Algebra II / Trigonometry
A Guided Inquiry

SHERMAN K. STEIN, CALVIN D. CRABILL, G.D. CHAKERIAN
University of California, Davis

SUNBURST COMMUNICATIONS, INC.
39 Washington Ave.
Pleasantville, NY 10570
(800) 431-1934
NY: (800) 221-5912
AK,Canada call collect: (914) 769-5030

International Standard Book Number: 1-55636-757-0
                      (formerly ISBN 0-916327-20-5)

Printed in the United States of America

Contents for Algebra II

This text is designed as a tool to aid the teaching and learning of mathematics in either the lecture, small-group, or individual approach. Unnecessary terms and symbols are avoided so that the average student can read the book. This encourages students to talk to each other about the material. After all, mathematics is not only a written language; it is a spoken language as well. It is important that students get ample opportunity to speak it, not just to see and hear it.

How this book is used in a particular class depends on the teacher and on the students. Some teachers lecture briefly on the material, then let the students work on it in small groups (usually four in a group). As the teacher sees obstacles develop, he or she then offers a brief explanation to get the class over the difficulty. Students may study the topic further, individually, at home. (The Solutions Key presents suggestions for teaching the material.)

We urge both teacher and student to experiment, to explore various ways of teaching and learning mathematics. The flexibility of the text encourages such explorations.

While the chapter titles indicate that the choice of topics is conservative and standard, the style is often new. We have usually used a specific problem to introduce a general concept and have preferred the geometric to the algebraic approach. For example, a bouncing ball introduces geometric series; a physical problem shows the need for logarithms. In the Complex Numbers chapter, the geometric definition of the addition and multiplication of complex numbers is treated first; only later is the algebraic view developed.

The first six chapters include a review of first year algebra as separate sections (optional).

Each chapter in its Project section offers an opportunity for further exploration.

Selected answers are given in the margins of the main section of each chapter. All answers for the self quizzes and tests, reviews, and first year algebra reviews are given at the end of each section.

Hints and learning suggestions are to be found in small type in the margins. Time may be saved by omitting some (or all) of the review exercises; however, most classes will find them helpful.

We wish to acknowledge the contributions of the many students and teachers who have used earlier versions of this book. In particular we wish to thank the following: Dr. Robert Van Dyne, Maria Wong, Archie White, Joanne Moldenhauer, Eugene Tashima, Valerie McIntyre, James O'Keefe, Ken Gelatt, Joan Graf, Ronnie James, and Craig Armstrong, Davis, Ca.; Mary Gorse, Palatine, Il.; Dr. Donald McKinley, Sacramento, Ca.; George Koonce, Dixon, Ca.; Ann Carroll, Santa Monica, Ca.; Gerry Benoit, Newport, R.I.; Henri Picciotto, San Francisco, Ca.; Sue Roosenvaad, Northfield, Mn.; Paul Zeitz, Colorado Springs, Co.; Jerry Weil, Gonzales Ca.; Donna Jean Wilson, Superior, Az., Hal Faulkner, Davis, Ca.

Sherman K. Stein
Calvin D. Crabill
G. D. Chakerian

Geometric Progressions

*** CENTRAL SECTION ***

In this chapter a bouncing ball will be used to introduce the mathematical concept of a "geometric series." Geometric series have a wide variety of applications within and outside of mathematics. For instance, they are used in finding the area under certain curves and in economic theory.

THE BOUNCING BALL

When a tennis ball, golf ball or super ball with high rebound qualities is dropped on a hard floor it rebounds and continues to bounce many times. The first exercise describes an experiment with such a ball.

Suppose a ball is dropped from 100 units above a hard level floor. Drop the ball several times. Observe the distance it rebounds on successive bounces.

EXERCISES

1. Drop a ball with high rebound qualities from a point 100 units (inches or centimeters) above the floor. Watch the experiment carefully in order to answer the following questions.
   (a) On the first drop of the ball, how high does it rebound? (For precision drop it several times from the same height. Average the heights of the rebounds.)
   (b) "Amount of rebound of the ball" is defined by the fraction

$$\frac{\text{height of rebound}}{\text{distance of fall}}$$

   On the basis of the results in (a), what is the amount of rebound of the ball in the experiment?
   (c) Use the results in (b) to <u>guess</u> the height of the second rebound if you were to let the ball bounce freely.
   (d) Let the ball **bounce tw**ice, and measure the height of the second rebound:

   Was your guess in (c) close to the results you observed?
   (e) What is the actual height of the second rebound?
   (f) What do you <u>guess</u> the height of the third rebound will be? How does it compare with the experimental results?
   (g) What is the total distance in a <u>downward</u> direction that the ball travels during the first fall? the first two falls? the first three falls? the first four falls?

If the class does this experiment in small groups, it may be convenient to have a meter stick for each small group. Thus, the distance is 100 centimeters. If it is a total class experiment directed by the teacher, it may be better to use a 100 inch scale of masking tape on the wall, so that everyone can see the results clearly. Labelling the tape every five inches is sufficient. A third alternative is to begin the chapter with the definition of exponential notation (following Exercise 2) and go directly to Exercise 3, which uses an "ideal" ball as a model of a geometric series.

THE EXPERIMENT IN EXERCISE 1 HELPS MOST STUDENTS.

2. (a) When the ball bounces freely, without being stopped, how many bounces can you count?

BECAUSE OF THE
STYLE OF WRITING
NO CENTRAL
SECTION EXERCISE
SHOULD BE
OMITTED

   (b) How far would you guess the ball drops during the fifth fall?   the ninth fall?   the tenth fall?
   (c) Compare your guesses in (b) with those of other students.

EXPONENTIAL NOTATION

In order to simplify further discussion of these problems, we pause to introduce an important tool, "exponential notation."

NOTATION.   Let r be a number and n be a positive integer.   Then the product,

$$\underbrace{r \cdot r \cdot r \cdots r}_{n \text{ factors}}$$

is denoted

$$r^n.$$

This is read

"r to the n"

or

"r to the nth power."

The integer n is called "the exponent."   The number r is called the "base." For instance, $(0.9)^4 = (0.9)(0.9)(0.9)(0.9) = 0.6561$ [base is 0.9; exponent is 4]

Exponential notation is useful in all branches of science and mathematics.

$$2^5 = 2 \cdot 2 \cdot 2 \cdot 2 \cdot 2 = 32 \text{ [base is 2; exponent is 5]}$$

$$\left(\tfrac{1}{2}\right)^6 = \tfrac{1}{2} \cdot \tfrac{1}{2} \cdot \tfrac{1}{2} \cdot \tfrac{1}{2} \cdot \tfrac{1}{2} \cdot \tfrac{1}{2} = \tfrac{1}{64} \text{ [base is } \tfrac{1}{2}\text{; exponent is 6]}$$

Exponential notation will be useful throughout this chapter.

AN IDEAL BALL

Imagine now an ideal ball that rebounds exactly 90 units when it is dropped from a height of 100 units.

 More generally, it always rebounds exactly 90 percent of the distance it falls.   For the rest of the chapter this "ideal ball" will be the model in exercises involving the bouncing ball.

3. Assume an ideal ball is dropped originally from a height of 100 units and that it will rebound 90% of the distance for each fall.

(a) Copy and fill in the following table:

number of fall	1st fall	2nd fall	3rd fall	4th fall	5th fall	25th fall
distance of fall	100	100 (0.9)				

(b) Indicate the total <u>downward distance</u> the ball falls for each of the following. (Do not simplify the expressions; use exponential notation.)

$$\text{1st fall} = 100$$
$$\text{first 2 falls} = 100 + 100(0.9)$$
$$\text{3 falls} = 100 + 100(0.9) + 100(0.9)^2$$
$$\text{4 falls} = \underline{\quad} + \underline{\quad} + \underline{\quad} + \underline{\quad}$$
$$\text{5 falls} = \underline{\quad} + \underline{\quad} + \underline{\quad} + \underline{\quad} + \underline{\quad}$$
$$\text{25 falls} = \underline{\quad} + \underline{\quad} + \underline{\quad} + \underline{\quad} + \underline{\quad} + \ldots + \underline{\quad}$$

(c) In your experiment in Exercise 2, about how many times did the real ball bounce before it stopped?

* * *

The ideal ball bounces an infinite number of times. To find the total distance it falls will require adding up an infinite number of positive numbers. The rest of this chapter is devoted to this type of problem.

A FORMULA FOR THE SUM OF FALLS

The total distance that the ideal ball falls during the first 25 falls is $100 + 100(0.9) + 100(0.9)^2 + \ldots + 100(0.9)^{23} + 100(0.9)^{24}$. There are 25 summands, of which only a few at the beginning and end are displayed above. The arithmetic needed to compute this sum is frightening. However, an algebraic trick, illustrated in Example 1, provides a shortcut for computing this sum. It concerns the first five falls, but the idea applies to any number of falls.

Example 1. Find a formula for the total downward distance that the ideal ball falls during the first five falls.
Solution. Denote the total downward distance by S_5 (read "S sub five," the sum of the first five falls).

From Exercise 3,
$$S_5 = 100 + 100(0.9) + 100(0.9)^2 + 100(0.9)^3 + 100(0.9)^4$$
Now multiply both sides of the equation by 0.9. Write the results of this operation below the original equation and line up vertically the terms of the same exponent (to prepare for a subtraction).

$$S_5 = 100 + 100(0.9) + 100(0.9)^2 + 100(0.9)^3 + 100(0.9)^4 \qquad \textit{given equation}$$
$$(0.9)S_5 = \qquad 100(0.9) + 100(0.9)^2 + 100(0.9)^3 + 100(0.9)^4 + 100(0.9)^5 \qquad \textit{resulting equation}$$

Now, <u>subtract both sides of the lower equation from the upper. Many of the summands on the right side cancel.</u> All that is left is the equation
$$S_5 - (0.9)S_5 = 100 - 100(0.9)^5.$$

Solving this equation for S_5 yields a "short formula" for S_5 :

$$(1 - 0.9)S_5 = 100 - 100(0.9)^5,$$

hence

$$S_5 = \frac{100 - 100(0.9)^5}{1 - 0.9}$$

4. In Example 1 a method was used to find a formula for S_5 (the sum of the first five falls). Use the same method to find

(a) S_6 (b) S_{10}

GEOMETRIC PROGRESSIONS

The bouncing ball problem is a model of what is called a "geometric progression." Recall that the distances the ball falls are:

1st fall:	100 units
2nd fall:	$100(0.9)$ units
3rd fall:	$100(0.9)^2$ units
4th fall:	$100(0.9)^3$ units
5th fall:	$100(0.9)^4$ units
	etc.
10th fall:	$100(0.9)^9$ units
	etc.

In Chapter 8 another type of sequence, called an "arithmetic progression," will be studied.

The sequence,

$$100, \ 100(0.9), \ 100(0.9)^2, \ 100(0.9)^3, \ ...$$

is an example of a "geometric progression." A geometric progression can be constructed with the aid of any two numbers, not just 100 and 0.9. A "geometric progression" is defined more generally as follows:

(Note that the nth number in the progression is ar^{n-1}, not ar^n. Make sure you understand why this is so.

> DEFINITION. A sequence of numbers is called a <u>geometric progression</u> if each number is obtained from the one just before it by multiplying by a fixed number, called the <u>ratio</u>. If the first number is <u>a</u> and the ratio is r, the geometric progression begins
>
> $$a, \ ar, \ ar^2, \ ar^3, \ ar^4, \ ..., \ ar^{n-1}, ...$$
>
> These numbers are called the <u>terms</u> of the geometric progression.

In the above definition, the first term is <u>a</u>, the second term is ar, and so on. The nth term is ar^{n-1}. The progression may or may not end with the nth term.

Example 2. Find the first five terms of the geometric progression that has first term a = 3 and ratio r = $\frac{1}{2}$.

(Note that the fifth term, $3(\frac{1}{2})^4$, has exponent 4, not 5.)

Solution. The geometric progression begins with the terms

$$3, \ 3\left(\frac{1}{2}\right), \ 3\left(\frac{1}{2}\right)^2, \ 3\left(\frac{1}{2}\right)^3, \ 3\left(\frac{1}{2}\right)^4, \ ...$$

When calculated, these terms are

$$3, \frac{3}{2}, \frac{3}{4}, \frac{3}{8}, \frac{3}{16}, \; \dots$$

* * *

5. (a) The first term in a geometric progression is 7 and the ratio is $\frac{1}{3}$. What is the 3rd term? 5th term? 25th term? nth term?

(a) 25th term is $7(\frac{1}{3})^{24}$, nth is $7(\frac{1}{3})^{n-1}$

 (b) The first term of a geometric progression is $\frac{3}{16}$ and the ratio is -2. What is the 3rd term? 5th term? nth term?

(b) 5th term is 3, nth is $\frac{3}{16}(-2)^{n-1}$

 (c) If $\underline{a}$ is the first term of a geometric progression and the ratio is r, what is the 3rd term? 5th term? 10th term? nth term?

(c) 3rd term is ar^2, 10th is ar^9.

6. State the first term, the ratio, and the number of terms in each of the following geometric progressions.

 (a) 1, 2, 4, 8, ... , 2048

(a) $\underline{a} = 1$, $r = 2$ $n = 12$

 (b) $5, \; 5\left(\frac{2}{3}\right), \; 5\left(\frac{2}{3}\right)^2, \; \dots, \; 5\left(\frac{2}{3}\right)^7$

 (c) $1, -\frac{1}{2}, \frac{1}{4}, -\frac{1}{8}, \; \dots \; -\frac{1}{512}$

(c) $\underline{a} = 1$, $r = -\frac{1}{2}$ $n = 10$

 (d) $\sqrt{3}, 3, 3\sqrt{3}, 9, \dots, 81$

7. Which of the following are geometric progressions?

 (a) 1, 2, 3, 4, 5, 6, ...
 (b) 1, 2, 4, 8, 16, ...
 (c) 16, -8, 4, -2, 1, ...
 (d) 1, 2, 3, 5, 8, 13, 21, ...

(a) no

(c) yes

 (e) $1, \frac{1}{2}, \frac{1}{3}, \frac{1}{4}, \frac{1}{5}, \; \dots$

 (f) 1, -1, 1, -1, 1, -1, ...

(e) no

THE SUM OF A GEOMETRIC PROGRESSION

Consider the first n terms of a geometric progression with first term $\underline{a}$ and whose ratio is r. Call their sum S_n (read "S sub n"):

$$S_n = a + ar + ar^2 + \dots + ar^{n-1}.$$

The next exercise shows how to find a short formula for this sum, S_n, for any geometric progression. This formula will be the key to finding the total distance that the ideal ball drops.

8.   (a) Copy and explain each step of the following:

$$S_n = a + ar + ar^2 + ar^3 + \ldots + ar^{n-1}$$

Observe in step 1 that like terms are lined up vertically so that they correspond to those in the original equation.

Step 1. $rS_n = \quad ar + ar^2 + ar^3 + \ldots + ar^{n-1} + ar^n$

Step 2. $S_n - rS_n = a - ar^n$

Step 3. $S_n(1 - r) = a - ar^n$

Step 4.

"SHORT FORMULA"

$$\boxed{S_n = \frac{a - ar^n}{1 - r} \quad \text{when } r \neq 1}$$ MEMORIZE!

(b) In Step 4 of (a) why must it be assumed that $r \neq 1$?

* * *

The <u>boxed formula</u> of step 4 is <u>important</u> and <u>should be memorized</u>. It will be used repeatedly in this chapter. But before considering its applications, examine Exercise 9, which treats the exceptional case where $r = 1$.

9.   Consider

$$S_n = a + ar + ar^2 + \ldots + ar^{n-1}$$

when $r = 1$.

Explain why $S_n = \underbrace{a + a + a + \ldots + a}_{n \text{ times}}$,

hence $S_n = na.$

* * *

Exercise 8 and 9 provide a formula for S_n when a, r, and n are known numbers. The special case when $r = 1$ is of little interest. As shown in Exercise 9, it is easy to compute without a formula. (The formula given in Exercise 8 would lead to division by zero if $r = 1$.)

Example 3.   Using the formula for the sum of a geometric progression, find $3 + 6 + 12 + 24 + 48$.
Solution.   This is the sum of 5 terms of the geometric progression with <u>a</u> = 3, r = 2.   Thus

It is wise to use parentheses in place of letters when substituting numbers in a formula.

$$S_n = \frac{a - ar^n}{1 - r}$$

$$= \frac{(3) - (3)(2)^5}{1 - (2)}$$

$$= \frac{3 - (3)(32)}{-1}$$

$$= \frac{3 - 96}{-1}$$

$$= \frac{-93}{-1}$$

$$= 93$$

10. Use the short formula in Exercise 8 to compute

$$S_n = a + ar + ar^2 + \ldots + ar^{n-1}$$

if (a) $n = 7$, $a = 2$, $r = -3$ (b) $a = 100$, $r = 0.9$, $n = 50$

(c) $r = 0.9$, $n = 25$, $a = 100$ (d) $a = \frac{1}{2}$ $r = \frac{2}{3}$, $n = 15$

(a) $\frac{2 - 2(-3)^7}{1 - (-3)}$

(c) $\frac{100 - 100(0.9)^{25}}{1 - (0.9)}$

11. Use the short formula for S_n and then simplify the expression to find a single fraction for each of the following sums:

(a) $1 - \frac{1}{2} + \frac{1}{4} - \frac{1}{8}$

(b) $\frac{3}{10} + \frac{3}{100} + \frac{3}{1000}$

(c) $1 + \frac{1}{3} + \frac{1}{9} + \frac{1}{27}$

(d) $1 + \frac{2}{3} + \frac{4}{9} + \ldots + \frac{32}{243}$

(e) $1 - \frac{2}{3} + \frac{4}{9} - \frac{8}{27} + \frac{16}{81} - \frac{32}{243}$

(f) $\frac{1}{2} + \frac{1}{4} + \ldots + \frac{1}{512}$

(a) $\frac{5}{8}$

(b) $\frac{333}{1000}$

(c) $\frac{40}{27}$

(d) $\frac{665}{243}$

(e) $\frac{133}{243}$

(f) $\frac{511}{512}$

* * *

Exercises 10 and 11 illustrate problems of varying difficulty. Generally if the process of simplifying the expression

$$S_n = \frac{a - ar^n}{1 - r}$$

is too tedious because n is a large number, the result is left as it is. For example,

$$S_{25} = \frac{100 - 100 (0.9)^{25}}{1 - (0.9)}$$

might be left as it is, for any more simplification (other than $1 - 0.9 = 0.1$) is tedious.

However, <u>you should simplify such expressions if it is not too difficult to do so or if the instructions for the problem specifically request it</u>, as in Exercise 11.

You may use a calculator at any time. However, this form of the answer for S_{25} is acceptable at this stage.

THE SHORT FORMULA FOR S_n APPLIED TO THE BALL

Return now to the ideal ball which has a 90% rebound. If it is dropped from 100 units, and it continues bouncing, could the total sum of the downward motions ever exceed 500 units? 1,000 units? 2,000 units? The next few exercises lead to the answers of these questions.

The sum of many terms of a geometric progression will involve the formula

$$S_n = \frac{a - ar^n}{1 - r}$$

Fortunately, when n is large and $-1 < r < 1$, this formula behaves so nicely that it will easily give the total downward motion of the ideal ball. Very little computation will be needed.

12. (a) When $r = 0.1$, what do you think happens to r^n as n gets larger?

   (b) Copy and fill in the table, using decimal notation in the last line.

n	1	2	3	4	5	6	7	8
$(\frac{1}{10})^n$	$(\frac{1}{10})$	$(\frac{1}{10})^2$						
$(0.1)^n$	0.1	0.01						

(b) $(0.1)^2 = 0.01$

$(0.1)^5 = 0.00001$

Exercises 12-15 are *important and worthy of a total class discussion.*

   (c) What happens to $(\frac{1}{10})^n = (0.1)^n$ as n gets larger in (b)?

* * *

The next table explores further the question raised in Exercise 12. The next table shows values or r^n when $n = \frac{1}{2} = 0.5$ for a few values of n.

n	1	2	3	4	7	16
$(\frac{1}{2})^n$	$\frac{1}{2}$	$(\frac{1}{2})^2 = \frac{1}{4}$	$(\frac{1}{2})^3 = \frac{1}{8}$	$(\frac{1}{2})^4 = \frac{1}{16}$	$(\frac{1}{2})^7 = \frac{1}{128}$	$(\frac{1}{2})^{16} = \frac{1}{65,536}$
$(0.5)^n$	0.5	0.25	0.125	0.0625	0.0078125	0.0000152

Observe that $(\frac{1}{2})^n = (0.5)^n$ gets closer and closer to zero as n gets larger.

The following table shows values for r^n when $r = \frac{9}{10} = 0.9$.

n	1	2	3	4	7	16
$(0.9)^n$	0.9	$(0.9)^2$ or 0.81	$(0.9)^3$ or 0.729	$(0.9)^4$ or 0.6561	$(0.9)^7$ or 0.4782969	$(0.9)^{16}$ or 0.185302

Observe that $(0.9)^n$ gets smaller (yet remains positive) as n gets larger.

Comparison of the two tables suggests that $(0.9)^n$ and $(0.5)^n$ both get close to 0 but that $(0.9)^n$ approaches 0 more slowly.

Use a calculator to experiment with $(.99)^n$ or $(0.999)^n$ as n gets large. Observe that they too approach 0, but much more slowly. Repeated use of the x^2 key is a quick method. Try $(0.999999999)^n$.

13. (a) Do Exercise 12 for $r = -0.1$. What number is r^n getting closer to as n gets larger?

   (b) Do Exercise 12 for $r = -\frac{1}{2} = -0.5$. What happens to r^n as n gets larger?

   (c) Do Exercise 12 for $r = -\frac{9}{10} = -0.9$ What happens to r^n as n gets larger?

14. Recall the tables in Exercises 12 and 13. What do you think happens to r^n when n is large if

(a) gets close to 0

(c) gets larger

(e) alternates between 1 and -1

   (a) $r = 0.8$ (b) $r = -0.8$ (c) $r = 2$

   (d) $r = 1$ (e) $r = -1$ (f) $r = 0.99$

In more advanced mathematics it is shown that if
$$-1 < r < 1$$
then r^n gets arbitrarily close to 0 when n is suitably large. (This is not the case for any other values of r.)

15. Return again to the ideal ball, which is dropped from 100 units and has a 90% rebound.

    (a) Write the formula for the sum, S_n, the sum of the first n <u>downward</u> falls of the ideal ball:
    $$S_n = \text{_____.}$$

    (b) From (a) show that $S_n < 1000$ for every value of n.

    (c) Show that when n is very large, S_n is very near 1000 (but still $S_n < 1000$)

    Hint: (b) the formula in (a) may be simplified and written as $S_n = 10[100-100(0.9)^n]$. Why?

* * *

Exercise 15 shows that even though the ideal ball bounces an endless number of times, the total distance it falls cannot be more than 1000 units.

SUM OF AN ENDLESS GEOMETRIC PROGRESSION

What happens to S_n in a geometric progression as n gets larger and larger? Study the following example carefully.

Example 4. Consider the geometric progression with <u>a</u> = 40 and r = ½. What happens to S_n if n is large?

Solution.
$$S_n = \frac{a - ar^n}{1 - r}$$

$$= \frac{(40) - (40)(\tfrac{1}{2})^n}{1 - \tfrac{1}{2}}$$

$$= \frac{40 - 40(\tfrac{1}{2})^n}{\tfrac{1}{2}}$$

$$= 80 - 80(\tfrac{1}{2})^n$$

When substituting in a formula it is wise to use ( ) in place of letters.

However, when n is large
$$(\tfrac{1}{2})^n \text{ is near } 0,$$

and therefore
$$80(\tfrac{1}{2})^n \text{ is near } 0.$$

Thus,
$$S_n \text{ is near } 80.$$

16. From Example 4,
$$S_n = 80 - 80(\tfrac{1}{2})^n.$$

    For what n is $S_n > 70$? That is, find a counting number, n, such that
    $$80 - 80(\tfrac{1}{2})^n > 70$$

    This exercise is worthy of a total class discussion.

    (a) Is the above inequality true for n = 1? n = 2? n = 3? n = 4?

    (b) For what value of n, a counting number, is $S_n > 75$?
    $$80 - 80(\tfrac{1}{2})^n > 75$$

    Experiment!

    (c) Does S_n ever get larger than 80? Explain.

* * *

Example 5.   Find what happens to the sum $S_n = 1 + \frac{1}{2} + \frac{1}{4} + \ldots + (\frac{1}{2})^{n-1}$ when n is large.

Solution.

$$S_n = \frac{a - ar^n}{1 - r}$$

Experiment with your calculator for larger and larger n, to show that
$(\frac{1}{2})^n = (0.5)^n$ *is close to 0.*

$$= \frac{(1) - (1)(\frac{1}{2})^n}{1 - (\frac{1}{2})}$$

Thus for large n, then $(\frac{1}{2})^n$ is close to 0 and S_n is near

$$\frac{(1) - (1)(0)}{1 - (\frac{1}{2})}$$

$$= \frac{1}{\frac{1}{2}}$$

$$= 2$$

17. Explain to a classmate why

$$S_n = \frac{a - ar^n}{1 - r}$$

Hint: See Exercises 12-16 and Example 5.

$$\doteq \frac{a}{1 - r}$$

when n is <u>large</u> and -1 < r < 1.

* * *

NOTE: Rather than say, "If -1 < r < 1, then

$$S_n = a + ar + ar^2 + \ldots + ar^{n-1}$$

is <u>close</u> to $\frac{a}{1 - r}$ when n is large," it is customary to say that

"a + ar + ar² + ... <u>equals</u> $\frac{a}{1 - r}$ when -1 < r < 1".

The expression $\frac{a}{1 - r}$ is called the "sum" of the endless (or infinite) geometric progression.

MEMORIZE!

$$\boxed{S = \frac{a}{1 - r}, \text{ when the geometric progression is } \underline{endless} \text{ and } -1 < r < 1}$$

Observe carefully the <u>two conditions</u> for using the formula, $S = \frac{a}{1 - r}$.

(1) The progression is endless.
(2) Also, -1 < r < 1.

Be careful! The equality sign does <u>not</u> have the usual meaning; no sum is equal to $\frac{a}{1 - r}$, but rather certain sums are <u>as near</u> $\frac{a}{1 - r}$ <u>as we please</u>. The dots tell us that we are dealing with sums having more and more summands.

Example 6. Find the sum of the endless geometric progression

$$1 - \frac{2}{3} + (\frac{2}{3})^2 - (\frac{2}{3})^3 + \ldots .$$

Solution.   This is the sum of an endless geometric progression with <u>a</u> = 1 and r = $-\frac{2}{3}$. Since -1 < r < 1,

$$S = \frac{a}{1-r}$$

$$= \frac{1}{(1) - (-\frac{2}{3})}$$

$$= \frac{1}{1 + \frac{2}{3}}$$

$$= \frac{1}{\frac{5}{3}}$$

$$= \frac{3}{5}$$

18. A ball is dropped from a height of 10 meters and bounces freely, rebounding 50% each time. Find the total distance of the downward falls.

19. Find the sum of each endless geometric progression if $\underline{a}$ is the first term and r is the ratio:

(a) $\underline{a} = 1$ and $r = \frac{1}{2}$ (b) $\underline{a} = 3$ and $r = -\frac{1}{2}$

(c) $\underline{a} = 10$ and $r = 0.9$ (d) $\underline{a} = 100$ and $r = 0.99$

20. Find the sum of each of these endless geometric progressions:

(a) $1 + \frac{2}{3} + \left(\frac{2}{3}\right)^2 + \left(\frac{2}{3}\right)^3 + \ldots$ (b) $1 - \frac{4}{5} + \left(\frac{4}{5}\right)^2 - \left(\frac{4}{5}\right)^3 + \ldots$

(c) $5 + 5\left(\frac{1}{11}\right) + 5\left(\frac{1}{11}\right)^2 + \ldots$ (d) $6 - 6\left(\frac{1}{11}\right) + 6\left(\frac{1}{11}\right)^2 - \ldots$

21. Compute:

(a) $7 + 7(0.6) + 7(0.6)^2 + 7(0.6)^3 + \ldots$
(b) $9 + 9(-\frac{1}{2}) + 9(-\frac{1}{2})^2 + 9(-\frac{1}{2})^3 + \ldots$
(c) $10 + 5 + \frac{5}{2} + \frac{5}{4} + \ldots$

* * *

Note. When $-1 < r < 1$, the formula for the "sum", S, of <u>all</u> the terms of an endless geometric progression, namely

$$S = \frac{a}{1-r}, \quad \text{(MEMORIZE!)}$$

is simpler than the formula for the sum, S_n, of the first n terms,

$$S_n = \frac{a - ar^n}{1-r}. \quad \text{(MEMORIZE!)}$$

Make sure you know the distinction between these two formulas and how they are used. Ask questions if you are not sure.

REPEATING DECIMALS

Geometric progressions have many applications. So far this chapter has applied them only to a bouncing ball. Next they will be applied to the arithmetic of decimals.

Consider the endless decimal

0.77777...   (the number 7 repeating without end)

The first 7 stands for

$$\frac{7}{10}.$$

The second 7 stands for

$$\frac{7}{100} \quad \text{or} \quad \frac{7}{10^2}.$$

The third 7 stands for

$$\frac{7}{1000} \quad \text{or} \quad \frac{7}{10^3}.$$

Thus 0.77777... is equal to

$$\frac{7}{10} + \frac{7}{10^2} + \frac{7}{10^3} + \frac{7}{10^4} + \frac{7}{10^5} + \cdots ,$$

the sum of a geometric progression with first term

$$\underline{a} = \frac{7}{10}$$

and ratio

$$r = \frac{1}{10}.$$

Thus, the sum is given by

$$S = \frac{a}{1 - r}$$

$$= \frac{\left(\frac{7}{10}\right)}{1 - \left(\frac{1}{10}\right)}$$

$$= \frac{\frac{7}{10}}{\frac{9}{10}}$$

$$= \left(\frac{7}{10}\right)\left(\frac{10}{9}\right)$$

$$= \frac{7}{9}.$$

Thus,

$$0.77777... = \frac{7}{9}.$$

Many other decimals are disguised geometric progressions, as illustrated in the next few exercises.

22. Show that $0.55555... = \frac{5}{9}$.

23. Show that $0.33333... = \frac{1}{3}$.

24. (a) Show that 0.13131313... (13's repeating without end) is equal to the sum of the geometric series with $\underline{a} = \frac{13}{100}$ and $r = \frac{1}{100}$.

   (b) Use (a) to show that

$$0.13131313... = \frac{13}{99} .$$

* * *

DEFINITION. (<u>repeating</u> decimal) A decimal expression is called a <u>repeating decimal</u> if, starting from some digit, it consists of a block of digits repeated without end.

For example, 8.378888...   (8's repeating without end) and

        16.04535353...   (53's repeating without end)

are repeating decimals.

NOTATION. The repeating block of digits in a repeating decimal may be shown by placing a horizontal bar over that block of digits. For instance

$$0.\overline{3} = 0.33333...$$
$$8.37\overline{8} = 8.3788888...$$
$$16.04\overline{53} = 16.04535353...$$

25. Write each of the following in the "bar" notation:
   (a)   10.99999...
   (b)   0.5414141...
   (c)   8.04139139139...

26. Using the formula for the sum of a geometric progression, show that

   (a) $10.\overline{9} = 11$ (b) $0.4\overline{9} = \frac{1}{2}$

27. (a) Express $0.7\underline{}$ as a fraction.
   (b) Express $0.0\overline{2}$ as a fraction.
   (c) Using (a) and (b), express $0.7\overline{2}$ as a fraction. $(c)\ \frac{13}{18}$

28. (a) Express $8.73\overline{}$ as a fraction.
   (b) Express $0.00\overline{4}$ as a fraction. $(c)\ \frac{7861}{900}$
   (c) Using (a) and (b), express $8.73\overline{4}$ as a fraction. $(a)\ \frac{5141}{999}$

29. Express as a fraction:

   (a) $5.\overline{146}$ (b) $13.\overline{874}$ (c) $1.45\overline{6}$

As the last few exercises show, a repeating decimal is always expressible as the quotient of two integers. In other words, a <u>repeating decimal is always a rational number.</u>

In the next chapter you will study more about rational and irrational numbers. A rational number is defined to be the quotient of two integers.

ANOTHER WAY TO CONSIDER ENDLESS DECIMALS

There is another way of finding the fraction equivalent to a repeating decimal. As Example 7 shows, it depends on the same method used to get the formula for S_n in Exercise 8.

Example 7. Find the fraction which equals 0.252525...

Solution. Let $X = 0.252525...$ Multiply each side of the equation by 100. The decimal point is moved two places to the right, thus

$$100 \ X = 25.252525...$$
$$X = 0.252525...$$

Subtract, $\qquad 99 \ X = 25$

hence $\qquad\qquad X = \dfrac{25}{99} \, .$

Comment. In Example 7 each side of the equation

$$X = 0.252525...$$

was multiplied by 100, or 10^2. Similarly, to compute 0.444... as a fraction, use the same method and write

$$X = 0.444... \, .$$

Then multiply each side by 10:

$$10 \ X = 4.444...$$
$$X = 0.444...$$

Subtract, $\qquad 9 \ X = 4$

hence $\qquad\qquad X = \dfrac{4}{9} \, .$

Observe in each of the above situations that you multiply each side of the equation by a power of 10 to eliminate the repeating block of decimals when you subtract. That is, multiply by 10^n, where n is the number of digits in the repeating block.

(a) $\dfrac{2}{3}$

(c) $\dfrac{41}{333}$

(e) $3\dfrac{64}{99}$

Hint: (f)
$2.1\overline{45} = 2\dfrac{1}{10}$ plus
$0.0\overline{45}$.

30. Use the method of Example 7 to find the fractional equivalent of each of the following:

(a) $0.\overline{6}$ $\qquad\qquad$ (b) $0.\overline{12}$ $\qquad\qquad$ (c) $0.\overline{123}$
(d) $0.\overline{9}$ $\qquad\qquad$ (e) 3.646464... $\qquad\quad$ (f) 2.1454545...

* * * SUMMARY * * *

A problem concerning a bouncing ball introduced a question concerning the sum of several numbers formed in a very special way.

Starting with a number $\underline{a}$ and repeatedly multiplying it by a fixed number r, the <u>ratio</u>, produces the following sequence of numbers:

$$a, ar, ar^2, ar^3, \ldots .$$

The nth number (or nth term) is ar^{n-1}. The sequence is called a <u>geometric progression</u>,

$$a + ar + ar^2 + \ldots + ar^{n-1}$$

is

$$S_n = \frac{a - ar^n}{1 - r} \quad \text{(where } r \neq 1\text{)}.$$

If $-1 < r < 1$, then the nth term of the progression approaches 0 as n gets larger. More important, r^n approaches 0 and the expression

$$\frac{a - ar^n}{1 - r} \quad \text{approaches} \quad \frac{a}{1 - r} .$$

That is, when the progression is endless and $-1 < r < 1$, then its "sum" is

$$S = \frac{a}{1 - r} .$$

The chapter concluded by showing how geometric progressions appear in the study of repeating decimals. In particular, every repeating decimal represents a fraction, in which the numerator and denominator are integers.

* * *   WORDS AND SYMBOLS   * * *

geometric progression $a, ar, ar^2, ar^3, \ldots$
first term, $\underline{a}$ nth term, ar^{n-1}
ratio, r   repeating decimal

* * * CONCLUDING REMARKS * * *

An example from economic theory will show that geometric series can be the model for many "real-world" problems other than that of the bouncing ball.

Assume that each person spends 90% of his income and saves 10%. Now, let us say that the government suddenly spends 1 billion dollars, which goes directly to the people. What is the total effect of this act on the economy.

Of the 1 billion dollars that people receive they turn around and spend 90%, or 0.9 billion dollars. Then recipients of those 0.9 billion dollars spend 90% of what they receive, or $(0.9)^2$ billion dollars. And so the effect continues to spread throughout the economy, as each person who receives some income in turn spends 90% of it. What is the total impact of that initial 1 billion dollars? One simply "adds" the numbers

$$1, 0.9, (0.9)^2, (0.9)^3, \ldots$$

to find out the amount (in billions of dollars) that is eventually spent. This graph indicates the total amount.

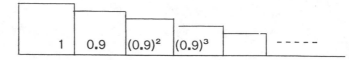

By the formula for the sum of a geometric progression, this sum is 10 billion dollars. The total impact is ten times as great as the initial amount spent. Economists say that the "multiplier" is ten.

There is also a "multiplier" effect in inflation. For instance, assume that the nation has to spend 1 billion dollars more for imported oil. First the price of oil goes up. Then the cost of transportation goes up. Then the cost of bread (or anything that is transported) goes up. Then wages go up to meet these rising costs. And so on.

Geometric series have also been used to analyse traffic delay at an intersection. It turns out that the delay increases much more rapidly than the number of cars in the streets.

For an application of geometric series in finding the area under a certain curve see Project 14.

* * *   CENTRAL SELF QUIZ   * * *

1. Define a geometric progression. Illustrate with an example of your own.
2. Find the sum:

(a) $\frac{1}{2} + \frac{1}{4} + \frac{1}{8} + ... + \frac{1}{1024}$

(b) $\frac{3}{8} - \frac{3}{4} + \frac{3}{2} - 3 + ... + 96$

(c) $1 - \frac{2}{3} + \frac{4}{9} - \frac{8}{27} + ...$

(d) $5 + 4 + \frac{16}{5} + \frac{64}{25} + ...$

3. Express 0.681681... as a fraction

(a) by showing first it is a geometric progression.
(b) by letting X = 0.681681..., multiplying a suitable power of 10, and subtracting.

4. Consider the geometric progression $2 - \frac{2}{3} + \frac{2}{9} - \frac{2}{27} + ...$

(a) Compute the sum of the first five terms.
(b) Find the sum of all the terms.

ANSWERS TO CENTRAL SELF QUIZ

1. See main section.   2. (a) $\frac{1023}{1024}$ (b) $\frac{513}{8}$ (c) $\frac{3}{5}$ (d) 25

3. $\dfrac{681}{999} = \dfrac{227}{333}$ 4. (a) $\dfrac{122}{81}$ (b) 1.5

* * * GEOMETRIC PROGRESSIONS REVIEW * * *

1. Suppose a flea is at one end in a box one foot wide. He jumps half the width on his first jump and one-half the remaining distance on each successive jump.

   (a) How far does he jump on the 3rd jump? the 10th jump? the nth jump?

   (b) How far has he traveled after the 3rd jump? after the 10th jump? after the nth jump?

   (c) Does he reach the other side? How close to the other side does he get?

2. (a) Compute $(0.7)^n$ decimally for n = 2, 3, 4, and 5.

   (b) When n is large what happens to $(0.7)^n$?

3. Let S_n be the sum of the first n terms of the geometric progression

   $$5 + \frac{5}{3} + \frac{5}{9} + \dots$$

   (a) Compute S_6 to two decimal places

   (b) Find a short formula for S_{100}.

   (c) Compute $5 + \dfrac{5}{3} + \dfrac{5}{9} + \dots$

4. Let S_n be the sum of the first n terms of the geometric progression

   $$3 - \frac{3}{2} + \frac{3}{4} - \frac{3}{8} + \dots \ .$$

   (a) What is $\underline{a}$? What is r?

   (b) Compute S_3 and S_4 to two decimal places.

   (c) Compute $3 - \dfrac{3}{2} + \dfrac{3}{4} - \dfrac{3}{8} + \dots$

5. (a) Show that 0.271271271... is the sum of a geometric series "in disguise."

   (b) Find its sum, as a fraction.

6. Solve Exercise 5(b) by multiplying the decimal by a suitable power of 10, and then subtracting.

7. Find the fractional equivalent of:

   (a) 0.333...   (b) 0.29999...   (c) 1.727272...   (d) 0.13444...

8.   Suppose we have a square of side 1. If we join the midpoints of each side, we get another square. We could continue this endlessly.

   (a)   What is the area of the 1st square (the largest square)? 2nd square? 3rd square? 10th square? nth square?
   (b)   What is the sum of the areas of the first 10 squares? 100 squares? n squares? (Ignore the fact that the areas overlap.)
   (c)   As in (b), can you find the sum of the areas of all the squares if the pattern continues endlessly? If not, why not? If so, what is the sum?

9.   Suppose we have a triangle of area 1. If we join the midpoints of the sides and form another triangle, how is the smaller triangle related to the original triangle? We can continue the pattern endlessly:

How are the dimensions of the smaller triangles related to the larger?

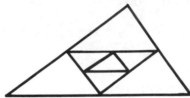 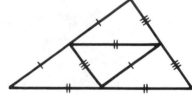

   (a)   What is the area of the 1st triangle? 2nd triangle? 3rd triangle? 10th triangle? nth triangle?
   (b)   What is the sum of the areas of the first 10 triangles? 100 triangles? n triangles? (Ignore the fact that the areas overlap.)
   (c)   As in (b), can you find the sum of the areas of all the triangles if the pattern continues endlessly? If not, why not? If so, what is the sum?

10.   Find the following products:

Note. Exercise 10 and 11 give a different view of a geometric progression.

   (a) $(1 - x)(1 + x)$
   (b) $(1 - x)(1 + x + x^2)$
   (c) $(1 - x)(1 + x + x^2 + x^3)$
   (d) $(1 - x)(1 + x + x^2 + x^3 + x^4)$
   (e) $(1 - x)(1 + x + x^2 + \ldots + x^{n-1})$

11.   (a)   Show that
$$1, x, x^2, \ldots, x^{n-1}$$
   is a geometric progression. What is the ratio?

   (b)   Write a compact expression for the sum: $1 + x + x^2 + \ldots + x^{n-1}$
   (c)   How is (b) related to Exercise 10(e)?

12.   A flea jumps
$$\frac{1}{2} \text{ ft., then } \frac{1}{4} \text{ ft., } \frac{1}{8} \text{ ft., } \frac{1}{16} \text{ ft., etc.}$$

However, each time he jumps, he turns around in the air so that the next jump is in the <u>opposite direction</u>. If we place him on the number line at $x = 0$, then the first jump is $x = \frac{1}{2}$. The second is to $x = \frac{1}{4}$. The third is to $\frac{3}{8}$, etc.

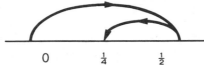

(a) How far does he jump and in what direction (to the right or left) on the 5th jump? the 10th jump? the 75th jump? the 100th jump? the nth jump?

Hint: 12(c)
Consider going to
the right as
positive and going
to the left as
negative.

(b) Draw the third and fourth jumps.

(c) What point is he getting near?

13. If you square each term in a geometric progression, do you obtain a geometric progression? Explain.

14. Which falls a greater distance on all its bounces? An ideal ball dropped from a height of 10 units that rebounds 50% or an ideal ball dropped from 5 units with a rebound of 80%?

* * * ANSWERS TO GEOMETRIC PROGRESSIONS REVIEW * * *

1. (a) $\frac{1}{8}$ ft., $\frac{1}{1024}$ ft., $\frac{1}{2^n}$ ft.   (b) $\frac{7}{8}$ ft., $\frac{1023}{1024}$ ft., $1 - (\frac{1}{2})^n$

(c) no, as close as he would like.

2. (a) 0.49, 0.342, 0.2401, 0.16807   (b) it gets close to 0

3. (a) 7.22   (b) $\dfrac{5 - 5(\frac{1}{3})^{100}}{1 - \frac{1}{3}}$ (c) 7.5

4. (a) $\underline{a} = 3$, $r = -\frac{1}{2}$ (b) 2.25, 1.88   (c) 2

5. (a) $0.271271271... = \dfrac{271}{10^3} + \dfrac{271}{10^6} + \dfrac{271}{10^9} + ...$; $\underline{a} = 0.271$, $r = \dfrac{1}{1000}$

(b) $\dfrac{271}{999}$ 6. $\dfrac{271}{999}$ 7. (a) $\dfrac{1}{3}$ (b) $\dfrac{3}{10}$ (c) $\dfrac{19}{11}$ (d) $\dfrac{121}{900}$

8. (a) 1, $\frac{1}{2}$, $\frac{1}{4}$, $\dfrac{1}{512}$, $\dfrac{1}{2^{n-1}}$

(b) $S_{10} = \dfrac{1023}{512}$, $S_{100} = \dfrac{1 - (\frac{1}{2})^{100}}{1 - \frac{1}{2}}$, $S_n = \dfrac{1 - (\frac{1}{2})^n}{1 - \frac{1}{2}}$ (c) 2

9. (a) 1, $\frac{1}{4}$, $\dfrac{1}{16}$, $\dfrac{1}{4^3}$, $\dfrac{1}{4^{n-1}}$

(b) $S_{10} = \dfrac{1 - (\frac{1}{4})^{10}}{1 - \frac{1}{4}}$ $S_{100} = \dfrac{1 - (\frac{1}{4})^{100}}{1 - \frac{1}{4}}$ $S_n = \dfrac{1 - (\frac{1}{4})^n}{1 - \frac{1}{4}}$ (c) 1.33...

10. (a) $1 - x^2$ (b) $1 - x^3$ (c) $1 - x^4$ (d) $1 - x^5$ (e) $1 - x^n$

11. (a) ratio = x   (b) $\dfrac{1 - x^n}{1 - x}$

    (c) Exercise 10(e) shows $1-x$ is a factor of $1 - x^n$

    (i.e., $\dfrac{1 - x^n}{1 - x} = 1 + x + x^2 + \ldots + x^{n-1}$).

12. (a) $\dfrac{1}{32}$ ft. to the right, $\dfrac{1}{1024}$ ft. left, $\dfrac{1}{2^{75}}$ ft. right, $\dfrac{1}{2^{100}}$ ft left, $\frac{1}{2}$ ft.

    left if n is even, right if n is odd.

    (b) 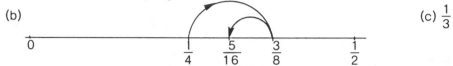   (c) $\dfrac{1}{3}$

13. yes   14. The 5 unit, 80% ball.

* * * REVIEW SELF TEST * * *

1. Find the sum:

    (a) $4 + 8 + 16 + \ldots + 256$ (b) $\dfrac{1}{2} - \dfrac{1}{3} + \dfrac{2}{9} - \dfrac{4}{27} + \dfrac{8}{81}$

2. Compute the fractional equivalent of 0.838383... in two different ways.

3. Which of the following are geometric progressions?

    (a) $\dfrac{1}{3}$, 1, $1\dfrac{1}{3}$, ...   (b) $\dfrac{2}{3}$, $-\dfrac{1}{2}$, $\dfrac{3}{8}$, $-\dfrac{9}{32}$, ...

    (c) 1, 4, 9, 16, ...   (d) $\dfrac{1}{9}$, $\dfrac{1}{3}$, 1, 3, ...

4. Find the first term of a geometric progression if the ratio is 2 and the 6th term is 96.

5. (a) Write the first five terms of a geometric progression if $\underline{a}$ = -3 and $r = -\dfrac{1}{3}$.   (b) Find the sum of the first five terms in (a).

6. (a) A ball dropped from 20 feet rebounds 25% of the distance from which it fell. What is the sum of the downward motions of the ball?

    (b) How far does the ball travel vertically (both up <u>and</u> down motion) before coming to rest?

7. (a) Is the following reasoning correct?

$$\text{Let } x = 1 + 2 + 2^2 + 2^3 + \ldots$$
$$= 1 + 2(1 + 2 + 2^2 + \ldots)$$
$$= 1 + 2\left(\dfrac{1}{1-2}\right)$$
$$= 1 - 2$$
$$x = -1$$

(b) Justify the reasoning in (a), or show where it is incorrect.

* * * ANSWERS TO REVIEW SELF TEST * * *

1. (a) 508   (b) $\frac{55}{162}$ 2. $\frac{83}{99}$

3. (a) and (c) are not geometric progressions, but (b) and (d) are.   4. 3

5. (a) -3, 1, $-\frac{1}{3}$, $\frac{1}{9}$, $-\frac{1}{27}$, (b) $-\frac{61}{27}$ 6. (a) $\frac{80}{3}$ ft., (b) $2\left(\frac{80}{3}\right) - 20 = \frac{100}{3}$ ft.

7. It is incorrect, for $\frac{a}{1-r}$ is the sum of an endless geometric progression where $-1 < r < 1$.

* * * PROJECTS * * *

You will greatly increase your understanding of geometric progressions if you work on these projects.

Hint: There is a very easy way!

1. An automobile depreciates as it gets older. Typically each year it loses a third of its value. After n years what is the value of an automobile which initially cost $10,000.

2. Two locomotives are 20 miles apart and approaching each other. Each is going 10 miles per hour. A busy bee buzzes back and forth between them until, alas, the bee is squashed when the locomotives collide. The bee flies 40 miles per hour. How far does the bee travel?

3. Counting only blood relatives, state the number of your
    (a) parents   (b) grandparents   (c) great grandparents
    (d) Assuming that there are 3 generations per century, how many ancestors have you had in the 300 years prior to your birth?
    (e) Assuming that there are 4 generations per century, answer (d).

4. Assume that all businesses and individuals in the U.S. spend 80% of all income or revenue, and invest or save the remaining 20%. Now, suppose the government spends an extra billion dollars without raising taxes.

    (a) What is the total increase in spending in the country due to the government action? [Hint: 80% money is passed on repeatedly from business to business (or individuals) over and over again.]

    (b) If the amount people and businesses spend increases a little, so that they spend $\frac{5}{6}$ of the money they receive (this increase is less than 4%) what is the total increase in spending in the country?   What is the percentage increase over that in part (a)?

5. A repeating decimal is always a fraction. Is the reverse true? If A and B are positive integers, is the decimal form $\frac{A}{B}$ always repeating?

    (a) Carry out the long division to show that the decimal forms of $\frac{2}{7}$ and $\frac{3}{17}$ are both repeating.

(b) Look carefully at the remainders met in (a) on each subtraction. They are always less than the denominator. From this show that $\frac{71}{1029}$

will have a repeating decimal and that the repeating block will have at most 1028 digits.

(c) Why will $\frac{A}{B}$ always have a repeating decimal? How long can the repeating block be?

(d) If you add two repeating decimals will you always get a repeating decimal?

6.   Suppose that you have a two pan balance:

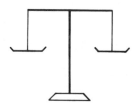

(You balance it by putting weights on one side and objects on the other.)

You want to be able to weigh objects weighing a whole number of pounds. How would you choose a fixed supply of weights in order to
   (a)   weigh objects from 1 to 13 pounds, using only three weights?
   (b)   weigh objects from 1 to 40 pounds, using only four weights?
   (c)   weigh objects from 1 to 121 pounds, using only five weights?
   (d)   There is a geometric progression in this problem; what is it?

7.   If x, y, and z are successive terms in a geometric progression, which of the following are always true?

   (a) $x^2 = yz$ (b) $y^2 = xz$
   (b) $x^3 = y^2z$ (d) $xy = yz$

8.   What number when added to each of 1, 10, and 46 give three consecutive terms of a geometric progression?

9.   (a)   Divide 1 by 1 - x,

$$1 - x \overline{\smash{\big)}\begin{array}{l} 1 + x + ... \\ 1 + 0{\cdot}x + 0{\cdot}x^2 + 0{\cdot}x^3 \\ \underline{1 - x} \\ x \\ \\ \underline{x - x^2} \\ \text{etc.} \end{array}}$$

   and convince yourself that

$$\frac{1}{1 - x} = 1 + x + x^2 + x^3 + ...$$

(b)   Have you seen an expression similar to the "equation" in (a) before?

(c) In (a), if we let x = -1, on the left side we get

$$\frac{1}{1 - x} = \frac{1}{1 - (-1)} = \frac{1}{2}$$

while on the right side, we get

$$1 + x + x^2 + x^3 + \ldots = 1 - 1 + 1 - 1 + \ldots$$

which is clearly not equal to $\frac{1}{2}$. Explain the apparent contradiction.

10. (a) A man starts 25 days before a national election to work for his candidate. The first day he convinces one person to work for his candidate. On the second day each of them convinces one new person to do the same. On the third day each of the workers convinces one new person. And so on for 25 days. How many voters were assured for the candidate on election day?

(b) If each man was able to convince two people each day instead of one, how many people were recruited on the 1st day? 2nd day? 3rd day? 4th day? 8th day? Could they continue this in the United States for 25 days? Explain. (Assume that the population of the United States is 240 million.)

11. (a) A frog near the edge of a cliff jumps one foot on his first jump toward the edge of the cliff. The closer he gets to the edge, the less he jumps. His second jump is $\frac{1}{2}$ foot; the third, $\frac{1}{3}$ foot, and so on. He should be placed far enough back so that he will not go over if he keeps jumping toward the cliff indefinitely. Will he be safe if he is placed 3 feet from the edge of the cliff? What about 10 feet? 100 feet?

(b) Clearly, part (a) is too tedious to be solved directly. To solve it let $S_n = 1 + \frac{1}{2} + \frac{1}{3} + \ldots + \frac{1}{n}$ and note that

$$S_1 = 1$$

$$S_2 = S_1 + \tfrac{1}{2}$$

$$S_4 = S_2 + \frac{1}{3} + \frac{1}{4} > S_2 + \frac{1}{4} + \frac{1}{4}$$

$$S_8 = S_4 + \frac{1}{5} + \frac{1}{6} + \frac{1}{7} + \frac{1}{8} > S_4 + \frac{1}{8} + \frac{1}{8} + \frac{1}{8} + \frac{1}{8}$$

Can you make a generalization about this? This should enable you to answer part (a).

(c) The frog swallows some lead and only jumps $\frac{1}{10}$th as far as he normally would. That is, he jumps in the pattern, $\frac{1}{10}$ ft., $\frac{1}{20}$ft., $\frac{1}{30}$ft., ... How far back must he be placed to be safe?

He will eventually go over the cliff. (Why?)

12. Consider again the bouncing ball from the beginning of this chapter.
    (a) Drop the ball again from 100 inches. Listen to the <u>sound</u> as the ball hits the floor. What do you observe about the times between sounds?

    In part (a) you heard the time intervals between bounces progressively getting shorter. In the end, they were like a "buzz" and then the ball was still. Let us look at this mathematically.

    Galileo discovered that the distance an object falls in t seconds, after starting from a position of rest, is
    $$16t^2 \text{ feet.}$$
    Thus we can compute the time it takes an object to fall if we know the distance it falls.

    (b) Assuming that the ball is dropped from 100 feet, fill in the table for the ball with the rebound ratio of 90%:

	1st fall	2nd fall	3rd fall	4th fall
distance (feet)	100			

    (c) Fill in the table:

	1st fall	2nd fall	3rd fall	4th fall
time (seconds)				

    (d) Show that the durations of the <u>falls</u> form a geometric progression.

    (e) Use (d) to explain why the times between <u>bounces</u> form a geometric progression.

    (f) How long does the ball bounce? Does it stop, or does it bounce forever, except that the bounces are so small that they are invisible?

13. In the right triangle ACB shown below, the lines E_1D_1, E_2D_2, ... are parallel to AC and the lines CD_1, E_1D_2, E_2D_3, ... are perpendicular to AB.

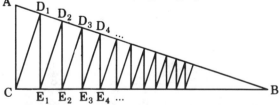

Hint: Consider corresponding sides of similar triangles.

    (a) Show that the lengths of CD_1, E_1D_2, E_2D_3, E_3D_4, ... form a geometric progression.

    (b) Find the sum of the progression in (a), using the formula for the sum of a geometric progression.

(c)   Find the sum in (b) without using the formula for the sum of a geometric progression. (Use geometry instead.)

14. In the 17th century, Pierre Fermat used geometric progressions to find the area under the parabola $y = x^2$.

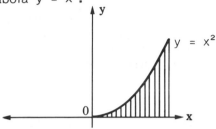

Select r such that $0 < r < 1$. (Later r will be near 1.)   Draw the points with x-coordinates $1, r, r^2, r^3, \ldots$. Construct rectangles as shown in this figure:

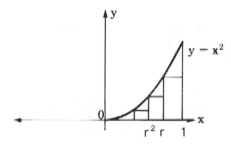

(a)   Show that the total area of the rectangles is

$$(1 - r)r^2 + (r - r^2)r^4 + (r^2 - r^3)r^6 + \ldots$$

(b)   Show that the sum of the geometric progression in (a) is

$$\frac{r^2}{1 + r + r^2}$$

(c)   When r is near 1, the rectangles are a better approximation of the area under the parabola. Use this fact to show that the area under the parabola is $\frac{1}{3}$.

Hint: (a) The rectangle on the right has $1 - r$ as the base and r^2 as the height.
(b) $1 - r^3 =$
$(1 - r)(1 + r + r^2)$

15. As far back as the 5th century B.C., great minds have wrestled with the infinite in mathematics. Read about "Zeno's Paradox" on pp. 24-25 in E. T. Bell's Men of Mathematics (Simon and Schuster) or on pp. 37-39 in Kasner and Newman's Mathematics and the Imagination (Simon and Schuster).

16. If you take every other term in a geometric progression do you get a geometric progression? Explain.

17. Suppose a person has a gold chain of 63 links and that the government demands one gold link per month for taxes. How can the person remove exactly 3 links and pay the taxes for 63 consecutive months? (Assume that the government can make change using previously paid links.)

* * * REVIEW OF FIRST YEAR ALGEBRA (OPTIONAL) * * *

Each of the first six chapters contains a review of first year algebra. These reviews are independent of the rest of the course. MOST STUDENTS SHOULD PROFIT FROM THESE REVIEWS.

Each of the first six chapters presents a separate review section of first year algebra. While occasionally there is some overlap with ideas that are presented in the main sections of the chapter, generally there is no direct relation to a particular chapter. The student who is less well prepared may discover his weaknesses and correct them as the course progresses.

1. Add:

      (a) 13   (b) 13   (c) -13   (d) -13
          8   -8   8   -8

2. (a) Subtract the lower number from the upper number in each part of Exercise 1.

   (b) Check your numbers in (a) by answering the question, "What must be added to the lower number in order to make the sum equal to the upper number?"

3. Compute:

   (a) $19 - (-8)$ (b) $7 - 8 + 9 - 11 - 14$

   (c) $(-13)(-7)$ (d) $\dfrac{-54}{6}$

   (e) $(-17)(3)$ (f) $\dfrac{-84}{-7}$

4. Compute:

   (a) $(-3)(-3)(-3)(-3)$ (b) $(-2)(-2)(-2)(-2)(-2)$

   (c) $\dfrac{(-13)(17)(-7)(0)}{(73)(-19)}$

EXPONENTS

If x is any number and n is any positive integer, then x^n signifies x is a factor n times:

$$x^n = \underbrace{x \cdot x \cdot \ldots \cdot x}_{n \text{ factors}}$$

The n is called the exponent and x is called the base:

$$\text{base} \longrightarrow x^n \longleftarrow \text{exponent}$$

For example,

$$5^3 = 5 \cdot 5 \cdot 5 = 125$$

RULES FOR EXPONENTS

Where x and y are numbers different from 0, and m and n are integers, the following rules for exponents are assumed:

I. $x^m \cdot x^n = x^{m+n}$ (product rule or "basic rule")

II. $x^0 = 1$ ("zero exponent rule")

III. $x^{-n} = \dfrac{1}{x^n}$ ("negative exponent rule")

IV. $\dfrac{x^m}{x^n} = x^{m-n}$ ("quotient rule")

V. $(x^m)^n = x^{mn}$ ("Power rule") *of a power*

VI. $\left(\dfrac{x}{y}\right)^n = \dfrac{x^n}{y^n}$ (Power of a quotient")

VII. $(x \cdot y)^n = x^n \cdot y^n$ ("Power of a product")

All the other rules depend on the "basic rule" as you will see in Chapter 2. Rules 6 and 7 will be useful in simplifying radicals. NOTE. THE RULES APPLY TO NEGATIVE INTEGERS AS WELL AS POSITIVE.

For example,

$3^4 \cdot 3^7 = 3^{11}$ ("basic rule")

$5^0 = 1$ ("zero rule")

$2^{-7} = \dfrac{1}{2^7} = \dfrac{1}{128}$ ("negative rule")

$\dfrac{6^9}{6^4} = 6^{9-4} = 6^5$ ("quotient rule")

$(3^4)^5 = 3^{20}$ ("Power rule") *of a power*

$\left(\dfrac{2^5}{3}\right) = \dfrac{2^5}{3^5}$ ("Power of quotient")

$(5 \cdot 7)^3 = 5^3 \cdot 7^3$ ("Power of product")

5. Find x.

(a) $(-1)^{2001} = x$

(b) $(-1)^{1492} = x$

(c) $(-1)^{1984} = x$

(d) $(-1)^0 = x$

(e) $(3^2)(3^4) = 3^x$

(f) $(5^4)^7 = 5^x$

(g) $\dfrac{7^{13}}{7^8} = 7^x$

(h) $\dfrac{4^3}{9^3} = \left(\dfrac{4}{9}\right)^x$

(i) $(3 \cdot 8)^5 = 3^x \cdot 8^x$

(j) $\dfrac{1}{32} = 2^x$

(k) $(3^{-4})(3^5) = 3^x$

(l) $(5^{-3})^2 = 5^x$

6. Multiply:

(a) $(3x^2)(-5x^4)$

(b) $(-2nx)(4x^3)$

(c) $(-x)(x^2)$

(d) $(-7x^6)(-5x^4)$

7. Divide:

(a) $\dfrac{39x^8}{13x^2}$

(b) $\dfrac{-18a^7 b^5}{-3a^4 b}$

(c) $\dfrac{-13x^4}{-13x^4}$

(d) $\dfrac{-x^3}{x}$

* * * ORDER OF OPERATIONS * * *

Expressions such as

$$6(7 - 2)^2 - 7(\tfrac{1}{3})^4 + 8$$

must be simplified as follows:

1. First, simplify within parentheses,

$$6(5)^2 - 7(\tfrac{1}{3})^4 + 8$$

2. Then consider the <u>exponents</u> and multiply as they indicate.

$$6(25) - 7(\tfrac{1}{81}) + 8$$

3. Then <u>multiply</u> (or <u>divide</u>) within each term <u>left to right</u>.

$$150 - \tfrac{7}{81} + 8$$

4. Finally, carry out the <u>additions</u> or <u>subtractions</u>.

$$158 - \tfrac{7}{81} = 157\,\tfrac{74}{81}$$

The order of operations described here must be followed in order to get the correct answer in simplifying an expression.

8. Compute:

   (a) $32(\tfrac{1}{2})^3 - 5(\tfrac{1}{2})^2 + 3$ (b) $2(-3)^4 - 5(-2)^5 + 6(4)^2$

   (c) $(2^2)(3^5) - (5^2)(7) + 4$ (d) $2(3^3) - 4(\tfrac{1}{5})^2 + 7$

9. Copy and fill in the table for $y = 2x^2 - x - 1$.

*IMPORTANT!
Use ( ) in place
of x when you
substitute, then
follow the
correct <u>order of
operations</u>
(exponents, then
products and
quotients, then
addition and
subtraction).*

x	$2x^2 - x - 1$	=	y	(x,y)
3	$2(3)^2 - (3) - 1$ $2(9) - 3 - 1$ $18 - 3 - 1$ 14		14	(3,14)
2	(Leave space here for computation, as was done for x = 3)			
1				
$\tfrac{1}{2}$				
0				
-1				
-2				

GRAPHING POINTS IN THE (x,y)-PLANE

Recall how points are graphed in the (x,y)-coordinate plane. For example, the points A = (3,4), B = (-2,7), C = (-4,-5), D = (5,-1) and the origin, (0,0) are plotted below:

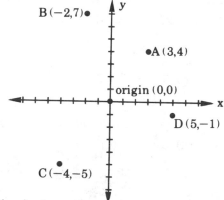

10. Plot each of the following points in the (x,y)-plane.

   (a)   (-7,5)   (b)   (-11,-2)   (c)   (6,-3)   (d)   (1,8)

11. (a) Graph the points from the table in Exercise 9 in the (x,y)-plane.

   (b) Draw a smooth curve through the points you plotted in (a).

12. (a) Copy and fill in the table for y = -x² - 2x + 3.

x	-x² - 2x + 3	=	y	(x,y)
2	(Show all computation steps here.)			
1				
0				
-1				
-2				
-3				
-4				

Observe
$(-x)^2 - (\infty)(-\infty) = x^2$
whereas
$-x^2 = -(x)^2$
Thus
x^2 is *not* $(-x)^2$

   (b) Graph the points in (a).

   (c) Draw a smooth curve through the points in (b).

13. (a) Find three points that lie on the line y = -5x + 3.

   (b) Plot the points in (a) on the (x,y)-coordinate plane.

   (c) Find three points that lie on the line y = 3x - 1.

   (d) Plot the points in (c) on the same coordinate system as (b).

   (e) Estimate the coordinates of the point where the two lines cross each other.

Two points determine a line. The third point is a check.

ANSWERS TO ALGEBRA REVIEW

1. (a) 21   (b) 5   (c) –5   (d) –21

2. (a) 5   (b) 21   (c) –21   (d) –5 (Check your answers as suggested.)

3. (a) 27   (b) –17   (c) 91   (d) –9   (e) –51   (f) 12

4. (a) 81   (b) –32   (c) 0

5. (a) –1   (b) 1   (c) 1   (d) 1   (e) 6   (f) 28
    (g) 5   (h) 3   (i) 5   (j) –5   (k) 1   (l) –6

6. (a) $-15x^6$ (b) $-8nx^4$ (c) $-x^3$ (d) $35x^{10}$

7. (a) $3x^6$ (b) $6a^3b^4$ (c) 1   (d) $-x^2$

8. (a) $5\frac{3}{4}$ (b) 418   (c) 801   (d) $60\frac{21}{25}$

9. (2,5),   (1,0), $(\frac{1}{2},-1)$,   (0,-1),   (-1,2),   (-2,9)

10. Plot points

11. (a) Plot points   (b) "_/" shaped curve (parabola)

12. (a) (2,-5),   (1,0),   (0,3),   (-1,4),   (-2,3),   (-3,0),   (-4,-5)
    (b) plot points
                                                    (c) "/\" shaped curve (parabola)

13. (a)   Find three points   (b) Plot points in (a)
    (c)   Find three points   (d) Plot points in (c)
    (e) $(\frac{1}{2},\frac{1}{2})$

*   ***   *

Exponents and Radicals

* * * CENTRAL SECTION * * *

An indulgent father told his son, "If you will deposit $1 with me now, in one year it will be worth $4. If you leave it with me for two years it will be worth $16. After three years, $64. Your money will keep growing all the time, or you can withdraw it at any time."

EXERCISES

1. (a) What do you think the boy's account will be worth after four years?
   (b) After 6 months?
   (c) After one and a half years?

At the end of this chapter we will return to Exercise 1. Problems of the type in which the father asks, "At what time will the value of your deposit be worth $5?" will be answered in Chapter 5.

BASES AND EXPONENTS

Recall the following definition:

DEFINITION. If n is a positive integer and b is a number, then b^n is the product,

$$b^n = \underbrace{b \cdot b \cdots b.}$$

$$\text{n factors}$$

The special case where $n = 1$ is simply

$$b^1 = b.$$

In the notation b^n, the number b is called the <u>base</u> and n is called the <u>exponent</u>.

$$\text{"base"} \leftarrow b^n \leftarrow \text{"exponent"}$$

Read b^n as "b to the nth power," or "b to the n".

2. Evaluate: (a) 5^2 (b) 2   (c) $(-3)^4$ (d) $(\frac{1}{2})^2$

* * *

<u>Henceforth assume that b, the base, is positive</u>, since this is the case that is important in Chapters 4 and 5 and throughout the text. Negative bases, which raise interesting questions, are discussed later in the chapter.

BASIC RULE OF EXPONENTS

The exponential b^n has so far been defined for positive integers n. In this chapter b^n will be defined for all real numbers n (positive, negative or zero). Thus $b^{\frac{1}{2}}$, b^{-3}, and even b^{π} will be defined. Throughout, the guiding principle will be the equation

$$b^m \cdot b^n = b^{m+n}.$$

For example,

$$2^3 \cdot 2^5 = (2 \cdot 2 \cdot 2)(2 \cdot 2 \cdot 2 \cdot 2 \cdot 2)$$

$$= 2^8$$

illustrates this principle.   All definitions of the exponential b^n will conform to the following rule:

Say the basic rule of exponents in your own words. MEMORIZE!

> ### BASIC RULE OF EXPONENTS
> For any positive base b and any real numbers m and n
> $$b^m \cdot b^n = b^{m+n}.$$

(a) 7
(c) 13
(e) 4
(g) 10
(i) 10

3.   Use the basic rule of exponents to find x:

(a) $2^3 \cdot 2^4 = 2^x$ (b) $5^6 \cdot 5^7 = 5^x$ (c) $9^5 \cdot 9^8 = 9^x$

(d) $(\frac{1}{2})^3(\frac{1}{2})^5 = (\frac{1}{2})^x$ (e) $11^2 \cdot 11^x = 11^6$ (f) $(0.7)^x(0.7)^4 = (0.7)^9$

(g) $b^7 \cdot b^3 = b^x$ (h) $n^3 \cdot n^x = n^6$ (i) $z^2 \cdot z^x = z^{12}$

* * *

In the next few exercises the basic rule of exponents is used to help define such expressions as b^0, $b^{\frac{1}{2}}$, and b^{-3}.

THE EXPONENT ZERO

The first exponent to be considered is 0. How should 2^0 be defined? It makes no sense to say "it is the product of zero 2's." The next exercise, using the basic rule of exponents, shows that 2^0 should be 1.

4.   (a) If the basic rule of exponents is always to be true, what must x be in the following equation?

$$2^0 \cdot 2^3 = 2^x$$

(b) From (a) conclude that 2^0 <u>must be</u> 1.

5.   (Review Exercise 4)
(a) Assume the basic rule of exponents and that b is positive; what must x be in the following equation?

$$b^0 \cdot b^n = b^x$$

(b) Explain why (a) suggests the following:

> ### THE ZERO EXPONENT RULE
> $$b^0 = 1$$

Express the meaning of the zero-exponent rule in your own words. MEMORIZE!

NEGATIVE EXPONENTS

The numbers 0, 1, 2, 3, ..., that is, all the nonnegative integers, now all have meaning as exponents. What meaning should be given to such expressions as

$$5^{-1}, \text{ or } 2^{-3}, \text{ or } b^{-7}?$$

where the exponents are negative?

If the <u>Basic Rule of Exponents</u> is to remain valid for negative integer exponents, there is only one possible choice for the value of such expressions as 5^{-1}, 2^{-3}, and b^{-7}. The next few exercises show why. First, recall an important concept, that of the "reciprocal".

DEFINITION. If the product of two numbers is 1, then each is called the reciprocal of the other. In symbols

---

RECIPROCAL

If $A \cdot B = 1$, then A is the <u>reciprocal</u> of B (written $A = \frac{1}{B}$),

and B is the reciprocal of A (written $B = \frac{1}{A}$).

Since $(5)(0.2) = 1$, the reciprocal of 5 is 0.2. Consequently, we write

$$0.2 = \frac{1}{5} .$$

Also, 5 is the reciprocal of 0.2. Thus,

$$5 = \frac{1}{0.2} .$$

Similarly, $\frac{4}{3}$ and $\frac{3}{4}$ are reciprocals, for $(\frac{4}{3})(\frac{3}{4}) = 1$.

6. Find N for each of the following:

(a) $(N)(2) = 1$ (b) $\frac{1}{2} \cdot N = 1$ (c) $2^4 \cdot N = 1$

(d) $(0.25)(N) = 1$ (e) $N \cdot (\frac{2}{3}) = 1$ (f) $\frac{1}{(\frac{3}{4})} N = 1$

(g) $A \cdot N = 1$, where $A \neq 0$,   (h) $b^x \cdot N = 1$, where $b > 0$.

(i) Write N as an integer if $(\frac{1}{2})^2 \cdot N = 1$.

(a) $\frac{1}{2}$

(c) $\frac{1}{2^4}$

(e) $\frac{1}{\frac{2}{3}}$ or $\frac{3}{2}$

(g) $\frac{1}{A}$

(i) $\frac{1}{(\frac{1}{2})^2}$ or 4

7. If the basic rule of exponents remains true for negative exponents, evaluate each of the following in the simplest form.

(a) $3^2 \cdot 3^{-2}$ (b) $5^3 \cdot 5^{-3}$ (c) $7^4 \cdot 7^{-4}$ (d) $b^5 \cdot b^{-5}$

(a) $3^0 = 1$
(c) $7^0 = 1$

8. Answer each of the following:

(a) Why is $(5^2)(5^{-2}) = 1$?   (See Exercise 7.)

(b) Why is $(5^2)(\frac{1}{5^2}) = 1$?   (c) Why is $5^{-2} = \frac{1}{5^2}$?

9. Give the reason for each step of the following proof that b^{-n} is the reciprocal of b^n. For $b > 0$ and any positive integer n,

Step 1. $b^n \cdot b^{-n} = b^{n + (-n)} = b^0$

Step 2.   But $b^0 = 1$.

Step 3.   Thus $b^n \cdot b^{-n} = 1$.

Step 4.   But $b^n \cdot \frac{1}{b^n} = 1$.

Step 5.   Hence $b^{-n} = \dfrac{1}{b^n}$.

* * *

In symbols,

> ### THE NEGATIVE EXPONENT RULE
>
> If n is a positive integer and $b > 0$, then
>
> $$b^{-n} = \frac{1}{b^n},$$
>
> the reciprocal of b^n.

The equation

$$b^n \cdot b^{-n} = 1$$

also asserts that b^n is the reciprocal of b^{-n}. Thus we also have

$$b^n = \frac{1}{b^{-n}}.$$

By the negative exponent rule,

$$2^{-3} = \frac{1}{2^3} = \frac{1}{8} \text{ or } 0.125,$$

$$\left(\frac{1}{3}\right)^{-1} = \frac{1}{\left(\frac{1}{3}\right)} = 3,$$

and

$$x^{-1} = \frac{1}{x}.$$

(a) 0.25

(c) 27

(e) 32

10. Use the negative exponent rule to help evaluate and express in decimal notation each of the following:

(a) 2^{-2} (b) $\left(\frac{1}{2}\right)^{-1}$ (c) $\left(\frac{1}{3}\right)^{-3}$ (d) $\left(\frac{3}{4}\right)^{-1}$ (e) $\left(\frac{1}{2}\right)^{-5}$

(a) 2

(c) -2

(e) 3

(f) -3

11. Find x:

(a) $\dfrac{1}{4^{-2}} = 4^x$ (b) $\dfrac{1}{3^2} = 3^x$ (c) $11^3 = \dfrac{1}{11^x}$ (d) $10^{-4} = \dfrac{1}{10^x}$

(e) $\dfrac{1}{2^{-3}} = 2^x$ (f) $\dfrac{1}{\left(\frac{1}{5^{-3}}\right)} = 5^x$

(a) -1

(c) -3

12. Find x:

(a) $2^x = 0.5$ (b) $2^x = 0.25$ (c) $2^x = 0.125$ (d) $2^x = 1$

13. (a)   Copy and fill in the table:

x	-4	-3	-2	-1	0	1	2	3	4
2^x			$\frac{1}{8}$				4		

(b)   Use the table in (a) to graph $y = 2^x$.

(c)   For any integer x, will 2^x ever equal zero?   Explain.

(d)   For any integer x, will 2^x ever be negative?   Explain.

(e)   Will the graph of $y = 2^x$ ever cross the x-axis?

(f)   Where does the graph of $y = 2^x$ cross the y-axis?

(g)   Compare your graph with that of other students.   Do you agree?

Make sure you
understand all of
Exercise 13.

(a) $2^{-4} = \frac{1}{16}$ for
instance

(c) no (Explain)

* * *

As the graph in Exercise 13(b) suggests, 2^x is positive for all x.   Similarly the following will be assumed:

$$\text{If } b > 0, \text{ then } b^x > 0 \text{ for all } x.$$

This important fact will be useful later in the chapter.

14. (a)   Copy and fill in the table:

x	-4	-3	-2	-1	0	1	2	3	4
2^{-x}									

(b)   Use the table in (a) to graph $y = 2^{-x}$.

(c)   How could the graph in (b) be obtained quickly from the graph in Exercise 13?

Hint: consider a
reflection across
a line.

THE RULE CONCERNING $\frac{b^m}{b^n}$ (THE DIVISION RULE)

The next three exercises show that there is another rule of exponents which is very much like the basic rule of exponents.   This rule concerns division, rather than multiplication.

Example 1.   Express in the form 3^x.   (a) $\frac{3^5}{3^2}$ (b) $\frac{3^3}{3^7}$

Solution.   (a) $\frac{3^5}{3^2} = \frac{3 \cdot 3 \cdot 3 \cdot 3 \cdot 3 \cdot 3}{3 \cdot 3} = 3^3$.

(b) $\frac{3^3}{3^7} = \frac{3 \cdot 3 \cdot 3}{3 \cdot 3 \cdot 3 \cdot 3 \cdot 3 \cdot 3 \cdot 3} = \frac{1}{3^4} = 3^{-4}$

15. Use the method of Example 1 to find x:

(a) $\dfrac{2^3}{2^5} = 2^x$ (b) $\dfrac{11^7}{11^4} = 11^x$ (c) $\dfrac{(\frac{1}{2})^9}{(\frac{1}{2})^6} = (\frac{1}{2})^x$ (d) $\dfrac{7^4}{7^9} = 7^x$

(e) $\dfrac{b^{10}}{b^2} = b^x$ (f) $\dfrac{b^{15}}{b^{10}} = b^x$

16. (a) Copy and explain each step of the following:

Step 1. $\dfrac{2^5}{2^3} = 2^5 \cdot \dfrac{1}{2^3}$

Step 2. $= 2^5 \cdot 2^{-3}$

Step 3. $= 2^{5 + (-3)}$

Step 4. $= 2^{5-3}$

Step 5. Thus, $\dfrac{2^5}{2^3} = 2^2.$

(b) Explain each step, where b > 0 and m and n are integers:

Step 1. $\dfrac{b^m}{b^n} = b^m \cdot \dfrac{1}{b^n}$

Step 2. $= b^m \cdot b^{-n}$

Step 3. $= b^{m + (-n)}$

Step 4. $= b^{m-n}$

Step 5. Hence $\dfrac{b^m}{b^n} = b^{m-n}.$

* * *

This reasoning leads to the following rule:

THE DIVISION RULE

If m and n are integers and
b > 0, then

$$\dfrac{b^m}{b^n} = b^{m-n}$$

Example 2 illustrates the division rule.

Example 2. Find x:

(a) $\dfrac{2^3}{2^8} = 2^x$ (b) $\dfrac{3^{-4}}{3^{-5}} = 3^x$ (c) $\dfrac{b^7}{b^4} = b^x$

Solution:

(a) $\dfrac{2^3}{2^8} = 2^{3-8}$ (b) $\dfrac{3^{-4}}{3^{-5}} = 3^{-4-(-5)}$ (c) $\dfrac{b^7}{b^4} = b^{7-4}$

thus x = -5   thus x = 1   thus x = 3

17. Find x in each case:

(a) $\dfrac{2^{-5}}{2^{-2}} = 2^x$ (b) $\dfrac{4^{-3}}{4^2} = 4^x$ (c) $\dfrac{11^5}{11^{-2}} = 11^x$

$(a)\ -3$

$(c)\ 7$

$(e)\ 6$

(d) $\dfrac{\left(\frac{1}{2}\right)^{-8}}{\left(\frac{1}{2}\right)^{-2}} = \left(\frac{1}{2}\right)^x$ (e) $\dfrac{(.3)^8}{(.3)^2} = (.3)^x$ (f) $\left(\frac{2}{3}\right)^x = 1.5$

18. The following exercise illustrates both the division rule and the negative exponents rule. Solve for x and y:

(a) $\dfrac{2^{-5}\cdot 2^{-3}}{2^{-4}\cdot 2^7} = 2^x = \dfrac{1}{2^y}$ (b) $\dfrac{11^{-6}\cdot 11^4\cdot 11^0}{11^5\cdot 11^{-10}} = 11^x = \dfrac{1}{11^y}$

$(a)\ x = -11$
$\qquad y = 11$

$(c)\ x = 7$
$\qquad y = 7$

(c) $\dfrac{10^2\cdot 10^{-7}\cdot 10^4}{10^3\cdot 10^{-5}\cdot 10^{-6}} = 10^x = \dfrac{1}{10^y}$ (d) $\dfrac{\left(\frac{1}{2}\right)^3\left(\frac{1}{2}\right)^0}{\left(\frac{1}{2}\right)^{-1}\left(\frac{1}{2}\right)^{-3}} = \left(\frac{1}{2}\right)^x = \dfrac{1}{\left(\frac{1}{2}\right)^y}$

THE RULE CONCERNING $(b^m)^n$

The next few exercises lead to a law of exponents concerned with a
"power of a power". Example 3 suggests how to simplify $(b^m)^n$ when $b>0$
and both m and n are integers.

Example 3. Find x in each case:

(a) $2^x = (2^3)^2$ (b) $3^x = (3^{-4})^3$ (c) $5^x = (5^2)^{-3}$

Solution. Use the basic rule for exponents.

(a) $2^x = (2^3)^2$ (b) $3^x = (3^{-4})^3$ (c) $5^x = (5^2)^{-3}$

$\quad = (2^3)(2^3)$ $= (3^{-4})(3^{-4})(3^{-4})$ $= \dfrac{1}{(5^2)^3}$

$\quad = 2^6$ $= 3^{-12}$

Thus x = 6.   Thus x = -12.

$= \dfrac{1}{5^2}\cdot\dfrac{1}{5^2}\cdot\dfrac{1}{5^2}$

$= \dfrac{1}{5^6}$ or 5^{-6}

thus x = -6.

19. Use the method of Example 3 to find x.

   (a) $(10^2)^3 = 10^x$ (b) $(2^{-3})^2 = 2^x$ (c) $(5^4)^3 = 5^x$ (d) $(7^2)^{-3} = 7^x$

   (e) $((\frac{1}{2})^5)^7 = (\frac{1}{2})^x$ (f) $(b^4)^{-3} = b^x$ $(b > 0)$

   The answers in Exercise 19 should agree with the following rule, which will be assumed valid for all exponents that are real numbers.

POWER OF A POWER RULE

If m and n are real numbers and b > 0, then

$$(b^m)^n = b^{mn}$$

---

20. Find x in each case:

   (a) $(2^7)^3 = 2^x$ (b) $(5^{-2})^3 = 5^x$ (c) $(10^{-4})^{-3} = 10^x$

   (d) $(7^{-4})^5 = 7^x$ (e) $(b^{-3})^4 = b^x$ (f) $(b^a)^c = b^x$

Example 4.   Sometimes it is convenient to write b^x in terms of a different base. For example, express 4^3 in the form 2^x for a suitable number x.

Solution. $2^x = 4^3$

$$= (2^2)^3$$

$$= 2^6$$

Thus $x = 6$

21. Use the method of Example 4 to find x in each case:

   (a) $9^3 = 3^x$ (b) $64^2 = 2^x$ (c) $25^{-3} = 5^x$

   (d) $(\frac{1}{8})^4 = 2^x$ (e) $9^4 = (\frac{1}{3})^x$ (f) $9^{-5} = (\frac{1}{3})^x$

   (g) $3^{-10} = (\frac{1}{3})^x$ (h) $125^2 = 25^x$ (i) $32^{-1} = 2^x$

EXPONENTS THAT ARE NOT INTEGERS

When n is an integer,

$$..., -3, -2, -1, 0, 1, 2, 3, ...$$

and b > 0, the expression b^n has meaning. But what meaning should be given to such expressions as $5^{\frac{1}{2}}$, $9^{\frac{1}{2}}$, $16^{-\frac{3}{4}}$, or $2^{\frac{3}{4}}$?

If the basic rule of exponents is to remain valid for fractional exponents, then there is only one possible value for such expressions. The next few exercises show why. We shall agree that for a positive base, b, the value of b^x will be positive.

22. If the basic rule of exponents is to remain valid when the exponents are fractions, what must be the value of the number x in each of the following?

(a) $(2^{\frac{1}{2}})(2^{\frac{1}{2}}) = 2^x$ (b) $(5^{\frac{1}{2}})(5^{\frac{1}{2}}) = 5^x$ (c) $(7^{\frac{1}{3}})(7^{\frac{1}{3}})(7^{\frac{1}{3}}) = 7^x$

* * *

The basic rule of exponents implies that

$$4^{\frac{1}{2}} \cdot 4^{\frac{1}{2}} = 4^1$$
$$= 4.$$

Since $(2)(2) = 4$ and $(-2)(-2) = 4$, it follows that $4^{\frac{1}{2}}$ is either 2 or -2. Since 4^x will be positive for all x, by agreement,

$$4^{\frac{1}{2}} = 2.$$

23. Assume that the basic rule of exponents is valid when the exponents are fractions, then find x in each case:

(a) $9^{\frac{1}{2}} \cdot 9^{\frac{1}{2}} = 9^Y$ (b) $25^{\frac{1}{2}} \cdot 25^{\frac{1}{2}} = 25^x$ (c) $27^{\frac{1}{3}} \cdot 27^{\frac{1}{3}} \cdot 27^{\frac{1}{3}} = 27^x$

(d) $16^{\frac{1}{4}} \cdot 16^{\frac{1}{4}} \cdot 16^{\frac{1}{4}} \cdot 16^{\frac{1}{4}} = 16^x$

(a) x = 1

(c) x = 1

(e) What integer is $9^{\frac{1}{2}}$? $25^{\frac{1}{2}}$? $27^{\frac{1}{3}}$? $16^{\frac{1}{4}}$? Explain.

ROOTS and the EXPONENT $\frac{1}{n}$

The following statements illustrate what we mean by "square root", "cube" root", and "fourth root".

If $r^2 = k$, then r is called a <u>square root</u> of k.
If $r^3 = k$, then r is called a <u>cube root</u> of k.
If $r^4 = k$, then r is called a <u>fourth root</u> of k.

In Chapter 6 when you study complex numbers, you will see that there are in the complex numbers two square roots for every number $k \neq 0$. Likewise, there are three cube roots, four fourth roots and so on.

Exercise 23 shows that

$9^{\frac{1}{2}}$ is a square root of 9, for $(9^{\frac{1}{2}})^2 = 9$

$27^{\frac{1}{3}}$ is a cube root of 27, for $(27^{\frac{1}{3}})^3$ is 27

and $16^{\frac{1}{4}}$ is a fourth root of 16, for $(16^{\frac{1}{4}})^4 = 16$. In general, for b > 0 and all positive integers n,

$$\left(b^{\frac{1}{n}}\right)^n = b.$$

Remember $b^{\frac{1}{n}}$ is always chosen to be <u>positive</u>.

24. Evaluate:

(a) $(\frac{1}{4})^{\frac{1}{2}}$ (b) $36^{\frac{1}{2}}$ (c) $64^{\frac{1}{3}}$ (d) $125^{\frac{1}{3}}$

(a) $\frac{1}{2}$

(b) 4

THE EXPONENT $\frac{m}{n}$

Consider what $b^{\frac{m}{n}}$ must be when m and n are integers and n is positive.

Example 5.   Use the basic rule of exponents to evaluate $4^{\frac{3}{2}}$.

Solution. $4^{\frac{3}{2}} = 4^{\frac{1}{2}+\frac{1}{2}+\frac{1}{2}}$ (since $\frac{3}{2} = \frac{1}{2} + \frac{1}{2} + \frac{1}{2}$)

$$= (4^{\frac{1}{2}})(4^{\frac{1}{2}})(4^{\frac{1}{2}}) \quad \text{(basic rule of exponents)}$$

$$= 2 \cdot 2 \cdot 2$$

$$= 8.$$

25.   Use the technique of Example 5 to evaluate:

(a) 125

(c) 16

(e) 4

(a) $25^{\frac{3}{2}}$ (b) $8^{\frac{2}{3}}$ (c) $64^{\frac{2}{3}}$ (d) $64^{\frac{5}{6}}$ (e) $32^{\frac{2}{5}}$ (f) $32^{\frac{6}{5}}$

* * *

The "power of a power" rule provides a better way to deal with $b^{\frac{m}{n}}$, as the next example illustrates.

Example 6.   Use the "power of a power" rule to find (a) $4^{\frac{3}{2}}$ and (b) $27^{-\frac{4}{3}}$

Solution.   (a) $4^{\frac{3}{2}} = (4^{\frac{1}{2}})^3$ [since $\frac{1}{2} = (3)(\frac{1}{2})$, "power of a power" rule]

$$= (2)^3$$

$$= 8.$$

(b) $27^{-\frac{4}{3}} = 27^{(\frac{1}{3})(-4)}$ [power of a power rule]

$$= (27^{\frac{1}{3}})^{-4}$$

$$= 3^{-4}$$

$$= \frac{1}{3^4}$$

$$= \frac{1}{81}$$

Of course you can write

$27^{-\frac{4}{3}} = (27^{-4})^{\frac{1}{3}}$

here, but that would make the problem more difficult.

26.   Using the "power of a power" rule, evaluate

(a) 9

(c) 32

(a) $27^{\frac{2}{3}}$ (b) $27^{-\frac{2}{3}}$ (c) $4^{\frac{5}{2}}$ (d) $4^{-\frac{3}{2}}$

27. Using the "power of a power" rule, evaluate $9^{\frac{3}{2}}$ in two ways:

   (a) writing it as $(9^{\frac{1}{2}})^3$

   (b) writing it as $(9^3)^{\frac{1}{2}}$.

28. Which of these quantities equal $(81)^{-\frac{3}{4}}$?

   (a) -27 (b) 27 (c) $\frac{1}{27}$ (d) $\frac{1}{81^{\frac{3}{4}}}$

29. Evaluate:

   (a) $64^{\frac{1}{6}}$ (b) $64^{\frac{1}{2}}$ (c) $64^{\frac{2}{3}}$ (d) $64^{-\frac{1}{2}}$

   (a) 2

   (c) 16

30. With the aid of Exercise 29, graph $y = 64^x$.

* * *

When working with $b^{\frac{m}{n}}$ keep in mind the following three equivalent statements which are all versions of the "power to a power" rule:

POWER OF A POWER RULE (FRACTIONAL EXPONENTS)

If m and n are integers with n > 0,

$$(b^{\frac{1}{n}})^m = b^{\frac{m}{n}} = (b^m)^{\frac{1}{n}}$$

MEMORIZE!

EXPONENTS NOT OF THE FORM $\frac{m}{n}$

So far in this chapter b^x has been defined for any exponent x that can be expressed as the quotient of two integers. In other words, b^x has been defined whenever x is a <u>rational number</u>, $\frac{m}{n}$. The definition of b^x when x is not rational is rather complicated, and is left for the Projects section. There it is shown how to define $2^{\sqrt{2}}$ or 2^π. A real number that is not rational is called <u>irrational</u>. For instance $\sqrt{2}$ and π are irrational.

POWER OF A PRODUCT $(ab)^n$

The basic rule concerning $b^m \cdot b^n = b^{m+n}$ refers to the product of two exponentials with the same base. The next rule concerns different bases but the same exponent. It is illustrated by the following two simple cases:

Observe that

$$(3 \cdot 5)^4 = (3 \cdot 5)(3 \cdot 5)(3 \cdot 5)(3 \cdot 5)$$

$$= (3 \cdot 3 \cdot 3 \cdot 3)(5 \cdot 5 \cdot 5 \cdot 5)$$

$$= 3^4 \cdot 5^4$$

Thus, $(3 \cdot 5)^4 = 3^4 \cdot 5^4$

Observe that

$$(7 \cdot 9)^{-1} = \frac{1}{7 \cdot 9}$$

$$= \frac{1}{7} \cdot \frac{1}{9}$$

$$= 7^{-1} \cdot 9^{-1}$$

Thus, $(7 \cdot 9)^{-1} = (7^{-1})(9^{-1})$

Note that the exponent is the same throughout each of the final equations. The following rule, suggested by these two examples, will be assumed for all exponents.

POWER OF A PRODUCT RULE

For any positive numbers a and b and any exponent n,
$$(ab)^n = a^n \cdot b^n.$$

---

31. If $a^4 = 9$ and $b^4 = 7$ find each of the following:

    (a) a^{-4} (b) a^8 (c) $(ab)^4$ (d) $(ab)^{-4}$ (e) a^2 (f) $a^2 b^4$ (g) a^6

    (h) $(ab)^8$.

(a) $\frac{1}{9}$

(c) 63
(e) 3
(g) 27

32. Find x and y.

    (a) $(2 \cdot 3)^4 = 2^x \cdot 3^y$ (b) $(3 \cdot 5)^{-2} = 3^x \cdot 5^y$

    (c) $(\frac{1}{4} \cdot 7)^3 = (\frac{1}{4})^x \cdot 7^y$ (d) $(ab)^6 = a^x \cdot b^y$

    (e) $(3^9)(7^9) = 21^x$ (f) $(2^{-5})(3^{-5}) = 6^y$

    (g) $15^3 = (3^x)(5^x)$ (h) $x^3 = (2^3)(5^3)$

(a) $x = y = 4$
(c) $x = y = 3$
(e) $x = 9$
(g) $x = 3$

POWER OF A QUOTIENT, $(\frac{a}{b})^n$

The next rule relates $(\frac{a}{b})^n$ to a^n and b^n. It is suggested by this typical illustration:

$$(\frac{2}{3})^4 = (\frac{2}{3})(\frac{2}{3})(\frac{2}{3})(\frac{2}{3})$$

$$= \frac{2 \cdot 2 \cdot 2 \cdot 2}{3 \cdot 3 \cdot 3 \cdot 3}$$

$$= \frac{2^4}{3^4}$$

Thus

$$(\frac{2}{3})^4 = \frac{2^4}{3^4}$$

More generally,

POWER OF A QUOTIENT RULE

For any positive numbers a and b and any exponent n,
$$(\frac{a}{b})^n = \frac{a^n}{b^n}$$

---

33. Find x and y.

(a) $(\frac{2}{3})^4 = \frac{2^x}{3^y}$

(b) $(\frac{4}{5})^{-1} = \frac{4^x}{5^y}$

(c) $(\frac{11}{7})^2 = \frac{11^x}{7^y}$

(d) $\frac{4^2}{9^2} = (\frac{4}{9})^x$

(e) $\frac{5^{-3}}{7^{-3}} = (\frac{5}{7})^x$

(f) $5^4 = \frac{15^x}{3^x}$

(g) $x^3 = \frac{3^3}{7^3}$

(h) $\frac{7^x}{10^x} = (\frac{7}{10})^5$

(i) $\frac{2^3}{5^3} = x^3$

(a) $x = y = 4$
(c) $x = y = 2$
(e) $x = -3$
(g) $x = \frac{3}{7}$
(i) $x = \frac{2}{5}$

34. Express each of these in the form 3^n:

(a) $\frac{6^5}{2^5}$

(b) $\frac{4^7}{12^7}$

(c) $(3^4)^5$

(d) $\frac{5^{-4}}{15^{-4}}$

(e) 1

(f) $\frac{1}{3}$

(g) $\frac{21^{-2}}{7^{-2}}$

(a) 3^5
(c) 3^{20}
(e) 3^0
(g) 3^{-2}

The rules concerning $(ab)^n$ and $(\frac{a}{b})^n$ may suggest that there is a rule concerning $(a + b)^n$. The next exercise shows that this is not the case.

35. (a) Is $(3 + 4)^2$ equal to $3^2 + 4^2$?

(b) Is $(9 + 16)^{\frac{1}{2}}$ equal to $9^{\frac{1}{2}} + 16^{\frac{1}{2}}$?

(c) Is $(a + b)^n$ equal to $a^n + b^n$?

Pause and study this exercise now. It will help to avoid many mistakes later.

A LIST OF RULES

Below is a short summary of the rules of exponents examined so far in this chapter.

$$b^n > 0$$

$$b^m \cdot b^n = b^{m+n} \quad \text{(The basic rule)}$$

$$b^0 = 1$$

$$b^{-n} = \frac{1}{b^n}$$

$$\frac{b^m}{b^n} = b^{m-n}$$

$$(b^m)^n = b^{mn}$$

$b^{\frac{1}{n}}$ is the positive nth root of b if n is a positive integer.

$$(ab)^n = a^n b^n$$

$$(\frac{a}{b})^n = \frac{a^n}{b^n}$$

The following exercises apply some of the above rules.

Make a list of these rules for your own notes and memorize them. They will be used again and again.

36. Use the laws of exponents to find x.

(a) $8 = 4^x$

(b) $9^x = 27$

(c) $25^{\frac{1}{3}} = 5^x$

(d) $25 = 125^x$

(e) $25^{\frac{1}{3}} = 125^x$

(f) $4 = (\frac{1}{2})^x$

(g) $8 = (\frac{1}{4})^x$

(h) $25 = (\frac{1}{5})^x$

(i) $27 = (\frac{1}{9})^x$

Exercises 36 and 37 will take time. Do not rush through them. Hint:
(a) $4^x = (2^2)^x$
$= 2^{2x}$
(a) $\frac{3}{2}$
(c) $\frac{2}{3}$
(e) $\frac{2}{9}$
(g) $-\frac{3}{2}$
(i) $-\frac{3}{2}$

37. Find x.

(a) $(2)^{\frac{2}{3}} = 4^x$ (b) $2^{\frac{2}{3}} = 16^x$ (c) $3^{\frac{2}{5}} = 9^x$

(d) $(\frac{1}{3})^{\frac{4}{5}} = 81^x$ (e) $(2^{\frac{1}{4}})^3 = 32^x$ (f) $4^{\frac{3}{2}} = 16^x$

(g) $8^{-\frac{2}{3}} = 2^x$ (h) $25^{-\frac{3}{2}} = 125^x$ (i) $9^x = 3^{\frac{3}{2}}$

* * *

For some values of b it is easy to evaluate $b^{\frac{1}{2}}$, the square root of b. For instance,

$$9^{\frac{1}{2}} = 3 \quad \text{and} \quad 100^{\frac{1}{2}} = 10.$$

But finding $3^{\frac{1}{2}}$, for instance, offers a challenge. Since

$$(1.7)^2 = 2.89 \text{ is less than } 3$$

and

$$(1.8)^2 = 3.24 \text{ is more than } 3,$$

the square root of 3 is between 1.7 and 1.8.

We could continue this guessing process for many decimal places. For example, the square root of 3 is between 1.7320 and 1.7321 because

$$(1.7320)^2 = 2.999824 \text{ and } (1.7321)^2 = 3.00017041$$

From this we may conclude that 1.732 is a good approximation of $3^{\frac{1}{2}}$, $3^{\frac{1}{2}} \doteq 1.732$.

38. Is 2.646 smaller or larger than $7^{\frac{1}{2}}$?

* * *

The next example shows that this information can be used to evaluate other exponentials.

Example 7. Similar experimenting shows

$$2^{\frac{1}{2}} \doteq 1.414.$$

Use this information to estimate $2^{\frac{5}{2}}$.

Solution. $2^{\frac{5}{2}} = 2^{\frac{4}{2}} \cdot 2^{\frac{1}{2}}$

$= 2^2 \cdot 2^{\frac{1}{2}}$

$\doteq 4(1.414)$

$\doteq 5.656.$

39. Using the estimates $2^{\frac{1}{2}} \doteq 1.414$ and $3^{\frac{1}{2}} \doteq 1.732$, estimate each of the following:

(a) $2^{\frac{3}{2}}$ (b) $2^{\frac{7}{2}}$ (c) $3^{\frac{5}{2}}$ (d) $3^{\frac{7}{2}}$ (e) $\left(\frac{3}{2}\right)^{\frac{1}{2}}$ (f) $6^{\frac{1}{2}}$

40. Using the estimate $5^{\frac{1}{2}} \doteq 2.236$, estimate

(a) $20^{\frac{1}{2}}$ (b) $45^{\frac{1}{2}}$ (c) $5^{\frac{3}{2}}$ (d) $5^{-\frac{1}{2}}$ (e) $125^{\frac{1}{2}}$ (f) $5^{\frac{5}{2}}$

Hint: Use the method of Example 7.

(a) 2.828
(c) 15.588
(e) 1.225

(a) 4.472
(c) 11.18
(e) 11.18

RADICALS

The exponents $\frac{1}{2}$, $\frac{1}{3}$, $\frac{1}{4}$... occur so frequently in mathematics that a special notation is sometimes used for them. It makes use of the symbol $\sqrt{}$, called the <u>radical sign</u>.

Study this section carefully!

For any positive number b,

$b^{\frac{1}{2}}$ is written also $\sqrt{b}$

$b^{\frac{1}{3}}$ is written also $\sqrt[3]{b}$

$b^{\frac{1}{4}}$ is written also $\sqrt[4]{b}$

In general, if n is any integer larger than 2,

$b^{\frac{1}{n}}$ is written also as $\sqrt[n]{b}$.

In radical form $b^{\frac{1}{2}}$ is written simply $\sqrt{b}$, not $\sqrt[2]{b}$. For instance,

$25^{\frac{1}{2}} = \sqrt{25} = 5$

$8^{\frac{1}{3}} = \sqrt[3]{8} = 2$

$16^{\frac{1}{4}} = \sqrt[4]{16} = 2$

$64^{\frac{1}{6}} = \sqrt[6]{64} = 2$

Keep in mind that $\sqrt{2}$ is the <u>positive number</u> whose square is 2. Similarly, $\sqrt[4]{2}$ is the <u>positive</u> fourth root of 2. In short, $\sqrt[n]{b}$ is <u>always positive</u> for b > 0.

To indicate the negative number whose square is 2 (approximately -1.414), put a minus sign in front of $\sqrt{2}$; that is

$-\sqrt{2}$ is the negative of the square root of 2.

Thus $-\sqrt{4} = -2$ and $-\sqrt{5} = -2.236$.

Remember:
$\sqrt{4} = 2$
But
$-\sqrt{4} = -2$

41. Evaluate:

(a) $\sqrt[3]{125}$ (b) $-\sqrt{64}$ (c) $-\sqrt[4]{81}$

(d) $\sqrt{49}$ (e) $-\sqrt{49}$ (f) $\sqrt[4]{16}$

(a)   5
(c)   -3
(e)   -7

42.   Evaluate:

(a) $\sqrt{\dfrac{4}{9}}$ (b) $\dfrac{\sqrt[3]{64}}{\sqrt[3]{125}}$ (c) $\sqrt[4]{16}$ (d) $\sqrt[5]{\dfrac{32}{243}}$

(a) $\dfrac{2}{3}$

(c) 2

* * *

The "power of a product" and "power of a quotient" rules lead to corresponding rules for radicals.

RULES FOR RADICALS

For positive numbers <u>a</u> and b and any integer n $\geq$ 2

$$\sqrt[n]{ab} = \sqrt[n]{a}\ \sqrt[n]{b} \qquad \text{(product rule)}$$

and

$$\sqrt[n]{\dfrac{a}{b}} = \dfrac{\sqrt[n]{a}}{\sqrt[n]{b}} \qquad \text{(quotient rule)}$$

For instance, the first of the two rules is just the rule

$$(ab)^{\frac{1}{n}} = a^{\frac{1}{n}}\ b^{\frac{1}{n}}$$

expressed in radicals.

MEMORIZE!

These are the only <u>rules</u> for radicals. Before inventing other rules, check them with specific numbers. In particular there is no rule relating $\sqrt{a + b}$ to $\sqrt{a}$ and $\sqrt{b}$

SIMPLIFYING RADICALS

The next example shows how the "product rule" is used in simplifying radicals.

Example 8.   Simplify:

(a) $\sqrt{8}$ (b) $\sqrt{48}$ (c) $\sqrt[3]{16}$ (d) $\sqrt{x^3}$ where $x \geq 0$

Solution.   (a) $\sqrt{8} = \sqrt{4 \cdot 2}$ (b) $\sqrt{48} = \sqrt{2^4 \cdot 3}$
$\phantom{Solution.   (a) \sqrt{8}} = \sqrt{4}\ \sqrt{2}$ $\phantom{(b) \sqrt{48}} = \sqrt{2^4}\ \sqrt{3}$
$\phantom{Solution.   (a) \sqrt{8}} = 2\sqrt{2}$ $\phantom{(b) \sqrt{48}} = 2^2\sqrt{3}$ or $4\sqrt{3}$

(c) $\sqrt[3]{16} = \sqrt[3]{2^3 \cdot 2}$ (d) $\sqrt{x^3} = \sqrt{x^2 \cdot x}$
$\phantom{(c) \sqrt[3]{16}} = \sqrt[3]{2^3}\ \sqrt[3]{2}$ $\phantom{(d) \sqrt{x^3}} = \sqrt{x^2}\ \sqrt{x}$
$\phantom{(c) \sqrt[3]{16}} = 2\sqrt[3]{2}$ $\phantom{(d) \sqrt{x^3}} = x\sqrt{x}$

43.   Simplify (use the product rule for radicals):

(a) $\sqrt{12}$ (b) $\sqrt{20}$ (c) $\sqrt{45}$ (d) $\sqrt{72}$ (e) $\sqrt{125}$ (f) $\sqrt[3]{81}$

(g) $\sqrt[3]{54}$ (h) $\sqrt[3]{128}$ (i) $\sqrt[4]{48}$ (j) $\sqrt[5]{64}$ (k) $\sqrt[4]{32}$ (l) $\sqrt{1000}$

(m) $\sqrt{98}$ (n) $\sqrt[3]{24}$ (o) $\sqrt{63}$ (p) $\sqrt[4]{32}$ (q) $\sqrt[6]{128}$ (r) $\sqrt[3]{10,000}$

(a) $2\sqrt{3}$
(c) $3\sqrt{5}$
(e) $5\sqrt{5}$
(g) $3\sqrt[3]{2}$
(i) $2\sqrt[4]{3}$
(k) $2\sqrt[4]{2}$
(m) $7\sqrt{2}$
(o) $3\sqrt{7}$
(q) $2\sqrt[6]{2}$

44. Simplify:

(a) $\sqrt{28x^2}$ (b) $2\sqrt{27x^5}$ (c) $\sqrt{32x^3}$ (d) $\sqrt{288x^9}$ (e) $\sqrt[3]{16x^4}$

(f) $\sqrt[4]{8x^5}$ (g) $\sqrt{x^2}$ (h) $x^{\frac{5}{3}}$ (i) $\sqrt[n]{x^n}$ (j) $\sqrt{x^{10}}$

(k) $\sqrt{x^9}$ (l) $\sqrt[3]{54x^4}$ (m) $\sqrt[3]{x^8}$ (n) $\sqrt[4]{625x^7}$ (o) $\sqrt[3]{72x^5}$

> Note.   The expression 0^x is defined to be 0 for $x > 0$.   For no other exponents is 0^x defined.

(a) $2x\sqrt{7}$
(c) $4x\sqrt{2x}$
(e) $2x\sqrt[3]{2x}$
(g) x
(i) x
(k) $x^4\sqrt{x}$
(m) $x^2\sqrt[3]{x^2}$
(o) $2x\sqrt[3]{9x^2}$
Of course assume $x > 0$.

45. (a)   For positive numbers a and b, does $\sqrt{a + b}$ always equal $\sqrt{a} + \sqrt{b}$?   Sometimes?   Never?   Give examples in support of your answer.

(b)   If another person told you that $\sqrt{a + b}$ equals $\sqrt{a} + \sqrt{b}$, how would you convince yourself that the person was wrong?

* * *

The quotient rule can also be used to help you simplify radicals.

Example 9.   Simplify:

(a) $\dfrac{\sqrt{8}}{\sqrt{2}}$ (b) $\dfrac{\sqrt{81}}{\sqrt{27}}$ (c) $\dfrac{\sqrt[3]{16x}}{\sqrt[3]{2x}}$ ($x > 0$)

Assume all letters denote positive numbers

Solution.

(a) $\dfrac{\sqrt{8}}{\sqrt{2}} = \sqrt{\dfrac{8}{2}}$ (b) $\dfrac{\sqrt{81}}{\sqrt{27}} = \sqrt{\dfrac{81}{27}}$ (c) $\dfrac{\sqrt[3]{16x^4}}{\sqrt[3]{2x}} = \sqrt[3]{\dfrac{16x^4}{2x}}$

$\qquad = \sqrt{4}$ or 2 $\qquad = \sqrt{3}$ $\qquad = \sqrt[3]{8x^3}$

$\qquad\qquad\qquad\qquad\qquad\qquad\qquad\qquad\qquad\qquad = 2x$

46. Using the quotient rule for radicals, write each of the following as an expression free of radicals in the denominator, then simplify as in Example 9.

(a) $\dfrac{\sqrt{12}}{\sqrt{2}}$ (b) $\dfrac{\sqrt{27}}{\sqrt{9}}$ (c) $\dfrac{\sqrt[3]{54}}{\sqrt[3]{2}}$ (d) $\dfrac{\sqrt[4]{32}}{\sqrt[4]{8}}$ (e) $\dfrac{\sqrt{125}}{\sqrt{5}}$

(f) $\dfrac{\sqrt[5]{128}}{\sqrt[5]{64}}$ (g) $\dfrac{\sqrt{4}}{\sqrt{2}}$ (h) $\dfrac{\sqrt{15}}{\sqrt{3}}$

(a) $\sqrt{6}$
(c) 3
(e) 5
(g) $\sqrt{2}$

47. Simplify as in Exercise 46.

(a) $\dfrac{\sqrt{8x}}{\sqrt{2x}}$ (b) $\dfrac{7\sqrt{63x^3}}{2\sqrt{7x}}$ (c) $\dfrac{5\sqrt{27x^3}}{6\sqrt{3x}}$ (d) $\dfrac{4\sqrt{8m^3}}{8\sqrt{2m}}$

(a) 2
(c) $\dfrac{5x}{2}$
(e) $6n\sqrt{2}$
(g) $3a\sqrt{2}$

$$\text{(e) } \frac{21\sqrt{8n^3}}{\sqrt{49n}} \qquad \text{(f) } \frac{\sqrt{x^2y}}{x\sqrt{y}} \qquad \text{(g) } \frac{\sqrt{54a^2}}{\sqrt{3}} \qquad \text{(h) } \frac{2\sqrt{15x^5}}{\sqrt{3x}}$$

* * *

The following example illustrates a typical multiplication of two radicals. Observe that it corresponds to the equation

$$2^{\frac{1}{2}} \cdot 2^{\frac{1}{3}} = 2^{\frac{5}{6}},$$

expressed in radicals.

Example 10.   Write the product $\sqrt{2} \cdot \sqrt[3]{2}$ as a single radical.

Solution. $\sqrt{2} \cdot \sqrt[3]{2} = 2^{\frac{1}{2}} \cdot 2^{\frac{1}{3}}$

$$= 2^{\frac{1}{2} + \frac{1}{3}}$$

$$= 2^{\frac{5}{6}}$$

$$= (2^5)^{\frac{1}{6}}$$

$$= \sqrt[6]{2^5}$$

48. Write each of the following as a single radical:

(a) $\sqrt{5} \ \sqrt[3]{5}$ (b) $\sqrt{3} \ \sqrt[3]{3}$ (c) $\sqrt[3]{9} \ \sqrt{3}$ (d) $\sqrt[4]{4} \ \sqrt{2}$

(e) $\sqrt[4]{4} \ \sqrt[3]{3}$ (f) $\sqrt{5} \ \sqrt[9]{16}$

(a) $\sqrt[6]{5^5}$
(c) $3\sqrt[3]{3}$
(e) $\sqrt[12]{72}$

The exponent $\frac{1}{2}$ is very common in algebra and its applications in trigonometry and physics.   The next few exercises concern manipulations with this exponent.

THE PYTHAGOREAN THEORM

Recall the Pythagorean Theorem from geometry: In a right triangle, the length of the hypotenuse squared equals the sum of the squares of the lengths of the other two sides:

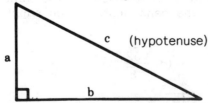

c (hypotenuse) $c^2 = a^2 + b^2$

a

b

49. Approximate.   x to three decimals places if $\sqrt{2} \doteq 1.414$, $\sqrt{3} \doteq 1.732$ and $\sqrt{5} \doteq 2.236$ (use no tables):

(a) 1.414

(a)

(b)

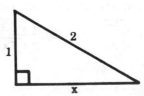

(c)

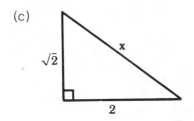

(d)

* * *

In the preceding exercise the irrational numbers $\sqrt{2}$, $\sqrt{3}$, and $\sqrt{5}$ appeared. In using the Pythagorean Theorem, irrational numbers frequently occur. For example, in finding the hypotenuse of this triangle, an irrational number appears.

$$x^2 = 2^2 + 1^2$$
$$x^2 = 5$$
$$x = \sqrt{5}$$

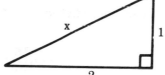

See the Projects for more information on irrational numbers.

As long ago as Euclid (300 B.C.) and Pythagoras (5th century B.C.) it was known that $\sqrt{5}$ is irrational. In the Projects it will be shown that whenever the positive integer N is not the square of an integer, then $\sqrt{N}$ is irrational. Thus, $\sqrt{1} = 1$, $\sqrt{4} = 2$, $\sqrt{9} = 3$, $\sqrt{16} = 4$, ... are rational, but $\sqrt{2}$, $\sqrt{3}$, $\sqrt{5}$, $\sqrt{6}$, $\sqrt{7}$, ... are irrational.

For instance, $\sqrt{54}$ is irrational, but it can be simplified:

$$\sqrt{54} = \sqrt{9 \cdot 6}$$
$$= \sqrt{9} \cdot \sqrt{6}$$
$$= 3\sqrt{6}$$

Because $\sqrt{54}$ is irrational a square root symbol will appear in any such simplification.

RATIONALIZING THE DENOMINATOR

Using the square root table, we could compute

$$\frac{1}{\sqrt{2}}$$

as follows:

$$\frac{1}{\sqrt{2}} \doteq \frac{1}{1.414},$$
$$\doteq 0.7072$$

While a calculator can do the tedious arithmetic, the technique of rationalizing the denominator is still very useful in certain problems as you will see.

Note the tedious division of 1.414 into 1. The next example illustrates a short cut, called <u>rationalizing the denominator</u>.

Example 11.   If $\sqrt{2} \doteq 1.414$, approximate

(a) $\dfrac{1}{\sqrt{2}}$ (b) $\dfrac{2}{\sqrt{27}}$

Recall N·1 = N for all numbers N. When rationalizing a denominator multiply by 1 in some suitable form so that the denominator becomes rational. Here the form $1 = \dfrac{\sqrt{2}}{\sqrt{2}}$ is used for (a) and $\dfrac{\sqrt{3}}{\sqrt{3}} = 1$ for (b)

Solution.

(a) $\dfrac{1}{\sqrt{2}} = \dfrac{1}{\sqrt{2}} \cdot \dfrac{\sqrt{2}}{\sqrt{2}}$ (b) $\dfrac{2}{\sqrt{27}} = \dfrac{2}{\sqrt{27}} \cdot \dfrac{\sqrt{3}}{\sqrt{3}}$

$\qquad = \dfrac{\sqrt{2}}{2}$ $= \dfrac{2\sqrt{3}}{\sqrt{81}}$

$\qquad = \dfrac{1.414}{2}$ $= \dfrac{2\sqrt{3}}{9}$

$\qquad = 0.707$ $\doteq 0.385$

Since $\sqrt{3} \doteq 1.732$ the answer is an approximation.

(Note the easy division by 2.)

50. Compute to three decimal places (refer to the square root table).

(a) 0.447
(c) 1.414

(a) $\dfrac{1}{\sqrt{5}}$ (b) $\dfrac{1}{\sqrt{3}}$ (c) $\dfrac{2}{\sqrt{2}}$ (d) $\dfrac{1}{\sqrt{12}}$

CONJUGATES

Even more complicated denominators can be treated in a similar way. To do this, the identity
$$(a + b)(a - b) = a^2 - b^2$$
is useful. For instance,
$$(1 + \sqrt{2})(1 - \sqrt{2}) = 1^2 - (\sqrt{2})^2$$
$$= 1 - 2$$
$$= -1.$$

The numbers $1 + \sqrt{2}$ and $1 - \sqrt{2}$ are each called the <u>conjugate</u> of the other.

Example 12.   Assume that $\sqrt{2} \doteq 1.414$.   Estimate
$$\dfrac{1}{1 + \sqrt{2}}$$
to three decimal places.

Solution.

$$\frac{1}{1 + \sqrt{2}} = \frac{1}{1 + \sqrt{2}} \cdot \frac{1 - \sqrt{2}}{1 - \sqrt{2}}$$

$$= \frac{1 - \sqrt{2}}{1^2 - (\sqrt{2})^2}$$

$$= \frac{1 - \sqrt{2}}{1 - 2}$$

$$= \frac{1 - \sqrt{2}}{-1}$$

$$= \sqrt{2} - 1$$

$$\doteq 1.414 - 1$$

$$\doteq 0.414$$

Note that the denominator has no square root.

Of course, $\frac{1}{1 + \sqrt{2}}$ can be estimated by division:

$$\frac{1}{1 + \sqrt{2}} \doteq \frac{1}{1 + 1.414}$$

$$\doteq \frac{1}{2.414}$$

$$\doteq 0.414$$

A fraction with a more complicated denominator can be simplified similarly. For instance,

$$\frac{1}{\sqrt{5} - \sqrt{2}} = \frac{1}{\sqrt{5} - \sqrt{2}} \cdot \frac{\sqrt{5} + \sqrt{2}}{\sqrt{5} + \sqrt{2}}$$

$$= \frac{\sqrt{5} + \sqrt{2}}{(\sqrt{5})^2 - (\sqrt{2})^2}$$

$$= \frac{\sqrt{5} + \sqrt{2}}{5 - 2}$$

$$\doteq \frac{2.236 + 1.414}{3}$$

$$\doteq 1.217$$

Again, the radical disappears from the denominator.   (The expression $\sqrt{5} + \sqrt{2}$ is called the conjugate of $\sqrt{5} - \sqrt{2}$).

51. Use the estimates $\sqrt{2} \doteq 1.414$, $\sqrt{3} \doteq 1.732$ and $\sqrt{5} \doteq 2.236$ and the technique of <u>rationalizing</u> the <u>denominator</u> to estimate each of the following to three decimal places.

Use a calculator only for arithmetic here. Do not use the square root key for Exercise 51.

(a) $\dfrac{1}{\sqrt{5} - \sqrt{2}}$ (b) $\dfrac{1}{\sqrt{3} + \sqrt{2}}$ (c) $\dfrac{\sqrt{2}}{\sqrt{6} - \sqrt{2}}$ (d) $\dfrac{2}{1 + \sqrt{3}}$

(e) $\dfrac{1}{\sqrt{5} + \sqrt{2}}$ (f) $\dfrac{1}{2\sqrt{5} - \sqrt{2}}$ (g) $\dfrac{\sqrt{3}}{1 + \sqrt{3}}$ (h) $\dfrac{\sqrt{6}}{\sqrt{3} - \sqrt{2}}$

* * *

The next example applies the distributive rule:

$$a(b + c) = ab + ac$$

to help simplify products of expressions that contain radicals.

Example 13.   Multiply:

(a) $\sqrt{3}\,(\sqrt{6} - \sqrt{5})$ (b) $(\sqrt{12} - \sqrt{18})(\sqrt{8} + \sqrt{27})$

Solution: (a) $\sqrt{3}(\sqrt{6} - \sqrt{5}) = \sqrt{18} - \sqrt{15}$

$$= \sqrt{9 \cdot 2} - \sqrt{15}$$

$$= 3\sqrt{2} - \sqrt{15}$$

(b) $(\sqrt{12} - \sqrt{18})(\sqrt{8} + \sqrt{27}) = (2\sqrt{3} - 3\sqrt{2})(2\sqrt{2} + 3\sqrt{3})$

$$= 4\sqrt{6} + 6\sqrt{9} - 6\sqrt{4} - 9\sqrt{6}$$

$$= 4\sqrt{6} + 6(3) - 6(2) - 9\sqrt{6}$$

$$= -5\sqrt{6} + 6$$

Observe in Example 13(b) that the <u>radicals were simplified before multi-plying</u>; thus the computation was less tedious.

52.   Simplify the product, but leave the answer in radical form.

(a) $\sqrt{2}(\sqrt{2} + \sqrt{6})$ (b) $(\sqrt{2} + 3)(3\sqrt{2} - 1)$

(c) $(\sqrt{3} - \sqrt{2})^2$ (d) $(\sqrt{5} - \sqrt{7})(\sqrt{5} + \sqrt{7})$

(e) $(3\sqrt{3} - 5\sqrt{2})(2\sqrt{3} + 2\sqrt{2})$ (f) $(\sqrt{27} - \sqrt{50})(\sqrt{12} + \sqrt{8})$

(g) $(\sqrt{98} - \sqrt{72})(\sqrt{48} + \sqrt{18})$ (h) $(\sqrt{3} + \sqrt{2})(\sqrt{8} - \sqrt{3})$

53.   Multiply, then simplify the radicals (assume a and b are positive).

(a) $(\sqrt{a} + \sqrt{b})(\sqrt{a} - \sqrt{b})$ (b) $(\sqrt{a} + \sqrt{b})^2$

(c) $(3\sqrt{a} + 2\sqrt{b})(\sqrt{a} - 2\sqrt{b})$ (d) $(\sqrt{a} - \sqrt{b})^2$

* * *

Sometimes it is necessary to rationalize the denominator when the expression is more complicated.

Example 14. Simplify:

(a) $\sqrt{2\tfrac{1}{3}}$

(b) $\sqrt{\tfrac{1}{x} - 1}$ (assume x between 0 and 1)

Solution:

(a) $\sqrt{2\tfrac{1}{3}} = \sqrt{2 + \tfrac{1}{3}}$

$= \sqrt{2 \cdot \tfrac{3}{3} + \tfrac{1}{3}}$

$= \sqrt{\tfrac{6}{3} + \tfrac{1}{3}}$

$= \sqrt{\tfrac{7}{3}}$

$= \sqrt{\tfrac{7}{3} \cdot \tfrac{3}{3}}$

$= \dfrac{\sqrt{21}}{3}$

(b) $\sqrt{\tfrac{1}{x} - 1} = \sqrt{\tfrac{1}{x} - 1(\tfrac{x}{x})}$

$= \sqrt{\tfrac{1}{x} - \tfrac{x}{x}}$

$= \sqrt{\tfrac{1-x}{x}}$

$= \sqrt{(\tfrac{1-x}{x}) \tfrac{x}{x}}$

$= \sqrt{\tfrac{x - x^2}{x^2}}$

$= \dfrac{\sqrt{x - x^2}}{x}$

Observe that $\sqrt{x - x^2}$ cannot be simplified because there is no rule for simplifying $\sqrt{a - b}$

54. Simplify (including "rationalizing the denominator"; assume that all letters denote positive numbers).

(a) $\sqrt{1\tfrac{1}{2}}$ (b) $\sqrt{1\tfrac{2}{5}}$ (c) $\sqrt{\tfrac{1}{x}}$ (d) $\sqrt{1 + \tfrac{1}{x}}$

(e) $\sqrt{x + \tfrac{1}{x}}$ (f) $\sqrt{2a + \tfrac{a}{4}}$ (g) $\sqrt{x^2 + \tfrac{x}{2}}$ (h) $\sqrt{\tfrac{1}{x + 1}}$

(i) $\sqrt{\tfrac{x + 1}{x - 1}}$ (where x > 1)   (j) $\sqrt{\tfrac{b^2}{4a^2} - \tfrac{c}{a}}$

Hint: Careful! In (d) first write $1 + \tfrac{1}{x}$ as a single fraction.

(a) $\dfrac{\sqrt{6}}{2}$

(c) $\dfrac{\sqrt{x}}{x}$

(e) $\dfrac{\sqrt{x^3 + x}}{x}$

(g) $\dfrac{\sqrt{4x^2 + 2x}}{2}$

(i) $\dfrac{\sqrt{x^2 - 1}}{x - 1}$

(j) $\dfrac{\sqrt{b^2 - 4ac}}{2a}$

NEGATIVE BASES

So far this chapter presented b^x for a positive base. The bases b = 0 or b < 0 have not been discussed.

Negative bases, such as -4, require some care. Integer exponents offer no new difficulty. For instance, they are treated as in the case of a positive base:

$(-4)^2 = (-4)(-4) = 16$

$(-4)^3 = (-4)(-4)(-4) = -64$

Similarly,

$(-4)^0$

is defined to be 1. Also $(-4)^{-3}$ is defined as $\dfrac{1}{(-4)^3} = -\dfrac{1}{64}$.

But the exponent $\frac{1}{2}$, for instance, causes trouble. We would expect
$$(-4)^{\frac{1}{2}}$$
to be a "square root" of -4. But there is no real number x whose square is -4. The equation
$$x^2 = -4$$
has no solution. Thus $(-4)^{\frac{1}{2}}$ is not defined.

Thus, if b is negative, $b^{\frac{1}{2}}$ is <u>not</u> defined. However, $b^{\frac{1}{3}}$ is defined. For instance, $(-8)^{\frac{1}{3}}$ is defined as the cube root of -8. In other words, the equation
$$x^3 = -8$$
has a solution, namely -2. Thus
$$(-8)^{\frac{1}{3}} = -2.$$
If n is an <u>odd</u> integer, then
$$b^{\frac{1}{n}}$$
is defined for any negative base b. It is the solution of the equation
$$x^n = b.$$
This solution is negative.

An exponential such as $(-8)^{\frac{2}{3}}$ is defined as
$$[(-8)^{\frac{1}{3}}]^2.$$

55. Find:

(a) $((-2)^3)^2$ (b) $((-3^2)^2$ (c) $((-3)^{-2})^2$ (d) $((-2)^{-3})^{-3}$

56. Find the value of

(a) $(-2)^{-3}$ (b) $(-3)^{-2}$ (c) $(-\frac{1}{2})^3$ (d) $(-27)^{\frac{1}{3}}$

(e) $(-1)^{\frac{1}{5}}$ (f) $(-64)^{\frac{1}{3}}$ (g) $(-64)^{\frac{2}{3}}$ **(h)** $(-3)^{-3}$

57. Compute (if possible):

(a) $(-8)^{\frac{2}{3}}$ (b) $(-32)^{\frac{4}{5}}$ (c) $(-4)^{\frac{3}{2}}$ (d) $(-64)^{\frac{4}{3}}$ (e) $(-27)^{\frac{5}{3}}$

(f) $(-16)^{\frac{3}{4}}$

58. It is now the appropriate time to return to the opening exercise of this chapter. Express the value of the boy's account in the form 4^x

(a) after four years.

(a) 64
(c) $\frac{1}{81}$

(a) $-\frac{1}{8}$

(c) $-\frac{1}{8}$

(e) -1
(g) 16

(a) 4
(c) impossible
(e) -243

(b) after six months.

(c) after one-and-a-half years.

(d) Evaluate the amounts in (a), (b), and (c).

*** SUMMARY ***

This chapter was concerned with the exponential expression b^x, primarily for $b > 0$. First, b^n was defined for positive integers n. The basic rule of exponents

$$b^m \cdot b^n = b^{m+n}$$

was taken as a guide in the definition of b^n when n was __not__ a positive integer. With the stipulation that b^n is positive, these definitions were made:

$b^0 = 1$ (zero exponent rule)

$b^{-n} = \dfrac{1}{b^n}$ (n is a positive integer)

$b^{\frac{1}{n}}$ is the solution of the equation $x^n = b$.

                                             (n is a positive integer)

$b^{\frac{m}{n}} = (b^{\frac{1}{n}})^m$ (m is an integer,
                                                  n is a positive integer)

(The definition of b^n when n is irrational is reserved for Projects.)

These rules apply to exponents:

$\dfrac{b^m}{b^n} = b^{m-n}$ (division rule)

$(b^m)^n = b^{mn}$ (power of a power)

$(ab)^n = a^n b^n$ (power of a product)

$(\frac{a}{b})^n = \dfrac{a^n}{b^n}$ (power of a quotient)

The radical sign ($\sqrt{}$) was introduced. For any positive number b,

$$\sqrt{b} = b^{\frac{1}{2}}$$

$$\sqrt[3]{b} = b^{\frac{1}{3}}$$

and so on. For positive integers n

$\sqrt[n]{ab} = \sqrt[n]{a}\,\sqrt[n]{b}$ (power of a product)

$$n \sqrt{\frac{a}{b}} = \frac{\sqrt[n]{a}}{\sqrt[n]{b}} \qquad\qquad \text{(power of a quotient)}$$

A brief discussion of negative bases b showed that they lead to trouble; for instance $b^{\frac{1}{2}}$ makes no sense if b is negative. Keep in mind that there are no convenient simplifications of

When tempted to invent such a "rule", test it first with some conventient numbers such as 9, 16, or 25.

$$(a + b)^n \qquad \text{or} \qquad (a - b)^n,$$

$$\sqrt{a + b} \qquad \text{or} \qquad \sqrt{a - b},$$

$$\sqrt{a^2 + b^2} \qquad \text{or} \qquad \sqrt{a^2 - b^2},$$

In particular, $\sqrt{a^2 + b^2}$ is not $\sqrt{a^2} + \sqrt{b^2}$.

* * * WORDS AND SYMBOLS * * *

exponent	radical sign $\sqrt{}$
base	root
"basic rule of exponents"	"rationalize the denominator"
zero exponent	conjugates
negative exponents	b^n
fractional exponents	$\sqrt[n]{b}$
	$\sqrt{b}$

* * * CONCLUDING REMARKS * * *

While b^n has been defined for every exponent n, the exponents 2 and $\frac{1}{2}$ are perhaps the most important in the "real" world. For instance, the formula for the distance that an object falls uses the exponents 2;

$$S = 16t^2$$

(an object falls $16t^2$ feet in t seconds.) The universal law of gravitation asserts that the force of gravitational attraction between two masses takes the form

$$\frac{k}{x^2},$$

where k is a constant and x is the distance between them. Early in the 18th century there were scientists who felt that the exponent in this equation may not be exactly 2, but very near 2. Later theory and experiments showed that '2' was the proper exponent.

The exponent 2 appears in the Einstein equation relating energy (E) and mass (m),

$$E = mc^2 ,$$

where c is the velocity of light.

Kepler early in the 17th century observed that the time it takes a planet to complete one revolution around the sun is proportional to $a^{\frac{3}{2}}$, where a is its average distance from the sun.

In computer use, we see such expressions as 16K, 32K, 64K, 128K, and 512K. The K is short for $2^{10} = 1024$ (which is close to a thousand, abbreviated K). The exponential 2^h, where h is a positive integer, is critical in computer design.

The exponents 2 and ½ both appear in the formula for the hypotenuse of a right triangle in terms of the other two sides:

$$c = (a^2 + b^2)^{\frac{1}{2}}$$

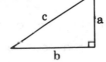

* * * CENTRAL SELF-QUIZ * * *

1. Define and demonstrate by an example:

    (a) base   (b) exponent   (c) basic rule of exponents
    (d) radical sign   (e) root   (f) "rationalize the denominator"

2. For what x is 2^x

    (a) negative?   (b) zero?   (c) positive?   (d) Explain.

3. Find x.

    (a) $\dfrac{b^{10}}{b^2} = b^x$ (b) $\dfrac{1}{2^6} = 2^x$ (c) $b^0 \cdot b^7 \cdot b^{-9} = b^x$

    (d) $(5^6)^{-4} = 5^x$ (e) $\dfrac{1}{b^{-3}} = b^x$ (f) $\dfrac{1}{b^{-n}} = b^x$

4. Simplify (include rationalizing denominator):

    (a) $\sqrt{75}$ (b) $\dfrac{\sqrt{88}}{\sqrt{11}}$ (c) $\dfrac{1}{\sqrt{x+1}}$

5. If $\sqrt{3} \doteq 1.732$ and $\sqrt{2} \doteq 1.414$, rationalize the denominator, then compute to 3 decimal places $\dfrac{2}{\sqrt{3} + \sqrt{2}}$.

6. Explain why $b^n > 0$ for $b > 0$ and n any interger. (Hint: See Exercise 2 above.)

7. Explain why $b < 0$ can lead to difficulties in b^n where n is a fractional exponent.

ANSWERS

1. See main section   2. 2^x is always positive.

3. (a) 8 (b) -6 (c) -2 (d) -24 (e) 3 (f) n

4. (a) $5\sqrt{3}$ (b) $2\sqrt{2}$ (c) $\dfrac{\sqrt{x-1}}{x-1}$ 5. 0.636

6. Consider n positive, zero, and negative.

7. $(-9)^{\frac{1}{2}}$ means $(-9)^{\frac{1}{2}}(-9)^{\frac{1}{2}} = (-9)^1$ or -9. Any real number times itself gives a number that is not negative.

* * *   EXPONENTS AND RADICALS REVIEW   * * *

1. *If you have not already made this list, do so now!* On a single sheet of paper list for handy reference all the rules of exponents for a positive base b.

2. Find x in each case:

    (a) $2^4 \cdot 2^7 = 2^x$ (b) $2^3 \cdot 2^5 = 2^x$ (c) $(2^{-3})(2^6) = 2^x$ (d) $2^{\frac{1}{3}} \, 2^{\frac{2}{7}} = 2^x$

    (e) $(2^{.03})(2^{.4}) = 2^x$ (f) $(2^{-4})(2^{-5}) = 2^x$

    (g) The above examples demonstrate "the basic rule of exponents." Describe it in your own words.

3. For any $b > 0$, how is b^0 defined?

4. For any $b > 0$ and $n > 0$, how is b^{-n} defined?

5. Find x in each case:

    (a) $\dfrac{5^6}{5^{-2}} = 5^x$ (b) $\dfrac{3^{-5}}{3^2} = \dfrac{1}{3^x}$ (c) $(2^4)^{-3} = 2^x$ (d) $(10)^3 = (2^x)(5^x)$

    (e) $(5 \cdot 7)^8 = 5^x \cdot 7^x$ (f) $(\frac{5}{7})^4 = \dfrac{5^x}{7^x}$ (g) $(2^{-4})^{-3} = 2^x$

    (h) $(.7)^3 = \dfrac{7^x}{(2^x)(5^x)}$

6. Say in your own words the following rule:

$$\frac{b^m}{b^n} = b^{m-n} \text{ where } b > 0.$$

7. Change to fractional exponent notation:

    (a) $\sqrt[5]{2}$ (b) $\sqrt[10]{73}$ (c) $\sqrt{11}$

8. (a) $5^{\frac{1}{4}} \cdot 5^{\frac{1}{4}} \cdot 5^{\frac{1}{4}} \cdot 5^{\frac{1}{4}} = ?$ (b) $\underbrace{2^{\frac{1}{n}} \cdot 2^{\frac{1}{n}} \cdot 2^{\frac{1}{n}} \cdot \ldots \cdot 2^{\frac{1}{n}}}_{n \text{ times}} = ?$

9. Which is larger (explain),

    (a) 27 or $\sqrt{728}$?   (b) 9 or $\sqrt[3]{729}$?   (c) 2 or $\sqrt[4]{71}$?   (d) 2 or $\sqrt[5]{31}$?

10. Find y

    (a) If $5 = 2^x$, then $10 = 2^y$.   (b) If $10 = 5^n$, then $2 = 5^y$.

    (c) If $5^n = 10$, then $5^{2n} = y$. (d) $(2^{x+1})(2^{x-1}) = 2^y$.

    (e) $\dfrac{2^{x+1}}{2^{x-1}} = 2^y$ (f) $(2^{x+1})^{x-1} = 2^y$ (g) $(2^x)^x = 2^y$ (h) $((2^2)^2)^2 = 2^y$

Some answers will be in terms of another letter.

11. Write as an expression involving a single radical sign:

    (a) $(\sqrt{2})(\sqrt[3]{2})$ (b) $(\sqrt[3]{4})(\sqrt[4]{8})$ (c) $(\sqrt[3]{3})(\sqrt[4]{3})$ (d) $(\sqrt[5]{9})(\sqrt{3})$

12. Simplify:

    (a) $\sqrt[3]{54}$ (b) $\sqrt{32}$ (c) $\sqrt[3]{24}$ (d) $\sqrt{63}$ (e) $\sqrt[4]{48}$ (f) $\sqrt[3]{40}$

    (g) $\sqrt{140}$ (h) $\sqrt{12}$ (i) $\sqrt[n]{2^n}$ (j) $\sqrt[5]{512}$

13. Simplify:

    (a) $\sqrt{12}\sqrt{3}$ (b) $\sqrt{18}\sqrt{5}$ (c) $\sqrt{18}\sqrt{15}$ (d) $\sqrt[3]{45}\sqrt[3]{3}$

    (e) $\sqrt{30}\sqrt{6}$ (f) $\sqrt[3]{300}\sqrt[3]{100}$

14. Multiply (simplify your answer):

    (a) $(\sqrt{2} - \sqrt{3})(\sqrt{6} + \sqrt{10})$ (b) $(\sqrt{18} + \sqrt{12})(\sqrt{3} - \sqrt{2})$

    (c) $(2\sqrt{5} + 3\sqrt{2})(\sqrt{5} - \sqrt{2})$ (d) $(\sqrt{2} + 1)(\sqrt{2} - 1)$

    (e) $(\sqrt{48} + \sqrt{72})(\sqrt{32} - \sqrt{54})$ (f) $(\sqrt{15} - \sqrt{12})(\sqrt{5} + \sqrt{8})$

Hint: simplify radicals before you multiply.

15. If $\sqrt{2} \doteq 1.414$, use conjugates to help you estimate to three decimal places:

$$\frac{2}{\sqrt{2}+1}$$

16. Rationalize the denominator, then approximate to three decimal places:

    (a) $\dfrac{1}{1 - \sqrt{3}}$ (b) $\dfrac{1}{\sqrt{5}}$ (c) $\dfrac{1}{\sqrt{5} - \sqrt{3}}$

17. Write in decimal notation without exponents, <u>if possible</u>:

    (a) $16^{\frac{3}{2}}$ (b) $16^{\frac{5}{4}}$ (c) $(.01)^{\frac{5}{2}}$ (d) $(.027)^{\frac{2}{3}}$ (e) $(-8)^{\frac{4}{3}}$ (f) $(-4)^{\frac{1}{2}}$

18. Explain why we do not define $(-25)^{\frac{1}{2}}$.

19. Between 1 and 1,000,000 inclusive, how many integers are

    (a) squares?   (b) cubes?   (c) sixth power of an integer?

Hint: This is an easy problem.

20. (a) Show by arithmetic or calculator that $(1.414)^2 < 2 < (1.415)^2$.
    (b) Explain why (a) shows that $1.414 < \sqrt{2} < 1.415$.

21. If $2^x = 700$, show that $9 < x < 10$.

22. (a) If $10^x = 3$, show that $0 < x < 1$.
    (b) If $10^x = 15$, show that $1 < x < 2$.
    (c) If $10^x = 120$, show that $2 < x < 3$.

23. Find integer k and $k + 1$ such that $k < x < k + 1$, if
    (a) $10^x = 1142$ (b) $10^x = 12,241$ (c) $10^x = .05$ (d) $10^x = .005$

24. Find x:
    (a) $\sqrt{\sqrt{3}} = 3^x$ (b) $\sqrt{\sqrt{9}} = 3^x$
    (c) Finding the "square root of the square root" of a number is the
        same as finding ___?___ root of the number.

25. What is the relationship between
    (a) $\dfrac{\sqrt{2}}{2}$ and $\dfrac{1}{\sqrt{2}}$?   (b) $\sqrt{3}$ and $\dfrac{3}{\sqrt{3}}$?   (c) $\sqrt{x}$ and $\dfrac{x}{\sqrt{x}}$?   Explain.

26. Simplify, rationalizing the denominator (assume all letters denote posi-
    tive numbers)

    (a) $\dfrac{\sqrt{27m^3n}}{\sqrt{3mn}}$ (b) $\dfrac{xy^2}{\sqrt{x^3y}}$ (c) $\sqrt{\dfrac{x^2}{4y^2} - \dfrac{z}{y}}$

    (d) $\sqrt{(\tfrac{x}{2})^2 + x^2}$ (e) $\sqrt{1\tfrac{3}{4}}$ (f) $\dfrac{\sqrt{99}}{\sqrt{11}}$

27. Write in the form b^x:
    (a) the cube of b.
    (b) the cube root of b.
    (c) the reciprocal of b.
    (d) the reciprocal of the square root of b.
    (e) the square root of the reciprocal of b.
    (f) the cube root of the square of b.
    (g) $\sqrt[5]{b^2}$

28. In the opening problem of the chapter, suppose that the boy's money
    (he has $1 right now) has been on deposit in the past at the same rate
    of interest that the father offers in the future. At what time was
    the account worth 25¢? $12\tfrac{1}{2}$¢? $6\tfrac{1}{4}$¢?   0¢?

ANSWERS TO EXPONENTS AND RADICALS REVIEW

1. See the summary.   2. (a) 11   (b) 8   (c) 3   (d) $\dfrac{13}{21}$ (e) .43   (f) –9
   (g) when you multiply powers of the same base, you add the exponents.

3. For $b > 0$, we define $b^0 = 1$.   4. For $b > 0$, we define $b^{-n} = \dfrac{1}{b^n}$.

5. (a) 8   (b) 7   (c) -12   (d) 3   (e) 8   (f) 4   (g) 12   (h) 3

7. (a) $2^{\frac{1}{5}}$ (b) $73^{\frac{1}{10}}$ (c) $11^{\frac{1}{2}}$ 8. (a) 5 (b) 2   9. (a) 27   (b) equal

9. (c) $\sqrt[4]{17}$ (d) 2.   10. (a) x+1   (b) n-1   (c) 100   (d) 2x   (e) 2   (f) x^2-1

10. (g) x^2 (h) 8   11. (a) $\sqrt[6]{2^5}$ (b) $2\sqrt[12]{2^5}$ (c) $\sqrt[12]{3^7}$ (d) $\sqrt[10]{3^9}$

12. (a) $3 \cdot \sqrt[3]{2}$ (b) $4\sqrt{2}$ (c) $2\sqrt[3]{3}$ (d) $3\sqrt{7}$ (e) $2\sqrt[4]{3}$ (f) $2\sqrt[3]{5}$ (g) $2\sqrt{35}$

    (h) $2\sqrt{3}$ (i) 2   (j) $2\sqrt[5]{16}$ 13. (a) 6 (b) $3\sqrt{10}$ (c) $3\sqrt{30}$ (d) $3\sqrt[3]{5}$

13. (e) $6\sqrt{5}$ (f) $10\sqrt[3]{30}$ 14. (a) $2\sqrt{3} + 2\sqrt{5} - 3\sqrt{2} - \sqrt{30}$ (b) $\sqrt{6}$

    (c) $4 + \sqrt{10}$ (d) 1   (e) $16\sqrt{6} - 36\sqrt{2} + 48 - 36\sqrt{3}$

    (f) $5\sqrt{3} + 2\sqrt{30} - 2\sqrt{15} - 4\sqrt{6}$ 15. 0.828

16. (a) $\dfrac{1 + \sqrt{3}}{-2} \doteq -1.366$ (b) $\dfrac{\sqrt{5}}{5} \doteq 0.447$ (c) $\dfrac{\sqrt{5} + \sqrt{3}}{2} \doteq 1.984$

17. (a) 64   (b) 32   (c) 0.00001   (d) .09   (e) 16   (f) not defined.

18. $(-25)^{\frac{1}{2}}(-25)^{\frac{1}{2}} = (-25)^1 = -25$ This is impossible for x^2 is never negative

    for all real numbers.   19. (a) 1000   (b) 100   (c) 10

21. $2^9 = 512,\ 2^{10} = 1024,\ 512 < 700 < 1024$ 22. (a) $10^0 = 1,\ 10^1 = 10,$
    $1 < 3 < 10$ (b) $10^1 = 10,\ 10^2 = 100,\ 10 < 15 < 100$ (c) $10^2 = 100,$
    $10^3 = 1000\quad 100 < 120 < 1000$

23. (a) k = 3, k+1 = 4 (b) k = 4, k+1 = 5   (c) k = -2, k+1 = -1
    (d) k = -3, k+1 = -2   24. (a) $\frac{1}{4}$ (b) $\frac{1}{2}$ (c) fourth

25. (a) equal   (b) equal (c) equal   26. (a) 3m   (b) $\dfrac{y\sqrt{xy}}{x}$ (c) $\dfrac{\sqrt{x^2 - 4yz}}{2y}$

    (d) $\dfrac{x}{2}\sqrt{5}$ (e) $\dfrac{\sqrt{7}}{2}$ (f) 3   27. (a) b^3 (b) $b^{\frac{1}{3}}$ (c) b^{-1} (d) $b^{-\frac{1}{2}}$ (e) $b^{-\frac{1}{2}}$

    (f) $b^{\frac{2}{3}}$ (g) $b^{-\frac{2}{5}}$ 28. (a) worth 25¢ 1 year ago. (b) worth 12½¢ 1½ yrs.

    ago. (c) worth 6¼¢ 2 yrs ago. (d) since $b^x > 0$ for all x, the amount

    was never 0.

* * * EXPONENTS AND RADICALS REVIEW SELF-TEST * * *

1. (a) $(2^{\frac{2}{3}})(2^{\frac{1}{4}}) = 2^x$ (b) $\dfrac{1}{2^{-7}} = 2^x$ (c) $\dfrac{2^{0.14}}{2^{0.3}} = 2^x$

    (d) $(2^{\frac{3}{10}})^{\frac{1}{2}} = 2^x$ (e) $\sqrt[3]{4} = 2^x$

2. Evaluate:

    (a) $(29)^0$ (b) $(-3)^0$ (c) $(\pi)^0$

3. Write as an equivalent expression with a positive exponent.

    (a) 2^{-7} (b) $\dfrac{1}{5^{-6}}$ (c) 10^{-29}

4. Which is larger, $\sqrt[3]{65}$ or 4? Explain.

5. Write as a single integer without an exponent, if possible.

    (a) $\sqrt[3]{64}$ (b) $-\sqrt{16}$ (c) $\dfrac{\sqrt{32}}{\sqrt{8}}$ (d) $\sqrt{8}\sqrt{18}$ (e) $(-4)^{\frac{1}{2}}$

    (f) $(1970)^0$ (g) $(64)^{\frac{2}{3}}$ (h) $(81)^{\frac{3}{4}}$ (i) $(100)^{\frac{3}{2}}$

6.   If $\sqrt{3} \doteq 1.732$, estimate $\dfrac{1}{2 - \sqrt{3}}$ in the easiest manner possible.

7.   Evaluate:

(a) $(-2)^2$ (b) $(-2)^{-2}$ (c) $(-3)^3$ (d) $(-3)^{-3}$

8.   Multiply (simplify your answer):

(a) $(\sqrt{12} - \sqrt{18})(\sqrt{3} + \sqrt{8})$ (b) $(\sqrt{5} + \sqrt{27})(\sqrt{125} - \sqrt{3})$

9.   Find x in terms of n:

(a) if $2^n = 5$, then $2^x = 10$ (b) if $2 = 5^n$, then $10 = 5^x$

(c) if $10 = 2^n$, then $5 = 2^x$ (d) if $10 = 3^n$, then $1000 = 3^x$

10. (a) If $10^x = 7$, then $\underline{\ ?\ } < x < \underline{\ ?\ }$.
     (b) If $10^x = .02$, then $\underline{\ ?\ } < x < \underline{\ ?\ }$.   (Answer with consecutive
     (c) If $10^x = 123$, then $\underline{\ ?\ } < x < \underline{\ ?\ }$.   integers)

11. Write the product using a single radical sign:

(a) $\sqrt{2}\ \sqrt[3]{2}$ (b) $\sqrt[5]{3}\ \sqrt[5]{4}$ (c) $\sqrt[3]{4}\ \sqrt[4]{8}$

ANSWERS

1   (a) $2^{\frac{11}{12}}$ (b) 2^7 (c) $2^{-0.16}$ (d) $2^{\frac{3}{20}}$ (e) $2^{\frac{2}{3}}$ 2. (a) 1 (b) 1 (c) 1   3. (a) $\dfrac{1}{2^7}$

3.   (b) 5^6 (c) $\dfrac{1}{10^{29}}$ 4. $\sqrt[3]{65}$ (why?)   5. (a) 4   (b) –4   (c) 2   (d) 12

(e) impossible   (f) 1   (g) 16   (h) 27   (i) 1000

6. $\dfrac{1}{(2-\sqrt{3})}\ \dfrac{(2 + \sqrt{3})}{(2 + \sqrt{3})} = \dfrac{2 + \sqrt{3})}{1} \doteq 3.732$.   7. (a) 4 (b) ¼ (c) –27 (d) $-\dfrac{1}{27}$

8.   (a) $\sqrt{6} - 6$ (b) $16 + 14\sqrt{15}$ 9. (a) n+1   (b) n+1   (c) n–1   (d) 3n

10. (a) $0 < x < 1$ (b) $-2 < x < -1$ (c) $2 < x < 3$

11. (a) $\sqrt[6]{2^5}$ (b) $\sqrt[5]{12}$ (c) $2\sqrt[12]{2^5}$

* * * EXPONENTS AND RADICALS PROJECTS * * *

1.   Where is the error in the following statement?

$1 = (1)^{\frac{1}{2}} = ((-1)(-1))^{\frac{1}{2}} = (-1)^{\frac{1}{2}}(-1)^{\frac{1}{2}} = -1$
Thus, $1 = -1$.

2.   The prime number 3 is of the form $N^2 - 1$ for some integer N.
$(3 = 2^2 - 1)$.

(a) Are there other primes of the form $N^2 - 1$?

(b) If so, how many?

3.   The prime number 5 is of the form $N^2 + 1$ $(5 = 2^2 + 1)$.

(a) Are there other primes of the form $N^2 + 1$?

(b) If so, how many?

4. Why is the Pythagorean Theorem true? The Babylonians of 4,000 years ago knew the theorem and used it, but there is no evidence that they knew why it is true. About 500 B.C. the first proof was discovered by Greek mathematicians. Here is a proof discovered by a Chinese mathematician some 1,000 years ago.

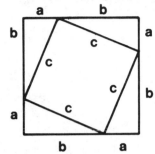 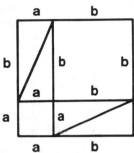

Compare the two large squares to show why

$$c^2 = a^2 + b^2.$$

What is the area of a square of side x?

5. Here is one way to define $2^{\sqrt{2}}$.

    (a) The decimal expansion of $\sqrt{2}$ begins 1.414... Is 2^1 defined? Is $2^{1.4}$? Is $2^{1.41}$? Is $2^{1.414}$?

    (b) The decimal expansion of $\sqrt{2}$ is endless. How can it be used to help define $2^{\sqrt{2}}$? (The same method can be used to define 2^x for any irrational exponent x.)

6. Which is larger, $3^{\frac{1}{3}}$ or $5^{\frac{1}{5}}$? Solve this question without using decimals.

Hint: 6. Take each to the 15th power.

7. Write in the form $\sqrt[n]{b}$

    (a) $\sqrt{\sqrt{b}}$

    (b) $\sqrt[3]{\sqrt{b}}$

    (c) $\sqrt[3]{\sqrt[3]{b}}$

8. (a) Estimate x^x for x = $\frac{1}{2}, \frac{1}{3}$, and $\frac{1}{6}$.

Use a calculator.

    (b) What do you think happens to x^x as x is chosen closer and closer to 0?

9. (a) Estimate $\sqrt[n]{n}$ for n = 2, 3 and 10.

    (b) About how large is $\sqrt[100]{100}$?

    (c) What do you think happens to $\sqrt[n]{n}$ as n becomes large? Experiment.

10. The integers 2 and 4 have the unusual property that

$$2^4 = 4^2.$$

Are there any other integers m and n, m ≠ n, such that

$$m^n = n^m ?$$

11. Recall the Pythagorean Theorem in Project 4, $a^2 + b^2 = c^2$.
    The world's most famous right triangle has sides of lengths 3, 4, and 5.
    Note that they are all integers.

    (a) Check that in a right triangle if a = 5, and b = 12, then the hypotenuse c is an integer.

    (b) Check that for any positive integer n, if $a = n^2 - 1$ and b = 2n, then c is an integer.

    (c) What right triangles result from (b), when n = 2? n = 3? n = 4?

12. Fermat (a 17th century mathematician and lawyer), motivated by the Pythagorean Theorem, asked whether there are three positive integers a, b, and c, such that $a^3 + b^3 = c^3$. Can you find any?

A number of the form x^3 for some nonnegative integer x is called a "cube."

13. G.H. Hardy, a famous number theorist, when visiting a friend, Ramanujan, who was quite ill, remarked that the license number of his taxi was quite dull; it was 1729. "That is a quite interesting number," Ramanujan replied, "it is the smallest integer that can be expressed in two different ways as the sum of two cubes." Show that he was right.

14. Which is larger

    (a) $5^{\frac{1}{2}}$ or $5^{\frac{1}{3}}$?

Hint: (a) Take each to the 6th power.

    (b) $(\frac{1}{9})^{\frac{1}{2}}$ or $(\frac{1}{16})^{\frac{1}{3}}$?

    (c) $(\frac{1}{9})^{\frac{1}{2}}$ or $(\frac{1}{9})^{\frac{1}{3}}$?

    (d) $(\frac{1}{64})^{\frac{1}{2}}$ or $(\frac{1}{64})^{\frac{1}{3}}$?

15. Using the definition of $x^{\frac{1}{2}}$ as the square root of x, prove that $a^{\frac{1}{2}}b^{\frac{1}{2}}$ is indeed $(ab)^{\frac{1}{2}}$.

16. It was mentioned that $\sqrt{2}$, for instance, is irrational. This outlines a proof. Assume that there are positive integers, M and N, such that
$$\sqrt{2} = \frac{M}{N}.$$

    (a) Deduce that $2N^2 = M^2$.

    (b) Write M and N as the product of primes. M is the product of, say, D primes; N is the product of, say, E primes.
$$M = P_1 P_2 \cdots P_D ,$$
$$N = Q_1 Q_2 \cdots Q_E .$$

    (c) Using (b) and the fact that factorization into primes is unique, show that
$$2E + 1 = 2D.$$

    (d) Why is 2E + 1 = 2D impossible? The same proof shows that for any prime P, $\sqrt{P}$ is irrational. Moreover, the argument can be slightly modified to show that the only positive integers with a rational square root are the squares.

17. This is the way that $\sqrt{b}$ is estimated by a modern calculator or by an ancient Babylonian.

   (a) Take a guess x for the square root of b.

   (b) Find $\frac{b}{x}$.

   (c) If $x = \frac{b}{x}$, then $x = \sqrt{b}$, and stop.

   (d) If $x \neq \frac{b}{x}$, let y be the average of x and $\frac{b}{x}$,

   $$y = \frac{x + \frac{b}{x}}{2}.$$

   This number, y, is a better approximation to $\sqrt{b}$ than x is.

   (e) Repeat the process as many times as desired. Carry out the process for estimating $\sqrt{10}$ starting with $x = 3$. (The next step will be approximately 3.17 and the following, 3.16, which is quite accurate.)

18. Which is larger $10^{\frac{1}{3}}$ or $7^{\frac{1}{2}}$?

19. This is Newton's procedure for estimating $\sqrt[3]{b}$.

   (a) Take a guess x for the cube root of b.

   (b) If $x = \frac{b}{x^2}$, stop; $x = \sqrt[3]{b}$.

   (c) Otherwise let $y = \frac{2}{3}x + \frac{1}{3} \cdot \frac{b}{x^2}$.

   Then y is the next estimate of $\sqrt[3]{b}$. Use this method to estimate $\sqrt[3]{10}$ to three decimal places, starting with $x = 2$.

20. (See Project 16.)

   (a) Show that $\sqrt{2}$ cannot be a terminating decimal.

   (b) Can $\sqrt{2}$ be a repeating decimal?

   (c) Do the same problem as (a) and (b) for $\sqrt{3}$.

21. Use a calculator to compute $(1 + \frac{1}{n})^n$ for n = 1, 2, 3, 5, 10, 20, 100, 500, 1000, and larger numbers. As n gets very large, $(1 + \frac{1}{n})^n$ approaches a number of great importance in the calculus, the number e.

22. Use a calculator to estimate x to three decimal places: $7^x + x = 2$.

* * * FIRST YEAR ALGEBRA REVIEW (OPTIONAL) * * *

To find the area of a rectangle, take the product of its length and width.

Some of this first year algebra review duplicates some of the material in the main section of Chapter 3 (Polynomials).

THE DISTRIBUTIVE IDENTITY

1. Use the area of the rectangles below in the figure to explain the distributive identity:

The distributive identity is very important.

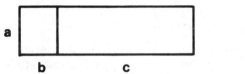

$$a(b + c) = ab + ac$$

This geometric argument works only when a, b, and c are positive numbers. The distributive identity holds also for negative or zero numbers.

The identity

$$a(b - c) = ab - ac$$

is also called the distributive identity.

2. Explain why Exercise 1 could have been written

$$(b + c)a = ba + ca.$$

3. Use the distributive identity to write each of the following as an equivalent expression without parentheses.

(a) $2(3x + 4)$ (b) $5(4x - 3)$ (c) $-3(6x + 1)$
(d) $-3x(5x - 3)$ (e) $-7x(-6x + 5)$ (f) $-x(x^2 - x)$
(g) $-x^2(3x^3 + 4x^2)$ (h) $-(x - 1)$ (i) $-x^3(-x^2 - x)$

4. Explain each step of the following:

(a) $3x + 9x = (3 + 9)x$ (b) $6x - 10x = (6 - 10)x$
$ = 12x$ $ = -4x$

(c) In parts (a) and (b), the distributive identity was used to simplify (or "collect like terms" in) an expression. Can you collect like terms in the expression

$$-3x^2 + 5x \ ?$$

Explain.

5. Simplify ("collect like terms"), if possible:

(a) $-12x + 3x$ (b) $-6a^2 + 15a^2$ (c) $15x^3 - 21x^3$ (d) $6x - 9x^2$

* * *

When simplifying an expression in which parentheses are inside other parentheses, first eliminate the inner parentheses. For example,

$$3[2x^2 - 4x(x - 5)] = 3[2x^2 - 4x^2 + 20x]$$
$$= 3[-2x^2 + 20x]$$
$$= -6x^2 + 60x$$

6.   Simplify ("collect like terms"):

   (a) $4x + [x - 3(x + 5)]$ (b) $2[3x - 2(x - 5)] - 4x - 9$

   (c) $\frac{x}{3} [3x + (x - 1)6]$ (d) $6 - (-4x + [-(x - 1)])$

POLYNOMIALS

In algebra polynomials are added, subtracted, multiplied, and divided in a way similar to that of arithmetic.

For addition and subtraction, powers of ten are lined up in arithmetic; likewise, powers of x are lined up in algebra:

Add:

```
     341                  2x³        - x + 5
      29.2                     x²  + 10x- 2
    4512.648           -5x³ + 9x²        + 6
   ──────────          ──────────────────────
    4882.848           -3x³ + 10x² + 9x + 9
```

Subtract (lower from upper):

```
49    16    14          2x³        - x + 5
15    79   -50               x² + 10x - 2
──   ───   ───          ──────────────────────
34   -63    64          2x³ - x²  - 11x + 7
```

Check your subtraction: does the lower number added to the answer give the upper number?

7.   Add:

(a) $x^3 - 5x^2 + 2x - 6$
        $6x^2 + 4x + 9$
  $4x^3$ $- 8x - 15$

(b) $6x^2 - 3x + 2$
      $4x^2 - 9x + 8$
         $5x - 11$

(c) $-x^3 + 4x^2 - x + 3$
  $6x^3 - 5x^2$ $- 7$
       $8x^2 - 4x + 14$

(d) $x^4 - 1$, $x^2 - x$, and $3x - 5$

8.   Subtract (lower from upper):

(a) $9x$
    $-14x$

(b) $6x - 7$
    $11x + 3$

(c) $5x^2 - 3x + 2$
   $-9x^2 + 2x - 5$

9. Subtract $3x^4 - 2x^2 + 5$ from $-2x^3 + x$.

* * *

Algebraic multiplying is similar to arithmetic multiplying. Powers of x are lined up in algebra, whereas powers of 10 are lined up in arithmetic:

$$\begin{array}{r} 325 \\ 36 \\ \hline 1950 \\ 975 \\ \hline 11,700 \end{array} \qquad \begin{array}{r} 3x^2 - 2x + 5 \\ 3x - 6 \\ \hline -18x^2 + 12x - 30 \\ 9x^3 - 6x^2 + 15x \\ \hline 9x^3 - 24x^2 + 27x - 30 \end{array}$$

In a sense, the algebraic case is easier, since there is no 'carrying'.

10. Multiply.

(a) $\begin{array}{r} x - 5 \\ x + 3 \end{array}$ 　　(b) $\begin{array}{r} 2x - 4 \\ x + 7 \end{array}$ 　　(c) $\begin{array}{r} 2x^2 + x - 3 \\ -3x^2 + 5x - 1 \end{array}$

(d) $\begin{array}{r} x^2 + x + 1 \\ x - 1 \end{array}$ 　　(e) $\begin{array}{r} x^3 - 3x^2 + 2x - 1 \\ 4x^2 - x + 5 \end{array}$

* * *

The method of dividing polynomials is similar to long division in arithmetic.

These quotients may also be written
$208\frac{27}{31}$ *and*
$2x^2 + 2x + 3 - \frac{2}{3x - 1}$

$$\begin{array}{r} 208 \text{ rem } 27 \\ 31 \overline{\smash{)}6475} \\ \underline{62} \\ 275 \\ \underline{248} \\ 27 \end{array} \qquad \begin{array}{r} 2x^2 + 2x + 3 \text{ rem } -2 \\ 3x - 1 \overline{\smash{)}6x^3 + 4x^2 + 7x - 5} \\ \underline{6x^3 - 2x^2} \\ 6x^2 + 7x \\ \underline{6x^2 - 2x} \\ 9x - 5 \\ \underline{3x - 3} \\ -2 \end{array}$$

Check:

$$(31)(208) + 27 = 6475$$

and $(3x - 1)(2x^2 + 2x + 3) - 2 = 6x^3 + 4x^2 + x - 5$

11. Divide.

Hint: (b) write
$x^4 + 0 \cdot x^3 + 0 \cdot x^2 + 0 \cdot x - 1$

(a) $2x - 3 \overline{\smash{)}2x^3 - 9x^2 + 11x - 3}$ 　　(b) $x + 1 \overline{\smash{)}x^4 - 1}$

(c) $2x - 1 \overline{\smash{)}8x^3 + 1}$ 　　(d) $x^2 - x + 1 \overline{\smash{)}x^4 + x^2 + 1}$

12. Use the areas of the figure to explain why

$$(a + b)^2 = a^2 + 2ab + b^2$$

13. Is $(a + b)^2$ equal to $a^2 + b^2$?   Explain.

14. Write as an expression free of parentheses.

   (a) $(x + 2)^2$ (b) $(x + 6)^2$ (c) $(3x + 1)^2$ (d) $(3x + 2)^2$

15. The areas of the rectangles in the figure can be used to justify the identity

$$(a - b)^2 = a^2 - 2ab + b^2$$

   Justify this identity using the figure to explain the first two steps.

   Step 1.   The shaded area equals $(a - b)^2$.
   Step 2. $(a - b)^2 = a^2 - b(a - b) - b(a - b) - b^2$
   Step 3. $(a - b)^2 = a^2 - ab + b^2 - ab + b^2 - b^2$
   Step 4. $(a - b)^2 = a^2 - 2ab + b^2$

16. Write as an expression free of parentheses.

   (a) $(x - 3)^2$ (b) $(x - 5)^2$ (c) $(2x - 1)^2$ (d) $(3x - 2)^2$

17. Use the areas of the following rectangles and squares to explain why
        if $a > b$, then

    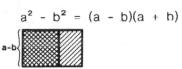

$$a^2 - b^2 = (a - b)(a + b)$$

18. Write as a simplified expression free of parentheses.

   (a) $(x - 2)(x + 2)$ (b) $(x + 5)(x - 5)$ (c) $(3x + 1)(3x - 1)$
   (d) $(3x + 2)(3x - 2)$ (e) $(4x - 3)(4x + 3)$ (f) $(4 - x)(4 + x)$

19. Using the pattern,
        $$(x + 1)(2x - 3) = 2x^2 - x - 3,$$

   multiply each of the following at sight.

   (a) $(x - 2)(x + 3)$ (b) $(2x + 3)(x + 4)$ (c) $(2x - 1)(3x - 5)$

20. Multiply each of the following at sight:

   (a) $(x - 1)(x + 1)$ (b) $(x + 1)(x + 1)$ (c) $(x - 1)(x - 1)$
   (d) $(2x - 5)(x + 3)$ (e) $(2x + 3)(2x + 3)$ (f) $(6x - 5)(x + 7)$
   (g) $(1 - x)(1 + x)$ (h) $(2x - 3)(3x - 2)$ (i) $(4x - 5)(4x - 5)$

21. Explain why the areas in the figure suggest the following:

   $(a + b)(c + d) = ac + ad + bc + bd$

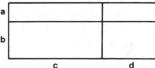

22. Use algebra to explain each step:

   Step 1. $(a + b)(c + d) = (a + b)c + (a + b)d$
   Step 2. $= ac + bc + ad + bd$
   Step 3. $= ac + ad + bc + bd$

* * * SUMMARY OF ALGEBRAIC IDENTITIES * * *

Make sure you remember these identities.

reciprocals:

$$(a)\left(\frac{1}{a}\right) = 1 \quad \text{(where } a \neq 0\text{)}$$

opposites:

$$a + (-a) = 0$$

commutative identities:

$$a + b = b + a \quad \text{(addition)}$$
$$ab = ba \quad \text{(multiplication)}$$

associative identities:

$$a + (b + c) = (a + b) + c \quad \text{(addition)}$$
$$a(bc) = (ab)c \quad \text{(multiplication)}$$

distributive identity:

$$a(b + c) = ab + ac \quad \text{(relates addition and multiplication)}$$

multiplication by 1 and -1:

$$-a = (-1)a$$
$$a = (1)a$$

other identities:

$$-a(b) = -ab$$
$$(-b)b = -b^2$$
$$-(a + b) = -a - b$$
$$-(a - b) = -a + b$$
$$a(b - c) = ab - ac$$
$$(a + b)(c + d) = ac + bc + ad + bd$$
$$(a + b)(a - b) = a^2 - b^2$$
$$(a + b)^2 = a^2 + 2ab + b^2$$
$$(a - b)^2 = a^2 - 2ab + b^2$$

* * *

ANSWERS

1. The area of the total figure equals the sum of the areas of the smaller rectangles.
2. Commutative property: $a(b + c) = (b + c)a$ and $ab = ba$.   3. (a) $6x+8$ (b) $20x-15$ (c) $-18x-3$ (d) $-15x^3+9x$ (e) $42x^2-35x$ (f) $-x^3+x^2$ (g) $-3x^5-4x^4$ (h) $-x+1$ (i) $x^5 + x^4$
4. (a) distributive identity (b) distributive identity (c) no. x and x^2 are not "like terms". 5.(a) $-9x$ (b) $9a^2$ (c) $-6x^3$ (d) can't simplify 6. (a) $2x-15$ (b) $-2x + 11$ (c) $3x^2 - 2x$ (d) $5x+5$ 7. (a) $5x^3+x^2-2x-12$ (b) $10x^2-7x-1$ (c) $5x^3+7x^2-5x+10$ (d) $x^4 + x^2 + 2x-6$ 8. (a) $23x$ (b) $-5x-10$ (c) $14x^2 -5x+7$ 9. $-3x^4 -2x^3 +2x^2 + x-5$
10. (a) $x^2 -2x-15$ (b) $2x^2 + 10x-28$ (c) $-6x^4 + 7x^3 + 12x^2 -16x+3$ (d) $x^3 -1$ (e)$4x^5 -13x^4 + 16x^3 -21x^2 + 11x-5$ 11.(a) $x^2 -3x + 1$ (b) $x^3 - x^2 + x-1$ (c)$4x^2 + 2x+1$ R2 (d) $x^2 +x+1$ 12. The area of the large square is $(a+b)^2$. The areas of the pieces

a^2, ab, ab, and b^2 . Thus, $(a+b)^2 = a^2 + ab + ab + b^2 = a^2 + 2ab + b^2$ 13. no. See #12.
14. (a) $x^2 + 4x + 4$ (b) $x^2 + 12x + 36$ (c) $9x^2 + 6x + 1$ (d) $9x^2 + 12x + 4$. 15. $(a - b)^2$ is the shaded area. You get the same area by subtracting areas of the unshaded rectangles from that of the total figure.   16. (a) $x^2 - 6x + 9$ (b) $x^2 - 10x + 25$ (c) $4x^2 - 4x + 1$
(d) $9x^2 - 12x + 4$ 17. Use areas of rectangles.   18. (a) $x^2 - 4$
(b) $x^2 - 25$ (c) $9x^2 - 1$ (d) $9x^2 - 4$ (e) $16x^2 - 9$ (f) $16 - x^2$ 19. (a) $x^2 + x - 6$
(b) $2x^2 + 11x + 12$ (c) $6x^2 - 13x + 5$ 20. (a) $x^2 - 1$ (b) $x^2 + 2x + 1$ (c) $x^2 - 2x + 1$
(d) $2x^2 + x - 15$ (e) $4x^2 + 12x + 9$ (f) $6x^2 + 37x - 35$ (g) $1 - x^2$ (h) $6x^2 - 13x + 6$
(i) $16x^2 - 40x + 25$ 21.   The area of the total figure equals the sum of the areas of the rectangles. 22. 1. distributive identity 2. distributive identity 3. addition commutative identity.

* * *

Polynomials

* * * CENTRAL SECTION * * *

Riddle: I am thinking of three numbers. For each of them, the following statement is true: If I add 3 to its cube I get the same number as when I add three times the number to its square. What are the three numbers?

* * *

An important application of algebra, especially in calculus, is solving an equation such as

$$x^3 - 3x^2 - 2x + 4 = 0$$

That means finding all numbers x that make the equation true. In this case there are three solutions; Exercises 1-6 help find them arithmetically. The goal of this chapter is to provide more rapid ways of solving equations.

THE SOLUTION OF $x^3 - 3x^2 - 2x + 4 = 0$

1. Copy and fill in the table for $y = x^3 - 3x^2 - 2x + 4$.

x	x^3 $3x^2$ $2x$ + 4	y
-2	$(-2)^3 - 3(-2)^2 - 2(-2) + 4$ $-8 - 3(4) - 2(-2) + 4$ $-8 - 12 + 4 + 4$ -12	-12
-1		
0		
1		
2		
3		
4		

In place of the variable use parentheses ( ), to help avoid mistakes. Start with $(\ )^3 - 3(\ )^2 - 2(\ ) + 4$ Then put "-2" inside the parentheses. Leave room in the middle column for computations.

When evaluating a polynomial, recall the order of computation:
1. exponents then
2. products (multiplying) and quotients lastly,
3. sums (adding) and differences.

2. Plot the seven points (x,y) described by the table in Exercise 1.

3. Sketch a graph of $y = x^3 - 3x^2 - 2x + 4$. It is a smooth curve passing through the seven points in Exercise 2.

4. Look at the graph in Exercise 3.

    (a) What value of x is definitely a solution of the opening equation, $x^3 - 3x^2 - 2x + 4 = 0$?

Hint: when does $y = 0$?

(b) Why is there also a solution between x = -2 and x = -1?

(c) Why is there also a solution between x = 3 and x = 4?

5. (a) With the aid of the graph in Exercise 3, <u>estimate</u> to one decimal place the solution between -2 and -1 and the solution between 3 4.

(b) Compare your estimates with those of your classmates.

(c) Check your estimates by sustituting in the original equation. Does your estimate seem reasonably close?

6. If you were willing to do the arithmetic, you could estimate the solutions of the equation $x^3 - 3x^2 - 2x + 4 = 0$ to as many decimal places as you wished. How could you go about doing this? Briefly describe your method in writing (but do not do the actual estimating).

You may assume the use of a calculator here.

* * *

The six opening exercises suggest one way of solving the equation

$$x^3 - 3x^2 - 2x + 4 = 0$$

<u>The rest of the chapter develops an easier way.</u>

POLYNOMIALS

An expression of the form ax^n, where <u>a</u> is a real number and n is a non-negative integer, is called a <u>monomial.</u> For instance,

$$2x^3, \qquad -\frac{\sqrt{2}}{2}x^2, \qquad \text{and} \qquad 5x^0$$

Read this section carefully.

are monomials. It is customary to write the monomial ax^0 simply <u>a</u>. The <u>coefficient</u> of ax^n is <u>a</u>. The <u>degree</u> of ax^n is the exponent n; however, the monomial $0x^0$ (= 0) is not assigned a degree.

$$\text{coefficient} \longrightarrow ax^n \longleftarrow \text{degree}$$

Thus $2x^3$ has coefficient 2 and degree 3. The monomial $-\frac{\sqrt{2}}{2}x^2$ has coefficient $-\frac{\sqrt{2}}{2}$ and degree 2. The monomial 5 has coefficient 5 and degree 0.

Think of 5 as $5x^0$, or -2 as $-2x^0$; thus 5 and -2 are polynomials of degree zero.

A <u>polynomial</u> is a sum of monomials. For instance,

$$2x + 1, \qquad x^2 - 3x + 3 \qquad \text{and} \qquad x^3 - 3x^2 - 2x + 4$$

are polynomials. The <u>degree</u> of a polynomial is the highest degree of the monomials that appear in it with a non-zero coefficient. Thus

$2x + 1$	has degree 1
$x^2 - 3x + 3$	has degree 2

and

$x^3 - 3x^2 - 2x + 4$	has degree 3.

The polynomial all of whose coefficients are 0 is called "the polynomial 0." It is not assigned a degree.   A polynomial of degree 2 is also called a "quadratic" or "quadratic polynomial".   A third degree polynomial is also called a "cubic" or "cubic polynomial".

7.   What is the degree of each of these polynomials?

(a) 6x - 8   (b) 5x^{10} + x^2 (c) -x^{100} + x^4 - x^2 (d) x^3 + 5x^8

(a)   1
(c)   100

8.   Give an example of your own of a polynomial of

(a) degree 1   (b) degree 2   (c) degree 3   (d) degree 60

THE GRAPHS OF POLYNOMIALS

Exercises 9-14 concern the graphs of polynomials.   Notice the influence of the degree of the polynomial upon the shape of the graph.

9.   Make a table of values, then graph:

Use a format for your table similar to Exercise 1.

(a) y = 3x - 2   (b) y = x + 5   (c) y = -2x + 4

(d)   If ax + b is a polynomial of degree 1, describe the shape of the graph of y = ax + b.

10. Make a table of values, then graph:

(a) y = x^2 + 2x + 2   (b) y = -x^2 - 2x - 2   (c) y = 2x^2 - x + 1

(d)   If ax^2 + bx + c is a polynomial of degree 2, describe the shape of the graph of y = ax^2 + bx + c.

11. (a)   Make a table of values, then graph y = x^3.
     (b)   In what way is it like the graph sketched in Exercise 3?
     (c)   How many times does the graph of y = x^3 meet the x-axis?
     (d)   How many solutions are there of the equation x^3 = 0?

12. Copy and fill in the table for y = x^3 - x^2 - 6x + 1:

(a)

If you compute directly by substitution, leave enough room on your paper in the middle column to record every step in the computation, as in Exercise 1.

x	x^3 - x^2 - 6x + 1	= y
-3		
-2		
-1		
0		
1		
2		
3		
4		
10		
-10		

(b)   Use the first eight values of the table in (a) to plot eight points (x,y) on the graph of y = x^3 - x^2 - 6x + 1.

(c)   Draw a smooth curve through the points in (b).

(d) Use the graph in (c) to help find the number of solutions of the equation $x^3 - x^2 - 6x + 1 = 0$.

(e) What does the graph look like when x is large and positive? When x is large and negative?

13. How does the shape of a graph of

$$y = \text{a polynomial of degree } 3$$

differ from the shape of a graph of

$$y = \text{a polynomial of degree } 2?$$

14. Each of these is the graph of

$$y = \text{polynomial}$$

and the polynomial has degree at most 3. What is the degree in each case? Why?

Hint: What happens to the graph as the degree increases from 1 up to degree 3?

The First Year Algebra Review (optional) of this chapter has more exercises concerning polynomials.

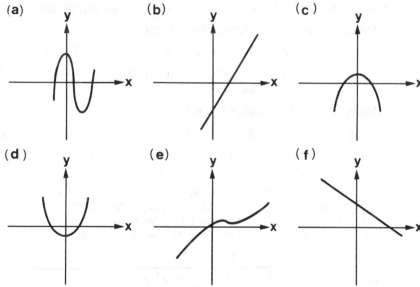

THE ALGEBRA OF POLYNOMIALS

There are three basic operations possible with two <u>integers a</u> and <u>b</u> that will produce another integer. (An "integer" is one of the numbers

$$\ldots -3, -2, -1, 0, 1, 2, 3, \ldots .)$$

You can add them, getting a + b; you can subtract them, getting a - b; or you can multiply them, getting ab. It is not always possible to divide one integer by another integer and obtain an integer for an answer. For instance "7 divided by 3" is not an integer; it is the fraction $\frac{7}{3}$. In this case we also say, "3 divides 7 twice with a remainder 1." Or

$$7 = 2 \cdot 3 + 1.$$

You will see a close correspondence between integers and polynomials.

On the other hand, 3 divides 15, for the remainder is 0:

$$15 = 5 \cdot 3.$$

Similarly, you can always add, subtract or multiply two polynomials. But, as with integers, you cannot always divide them so that the remainder is 0.

Example 1.   Find the sum of the polynomials $3x^2 + 5x + 1$ and $x^3 + 6x^2 + 8x - 1$.

Solution. Add them by adding monomials of the same degree.

$$x^3 + 6x^2 + 8x - 1$$

(add) $$\underline{\qquad 3x^2 + 5x + 1 \qquad}$$

$$x^3 + 9x^2 + 13x + 0$$

Line up like powers of x just as you line up digits with like powers of ten when adding integers:
$$1642$$
$$\underline{137}$$
$$\underline{1779}$$
Note that with polynomials you never have to "carry" or "borrow."

The sum of $x^3 + 6x^2 + 8x - 1$ and $3x^2 + 5x + 1$ is therefore the polynomial $x^3 + 9x^2 + 13x + 0$, or simply $x^3 + 9x^2 + 13x$.

15. Find the sum of the two polynomials:

(a) $5x + 1$ and $3x + 2$ (b) $6x^2 - 3x + 2$ and $-3x^2 - 3x + 7$

(c) $6x^2 + 8x + 2$ and $-6x^2 + 7x + 3$ (d) x^5 and $6x^4$

15. (a) $8x + 3$
  (c) $15x + 5$

16. When you add two polynomials of degree 4 can you ever get

(a) a polynomial of degree 5?   (b) a polynomial of degree 4?
(c) a polynomial of degree 3?   (d) a polynomial of degree 2?
(e) a polynomial of degree 1?   (f) a polynomial of degree 0?
(g) the polynomial 0?   (h) Explain in each case, giving an example if the answer is "yes".

* * *

Now consider the subtraction of polynomials.

Example 2. Find $(6x^2 - 3x + 2) - (4x^2 + 5x - 6)$.

Solution. Write the second polynomial beneath the first, lining up monomials of the same degree, and subtract term by term.

$$6x^2 - 3x + 2$$

(subtract) $$4x^2 + 5x - 6$$

$$\overline{2x^2 - 8x + 8}$$

Of course polynomials may be added (or subtracted horizontally as you see at the beginning of Exercises 1 and 2. Remove parentheses and collect terms. The vertical method shows the close relationship between polynomials and integers.

Subtraction can be done by "adding the opposite."

The result is $2x^2 - 8x + 8$.

When some monomials are missing in one or both of a pair of polynomials, adding or subtracting requires careful bookkeeping, as Example 3 illustrates.

Example 3. Compute

$$(5x^3 + 6x + 2) - (x^4 + x^2 - 8x).$$

Solution. Write out each polynomial in full, using 0 for the coefficient for each missing power of x:

$$0x^4 + 5x^3 + 0x^2 + 6x + 2$$

(subtract) $1x^4 + 0x^3 + 1x^2 - 8x + 0$

$$-1x^4 + 5x^3 - 1x^2 + 14x + 2$$

The result is written more simply as

$$-x^4 + 5x^3 - x^2 + 14x + 2.$$

Line up monomials
of the same
degree carefully.
(a) $6x^3 - 5x^2 - x + 12$
(c) $6x^4 + 2x^3 - 5x^2$
        $-4x + 4$

17. Subtract:

   (a) $-2x^3 + 2x^2 + x - 5$ from $4x^3 - 3x^2 + 7$ (b) $x + 2$ from $x^2 - 9$

   (c) $-2x^3 + 3x - 1$ from $6x^4 - 5x^2 - x + 3$

   (d) $2x^3 + 3x^2 + x + 5$ from $x^3 - 4x^2 + 3x - 1$

(a) $5x^3 - 5x^2 + 5x + 1$

18. Carry out the indicated subtraction:

   (a) $6x^3 - 5x^2 + 0x + 3$

       (subtract) $x^3 + 0x^2 - 5x + 2$

   (b) $8x^4 + 0x^3 + 0x^2 + 3x + 0$

       (subtract) $8x^4 - 6x^3 + 5x^2 + 0x + 4$

(a) $2x^3 + 2x^2 - 9x + 5$

19. Carry out the indicated subtraction:

   (a)
                                $0x^3 + 2x^2 - 0x + 3$

       (subtract) $-2x^3 - 0x^2 + 9x - 2$

   (b)
                                $7x^3 - 6x^2 + 5x - 4$

       (subtract) $-7x^3 + 6x^2 - 5x + 4$

MULTIPLICATION OF POLYNOMIALS

The multiplication of polynomials is similar to the multiplication of integers. Consider the computation of (483)(25) from arithmetic.

$$
\begin{array}{r}
483 \\
25 \\
\hline
\end{array}
$$

2415	multiplying by 5
966	multiplying by 2
12075	adding the results

Observe that $483 = 4(10^2) + 8(10) + 3$ and $25 = 2(10) + 5$. Compare this computation to the easier task of multiplying the polynomials, illustrated in the next example.

Example 4.   Multiply $4x^2 + 8x + 3$ and $2x + 5$.

Solution.

$$4x^2 + 8x + 3$$
$$\underline{\qquad\qquad 2x + 5}$$
$$20x^2 + 40x + 15 \qquad \text{multiplying by 5}$$
$$\underline{8x^3 + 16x^2 + \ \ 6x \qquad\qquad} \qquad \text{multiplying by 2x}$$
$$8x^3 + 36x^2 + 46x + 15 \qquad \text{adding the results}$$

You may use a calculator to save time here.

20. This exercise is a check by arithmetic that when x = 10

$(4x^2 + 8x + 3)(2x + 5)$ equals $8x^3 + 36x^2 + 46x + 15$ in agreement with the result in Example 4.

(a) Compute $4x^2 + 8x + 3$ when x = 10.

(b) Compute $2x + 5$ when x = 10.

(c) Compute $8x^3 + 36x^2 + 46x + 15$ when x = 10.

(d) Is the number found in (c) the product of the numbers found in (a) and (b)?

As Exercise 20 suggests, you can check a multiplication of polynomials by putting in a number for x.

21. (a) Compute the product $(3x + 5)(6x + 8)$.

(b) Check your answer when x = 2.

22. (a) Compute the product $(x^3 + 6x^2 + x + 3)(x^2 - 3x + 2)$.

(b) Check your answer when x = 1.

23. Compute the following products (be sure to put $0x^n$ during the computations to record a "missing" term, if necessary).

(a) $(x^3 + x^2 + x + 1)(x - 1)$ (b) $(2x^4 + 1)(x^3 - 5)$

(c) $(x^3 + x - 1)(x^3 + x^2 + 5)$ (d) $(x^2 + \sqrt{2}\,x + 1)(x^2 - \sqrt{2}\,x + 1)$

(a) $x^4 - 1$
(c) $x^6 + x^5 + x^4 + 5x^3 - x^2 + 5x - 5$

24. (a) When you add a polynomial of degree 3 and a polynomial of degree 5, you <u>always</u> get a polynomial of degree _____.

(b) When you multiply a polynomial of degree 3 and a polynomial of degree 5, you <u>always</u> get a polynomial of degree _____.

Explain and check your answers with those of other students.

25. Compute the products:

(a) $(x^4 - x^3 + x^2 - x + 1)(x + 1)$ (b) $(3x^2 - 2x + 1)(4x - 3)$

(c) $(x^2 + x + 1)(x^2 - x + 1)$ (d) $(x^3 - 2x + 1)(x^2 + x - 3)$

(a) $x^5 + 1$
(c) $x^4 + x^2 + 1$

DIVISION OF POLYNOMIALS

The last operation in the arithmetic of polynomials is the division of polynomials. It closely resembles "long division" in arithmetic but, surprisingly, the computations are easier.

Recall from arithmetic how to divide:

$$
\begin{array}{r}
28 \\
32\overline{\smash{\big)}896} \\
\underline{64} \\
256 \\
\underline{256} \\
0
\end{array}
$$

Check:

$$
\begin{array}{r}
28 \\
\times\ 32 \\
\hline
56 \\
84 \\
\hline
896
\end{array}
$$

Similar steps are followed when dividing polynomials as Example 5 illustrates.

Example 5.   Divide the polynomial x + 2 into the polynomial $2x^2 - x - 10$.

Solution.

$$
\begin{array}{r}
2x\ -\ 5 \\
x + 2\overline{\smash{\big)}2x^2 - x - 10} \\
\underline{2x^2 + 4x} \\
-5x - 10 \\
\underline{-5x - 10} \\
0
\end{array}
$$

$\longleftarrow$ this is (2x)(x + 2)

$\longleftarrow$ result of subtracting
and bringing down -10.

$\longleftarrow$ this is (-5)(x + 2)

$\longleftarrow$ result of subtraction

In this case the remainder is 0. (This seldom happens.) The quotient is
2x - 5.

26. Just as in arithmetic, division can always be checked by multiplying.
    To check Example 5, multiply (x + 2) by (2x - 5). Do you get $2x^2 - x - 10$?

Often in arithmetic there is a non-zero remainder in a division problem
involving integers.

$$
\begin{array}{r}
34 \\
29\overline{\smash{\big)}997} \\
\underline{87} \\
127 \\
\underline{116} \\
11
\end{array}
$$

The division process in arithmetic stops when the remainder, 11 in this case,
is smaller than the divisor, which is 29.

The next example illustrates a typical division of polynomials. Note
the similarity to the division of integers.

Example 6.   Divide the polynomial $x^2 + 2x + 1$ into the polynomial $5x^4 - x^2 + 2$ and find the quotient and the remainder.

Solution.

$$
\begin{array}{r}
5x^2 - 10x + 14 \\
x^2 + 2x + 1 \overline{\smash{\big)}\ 5x^4 + 0x^3 - x^2 + 0x + 2} \\
5x^4 + 10x^3 + 5x^2 \\
\hline
-10x^3 - 6x^2 + 0x \\
-10x^3 - 20x^2 - 10x \\
\hline
14x^2 + 10x + 2 \\
14x^2 + 28x + 14 \\
\hline
-18x - 12
\end{array}
$$

Note the use of zero for the coefficient of the missing x^3 term.

In this case the quotient is $5x^2 - 10x + 14$ and the remainder is $-18x - 12$. The division process stops because the degree of $-18x - 12$ is smaller than the degree of $x^2 + 2x + 1$.

Note also that division here does not lead to a polynomial quotient because the remainder is nonzero.

27. Determine the quotient and the remainder when you divide

(a) $x + 2$ into $x^2 - 6x + 9$
(b) $x - 3$ into $5x^3 - 6x^2 - 25x + 6$
(c) $x^2 + 3x - 1$ into $2x^2 + 6x - 2$
(d) $x^2 + 3x - 1$ into $3x^2 + 9x$
(e) $x^2 + 3x - 1$ into $3x^2 + 5x + 2$.

(a) $x -8$; rem. 25
(b) $5x^2 + 9x + 2$; rem. 12
(c) 2
(d) 3; rem. 3
(e) 3; rem. $-4x + 5$

28. Determine the quotient and remainder when you divide

(a) $x - 1$ into $x^4 - 1$ (b) $x + 1$ into $x^3 + 1$

(c) $x + 3$ into $2x^5 + x^3 - 6x^2 + 5x - 2$ (d) $x^2 + 6x$ into $x^5 + x^2 - 3$

(c) $x^2 - x + 1$ into $x^4 + x^2 + 1$ (f) $x - 2$ into $x^3 - 8$

(a) $x^3 + x^2 + x + 1$
(c) $2x^4 - 6x^3 + 19x^2$
$-63x + 194$
rem. -584
(e) $x^2 + x + 1$

29. Whenever you divide a polynomial of degree 5 by a polynomial of degree I

(a)   the quotient has degree _____.

(b)   the remainder is either 0 or has degree at most _____.

30. Whenever you divide a polynomial of degree 5 by a polynomial of degree 2.

(a)   the quotient has degree _____.

(b)   the remainder is either 0 or has degree at most _____.

* * *

When dividing one <u>integer</u> into another <u>integer</u>, the quotient and remainder are checked as follows:

$$\begin{array}{r} 5 \quad\quad\text{quotient} \\ 15 \overline{\smash{\big)}\, 87} \\ 75 \\ \hline 12 \quad\text{remainder} \end{array}$$

Check:

Since $(5)(15)+12=75+12$
$\qquad\qquad\qquad =87$

the result is correct.

The next example shows a similar check in the division of polynomials.

Example 7.   Divide $x - 3$ into $2x^3 + 5x^2 - 6x + 9$ and then check.

Solution.

$$\begin{array}{r} 2x^2 + 11x \ + 27 \longleftarrow \text{quotient} \\ x - 3 \overline{\smash{\big)}\, 2x^3 + 5x^2 - 6x + 9} \\ 2x^3 - 6x^2 \\ \hline 11x^2 - 6x \\ 11x^2 - 33x \\ \hline 27x + 9 \\ 27x - 81 \\ \hline 90 \longleftarrow \text{remainder} \end{array}$$

Check: Compute $(x - 3)(2x^2 + 11x + 27) + 90$:

$$\begin{array}{r} 2x^2 + 11x + 27 \\ x - \ \ 3 \\ \hline - 6x^2 - 33x - 81 \\ 2x^3 + 11x^2 + 27x \\ \hline 2x^3 + 5x^2 \ - 6x - 81 \end{array}$$

Then add the remainder 90:

$(2x^3 + 5x^2 - 6x - 81) + (90) = 2x^3 + 5x^2 - 6x + 9$, as desired.

31. Divide the first polynomial into the second.

(a) $2x+5$; rem. -1

(b) $2x^2-5x$; rem. 0

(c) $2x^2-7x + 19$
     rem. $-50x-13$

   (a) $x - 2$ into $2x^2 + x - 11$

   (b) $x + 3$ into $2x^3 + x^2 - 15x$

   (c) $x^2 + 3x + 1$ into $2x^4 - x^3 + 6$

32. Divide $2x^3 + x - 5$ by $x^2 - 7$.

33. Divide $2x^3 + 3x^2 - 4x + 12$ by $x + 2$

THE QUOTIENT AND REMAINDER

If A and B are positive integers and we divide A into B, we get quotient Q and a remainder R, smaller than A, and

$$B = QA + R.$$

If the remainder is 0 then we say that "A <u>divides</u> B", or "A is a <u>factor</u> of B." For instance, when we divide 5 into 30, the quotient is 6 and the remainder is 0.

$$30 = 6 \cdot 5$$

We say "5 divides 30." But 5 does not divide 32, since the remainder is 2.

The same idea carries over without change to polynomials A(x) and B(x). When we divide A(x) into B(x) there is a quotient Q(x) and remainder R(x), and

$$B(x) = Q(x) \cdot A(x) + R(x)$$

If R(x) is 0, then we say that "A(x) divides B(x)." For instance, division of x - 2 into $2x^2$ - x - 6 yields a quotient Q(x) = 2x + 3 and remainder R(x) = 0. (Carry out the division.) Then

$$2x^2 - x - 6 = (2x + 3)(x - 2),$$

as may be checked by multiplying. In each division, Q(x) is a polynomial and R(x) is a polynomial that has degree <u>less</u> than that of A(x) or is simply the polynomial 0.

34. Show all work in each case:

    (a) Does x - 2 divide $x^3 - 17x^2 + 11x + 38$?

    (b) Does $x^2 + 2x + 5$ divide $x^3 - 17x^2 + 11x + 38$?

    (c) Does x + 2 divide $x^3 - 17x^2 + 11x + 38$?

For bookkeeping purposes, polynomials are usually written in descending (or ascending) powers.

34. (a) yes
(b) no
(c) no

In general, the way to tell whether a polynomial A(x) divides another polynomial B(x) is to carry out the division ("long" or "synthetic") and see if the remainder is 0. But there is a shortcut in case the polynomial A(x) has degree 1 and the coefficient of x in A(x) is 1. In other words, there is a quick way to tell whether a polynomial of the form x + a (where <u>a</u> is a number, either positive or negative) divides a polynomial B(x). Later in this chapter, Exercise 69 develops this shortcut, known as "the factor theorem."

A NOTATION FOR SUBSTITUTION IN POLYNOMIALS

Every time x in $x^3 - 3x^2 + 2x - 4$ is replaced by a number, another number is obtained. For example, when x is replaced by 3, the computations look like this:

$$(3)^3 - 3(3)^2 + 2(3) - 4 = (27) - 3(9) + 2(3) - 4$$
$$= 27 - 27 + 6 - 4$$
$$= 2$$

In order to talk about this procedure simply, some new notation is needed.

Recall the order of operations here:
1. Parentheses
2. Exponents
3. Multiplication (Division)
4. Addition (Subtraction)

The symbol P(x) shall denote a polynomial. Thus P(x) may be $x^3 - 3x^2 + 2x - 4$ or $6x + 5$ or $x^6 - 2$, or any other polynomial. But A(x), B(x), Q(x), or R(x) may also denote polynomials.

When you replace x in a polynomial P(x) by some number, k, another number is obtained, which shall be denoted P(k). (This is read "P of k".) Thus, if

$$P(x) = 3x^2 - 3x + 2$$

then

$$P(k) = 3k^2 - 3k + 2$$

More specifically, if $P(x) = 2x^2 - 3x + 2$, then

$$P(1) = 2(1)^2 - 3(1) + 2$$

$$= 2(1) - 3(1) + 2$$

$$= 2 - 3 + 2$$

$$= 1$$

Skipping steps in the order of operations can cause errors. (see the marginal note for Exercise 1.)

and

$$P(2) = 2(2)^2 - 3(2) + 2$$

$$= 2(4) - 3(2) + 2$$

$$= 8 - 6 + 2$$

$$= 4$$

The number P(k) is called the <u>value of</u> P(x) <u>at</u> k (the value of the polynomial when x is replaced by k).

Show all steps in Exercises 35-40.
(a) 14 (b) 4 (c) 0 (d) 2 (e) 10

35. Let $P(x) = 3x^2 - 5x + 2$. Showing all arithmetic, compute

   (a) P(3)   (b) P(2)   (c) $P(\frac{2}{3})$ (d) P(0)   (e) P(-1)

(a) -3 (b) 2 (c) 0 (d) -8

36. Let $P(x) = 5x - 3$. Showing all arithmetic, compute

   (a) P(0)   (b) P(1)   (c) $P(\frac{3}{5})$ (d) P(-1)

(a) -1 (c) -5 (e) -10

37. Let $P(x) = x - 5$. Compute P(k) for k equal to

   (a) 4   (b) -4   (c) 0   (d) 5   (e) -5

38. Let $P(x) = x + 3$. For what value of k does P(k) = 0?

(a) -2 (b) $-\frac{3}{4}$ (c) 0 (d) 0

39. Let $P(x) = x^2 - 4x + 1$. Compute

   (a) P(1)   (b) $P(\frac{1}{2})$ (c) $P(2 - \sqrt{3})$ (d) $P(2 + \sqrt{3})$

40. Let $P(x) = 3x^2 + 7x - 20$. Compute

(a) -20 (b) 0 (c) 56 (d) 0

   (a) P(0)   (b) $P(\frac{5}{3})$ (c) P(4)   (d) P(-4)

* * *

The preceding six exercises gave practice in using the notations P(x) and P(k). Of particular interest are those numbers k such that P(k) is 0. The opening problem of the chapter is one instance of this. If k is a number such that P(k) = 0, then k is called a <u>solution</u> of the equation P(x) = 0.

The rest of the chapter develops methods for finding solutions of equations of the type

$$P(x) = 0$$

where P(x) is a polynomial of degree one, two, or three.

HOW TO SOLVE THE EQUATION ax + b = 0

A polynomial P(x) of degree one has the form ax + b where <u>a</u> is not 0. The equation

$$P(x) = 0$$

is just

$$ax + b = 0.$$

The next example shows how to find x, if ax + b = 0.

Example 8. Solve the equation: 3x + 7 = 0

Solution.

$$3x + 7 = 0$$

$$3x = -7$$

$$\frac{3x}{3} = -\frac{7}{3} \quad \text{(divide by 3)}$$

$$x = -\frac{7}{3}$$

As a check:

$$3(-\frac{7}{3}) + 7 \stackrel{?}{=} 0$$

$$-7 + 7 \stackrel{?}{=} 0 \qquad \text{yes}$$

41. Solve for x in the equation P(x) = 0 if P(x) is

   (a) 3x - 7   (b) 6x   (c) -2x + 5   (d) x + 5

   Check the results by substitution.

   $(a) \frac{7}{3} \quad (c) \frac{5}{2}$

42. Solve and check by substitution:

   (a) 2x - ½ = 0   (b) 6x + 3 = 0   (c) -5x + 8 = 0   (d) 7x + 4 = 0

   $(a) \frac{1}{4} \quad (c) \frac{8}{5}$

* * *

As Example 8 and Exercises 41-42 show, an equation of the type $P(x) = 0$, where $P(x)$ has degree 1, can be solved quickly. It always has exactly one solution.

HOW TO SOLVE THE EQUATION $ax^2 + bx + c = 0$

A polynomial $P(x)$ of degree two has the form $ax^2 + bx + c$, where <u>a</u> is not 0. The equation

$$P(x) = 0$$

takes the form

$$ax^2 + bx + c = 0.$$

Such an equation may have no real solutions (the solutions are not on the number line). For instance, the equation

$$x^2 + 4 = 0$$

has none, since the left hand side, $x^2 + 4$, is always at least 4 whenever x is replaced by a real number.

43. How many solutions has the equation $x^2 = 0$?

44. How many solutions has the equation $x^2 - 9 = 0$?

* * *

There are two ways to solve the equation,

$$ax^2 + bx + c = 0$$

(if there are any real solutions). One method is "factoring" the expression $ax^2 + bx + c$. It is useful when the coefficients a, b, and c are small integers and the expression $ax^2 + bx + c$ is easy to factor. The other method is "completing the square", which is the basis of the "quadratic formula." This latter method always works for second degree equations, no matter how complicated the coefficients may be.

SOLVING THE EQUATION $ax^2 + bx + c = 0$ BY FACTORING

The next example illustrates how solutions are found by factoring.

Example 9. Solve the equation

$$6x^2 - 11x - 10 = 0.$$

Solution. First try to write
$$6x^2 - 11x - 10$$
as the product of two first degree polynomials

$$6x^2 - 11x - 10 = (?) \, (?)$$

The coefficient of x^2 is 6. Now $6 = (1)(6)$ and $6 = (2)(3)$. Following up the $6 = (1)(6)$ case first, try

$$6x^2 - 11x - 10 = (1x + \underline{})(6x + \underline{})$$

The two blanks are to be filled in with numbers whose product is -10,

2 and -5 or -2 and 5, or 5 and -2, or -5 and 2.

Try each of the four cases

$$(1x + 2)(6x - 5) = 6x^2 + 7x - 10$$
$$(1x - 2)(6x + 5) = 6x^2 - 7x - 10$$
$$(1x + 5)(6x - 2) = 6x^2 + 28x - 10$$
$$(1x - 5)(6x + 2) = 6x^2 - 28x - 10$$

None of the four cases results in $6x^2 - 11x - 10$.

So turn to the $6 = (2)(3)$ possibility, testing the products in a similar manner:

$$(2x + 2)(3x - 5) = 6x^2 - 4x - 10$$
$$(2x - 2)(3x + 5) = 6x^2 + 4x - 10$$
$$(2x + 5)(3x - 2) = 6x^2 + 11x - 10$$
$$(2x - 5)(3x + 2) = 6x^2 - 11x - 10$$

Fortunately, the last case works. Thus the original equation

$$6x^2 - 11x - 10 = 0$$

now reads

$$(2x - 5)(3x + 2) = 0.$$

"The product of the numbers 2x - 5 and 3x + 2 is 0."

Now, <u>the only way the product of two numbers is 0 is when one of them is 0</u>. Thus $(2x - 5) = 0$ or $(3x + 2) = 0$. So we ask for solutions of either of these first degree equations

<table>
<tr><td>2x - 5 = 0</td><td>or</td><td>3x + 2 = 0</td></tr>
<tr><td>2x = 5</td><td></td><td>3x = -2</td></tr>
<tr><td>$x = \dfrac{5}{2}$</td><td></td><td>$x = -\dfrac{2}{3}$</td></tr>
</table>

If ab = 0, then a = 0 or b = 0. This is sometimes called the "zero product rule." (See First Year Algebra Review of this chapter)

The technique of solving first degree equations provides the solutions, $\dfrac{5}{2}$ and $-\dfrac{2}{3}$

Check: (Always substitute in the original equation)

$$6(\tfrac{5}{2})^2 - 11(\tfrac{5}{2}) - 10 \overset{?}{=} 0 \qquad\qquad 6(-\tfrac{2}{3})^2 - 11(-\tfrac{2}{3}) - 10 \overset{?}{=} 0$$

$$6(\tfrac{25}{4}) - 11(\tfrac{5}{2}) - 10 \overset{?}{=} 0 \qquad\qquad 6(\tfrac{4}{9}) + \tfrac{22}{3} - 10 \overset{?}{=} 0$$

$$\frac{75}{2} - \frac{55}{2} - 10 \overset{?}{=} 0 \qquad\qquad \frac{8}{3} + \frac{22}{3} - 10 \overset{?}{=} 0$$

$$\frac{20}{2} - 10 \overset{?}{=} 0 \qquad\qquad \frac{30}{3} - 10 \overset{?}{=} 0$$

$$10 - 10 \overset{?}{=} 0 \qquad\qquad 10 - 10 \overset{?}{=} 0$$

$$0 \overset{?}{=} 0 \qquad\qquad 0 \overset{?}{=} 0$$

$$\text{yes} \qquad\qquad\qquad \text{yes}$$

Some students may not write all of the work shown in Example 9. The next example shows how the problem might look on paper if the student does not need to experiment so much.

Example 10. Solve for x:

$$6x^2 - 11x - 10 = 0$$

Solution: $(2x - 5)(3x + 2) = 0$

This format is recommended for solving quadratic equations by factoring.

then

$2x - 5 = 0$	or	$3x + 2 = 0$
$2x = 5$		$3x = -2$
$x = \dfrac{5}{2}$		$x = -\dfrac{2}{3}$

Check: The check is the same as shown in Example 9.

Clearly, solving by factoring can take a lot of time if the coefficient a and c in $ax^2 + bx + c$ have many divisors.

(a) 3, $-\dfrac{1}{3}$

(c) 0, $\dfrac{2}{3}$

45. Solve by factoring and check your answers for (b) and (d):

(a) $3x^2 - 8x - 3 = 0$ (b) $2x^2 - x - 1 = 0$

(c) $3x^2 - 2x = 0$ (d) $3x^2 + 7x + 2 = 0$

46. Solve by factoring:

(a) $\dfrac{1}{3}$, -2

(c) -1, $-\dfrac{2}{3}$

(a) $3x^2 + 5x = 2$ (b) $2x^2 - 3x = 5$

(c) $3x^2 + 5x + 2 = 0$ (d) $8x^2 - 2x - 3 = 0$

Hint: (a) Write $3x^2 + 5x - 2 = 0$

SOLVING THE EQUATION $x^2 + bx + c = 0$ BY COMPLETING THE SQUARE

The next few exercises introduce a more general way of solving second degree ("quadratic") equations.

47. Consider the equation

$$x^2 - 6x + 7 = 0$$

(a) Copy and fill in this table for $y = x^2 - 6x + 7$

x	$x^2 - 6x + 7$	y
1		
2		
3		
4		
5		

Allow sufficient space for computations in the middle column.

(b) Plot the five points (x,y) provided by the table in (a).

(c) Sketch a smooth curve through the points in (b).

(d) On the basis of (c), show that the equation $x^2 - 6x + 7 = 0$ has at least two solutions.

48. Show that the polynomial $x^2 - 6x + 7$ is <u>not</u> the product of two polynomials of degree one whose coefficients are integers.   *Try all possibilities.*

* * *

As Exercise 47 and 48 show, the equation

$$x^2 - 6x + 7 = 0$$

has solutions, but the "factoring method" is useless in finding them.

There is a second and more powerful method, called "completing the square". Example 11 illustrates the general procedure in the case of the equation

$$x^2 - 6x + 7 = 0.$$

Example 11. Solve the equation

$$x^2 - 6x + 7 = 0$$

Solution. $x^2 - 6x + 7 = 0$

$$x^2 - 6x + 7 - 7 = 0 - 7$$

$$x^2 - 6x \qquad = -7 \qquad \text{(Leave a blank space for a constant term)}$$

Now $x^2 - 6x$ is the first two terms of the square of $x - 3$ since

$$(x - 3)^2 = x^2 - 6x + 9$$

To make use of this, add 9 to both sides of the equation

$$x^2 - 6x \qquad = -7$$

obtaining

$$x^2 - 6x + 9 = -7 + 9$$

or

$$(x - 3)^2 \qquad = 2.$$

This says that "x - 3, when squared, gives 2."   Thus

$$x - 3 = \sqrt{2} \qquad \text{or} \qquad x - 3 = -\sqrt{2}$$
$$x = 3 + \sqrt{2} \qquad\qquad\qquad x = 3 - \sqrt{2}$$

Hence the equation,

$$x^2 - 6x + 7 = 0,$$

has two solutions,

$$3 + \sqrt{2} \qquad \text{and} \qquad 3 - \sqrt{2}.$$

Check: Verify that $3 + \sqrt{2}$ is solution of $x^2 - 6x + 7 = 0$ as follows:

$$(3 + \sqrt{2})^2 - 6(3 + \sqrt{2}) + 7 \overset{?}{=} 0$$
$$(9 + 6\sqrt{2} + 2) - 18 - 6\sqrt{2} + 7 \overset{?}{=} 0$$
$$(11 + 6\sqrt{2}) - 18 - 6\sqrt{2} + 7 \overset{?}{=} 0$$
$$11 - 18 + 7 + 6\sqrt{2} - 6\sqrt{2} \overset{?}{=} 0$$
$$0 \overset{?}{=} 0 \quad \text{yes}$$

A similar check may be made for $3 - \sqrt{2}$.

Example 12 is similar to Example 11, but fewer steps are shown.

Example 12.   Solve for x:

$$x^2 + 2x - 9 = 0$$
$$x^2 + 2x \qquad = 9 \qquad \text{(prepare to complete the square;}$$
$$\text{leave space for the constant term)}$$
$$x^2 + 2x + 1 = 9 + 1 \qquad \text{(complete the square)}$$
$$(x + 1)^2 = 10$$

Study Example 12 carefully.

Thus,

$$x + 1 = \sqrt{10} \qquad \text{or} \qquad x + 1 = -\sqrt{10}$$
$$x = -1 + \sqrt{10} \qquad\qquad\qquad x = -1 - \sqrt{10}$$

Check:

$$(-1 + \sqrt{10})^2 + 2(-1 + \sqrt{10}) - 9 \overset{?}{=} 0$$
$$(1 - 2\sqrt{10} + 10) - 2 + 2\sqrt{10} - 9 \overset{?}{=} 0$$
$$(11 - 2\sqrt{10}) - 11 + 2\sqrt{10} \overset{?}{=} 0$$
$$0 \overset{?}{=} 0$$
$$\text{yes}$$

A check will show that $-1 - \sqrt{10}$ is also a solution.

49. (a) Check that $3 - \sqrt{2}$ is a solution of the equation $x^2 - 6x + 7 = 0$.

   (b) Check that $-1 - \sqrt{10}$ is a solution of the equation $x^2 + 2x - 9 = 0$.

50. Solve each of these equations by completing the square. <u>Show all steps.</u>

   (a) $x^2 - 6x + 4 = 0$ (b) $x^2 - 6x + 2 = 0$

   (c) $x^2 - 6x + 1 = 0$ (d) $x^2 + 10x - 1 = 0$

   (e) $x^2 + 4x + 2 = 0$ (f) $x^2 - 10x + 19 = 0$

51. Solve the equation $x^2 - 6x - 7 = 0$ by

   (a) factoring   (b) completing the square

(a) $3 + \sqrt{5}$ and $3 - \sqrt{5}$
(c) $3 + 2\sqrt{2}$ and $3 - 2\sqrt{2}$
(e) $-2 + \sqrt{2}$ and $-2 - \sqrt{2}$

* * *

To "complete the square", take half of the x-coefficient, <u>square it, and add this number to both sides of the equation.</u> For instance,

$$x^2 - 6x - 2 = 0$$

or

$$x^2 - 6x \quad = 2$$

becomes

$$x^2 - 6x + (-3)^2 = 2 + (-3)^2$$

$$x^2 - 6x + 9 = 2 + 9$$

$$(x - 3)^2 = 11$$

and so on.

Make sure you understand this key step.

This method still works when the x-coefficient is an odd number such as -5, 3, or 7, of indeed any real number. <u>Notice that in this discussion the x^2-cofficient is 1.</u> The case where the coefficient of the x^2 term is not 1 is treated later in Exercise 54.

Example 13. Solve the equation by completing the square:

$$x^2 - 5x + 3 = 0$$

Solution.

$$x^2 - 5x \quad = -3$$

$$x^2 - 5x + \left(-\frac{5}{2}\right)^2 = -3 + \left(-\frac{5}{2}\right)^2$$

$$\left(x - \frac{5}{2}\right)^2 = -3 + \frac{25}{4}$$

$$\left(x - \frac{5}{2}\right)^2 = \frac{-12 + 25}{4}$$

$$\left(x - \frac{5}{2}\right)^2 = \frac{13}{4}$$

$\left(x - \frac{5}{2}\right)^2 =$
$x^2 - 5x + \left(-\frac{5}{2}\right)^2$
so $\left(-\frac{5}{2}\right)^2$ is added to both sides of the equation $x^2 - 5x = -3$

Hence

$$x - \frac{5}{2} = \sqrt{\frac{13}{4}} \qquad \text{or} \qquad x - \frac{5}{2} = -\sqrt{\frac{13}{4}}$$

$$x = \frac{5}{2} + \sqrt{\frac{13}{4}} \qquad \Big| \qquad x = \frac{5}{2} - \sqrt{\frac{13}{4}}$$

The two solutions are therefore

$$x = \frac{5}{2} + \sqrt{\frac{13}{4}} \qquad \text{or} \qquad x = \frac{5}{2} - \sqrt{\frac{13}{4}},$$

usually simplified to

$$x = \frac{5 + \sqrt{13}}{2} \qquad \text{or} \qquad x = \frac{5 - \sqrt{13}}{2}$$

52. (a) Check that $\dfrac{5 + \sqrt{13}}{2}$ is a solution of the equation in Example 13,

$$x^2 - 5x + 3 = 0.$$

(b) Check that $\dfrac{5 - \sqrt{13}}{2}$ is also a solution of the equation.

$$* * *$$

Example 14. Solve x in the equation

$$x^2 - 3x + 1 = 0.$$

Solution.

$$x^2 - 3x + 1 = 0.$$

$$x^2 - 3x \qquad = -1 \qquad \text{(prepare to complete the square)}$$

$$x^2 - 3x + \left(-\frac{3}{2}\right)^2 = -1 + \left(-\frac{3}{2}\right)^2 \quad \text{(complete the square)}$$

$$\left(x - \frac{3}{2}\right)^2 = -1 + \frac{9}{4}$$

$$\left(x - \frac{3}{2}\right)^2 = \frac{5}{4}$$

$$x - \frac{3}{2} = \sqrt{\frac{5}{4}} \qquad \text{or} \qquad x - \frac{3}{2} = -\sqrt{\frac{5}{4}}$$

$$x = \frac{3}{2} + \sqrt{\frac{5}{4}} \qquad \qquad x = \frac{3}{2} - \sqrt{\frac{5}{4}}$$

$$x = \frac{3}{2} + \frac{\sqrt{5}}{2} \qquad \qquad x = \frac{3}{2} - \frac{\sqrt{5}}{2}$$

$$x = \frac{3 + \sqrt{5}}{2} \qquad \qquad x = \frac{3 - \sqrt{5}}{2}$$

NOTATION. The solutions $\dfrac{3 + \sqrt{5}}{2}$ and $\dfrac{3 - \sqrt{5}}{2}$ are sometimes written $\dfrac{3 \pm \sqrt{5}}{2}$. However, the two solutions should be checked separately.

53. Solve for x:

   (a) $x^2 - x - 1 = 0$ (b) $x^2 - 3x + 2 = 0$ (a) $\frac{1 \pm \sqrt{5}}{2}$

   (c) $x^2 - 3x - 2 = 0$ (d) $x^2 + 5x - 1 = 0$ (c) $\frac{3 \pm \sqrt{17}}{2}$

   (e) $x^2 - 7x - 11 = 0$ (f) $x^2 + 9x + 3 = 0$ (e) $\frac{7 \pm \sqrt{93}}{2}$

* * *

SOLVING $ax^2 + bx + c = 0$ WHERE $a \neq 1$ BY COMPLETING THE SQUARE

54. Explain why the equation
$$2x^2 - 10x + 6 = 0$$

has the same solutions as the equation
$$x^2 - 5x + 3 = 0$$

55. Solve each of the following equations by completing the square. Some
may have no solutions or only one solution.

   (a) $x^2\ \ 4x + 2 = 0$ (b) $x^2 - 4x + 5 = 0$ *Hint: (d) first*
   (c) $x^2 - 4x + 4 = 0$ (d) $3x^2 - 12x + 15 = 0$ *divide each side*
                                                                     by 3.

* * *

   Suppose you wish to solve the equation
$$25x^2 - 10x - 48 = 0$$

by completing the square. Note that the coefficient of x^2 is not 1. First
to make the coefficient of x^2 equal to 1, divide both sides by 25, obtaining

$$x^2\ \frac{10}{25}x\ \ \frac{48}{25} = 0.$$

then prepare to complete the square,

$$x^2 - \frac{2}{5}x\ \ = \frac{48}{25}.$$

   Take half of the coefficient of the x term and add its square to both
sides:

$$x^2 - \frac{2}{5}x + \left(-\frac{1}{5}\right)^2 = \frac{48}{25} + \left(-\frac{1}{5}\right)^2$$

or

$$\left(x - \frac{1}{5}\right)^2 = \frac{48}{25} + \frac{1}{25}$$

or

$$\left(x - \frac{1}{5}\right)^2 = \frac{49}{25}$$

Thus

$$x - \frac{1}{5} = \frac{7}{5} \qquad \text{or} \qquad x - \frac{1}{5} = -\frac{7}{5}$$

$$x = \frac{8}{5} \qquad\qquad\qquad x = -\frac{6}{5}$$

In this case, the solutions are relatively simple since they involve only quotients of integers, and no square root is required. (Only when the solutions are of this simple rational number form can they be found by the the method of factoring.)

56. Check that $\frac{8}{5}$ and $-\frac{6}{5}$ are solutions of the equation

$$25x^2 - 10x - 48 = 0.$$

57. Solve the equation $25x^2 - 10x - 48 = 0$ by factoring.

58. Solve the equation $9x^2 + 14x - 8 = 0$

Hint: (a) First divide by 9

    (a) by completing the square.

    (b) by factoring.

Hint: First divide, making 1 the x^2 coefficient.

(a) $1, \frac{1}{2}$

(c) $\frac{-2 \pm \sqrt{19}}{3}$

59. Solve for x by completing the square:

    (a) $2x^2 - 3x + 1 = 0$ (b) $3x^2 - 7x - 6 = 0$

    (c) $3x^2 + 4x - 5 = 0$ (d) $4x^2 - 2x - 3 = 0$

THE QUADRATIC FORMULA FOR SOLVING $ax^2 + bx + c = 0$

The method of "completing the square" applies to any equation in the form

$$ax^2 + bx + c = 0.$$

STUDY CAREFULLY!

First divide by <u>a</u>, then complete the square. Rather than go through the whole procedure each time, it is much quicker to do the <u>general case</u> once and obtain a formula which describes the solutions (if there are any) for all cases. This will now be done.

Consider the general equation,

$$ax^2 + bx + c = 0.$$

where <u>a</u> is not 0. First of all, it can be assumed that <u>a</u> is positive. (If it were not, multiply both sides of the equation by -1.) Now proceed to solve for x, as in the previous cases where a, b, and c were specific numbers.

$$ax^2 + bx + c = 0$$

$$x^2 + \frac{b}{a}x + \frac{c}{a} = 0 \qquad \text{(divide by a)}$$

$$x^2 + \frac{b}{a}x \qquad\qquad = -\frac{c}{a} \qquad \text{(add } -\frac{c}{a} \text{ to both sides)}$$

$$x^2 + \frac{b}{a}x + \left(\frac{b}{2a}\right)^2 = -\frac{c}{a} + \left(\frac{b}{2a}\right)^2 \qquad \text{(complete the square)}$$

$$\left(x + \frac{b}{2a}\right)^2 = \frac{b^2}{4a^2} - \frac{c}{a} \qquad \text{(a more useful form)}$$

$$\left(x + \frac{b}{2a}\right)^2 = \frac{b^2 - 4ac}{4a^2} \qquad \text{(add fractions)}$$

So

$$x + \frac{b}{2a} = \pm\sqrt{\frac{b^2 - 4ac}{2a}} \qquad \text{(square root of each side)}$$

$$x + \frac{b}{2a} = \pm\frac{\sqrt{b^2 - 4ac}}{2a} \qquad \text{(simplify radical)}$$

$$x = -\frac{b}{2a} \pm \frac{\sqrt{b^2 - 4ac}}{2a} \qquad \text{(add } -\frac{b}{2a} \text{ to both sides)}$$

$$x = \frac{-b \pm \sqrt{b^2 - 4ac}}{2a} \qquad \text{(add fractions)}$$

Thus

$$\boxed{x = \frac{-b \pm \sqrt{b^2 - 4ac}}{2a} \qquad \text{The Quadratic Formula}}$$

MEMORIZE!

(The formula is correct whether <u>a</u> is positive or negative.) This formula is a generalization of "completing the square" It should be <u>memorized</u> and used, for it solves the equation $ax^2 + bx + c = 0$ rapidly and with <u>less chance to make errors in arithmetic</u>, as Example 15 shows.

Write the quadratic formula from memory. Say it aloud without looking at it.

Comment. In using the quadratic formula, the value of a, b, and c are given by the equation in the form

$$ax^2 + bx + c = 0.$$

<u>Change the equation to this form before using the quadratic formula.</u>

Example 15. Solve the equation

$$25x^2 - 10x - 48 = 0.$$

Solution: Observe that

$$a = 25, \qquad b = -10 \qquad \text{and} \qquad c = -48.$$

By the quadratic formula,

$$x = \frac{-b \pm \sqrt{b^2 - 4ac}}{2a}$$

*Write the
quadratic formula.*

the solutions are

$$x = \frac{-(-10) \pm \sqrt{(-10)^2 - 4(25)(-48)}}{2(25)}$$

*Use parentheses
( ), in place of
a, b and c when
substituting in
the formula to
reduce the chance
for errors. Do
not skip steps.*

$$x = \frac{10 \pm \sqrt{100 + 4800}}{50}$$

$$= \frac{10 \pm \sqrt{4900}}{50}$$

$$= \frac{10 \pm 70}{50}$$

*Do not stop too
soon, for $\frac{1 \pm 7}{5}$
can be simplified.*

$$= \frac{1 \pm 7}{5}$$

$$= \frac{8}{5} \text{ or } -\frac{6}{5}$$

60. Use the quadratic formula to solve the equations:

(a) $x^2 - 5x + 3 = 0$ of Example 13.

(b) $x^2 - 3x + 1 = 0$ of Example 14.

*Hint: (b) First
write the equation
in the form
$ax^2 + bx + c = 0$
(a) $1, \frac{1}{2}$
(c) $\frac{3 \pm \sqrt{13}}{4}$
(e) $3, -\frac{2}{3}$*

61. Solve for x, using the quadratic formula:

(a) $2x^2 - 3x + 1 = 0$ (b) $2x^2 - 5x = 1$

(c) $4x^2 = 6x + 1$ (d) $5x^2 - 9x + 4 = 0$

(e) $3x^2 - 7x = 6$ (f) $6x + 7 = 2x^2$

*(a) $1 \pm \sqrt{3}$
(b) no solution,
   why?
(c) 1
Make sure you
understand this
important
exercise.*

62. Solve the equation if possible (use the quadratic formula):

(a) $x^2 - 2x - 2 = 0$ (b) $x^2 - 2x + 2 = 0$

(c) $x^2 - 2x + 1 = 0$

63. How many real-number solutions does the equation

$$ax^2 + bx + c = 0$$

have if $b^2 - 4ac$ is

*(a) two
(c) one*

(a) positive?   (b) negative?   (c) 0?   (d) Explain, using the quadratic
                                             formula.

THE DISCRIMINANT

The expression "$b^2 - 4ac$" is called the <u>discriminant</u> of the polynomial $ax^2 + bx + c$. As Exercises 62 and 63 show, the discriminant tells much about the solutions of the equation $ax^2 + bx + c = 0$.

64. Use the quadratic formula to find the solutions (if any) of these equations:

    (a) $x^2 + 5x - 3 = 0$ (b) $4x^2 - 12x + 9 = 0$

    (c) $2x^2 - 3x + 4 = 0$

65. Write the quadratic formula from memory.

66. Say the quadratic formula aloud without looking at it.

Have a classmate check your memorization of it.

THE GRAPH OF $y = ax^2 + bx + c$

The graph of an equation of the form

$$y = ax^2 + bx + c$$

is shaped as follows:

If $a > 0$:   If $a < 0$:

In the first case, the <u>lowest point</u> is called the <u>vertex</u>; in the second case, the <u>highest point</u> is called the <u>vertex</u>. In both cases, <u>the curve is symmetric with respect to a vertical line through the vertex.</u>

The vertex is sometimes called the "turning point."

Exercise 67 shows how to use the quadratic formula to find the x-coordinate of the vertex of the parabola. This knowledge is an aid in graphing the curve.

Make sure you understand Exercise 67. Part (c) makes graphing easier.

67. (a) Consider the solutions to

$$ax^2 + bx + c = 0,$$

    namely,

$$x = \frac{-b + \sqrt{b^2 - 4ac}}{2a} \quad \text{and} \quad x = \frac{-b - \sqrt{b^2 - 4ac}}{2a}$$

    (b) Show that the average of the two solutions in (a) is

$$-\frac{b}{2a}.$$

    (c) Use the symmetry of the curve to explain why $-\frac{b}{2a}$ is the <u>x-coordinate of the vertex of the parabola</u>, $y = ax^2 + bx + c$.

* * *

The results of Exercise 67 aid in graphing a parabola, as the following example shows.

Example 16.   Graph $y = x^2 - 6x + 5$.

Solution.   First find the x-coordinate of the vertex:

$$-\frac{b}{2a} = \frac{-(-6)}{2} = 3.$$

Now use the symmetry of the parabola to help sketch the graph:

Note the symmetry as pairs of x-values are chosen equidistant from the x-value of the vertex.

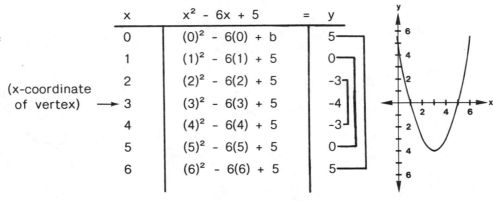

x	$x^2 - 6x + 5$	=	y
0	$(0)^2 - 6(0) + b$		5
1	$(1)^2 - 6(1) + 5$		0
2	$(2)^2 - 6(2) + 5$		-3
(x-coordinate of vertex) → 3	$(3)^2 - 6(3) + 5$		-4
4	$(4)^2 - 6(4) + 5$		-3
5	$(5)^2 - 6(5) + 5$		0
6	$(6)^2 - 6(6) + 5$		5

*A less precise (quicker) method may be used:
(1) Find the vertex.
(2) Find the x-axis intercept (where y = 0).
(3) Find the y-axis intercept (where x = 0).
(4) Use symmetry and rough sketch.*

Observe the symmetry in the table with reference to the vertex of the parabola.   This serves as a check for errors in computing y-values.

68. Use the method of Example 16 to graph each of the following:

(a) $y = x^2 - 2x - 2$ (b) $y = x^2 - 2x + 2$ (c) $y = x^2 - 2x + 1$

(d) Use the graphs to discuss the number of solutions for each equation in Exercise 62.

69. Use the method of Example 16 to help you graph

Hint: (c) as before pick values of x symmetric with respect to the x-coordinate of the vertex.

(a) $y = -2x^2 + 4x + 5$ (b) $y = -x^2 + 8x + 2$

(c) $y = x^2 - 3x + 3$

(d) Estimate where each of the above graphs crosses the x-axis.

70. (a) Devise an equation of your own of the form $y = ax^2 + bx + c$.

(b) Graph the equation in (a).

Compare your work with that of other students.

HOW TO SOLVE THE EQUATION $ax^3 + bx^2 + cx + d = 0$

So far the chapter has presented techniques for solving any first degree equation,

$$ax + b = 0,$$

and any second degree equation,

$$ax^2 + bx + c = 0.$$

The remainder of the chapter is devoted to solving equations of the third degree

$$ax^3 + bx^2 + cx + d = 0.$$

Though the methods will not take care of every case, they will be quite useful. In particular they will provide all the solutions of the opening equation,

$$x^3 - 3x^2 - 2x + 4 = 0.$$

THE FACTOR THEOREM

For a moment make a slight detour, and consider the following question. Suppose that $P(x)$ is a polynomial of degree at least one, and r is a number. How can we tell whether the polynomial

$$x - r$$

divides the polynomial

$$P(x)$$

without going through the labor of long division? The next exercise may suggest the shortcut.

71. Fill in this table with the words "yes" or "no", and show all work.

(Check the first line, which is already filled in.)

r	x-r	P(x)	Does x-r divide P(x)?	Is r a solution of the equation P(x)=0?
3	x - 3	$x^2 + 2x - 5$	no	no
3		$x^2 + 2x - 15$		
2		$x^3 + 8$		
-2	x + 2	$x^3 + 8$		
1		$x^3 - 3x^2 - 2x + 4$		
-1		$x^3 - 4x^2 + 5x + 9$		

Do not rush through Exercise 71. Study the results after you have completed it.

If your arithmetic in Exercise 71 was accurate, you found that in each case the two questions,

"Does x - r divide P(x)?"

and

"Is r a solution of the equation P(x) = 0?"

<u>always</u> have the same answers, both "yes" or both "no."

First of all, assume that x - r <u>does</u> divide P(x). That means there

is a polynomial Q(x) such that
$$P(x) = (x - r)Q(x).$$

Substituting the number r for x (which is perfectly legal) yields
$$P(r) = (r - r)Q(r).$$

But
$$r - r = 0.$$

Thus, no matter what Q(r) may be,
$$P(r) = 0 \cdot Q(r) = 0.$$

In other words, <u>if</u> x - r <u>divides</u> P(x), <u>then</u> r <u>must be a solution of the equation</u> P(x) = 0.

Next consider the converse: Whenever r is a solution of the equation P(x) = 0, must x - r divide P(x)?

In any case, when you divide x - r into P(x) by long division the remainder is either 0 or some other constant that does not involve x. (See Exercise 71.) <u>The remainder does not involve x.</u> Thus

Recall that the remainder will be of lower degree than the degree of x - r (or else the remainder is 0).

$$P(x) = (x - r)Q(x) + R$$

where the remainder, R, is a fixed number, not involving x.   Since the equation
$$P(x) = (x - r)Q(x) + R$$

holds for all values of x, it holds when we replace x by the particular number r. Thus
$$P(r) = (r - r)Q(r) + R.$$

But P(r) = 0 since r is a solution of the equation P(x) = 0. Consequently

$$0 = 0 \cdot Q(r) + R$$

and

$$R \text{ must be } 0.$$

Thus

$$P(x) = (x - r)Q(x) + 0.$$

In other words, x - r divides P(x).

All this reasoning is summarized in the very useful Factor Theorem.

Note that the factor theorem has two parts. Each is the converse of the other.

THE FACTOR THEOREM

If P(r) = 0, then x - r divides P(x).

If x - r divides P(x), then P(r) = 0.

---

72. (a) Show that 1 is a solution of the equation
$$2x^4 - 5x^3 + 6x^2 - 8x + 5 = 0.$$

(b) Carry out the division of x - 1 into $2x^4 - 5x^3 + 6x^2 - 8x + 5$.
What is the quotient, Q(x)? What is the remainder?

(c) By multiplying check that $(x - 1) \cdot Q(x)$ equals $2x^4 - 5x^3 + 6x^2 - 8x + 5$.

Example 17.   Does x - 2 divide $x^5 - 6x^2 + x$?

Solution. Use the factor theorem to decide.

In this case r = 2 and $P(x) = x^5 - 6x^2 + x$. Replacing x by 2 yields

$$P(2) = (2)^5 - 6(2)^2 + (2)$$
$$= (32) - 6(4) + 2$$
$$= 32 - 24 + 2$$
$$= 10$$

Since P(2) is <u>not</u> 0, it follows that x - 2 does <u>not</u> divide P(x). (For if x - 2 did <u>divide</u> P(x), the factor theorem would <u>tell</u> us that P(2) would be 0.)

73. Use the factor theorem to answer these questions. (Show your work.)

(a) Does x - 1 divide $x^{30} - 1$?   (b) Does x + 1 divide $x^{40} - 1$?

(c) Does x - 1 divide $x^6 - x^5 + 3x^4 - 6x^3 + x^2 + x + 1$?

(d) Does x + 2 divide $2x^4 + 5x^0 + x^2 + 4$?

Hint: (b) Write x + 1 in x - r form as x -(-1). Thus, r = -1 in the factor theorem.

74. (a) Is x + 2 a factor of $x^0 + 2x^2 - x - 2$?

(b) Is x - 1 a factor?   (c) What is another factor?

(c) x + 1 = x-(-1)

75. (a) Is x - 3 a factor of $x^3 - 7x - 6$?   (b) Find two other factors.

(b) one factor is x + 2

76. Find three first-degree factors of $x^3 + 4x^2 + x - 6$.

One factor is x + 2

77. What are the solutions of the equation

$$(x - 3)(x + 5)(x - 7) = 0?$$

3, -5, and 7

78. Use the factor theorem to find three factors $x^3 + x^2 - 4x - 4$.

79. (a) Find four solutions of the equation $x^4 + 2x^3 - 7x^2 - 8x + 12 = 0$.

(a) one solution is -3

(b) Find four factors of $x^4 + 2x^3 - 7x^2 - 8x + 12$.

(b) one factor is x - 2

80. Find all the solutions for each equation:

(a) $x^3 - 7x - 6 = 0$ (b) $x^4 + x^3 - 7x^2 - x + 6 = 0$.

(a) one solution is -1
(b) one solution is 2

HOW TO FIND THE RATIONAL SOLUTIONS OF P(x) = 0

The factor theorem shows that the search for a first degree divisor of the polynomial P(x) is related to the search for a solution of the equation P(x) = 0. There is no simple general procedure for solving the equation P(x) = 0. However, there is a procedure for finding all the rational solutions of the equation P(x) = 0 if P(x) has integer coefficients. This will now be described.

Observe that we are only consider-ing polynomials whose coefficients are integers.
Consider only polynomials, $P(x)$, whose coefficients are integers. If $P(x)$ is of degree 1, that is, of the form $ax + b$, then the equation

$$P(x) = 0$$

is simply

$$ax + b = 0.$$

This equation has the solution

$$-\frac{b}{a},$$

which is a <u>rational number</u> (both a and b are integers and $a \neq 0$).

If $P(x)$ is of degree two, that is, of the form $ax^2 + bx + c$, then the equation

$$P(x) = 0$$

is

$$ax^2 + bx + c = 0.$$

It may have solutions that are not rational. The same is true of equations of higher degree. Nevertheless, it is useful to be able to search for the rational solutions, if there are any. Example 18 illustrates how to do this in an orderly way.

Example 18. Find the rational solutions of the equation

$$2x^3 + x^2 - 7x + 3 = 0$$

(if there are any).

Solution. Assume that there is a rational number

$$\frac{m}{n}$$

Observe the assumption that the rational number $\frac{m}{n}$ is in reduced form.
that is a solution of the equation where m and n are integers. For con-venience assume that the fraction has been reduced; thus m and n have no divisor in common other than 1. It is no restriction also to assume n to be positive. If $\frac{m}{n}$ is a solution of the equation, then

$$2\left(\frac{m}{n}\right)^3 + \left(\frac{m}{n}\right)^2 - 7\left(\frac{m}{n}\right) + 3 = 0$$

or

$$\frac{2m^3}{n^3} + \frac{m^2}{n^2} - \frac{7m}{n} + 3 = 0$$

Clearing denominators by multiplying both sides by n^3 gives

$$2m^3 + m^2n - 7mn^2 + 3n^3 = 0$$

Now, m divides each of the first three terms

$$2m^3, \quad m^2n, \quad \text{and} \ -7mn^2,$$

so m divides

$$2m^3 + m^2n - 7mn^2$$

Since
$$2m^3 + m^2n - 7mn^2 = -3n^3$$
it follows that

$$m \text{ divides } 3n^3$$

Since m divides
the left side of
the equation it
must also divide
the right side.

Furthermore, since m and n have no common divisor other than 1 it follows that $$m \text{ divides } 3.$$

Thus the numerator, m, is either 3 or 1 or -3 or -1.

A similar approach shows what n, the denominator, may be. Since n divides each of the last three terms on the left side of
$$2m^3 + m^2n - 7mn^2 + 3n^3 = 0$$
it must divide $2m^3$, for

$$2m^3 = -m^2n + 7mn^2 - 3n^3$$

Thus
$$n \text{ divides } 2.$$
The denominator, n, is either 2 or 1 or -2 or -1.

Hence the only candidates for rational solutions $\frac{m}{n}$ of the equation

$$2x^3 + x^2 - 7x + 3 = 0$$

are

$$\frac{m}{n} = \frac{3}{2}, \frac{1}{2}, -\frac{3}{2}, -\frac{1}{2}, \frac{3}{1}, \frac{1}{1}, -\frac{3}{1}, \text{ or } -\frac{1}{1}$$

These must be tested one by one. For instance, is $\frac{3}{2}$ a solution? Check by substitution:

$$2\left(\frac{3}{2}\right)^3 + \left(\frac{3}{2}\right)^2 - 7\left(\frac{3}{2}\right) + 3 \overset{?}{=} 0$$

$$2\left(\frac{27}{8}\right) + \frac{9}{4} - \frac{21}{2} + 3 \overset{?}{=} 0$$

$$\frac{27}{4} + \frac{9}{4} - \frac{21}{2} + 3 \overset{?}{=} 0$$

$$\frac{27}{4} + \frac{9}{4} - \frac{42}{4} + \frac{12}{4} \overset{?}{=} 0$$

$$\frac{6}{4} \overset{?}{=} 0$$

$$\text{no}$$

Thus $\frac{3}{2}$ is _not_ a solution.

Is $\frac{1}{2}$ a solution?   Check by substitution:

$$2\left(\frac{1}{2}\right)^3 + \left(\frac{1}{2}\right)^2 - 7\left(\frac{1}{2}\right) + 3 \stackrel{?}{=} 0$$

$$2\left(\frac{1}{8}\right) + \frac{1}{4} - \frac{7}{2} + 3 \stackrel{?}{=} 0$$

$$\frac{1}{4} + \frac{1}{4} - \frac{14}{4} + \frac{12}{4} \stackrel{?}{=} 0$$

$$\frac{0}{4} \stackrel{?}{=} 0$$

$$\text{yes}$$

Thus $\frac{1}{2}$ is a solution.

More checks show that none of the other six possibilities

$$-\frac{3}{2} \quad -\frac{1}{2} \quad \frac{3}{1} \quad \frac{1}{1} \quad -\frac{3}{1} \quad -\frac{1}{1}$$

is a solution.

The same reasoning as in Example 18 establishes the following theorem, which holds for polynomials of any degree, but will be needed only for the third degree in this course.

RATIONAL SOLUTION THEOREM

Let $P(x)$ be a polynomial of degree three with integer coefficients,

$$ax^3 + bx^2 + cx + d.$$

Let m be an integer and n a positive integer.   Assume that the fraction $\frac{m}{n}$ is reduced.   If $\frac{m}{n}$ is a solution of the equation

$$ax^3 + bx^2 + cx + d = 0$$

then

the numerator, m, divides d

and

the denominator, n, divides a.

Some equations with integer coefficients have only irrational solutions, as the quadratic formula shows.   The next example shows that a third degree polynomial equation may not have any rational solutions.

Example 19.   Find all rational number solutions, $\frac{m}{n}$, for the equation

$$x^3 - 6x^2 + 5x + 2 = 0$$

Solution.   Rather than use the "rational solution theorem", start from scratch.   Doing so is not much harder but helps avoid mistakes.   If $\frac{m}{n}$ is a solution that is reduced, then

$$\left(\frac{m}{n}\right)^3 - 6\left(\frac{m}{n}\right)^2 + 5\left(\frac{m}{n}\right) + 2 = 0$$

or

$$1m^3 - 6m^2n + 5mn^2 + 2n^3 = 0$$

Inspection of this equation shows that

$$m \text{ divides } 2$$

and

$$n \text{ divides } 1.$$

Thus

$$m \text{ is } 1 \text{ or } 2 \text{ or } -1 \text{ or } -2$$

and

$$n \text{ (which is positive) is } 1.$$

The only possible rational solutions are

$$\frac{m}{n} = \frac{1}{1}, \quad \frac{2}{1}, \quad \frac{-1}{1}, \quad \text{or} \quad \frac{-2}{1}$$

that is, the integers,

$$1, \qquad 2, \qquad -1, \text{ or} \qquad -2.$$

As a little arithmetic shows, none of these four cases is a solution.

Thus <u>the equation</u>

$$x^3 - 6x^2 + 5x + 2 = 0$$

<u>has no rational solutions.</u>

81. Show that the following equation has no rational solutions.

$$x^3 + 4x^2 - 6x + 2 = 0$$

Hint: See the rational solution theorem on page 104.

82. Show that the following equation has exactly one rational solution, namely -3.

$$x^3 + 4x^2 + 6x + 9 = 0$$

83. Find and check all the rational solutions of the following equation:

$$x^3 - 4x^2 + x + 6 = 0$$

There are 3 rational solutions.

84. Find all rational solutions of the following equation:

$$2x^3 - x^2 + 5x + 8 = 0$$

-1

85. If $\frac{m}{n}$ is a reduced fraction and a solution of the equation

$$8x^3 - 5x^2 - 6x + 9 = 0.$$

then m divides _____ and n divides _____.

86. Find and check by substitution the three rational solutions of the equation

$$6x^3 - 7x^2 - x + 2 = 0.$$

* * *

SOLVING $ax^3 + bx^2 + cx + d = 0$ IF ONE SOLUTION IS KNOWN

To solve the equation

$$x^3 - x^2 - 5x + 2 = 0,$$

first look for rational roots $\frac{m}{n}$:

$$\left(\frac{m}{n}\right)^3 - \left(\frac{m}{n}\right)^2 - 5\left(\frac{m}{n}\right) + 2 = 0$$

or

$$m^3 - m^2 n - 5mn^2 + 2n^3 = 0$$

Hence

m divides 2

and

n divides 1.

Thus m = 1, 2, -1, or -2 while n = 1 (since n may be assumed positive). The only possibilities are

$$\frac{m}{n} = 1, 2, -1, -2$$

A little arithmetic shows that only -2 <u>is a solution.</u>

The factor theorem then says that x - (-2), that is, x + 2, divides $x^3 - x^2 - 5x + 2$. Division determines that

$$x^3 - x^2 - 5x + 2 = (x + 2)(x^2 - 3x + 1).$$

The original equation

$$x^3 - x^2 - 5x + 2 = 0$$

is therefore equivalent to the equation

$$(x + 2)(x^2 - 3x + 1) = 0.$$

But the product of two numbers is 0 only when one of them is 0. Thus any solution of the original equation other than x = -2 is also a <u>solution of the equation</u>

$$x^2 - 3x + 1 = 0.$$

The quadratic formula provides the solutions of the equation

$$x^2 - 3x + 1 = 0.$$

They are

$$x = \frac{-(-3) \pm \sqrt{(-3)^2 - 4(1)(1)}}{2(1)}$$

or

$$x = \frac{3 \pm \sqrt{5}}{2}$$

Thus, the solutions of the equation

$$x^3 - x^2 - 5x + 2 = 0$$

are 2 and $\frac{3 \pm \sqrt{5}}{2}$.

Observe that not all of the solutions of the equation

$$x^3 - x^2 - 5x + 2 = 0$$

are rational. But, note the role played by the rational solution theorem. It provided a solution, -2. By the factor theorem, $x + 2$ is a factor of $x^3 - x^2 - 5x + 2$. The quotient was a quadratic polynomial whose solutions could be found with the aid of the quadratic formula. This single problem has thus used all the techniques presented in the chapter.

87. Check that

$$\frac{3 + \sqrt{5}}{2}$$

is a solution of the equation

$$x^3 - x^2 - 5x + 2 = 0.$$

SUMMARY OF HOW TO SOLVE EQUATIONS OF DEGREE THREE

The steps in solving

$$ax^3 + bx^2 + cx + d = 0$$

are as follows:

1. Use the "rational solution theorem" to find all rational solutions. (There may be none!)
2. If r is a solution, divide $x - r$ into the given polynomial, obtaining a quotient Q(x), which is of degree two.
3. Use the quadratic formula to find the solutions of
$$Q(x) = 0.$$

They are also solutions of the original equation.

See Project 21 for a third degree polynomial with a real solution but no rational solutions. Can you devise another example?

Even though an equation of degree three has no rational solutions, it is still possible to find its solutions. (See Projects 18 and 21.)

$\dfrac{3 \pm \sqrt{41}}{4}$

88. One solution of the equation

$$2x^3 + x^2 - 10x - 8 = 0$$

is -2. There are two more solutions. Find them, showing your work.

89. Find the solutions for each of the following. (Show all work.)

(a) 1

(c) ½, $\dfrac{-1 \pm \sqrt{13}}{2}$

(a) $x^3 - x^2 + 2x - 2 = 0$ (b) $x^3 + 3x^2 - 2x - 6 = 0$

(c) $2x^3 + x^2 - 7x + 3 = 0$

90. The equation that opened the chapter was

$$x^3 - 3x^2 - 2x + 4 = 0.$$

As Exercise 1 showed, 1 is a solution of the equation:

(a) Divide x - 1 into $x^3 - 3x^2 - 2x + 4$.
(b) What is the quotient, Q(x)?
(c) Use the quadratic formula to find the solutions of the equation
     Q(x) = 0.
(d) Find all solutions of the equation $x^3 - 3x^2 - 2x + 4 = 0$.

* * *

Exercise 90 shows the solution that the arithmetic of Exercises 1-6 only suggested. Notice how Exercise 90 requires the factor theorem and the quadratic formula. It does not use the rational solution theorem since the solution 1 was found by sheer good luck in Exercise 1.

91. Solve the riddle at the beginning of the chapter.

$1, \pm \sqrt{3}$

* * *   SUMMARY   * * *

This chapter opened and closed with the problem of finding all solutions of the equation

$$x^3 - 3x^2 - 2x + 4 = 0.$$

The body of the chapter developed techniques for solving such equations.

It treated the addition, subtraction, multiplication, and division of polynomials, including the idea of _quotient_ and _remainder_.

Solutions of equations of degree one,

$$ax + b = 0$$

can be found quite easily.

Equations of degree two,

$$ax^2 + bx + c = 0,$$

sometimes can be solved by _factoring_, but the "completing the square" method always works. To complete the square for

$$x^2 + \frac{b}{a}x,$$

simply add $\left(\dfrac{b}{2a}\right)^2$.

The <u>quadratic formula</u>,

$$x = \frac{-b \pm \sqrt{b^2 - 4ac}}{2a}$$

is the general case of the completing the square method and provides the solutions directly for equations of the form,

$$ax^2 + bx + c = 0.$$

There are
$$\begin{cases} \text{two real solutions if } b^2 - 4ac > 0 \\ \text{one real solution if } b^2 - 4ac = 0 \\ \text{no real solutions if } b^2 - 4ac < 0 \end{cases}$$

The graph of

$$y = ax^2 + bx + c$$

is a <u>parabola</u> with the x-coordinate of the vertex equal to $-\dfrac{b}{2a}$.

Other observations were made, such as:

When a polynomial A(x) Is divided into a polynomial B(x), the remainder is either 0 or else a polynomial of degree loss than the degree of A(x).

When x is replaced in a polynomial P(x) by some number k, then another number is obtained which Is denoted P(k), which is read "P of k." Thus, if

$$P(x) = 3x^2 - 4x + 2$$

then

$$P(k) = 3k^2 - 4k + 2.$$

The solution of equations of degree three,

$$ax^3 + bx^2 + cx + d = 0$$

depends on two theorems. First, the <u>factor theorem</u>, which asserts that if P(x) is a polynomial and P(r) = 0, then x - r is a factor of P(x). Second, the <u>rational solution theorem</u>, which simplifies the search for <u>rational solutions</u> of an equation P(x) = 0, where P(x) is a polynomial with integer coefficients. (If the reduced fraction $\dfrac{m}{n}$ is a solution, then n must divide the coefficient of the highest power of x appearing in P(x), and m must divide the constant term.)

These tools solve many equations of degree three, P(x) = 0, as follows:

1. Find a rational number, r, such that P(r) = 0.
2. Divide x - r into P(x), obtaining a quotient Q(x), (and remainder 0)

$$P(x) = (x - r)Q(x).$$

3. Find the solutions of the equation

$$Q(x) = 0$$

by the quadratic formula. These solutions are also solutions of the equation

$$P(x) = 0.$$

If the equation P(x) = 0 has no rational solutions, the method does not work. (See Projects 1 and 2 for a more general method.)

* * *   WORDS AND SYMBOLS   * * *

monomial	P(x)
coefficient	P(r) where r is a number
polynomial	completing the square
degree	quadratic formula
quotient and remainder	discriminant (b^2 - 4ac)
factor	factor theorem
	rational solutions theorem

* * *   CONCLUDING REMARKS   * * *

Why solve equations?   As a specific example, the equation

$$x^3 - 3x^2 - 2x + 4 = 0$$

occurs in elementary calculus.   That equation must be solved to find the lowest point on the graph of

$$y = x^4 - 4x^3 - 4x^2 + 16x.$$

For many reasons scientists have sought the solutions of polynomial equations.   The formulas developed in this chapter for solving any equation of degree one or two are at least three thousand years old, going back to ancient Egypt and Babylon.

Notice that the solution of the second degree equation

$$ax^2 + bx + c = 0$$

is expressed in terms of the solution of another second degree equation,

$$y^2 = b^2 - 4ac$$

for $\sqrt{b^2 - 4ac}$ appears in the quadratic formula.   The advantage of this step is that a calculator gives the solutions of this second type of equation conveniently.

In the Renaissance, Italian mathematicians developed formulas for solving <u>any</u> equation of degree three or four. (The solutions were expressed in terms of square roots and cube roots.)   Mathematicians struggled in vain for two centuries to find a corresponding formula for equations of degree five. Finally, Ruffini and Abel, early in the 19th century, proved that no such formula could <u>ever</u> be found.

Perhaps, if the rapid electronic computers of the twentieth century had been available five hundred years ago, the need to express the solutions of all polynomial equations in terms of square roots, cube roots, and so on, would never have existed.   For today's computers obtain the solution of equations to many decimal places in a split second.   The tedious approach illustrated in Exercises 1 - 6 of this chapter would be most appropriate for an electronic computer. In fact, a computer solves equations by methods which are just improvements on this approach.

*** CENTRAL SELF QUIZ ***

1. (a) Solve the first degree equation $7x + 2 = 0$.
   (b) Check your answer.

2. (a) Solve the second degree equation $2x^2 - 7x = 4$ by factoring.
   (b) Solve (a) using the Quadratic Formula.

3. (a) Show that the third degree equation $3x^3 + 8x^2 - 9x + 2 = 0$ has one rational solution.
   (b) Find any other real solutions.

4. Give an example of

   (a) an equation of degree one whose only solution is $\frac{2}{7}$.
   (b) an equation of degree two whose solutions are 3 and 4.
   (c) an equation of degree three whose solutions are 1, 2, and 3.

5. Graph:
$$y = -x^2 - 2x + 5.$$

ANSWERS TO SELF QUIZ

1. (a) $-\frac{2}{7}$ 2. (a),(b) $4, -\frac{1}{2}$ 3. (a) $\frac{1}{3}$ (b) $\dfrac{-3 \pm \sqrt{17}}{2}$

4. Different examples   5. Graph

*** POLYNOMIALS REVIEW ***

1. If $P(x) = 3x^2 - 2x + 5$, find

   (a) P(0)   (b) P(2)   (c) P(-1)   (d) P(-3)

2. If $P(x) = 4x^3 - 2x^2 + 3x - 1$ and $Q(x) = 3x^2 + 4x + 7$, find

   (a) $P(x) + Q(x)$ (b) $P(x) - Q(x)$
   (c) $Q(x) - P(x)$ (d) $P(x) \bullet Q(x)$

3. Divide

   (a) $4x^3 + 3x^2 + 0x - 8$ by $x - 3$
   (b) $x^5 - 32$ by $x - 2$

   Hint (b) Write
   $x^5 + 0 \cdot x^4 + 0 \cdot x^3 + 0 \cdot x^2 + 0 \cdot x - 32$

4. Write an equation in the form $ax + b = 0$ whose solution is $\frac{3}{5}$.

5. Solve for x:

   (a) $(x - 7)(x + 9) = 0$ (b) $(x - 3)(x + 4)(x - 8) = 0$
   (c) $x(x + 1)(x - 5) = 0$

6. Solve for x by factoring:

   (a) $x^2 - 9x = 0$ (b) $6x^2 - x - 2 = 0$
   (c) $3x^2 - 8x - 3 = 0$ (d) $3x^2 + 5x - 2 = 0$
   (e) $2x^2 - 7x = 15$ (f) $3x^2 - 7x + 2 = 0$

   Hint: (e) Write
   $P(x) = 0$

7. Solve for x by completing the square:

(a) $x^2 + 6x - 4 = 0$

(b) $x^2 - 2x + \frac{7}{16} = 0$

(c) $x^2 - 5x - 2 = 0$

(d) $x^2 - 7x + 3 = 0$

8. Write the quadratic formula.

9. Solve $x^2 + 5x - 3 = 0$ for x by

(a) completing the square

(b) the quadratic formula

10. Solve for x using the quadratic formula:

Hint: (a) Write P(x) = 0

(a) $x^2 - 3x = 5$

(b) $2x^2 + x - 6 = 0$

(c) $3x^2 - 5x + 2 = 0$

(d) $2x^2 - 5x - 1 = 0$

11. (a) Does $x^2 + x + 1 = 0$ have a solution?   Explain.

(b) What does the discriminant, $b^2 - 4ac$, in the quadratic formula have to do with part (a)?

12. Use the x-coordinate of the vertex ($-\frac{b}{2a}$ from $y = ax^2 + bx + c$)

to help you graph each of the following parabolas:

(a) $y = x^2 - 2x + 4$

(b) $y = x^2 - 6x + 9$

(c) $y = -x^2 + 2x + 4$

(d) $y = -2x^2 + 4x + 8$

13. (a) x - 3 is a factor of $x^3 - 27$.   Use division of polynomials find the other factor.

Hint: Write $x^3 + 0 \cdot x^2 + 0 \cdot x - 27$

(b) Explain how you can use the factor theorem to show that x - 3 is a factor of $x^3 - 27$.

14. State the factor theorem aloud to a classmate.

15. Explain to a classmate how the factor theorem is used to help solve equations.

16. (a) Is 0 a solution of $x^5 - 6x^4 + x^3 - x + 2 = 0$?

(b) Is 0 a solution of $x^7 - x^3 - x^2 + 3x = 0$?

(c) How can you tell at a glance whether 0 is a solution of the equation $P(x) = 0$?

17. (a) Factor:
$$x^3 + 3x^2 - 4x - 12$$

(b) Solve:
$$x^3 + 3x^2 - 4x - 12 = 0$$

18. Solve for x:
$$x^3 - 2x^2 - 5x + 6 = 0$$

19. Show that $1 + \sqrt{5}$ is a solution of the equation $x^3 - 8x - 8 = 0$.

20. (a) Show that x + 1 is a factor of $x^3 + 2x^2 - 1$.

(b) What is the other factor in (a)?

(c) Show that the solutions to the equation $x^3 + 2x^2 - 1 = 0$ are -1 and $\frac{-1 \pm \sqrt{5}}{2}$.

21. Find the three real solutions of $x^3 + x^2 - 3x + 1 = 0$.

22. Suppose $P(x)$ is a polynomial with $P(3) = 8$ and $P(4) = -2$. Then it follows that the equation $P(x) = 0$ has a solution between __?__ and __?__.

23. How many solutions does an equation of degree one have?

24. How many real number solutions may an equation of degree two have? Which of these statements are correct?

    (a) always two solutions,   (b) sometimes three solutions,
    (c) sometimes one solution,   (d) sometimes no solution,
    (e) at least one solution,   (f) sometimes two solutions.

25. Consider the following proposed solution of the equation

$$x^2 - 5x + 6 = 1.$$

    Proposed solution:

$$(x - 3)(x - 2) = 1$$

    so

$x - 3 = 1$	or	$x - 2 = 1$
$x = 4$		$x = 3$

    (a) Show by substitution that 4 and 3 are <u>not</u> solutions.
    (b) Find the error in the reasoning of the proposed solution.

26. Use the rational solution theorem to solve for x:

    (a) $2x^3 - x^2 - 2x + 1 = 0$ (b) $6x^3 + 13x^2 + x - 2 = 0$

27. Show that any rational solution of an equation of the type

$$x^3 + bx^2 + cx + d = 0,$$

    where b, c, and d are integers, must itself be an integer.

28. Write a third degree equation with -1, 5, and 3 as solutions.

29. Write a third degree equation that has integer coefficients and $\frac{1}{2}$, $-\frac{1}{3}$ and 1 as solutions.

30. Write an equation of degree two with solutions $2 + \sqrt{5}$ and $2 - \sqrt{5}$. (Check your answer with other classmates.)

31. What is the value (positive, negative or zero) of the discriminant, $b^2 - 4ac$, in the quadratic formula if the number of rational solutions to a quadratic equation is

    (a) none   (b) one   (c) two

32. Use Exercise 31 to devise a quadratic equation that has

    (a) no solutions   (b) exactly one solution   (c) two solutions

ANSWERS TO POLYNOMIALS REVIEW

1. (a) 5   (b) 13   (c) 10   (d) 38   2. (a) $4x^3 + x^2 + 7x + 6$

    (b) $4x^3 - 5x^2 - x - 8$ (c) $-4x^3 + 5x^2 + x + 8$

    (d) $12x^5 + 10x^4 + 29x^3 - 5x^2 + 17x - 7$

3. (a) $4x^2 + 15x + 45$; rem.,127   (b) $x^4 + 2x^3 + 4x^2 + 8x + 16$.

4. $5x - 3 = 0$ 5. (a) 7, -9 (b) 3, -4, 8   (c) 0, -1, 5

6. (a) 0, 9 (b) $-\frac{1}{2}$, $\frac{2}{3}$ (d) $-\frac{1}{3}$, 3   (d) -2, $\frac{1}{3}$ (e) $-\frac{3}{2}$ 5 (f) $\frac{1}{3}$, 2

7. (a) $-3 \pm \sqrt{13}$ (b) $\frac{7}{4}$, $\frac{1}{4}$ (c) $\frac{5 \pm \sqrt{33}}{2}$ (d) $\frac{7 \pm \sqrt{37}}{2}$

8. $x = \frac{-b \pm \sqrt{b^2 - 4ac}}{2a}$ 9. (a) (b) $\frac{-5 \pm \sqrt{37}}{2}$

10. (a) $\frac{3 \pm \sqrt{29}}{2}$ (b) -2, $\frac{3}{2}$ (d) $\frac{2}{3}$, 1   (d) $\frac{5 \pm \sqrt{33}}{4}$

11. (a) No real number solution   (b) $b^2 - 4ac < 0$ 12. Graphs

13. (a) $x^2 + 3x + 9$ (b) If $x^3 - 27 = 0$, then $P(3) = 0$ and $x - 3$ divides $x^3 - 27$.

14. - 15. See the Summary of the main section. 16. (a) No (b) Yes

(c) If every term contains x.   17. (a) $(x-2)(x+2)(x+3)$ (b) ± 2, -3

18. 1, 3, -2   19. Check (other solutions are $1 - \sqrt{5}$ and 2).

20. (a) Use the factor theorem. (b) $x^2 + x - 1$

(c) $P(-1) = 0$; $(x + 1)(x^2 + x - 1) = 0$. Solving $x^2 + x - 1 = 0$, get

$\frac{-1 \pm \sqrt{5}}{2}$ 21. 1 and $-1 \pm \sqrt{2}$ 22. 3 and 4   23. 1

24. (c), (d), (f)   25. (b) If the product of two numbers is 1, neither of the numbers has to be 1. (The equation should first be put in the form $x^2 - 5x + 5 = 0$.)

26. (a) $\frac{1}{2}$, ± 1 (b) $\frac{1}{3}$, $-\frac{1}{2}$, -2   27. If $\frac{m}{n}$ is a rational solution of

$x^3 + bx^2 + cx + d = 0$ then m divides d and n divides 1, or $\frac{m}{n}$ is an integer.

28. $(x + 1)(x - 5)(x - 3) = 0$ 29. $6x^3 - 7x^2 + 1 = 0$

30. $x^2 - 4x - 1 = 0$.   31. (a) negative (b) zero (c) positive

32. Student examples.

* * * POLYNOMIALS REVIEW SELF TEST * * *

1. Write an equation with integer coefficients that has the solution(s)
    (a) $-\frac{3}{4}$ and degree one   (b) $\frac{2}{3}$ and $\frac{1}{4}$ and degree two

    (c) $1 - \sqrt{3}$ and $1 + \sqrt{3}$ and degree two. (d) -2, 2 and 3 and degree three.

2. Solve for x (show all work):
$$2x^2 - 5x + 3 = 0$$
    (a) by factoring   (b) by the quadratic formula

3. Solve by completing the square

$$x^2 - 3x - 9 = 0$$

4. One factor of $x^3 + x^2 + x + 1$ is $x + 1$. What is the other factor?

    Hint: When does $x^3 + x^2 + x + 1 = 0$?

5. How many times does the graph of $y = x^3 + x^2 + x + 1$ cross the x-axis?

6. Factor:
    (a) $x^2 - 5x$ (b) $x^3 - 6x^2 + 11x - 6$

7. Solve:

    (a) $x^2 - 5x = 0$ (b) $x^3 - 6x^2 + 11x - 6 = 0$

8. Solve: $3x^3 - 5x^2 - 4x + 4 = 0$.

ANSWERS TO POLYNOMIALS REVIEW SELF TEST

1. (a) $4x + 3 = 0$ (b) $12x^2 - 11x + 2 = 0$ (c) $x^2 - 2x - 2 = 0$
    (d) $x^3 - 3x^2 - 4x + 12 = 0$;   2. $\frac{3}{2}$, 1;   3. $\frac{3 \pm 3\sqrt{5}}{2}$ 4. $x^2 + 1$

5. once, $x^2 + 1 \neq 0$;   6. (a) $x(x - 5)$ (b) $(x - 1)(x - 2)(x - 3)$;

7. (a) 0, 5   (b) 1, 2, 3;   8. $\frac{2}{3}$, -1, 2

* * * POLYNOMIALS PROJECTS * * *

1. Read the history of solutions of equations of degree greater than 2 in Morris Kline's Mathematical Thought from Ancient to Modern Times, Oxford, 1972. See, in particular, pp. 263-270.

2. Look up the formula (see Exercise 1) for solving a cubic equation in a book of mathematical tables, such as Standard Mathematical Tables, published by the Chemical Rubber Co.

3. Give an example of a polynomial, $P(x)$, of degree two such that the equation $P(x) = 0$ has no real solutions.

4. Give an example of a polynomial $P(x)$ of degree three such that the equation $P(x) = 0$
    has exactly
    (a) one solution   (b) two solutions   (c) three solutions

5. Let $ax^2 + bx + c$ be a second degree polynomial. Show that if $b^2 - 4ac$ is the square of an integer, then $ax^2 + bx + c$ can be factored as the product of two polynomials of degree one that have integer coefficients.

6. Give an example of a second degree equation that cannot be solved by factoring, but does have two real solutions.

7. How can the discriminant, $b^2 - 4ac$, be used to tell how many points on the graph of $y = ax^2 + bx + c$ are on the x-axis?

8. Let <u>a</u> be positive, and consider the graph of

    $$y = ax^2 + bx + c.$$

    (a) When x is positive and large, what can be said about y?

    (b) When x is negative and large, what can be said about y?

(c) Show that

$$y = a \left(x^2 + \frac{b}{a} x + \frac{c}{a} \right)$$

(d) Complete the square in "$x^2 + \frac{b}{a} x$" to show that

$$y = a\left[\left(x + \frac{b}{2a} \right)^2 + \frac{c}{a} - \frac{b^2}{4a^2} \right]$$

(e) Using (d), show that the lowest point on the graph of
$$y = ax^2 + bx + c$$
occurs when

$$x = -\frac{b}{2a}$$

(f) Show that the lowest point occurs midway between the two solutions of the equation
$$ax^2 + bx + c = 0$$
if there are two solutions.

9. (See Exercise 8.) Let $a \neq 0$. (a) Show that

$$ax^2 + bx + c = a\left[\left(x + \frac{b}{2a} \right)^2 + \frac{c}{a} - \frac{b^2}{4a^2} \right]$$

(b) Show that at $x = -\frac{b}{2a} + u$ and $x = -\frac{b}{2a} - u$, the expression $ax^2 + bx + c$ takes equal values. (Here the "u" denotes any real number.)

(c) From (b) conclude that the graph of $y = ax^2 + bx + c$ is symmetric with respect to the line $x = -\frac{b}{2a}$.

10. Use experimentation and the factor theorem to convince yourself that any polynomial of degree 3 has a factor of degree 1.

11. Give an example of a second degree polynomial $P(x)$ whose coefficients are integers, such that the equation $P(x) = 0$ has $7 - 4\sqrt{11}$ as a solution.

12. The equation $x^2 + bx + c = 0$ has the solutions r_1 and r_2.

(a) Show that $x^2 + bx + c = (x - r_1)(x - r_2)$.

(b) Show that $r_1 r_2 = c$ and $r_1 + r_2 = -b$.

13. The equation $x^3 + bx^2 + cx + d = 0$ has the solutions r_1, r_2, and r_3.

(a) Show that $(x - r_1)(x - r_2)(x - r_3) = x^3 + bx^2 + cx + d$.

(b) Show that $r_1 r_2 r_3 = -d$, $r_1 r_2 + r_1 r_3 + r_2 r_3 = c$, $r_1 + r_2 + r_3 = -b$.

14. Give an example of a polynomial $P(x)$ of degree 4 such that the equation $P(x) = 0$ has

(a) no real solutions.   (b) exactly one real solution.

The next three exercises show that an equation of the form P(x) = 0, where P(x) has degree n, has at most n solutions.

15. Let A(x) and B(x) be polynomials. Let C(x) = A(x) • B(x). Prove that any solution of the equation

$$C(x) = 0$$

is a solution of

$$A(x) = 0 \qquad \text{or} \qquad B(x) = 0.$$

16. Let r be a number and B(x) be a polynomial. Let C(x) = (x - r)B(x).

(a) Prove that r is a solution of the equation

$$C(x) = 0.$$

(b) Prove that if s is any solution of the equation

$$C(x) = 0$$

<u>other than</u> r, then s is also a solution of the equation

$$B(x) = 0.$$

17. Prove that if P(x) has degree

(a) one, then the equation P(x) = 0 has exactly one solution.

(b) two, then the equation P(x) = 0 has at most two solutions.

(c) three, then the equation P(x) = 0 has at most three solutions.

(d) four, then the equation P(x) = 0 has at most four solutions.

18. (a) Show that if n is even and at least two, then there is a polynomial P(x) of degree n such that the equation P(x) = 0 has no real solution.   *Hint: consider the graph.*

(b) Show that if n is odd, then for any polynomial P(x) of degree n the the equation P(x) = 0 has at least one real solution.

19. This exercise concerns a classic ladder problem that requires the solution of a fourth degree equation. The diagram shows a 30 foot ladder and a 50 foot ladder crossing at a height of 10 feet. Find the distance between the bases of the ladders.

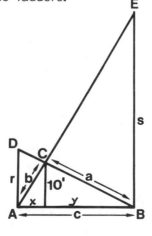

AE = 50
BD = 30
c = x + y = AB

(a) Estimate the answer by making an accurate drawing with the aid of a centimeter ruler.

(b) Using geometry (e.g., similar triangles and the Pythagorean theorem) show that

$$b^4 - 100b^3 + 2500b^2 + 15,610b - 391,000 = 0$$

(c) The solution of (b) that applies to this problem is about 11.9. Deduce that $a \doteq 22.9$, $x \doteq 6.5$, $y \doteq 20.6$, hence $AB \doteq 27.0$.

20. Solve for x as an <u>integer</u>.

(a) Hint: let

$$x = \sqrt{2 + \sqrt{2 + \sqrt{2}}}$$

inside the first radical, then solve.

(a) $x = \sqrt{2 + \sqrt{2 + \sqrt{2} + \ldots}}$ (The pattern continues.)

(b) $x = \sqrt{6 + \sqrt{6 + \sqrt{6} + \ldots}}$ (The pattern continues.)

(c) Devise a similar problem of your own.

21. Consider the function,

$$f(x) = x^3 - x^2 - x - 1.$$

(b) That is, find where the curve crosses the x-axis.

(a) Find $f(1)$ and $f(2)$. Between what two integers does the curve cross the x-axis?

(b) Use a calculator to estimate x to three decimal places such that $f(x) = 0$. (Observe that there are no rational solutions.)

* * *   FIRST YEAR ALGEBRA REVIEW (OPTIONAL)   * * *

PROPERTIES OF EQUALITY

If

$$a = b,$$

the properties of equality state that

$$a + c = b + c \qquad a - c = b - c \qquad ac = bc \qquad \frac{a}{c} = \frac{b}{c}, \ (c \neq 0)$$

These properties are useful in solving equations, such as those in Exercise 1.

1. Solve for x. (Show all steps.)

(a) $x + 2 = 3$ (b) $x - 5 = 7$ (c) $3x = 7$

(d) $\frac{x}{5} = 9$ (e) $\frac{x}{2} - 3 = 11$ (f) $x + b + 1$

(g) $x - a = b$ (h) $-bx = a$ (i) $\frac{x}{a} - b = c \ (a \neq 0)$

2. Solve for x. (Show all steps.)

(a) $2x - 5 + 6x = 7x - 9$ (b) $2(3x - 5) - 5(2x + 7) = 6(x - 1)$

(c) $-4[x - 3(5 - x) + 8] = 1$

(d) $-3[-(x - 1) - (3x + 2) - (7 - 2x)] = 0$

PRIMES

DEFINITION. A positive integer greater than 1 that is not a product of two smaller positive integers is called a <u>prime</u>.

Thus,

$$2, 7, \text{ and } 13 \text{ are prime.}$$

However, 10, 18, and 33 are not prime, for

$$10 = 2 \cdot 5 \qquad\qquad 18 = 2 \cdot 3^2 \qquad\qquad 33 = 3 \cdot 11$$

$$* * *$$

3. What are the prime integers

(a) between 50 and 70?   (b) between 90 and 100?

4. Write the factors of each of the following if they are not prime.

(a) 103   (b) 203   (c) 303   (d) 630

$$* * *$$

Just as there are prime integers, there are prime ("irreducible") polynomials. Their definition depends on the notion of "degree" of a polynomial, with nonzero coefficient <u>a</u>. Thus

$$
\begin{array}{ll}
x - 2 & \text{has degree 1} \\
7x^2 + x - 1 & \text{has degree 2} \\
14x^5 - x & \text{has degree 5} \\
\text{and } -3 & \text{has degree 0 (for } -3 = -3x^0)
\end{array}
$$

(The polynomial in which all coefficients are 0 is not assigned a degree.)

In the polynomial $7x^2 + x - 1$ the coefficient of x^2 is 7, the coefficient of x is 1 and the "constant" coefficient is -1.

DEFINITION. A polynomial with integral coefficients is called prime (with respect to integral coefficients) if

(a) it has degree at least one, and
(b) it is not the product of polynomials of lower degree with integral coefficients.

According to the definition, the polynomials
$$2x^2 + 5x - 3 = (2x - 1)(x + 3)$$
and
$$x^3 - x = x(x - 1)(x + 1)$$

are <u>not</u> prime with respect to integer coefficients, but

$$x^2 - 7$$

is prime. Observe that while

$$x^2 - 7 = (x - \sqrt{7})(x + \sqrt{7})$$

it is nevertheless prime with respect to integer coefficients.

5. Factor with integer coefficients, if possible:

(a) $x^2 - 81$ (b) $4x^2 + 4x + 1$ (c) $x^2 + x - 6$

$\qquad$ (d) $x^2 - 14x + 49$ $\qquad$ (e) $3x^2 - 3x$ $\qquad$ (f) $9x^2 - 12x + 4$

$\qquad$ (g) $5x^2 - 7x$ $\qquad$ (h) $16 - x^2$ $\qquad$ (i) $16x^2 - 25$

6. Factor:

(a) $x^3 - x^2 - 6x$ $\qquad$ (b) $x - x^3$ $\qquad$ (c) $x^2 + 4x + 4$

Hint: (a) First find a common factor.

7. (a) $x^2 - 4$ $\qquad$ (b) $x^2 + 4$ $\qquad$ (c) $x^2 + 4 + 4$

8. Factor with integer coefficients, if possible:

(a) $x^2 - x - 2$ $\qquad$ (b) $x^2 - x + 2$ $\qquad$ (c) $2x^2 + 5x + 2$

(d) $2x^2 + 5x - 3$ $\qquad$ (e) $2x^2 - 3x - 14$ $\qquad$ (f) $6x^2 - 5x - 6$

(g) $6x^2 + 13x + 6$ $\qquad$ (h) $8x^2 + 10x - 3$ $\qquad$ (i) $8x^2 + 2x - 3$

* * *

Example 1. Factor $ax + ay + bx + by$.

Solution.

$$ax + ay + bx + by = a(x + y) + b(x + y)$$
$$= (a + b)(x + y)$$

9. Factor with integer coefficients, if posible.

(a) $ac + bc + ad + bd$ $\qquad$ (b) $2x - 2y + ax - ay$

(d) $3x - ay + ax - 3y$ $\qquad$ (d) $ax + bx - a - b$

Example 2. Factor:

(a) $x^3 + y^3$ $\qquad$ (b) $x^3 - y^3$

Solution. (Verify each solution by multiplying)

(a) $(x + y)(x^2 - xy + y^2)$ $\qquad$ (b) $(x - y)(x^2 + xy + y^2)$

10. Factor.

(a) $x^3 - 1$ $\qquad$ (b) $x^3 + 1$

(c) $27 + 8a^3$ $\qquad$ (d) $64m^3 - n^3$

Recall the zero product rule:

> ### ZERO PRODUCT RULE
> If $ab = 0$, then $a = 0$, or $b = 0$, or both may be 0.

In using the zero product rule the equation must be put in the form: polynomial = 0.

The zero product rule helps in the solution of certain quadratic equations, illustrated in the next example.

Example 3. Solve for x:

$$2x^2 + 5x = 3$$

Solution. Write

$$2x^2 + 5x - 3 = 0$$

which is

$$(2x - 1)(x + 3) = 0$$

Thus the zero product rule says

$$2x - 1 = 0 \qquad \text{or} \qquad x + 3 = 0$$
$$2x = 1 \qquad\qquad\qquad x = -3$$
$$x = \tfrac{1}{2}$$

The solutions, $\frac{1}{2}$ and -3, may be checked in the original equation:

Check:

$$2(\tfrac{1}{2})^2 + 5(\tfrac{1}{2}) \overset{?}{=} 3 \qquad\qquad 2(-3)^2 + 5(-3) \overset{?}{=} 3$$
$$2(\tfrac{1}{4}) + 5(\tfrac{1}{2}) \overset{?}{=} 3 \qquad\qquad 2(9) - 15 \overset{?}{=} 3$$
$$\tfrac{2}{4} + \tfrac{5}{2} \overset{?}{=} 3 \qquad\qquad 18 - 15 \overset{?}{=} 3$$
$$3 = 3 \qquad\qquad\qquad 3 = 3$$

11. Solve for x using factoring and the zero product rule. (Show all work.)

   (a) $36x^2 - 49 = 0$ (b) $3x^2 - 7x = 20$

12. Is the following problem worked correctly? Explain.

   Solve for x: $\qquad 2x^2 - 3x = 1$

   $$x(2x - 3) = 1$$

   Then

   $$x = 1 \qquad \text{or} \qquad 2x - 3 = 1$$
   $$2x = 4$$
   $$x = 2$$

13. Solve for x. (Show all work.)

   (a) $2x^2 - 7x = 0$ (b) $2x^2 - 7x = 15$
   (c) $6x^2 - 13x - 5 = 0$ (d) $4x^2 + 19x - 5 = 0$
   (e) $x - x^2 = 0$ (f) $2x^2 + 3x - 35 = 0$

14. Solve for x:

   $$x(x - 5)(x + 4) = 0$$

15. Solve for x:

   (a) $(x - 2)(x + 3)(x - 7) = 0$ (b) $(x^2 - 1)(x + 9) = 0$

ANSWERS TO ALGEBRA REVIEW

1. (a) $x = 1$ (b) $x = 12$ (c) $x = \frac{7}{3}$ (d) $x = 45$ (e) $x = 28$

   (f) $x = a - b$ (g) $x = a + b$ (h) $x = -\frac{a}{b}$ (i) $x = a(b + c)$

2. (a) $x = -4$ (b) $x = -\frac{39}{10}$ (c) $x = \frac{27}{16}$ (d) $x = -4$

3. (a) 53, 59, 61, 67   (b) 97, 101, 103, 107, 109

4. (a) prime (b) $7 \cdot 29$ (c) $3 \cdot 101$ (d) $2 \cdot 3 \cdot 3 \cdot 5 \cdot 7$

5. (a) $(x - 9)(x + 9)$ (b) $(2x + 1)^2$ (c) $(x + 3)(x - 2)$ (d) $(x - 7)^2$
    (e) $3(x - 1)(x + 1)$ (f) $(3x - 2)^2$ (g) $x(5x - 7)$ (h) $(4 - x)(4 + x)$
    (i) $(4x - 5)(4x + 5)$

6. (a) $x(x - 3)(x + 2)$ (b) $x(1 - x)(1 + x)$ (c) $x(x + 1)(x + 2)$

7. (a) $(x - 2)(x + 2)$ (b) prime   (c) $(x + 2)^2$

8. (a) $(x - 2)(x + 1)$ (b) prime   (c) $(2x + 1)(x + 2)$ (d) $(2x - 1)(x + 3)$
    (e) $(2x - 7)(x + 2)$ (f) $(2x - 3)(3x + 2)$ (g) $(2x + 3)(3x + 2)$
    (h) $(2x + 3)(4x - 1)$ (i) $(2x - 1)(4x + 3)$

9. (a) $(a + b)(c + d)$ (b) $(2 + a)(x - y)$ (c) $(3 + a)(x - y)$
    (d) $(x - 1)(a + b)$

10. (a) $(x - 1)(x^2 + x + 1)$ (b) $(x + 1)(x^2 - x + 1)$ (c) $(3 + 2a)(9 - 6a + 4a^2)$
    (d) $(4m - n)(16m^2 + 4mn + n^2)$

11. (a) $x = \pm \frac{7}{6}$ (b) $x = -\frac{5}{3}$, $x = 4$

12. No. There is no rule that says if $a \cdot b = 1$, then either $a = 1$ or $b = 1$

13. (a) $x = 0$ or $x = \frac{7}{2}$ (b) $x = 5$ or $x = -\frac{3}{2}$ (c) $x = -\frac{1}{3}$ or $x = \frac{5}{2}$

    (d) $x = -5$ or $x = \frac{1}{4}$ (e) $x = 0$ or $x = 1$ (f) $x = -5$ or $x = \frac{7}{2}$

14. $x = 0$ or $x = 5$ or $x = -4$

15. (a) $x = 2$ or $x = -3$ or $x = 7$ (b) $x = -9$ or $x = 1$ or $x = -1$

    * * * CUMULATIVE REVIEW CHAPTERS 1 - 3 * * *

1. Find the sum:

    (a) $7 - \frac{7}{3} + \frac{7}{9} - \frac{7}{27} + \frac{7}{81} - \frac{7}{243}$ (b) $3 + 2 + \frac{4}{3} + \frac{8}{9} + \ldots$

2. Evaluate as a common fraction:

    (a) $0.4999\ldots$ (b) $2.676767\ldots$

3. For what x will 3^x be

    (a) negative?   (b) zero?   (c) positive?

4. Compute to three decimal places: $\frac{2}{1 + \sqrt{5}}$

5. Simplify:

    (a) $\sqrt{1 + \frac{1}{n}}$ (b) $(\sqrt{32} - \sqrt{27})(\sqrt{8} + \sqrt{48})$

6. Write the product $\sqrt[3]{2}\sqrt{3}$ using a single radical.

7. Explain why the expression $\sqrt{-36}$ is meaningless.

8. Find the sum of each of the following:

(a) $1 + \sqrt{2} + 2 + \sqrt{8} + 4 + \sqrt{32} + 8 + \sqrt{128}$

(b) the first eight terms of the geometric progression $\frac{3}{4}, \frac{3}{8}, \frac{3}{16}, \cdots,$

(c) the first six terms of the geometric progression $\frac{1}{\sqrt{2}}, -2, \frac{8}{\sqrt{2}}, \ldots,$

(d) Show the answer to (a) can also be written $15\sqrt{2} + 15.$

9. Which of the following is a geometric progression?

(a) $2 + \sqrt{3}, 3 + 2\sqrt{3}, 6 + 3\sqrt{3}, \ldots$
(b) $1 + \sqrt{2}, 2 + \sqrt{2}, 2 + 2\sqrt{2}, \ldots$
(c) $\sqrt{5}, 2\sqrt{5}, 3\sqrt{5}, 4\sqrt{5}, \ldots$

(d) $1 + \frac{1}{\sqrt{2}}, \frac{1}{2} + \frac{1}{\sqrt{2}}, \frac{1}{2} + \frac{1}{2\sqrt{2}}, \ldots$

10. If $P(x)$ is fifth degree and $Q(x)$ is third degree, state the degree of

(a) $P(x) + Q(x)$ (b) $\frac{P(x)}{Q(x)}$

(c) $P(x) \cdot Q(x)$ (d) $Q(x) - P(x)$

11. If $P(x) = 4x^2 + 2x + 1$ and $Q(x) = 2x - 1$, compute

(a) $P(x) \cdot Q(x)$ (b) $Q(x) - P(x)$

(c) $\frac{P(x)}{Q(x)}$ (d) $P(x) + Q(x)$

12. Solve for x:

(a) $3x - 7 = 11$ (b) $-2x + 5 = 6$

13. Solve $2x^2 + 5x - 3 = 0$ for x by

(a) factoring   (b) the quadratic formula

14. Solve $x^2 - 3x - 1 = 0$ for x by

(a) completing the square   (b) the quadratic formula

15. State the factor theorem.

16. Use the factor theorem to solve for x:

$$x^3 + 3x^2 - x - 3 = 0$$

17. Use the rational solution theorem and the quadratic formula to help solve for x:

$$2x^3 + x^2 - 3x + 1 = 0$$

18. A car purchased for $12,000 depreciates in value 25% every year. What is its value

(a) after 1 year?   (b) after 2 years?

(c) after 5 years?   (d) after 10 years?

19. Read in an encyclopedia how geometric progressions are found in the notes produced on a piano. For example, an orchestra in the United States tunes up to a "concert A" (440 cycles).

  (a) What note has 220 cycles? 110 cycles? 880 cycles?

  (b) What is the mathematical relation in (a)?

  (c) What is the musical relation in (a)?

ANSWERS TO CUMULATIVE REVIEW

1. (a) $\dfrac{1274}{243}$ (b) 9   2. (a) $\frac{1}{2}$ (b) $\dfrac{265}{99}$

3. (a) for no x   (b) for no x   (c) x any real number

4. 0.618   5. (a) $\dfrac{\sqrt{n^2 + n}}{n}$ (b) $-20 + 10\sqrt{6}$ 6. $\sqrt[6]{108}$

7. Let $x = (-36)^{\frac{1}{2}}$, then $(-36)^{\frac{1}{2}}(-36)^{\frac{1}{2}} = $?

8. (a) $\dfrac{15}{\sqrt{2} - 1}$ (b) $\dfrac{765}{512}$ (c) $\dfrac{-511}{4 + \sqrt{2}}$ (d) Multiply by $\dfrac{\sqrt{2} + 1}{\sqrt{2} + 1}$

9. (a) yes   (b) yes   (c) no   (d) yes

10. (a) fifth   (b) second   (c) eighth   (d) fifth

11. (a) $8x^3 - 1$ (b) $-4x^2 - 2$ (c) $2x + 2$; rem. 3   (d) $4x^2 + 4x$

12. (a) $x = 6$ (b) $x = -\frac{1}{2}$ 13. (a) (b) $\frac{1}{2}$, -3   14. (a) (b) $\dfrac{3 \pm \sqrt{13}}{2}$

15. For any polynomial P(x): If P(r) = 0, then x - r is a factor of P(x).
      If x - r is a factor of P(x), then P(r) = 0.

16. ±1, -3   17. $\frac{1}{2}$, $\dfrac{-1 \pm \sqrt{5}}{2}$

18. (a) $9,000   (b) $6,750   (c) $2,848   (d) $676 (ans: rounded off)

19. (a) one octave lower, two octaves lower, one octave higher.
    (b) An octave higher doubles the frequency.   (c) See (b).

* * * CENTRAL SECTION * * *

When Robinson Crusoe, whose hobby was mathematics, was marooned on an island, he wanted to keep his goat from straying. He tied one end of a rope to the goat and wound the other around a tree. The more times he wrapped it around the tree the stronger was the pull necessary to make the rope slip because of friction. So the more times he wrapped the rope around the tree the bigger was the pull it could resist. When one foot of the rope was wound around the tree, it could hold 10 lbs. Generally, when d feet of rope were wound around the tree it could hold $5 \cdot 2^d$ pounds.

How many feet should be wound around the tree in order to hold

    (a) 40 pounds?   (b) 80 pounds?   (c) 120 pounds?

Logarithms allow us to solve for an unknown exponent.

One of the above three questions is hardest to answer. Which one is it? With the aid of logarithms, all three parts will be easy to answer.

THE GRAPH OF $y = 2^x$

The next few exercises concern the exponential equation $y = 2^x$ and introduce the concept of logarithms.

1. Copy and fill in this table, using $\sqrt{2} \doteq 1.4$.

x	-3	-2	-1	$-\frac{1}{2}$	0	$\frac{1}{2}$	1	2	3
2^x									

2. Use the table in Exercise 1 to graph <u>very carefully</u> the equation
$$y = 2^x.$$

Label and keep the scales equal for both axes. Save this graph; you will use it again in later exercises.

3. Consider the graph of $y = 2^x$. What happens to y as
    (a) x gets larger?   (b) x gets near zero?   (c) x gets near -1000?
    (d) Does the graph cross the x-axis? If so, where?
    (e) Does the graph cross the y-axis? If so, where?
    (f) Is there a vertical line that the graph does not cross? Explain.
    (g) Is there a horizontal line that the graph does not cross? Explain.

Exercises 3-5 are worthy of much discussion.

4. Use your graph in Exercise 2 to estimate x in each equation:
    (a) $3 = 2^x$ (b) $0.8 = 2^x$ (c) $9.5 = 2^x$ (d) $6 = 2^x$

(a) 1.6
(c) 3.2

5. Use your graph in Exercise 2 to estimate y in each equation:
    (a) $y = 2^{1.8}$ (b) $y = 2^{-0.3}$ (c) $y = 2^{2.3}$ (d) $y = 2^{-1.6}$
    (e) $y = 2^{0.5}$

(a) 3.5
(c) 4.9
(e) 1.4

THE GRAPH OF $x = 2^y$

In Exercises 2 - 5, you sketched and examined the graph of $y = 2^x$. The next few exercises lead to the graph of $x = 2^y$, which is not the same graph. First, observe that the point (4,16) is on the graph of $y = 2^x$. Thus, it follows that the point (16,4) is on the graph of $x = 2^y$, because $16 = 2^4$. Similarly since the point (1,2) is on the graph of $y = 2^x$, it follows that the point (2,1) is on the graph of $x = 2^y$ ($2 = 2^1$). Exercise 6 develops this idea further.

6. (a) Since (3,8) is on the graph of $y = 2^x$, then (?,3) is on the graph of $x = 2^y$.

    (b) If each of the following points is on the graph of $y = 2^x$, what corresponding point will be on the graph of $x = 2^y$?

point on $y = 2^x$	corresponding point on $x = 2^y$
(6,64)	(64,4)
(0,1)	(,)
(-2,0.25)	(,)
(2,4)	(,)
(-1,0.5)	(,)

7. Using the idea in Exercise 6 and your data from the table in Exercise 1, copy and fill in the following table:

Use $\sqrt{2} \doteq 1.4$

y	-3	-2	-1	$-\frac{1}{2}$	0	$\frac{1}{2}$	1	2	3
$x = 2^y$				$\frac{1}{2}$					

Keep the y-axis vertical, the x-axis horizontal, and the scales equal for both axes. Save this graph; you will use it again.

8. Using the table in Exercise 7, graph $x = 2^y$ very carefully.

9. Consider the curves:

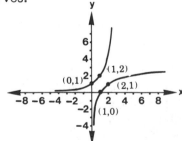

    (a) Which is the graph of $x = 2^y$?

    (b) Which is the graph of $y = 2^x$?

LOGARITHMIC NOTATION

There are only two new things to learn in the study of logarithms: one is <u>logarithmic notation</u> and the other is the use of <u>logarithms for computing.</u>

In the equation

$$8 = 2^3,$$

the 3 is called the <u>exponent</u>.   Now it shall also be called "the <u>logarithm</u> of 8 to the base 2."   This is written as

$$\log_2 8 = 3$$

Stop here. Read this notation aloud.

DEFINITION.   Let N be a positive number.   The number $\log_2 N$ is the exponent to which 2 must be raised to get the number N.   That is,

$$\log_2 N = x$$

means

$$N = 2^x.$$

MEMORIZE this definition now. The content of the chapter is based on it. <u>You must be able to go back and forth</u> between exponential and logarithmic notations comfortably. Observe a <u>LEFT-RIGHT</u> sense as you do this.

This is <u>not a new idea</u>, but merely a new way of writing an old idea. The next paragraph compares the relation between logarithmic and exponential notations to a common relationship in daily life.

There are two ways of stating the father-daughter relationship.   We may say

"John is the father of Mary."

or

"Mary is the daughter of John."

Also, the sentence,

"Bill is older than Mike."

says the same thing as the sentence

"Mike is younger than Bill."

Similarly with exponential and logarithmic notation, the equation

$$8 = 2^3$$

is also expressed by the equation

$$\log_2 8 = 3.$$

Example 1.   Find $\log_2 32 = x$.

Solution.   The problem is to find x such that

$$\log_2 32 = x$$

that is,

$$32 = 2^x.$$

Experience with exponentials tells us that

$$32 = 2^5$$

Hence

$$x = 5.$$

Practice reading aloud both exponential and logarithmic notations.

Thus,

$$\log_2 32 = 5.$$

The equations,

$$32 = 2^5 \quad \text{and} \quad \log_2 32 = 5,$$

both say the same thing.

(a) 4
(c) 6

10. Say each of the following aloud. Then find x:

    (a) $\log_2 16 = x$ (b) $\log_2(\frac{1}{8}) = x$ (c) $\log_2 64 = x$ (d) $\log_2 1 = x$

(a) $\log_2 \frac{1}{4} = -2$

(c) $\log_2 1 = 0$

11. Write in logarithmic notation:

    (a) $\frac{1}{4} = 2^{-2}$ (b) $8 = 2^3$ (c) $1 = 2^0$ (d) $2 = 2^1$

12. Find x:
    (a) $\log_2(2^3) = x$ (b) $\log_2(2^{\frac{1}{2}}) = x$ (c) $\log_2(2^{45}) = x$

(a) 3
(c) 45

    (d) $\log_2(2^{-100}) = x$

* * *

The following is the definition for logarithmic notation for any <u>positive</u> base (other than 1):

Again, you should note that the idea of "logarithm" is not new, but just another name for an exponent.

LOGARITHMIC NOTATION

Let b > 0, and b ≠ 1, and N and x be numbers such that

$$N = b^x.$$

We write

$$\log_b N = x$$

and say,

"the logarithm of N to the base b is x".

Note: Observe N>0 always. Why? Also, if we let b = 1, then x is not unique.

The next few exercises illustrate logarithmic notation.

(a) 3
(c) 3
(e) $\frac{1}{2}$
(f) $\frac{2}{3}$

13. Write the corresponding exponential equation and then solve for x:

    (a) $\log_2 8 = x$ (b) $\log_5 25 = x$ (c) $\log_3 27 = x$

    (d) $\log_{10} 10,000 = x$ (e) $\log_{16} 4 = x$ (f) $\log_8 4 = x$

14. Translate into logarithmic notation:

(a) $\log_3 9 = 2$
(c) $\log_{10} 10000 = 4$
(e) $\log_5 2 = x$
(g) $\log_{10} m = -1$
(i) $\log_2 1 = 0$

    (a) $9 = 3^2$ (b) $125 = 5^3$ (c) $10,000 = 10^4$ (d) $16 = 2^4$

    (e) $2 = 5^x$ (f) $4 = n^3$ (g) $m = 10^{-1}$ (h) $3^x = 4$

    (i) $2^0 = 1$

There are a lot of these exercises here because drill is important: translating between exponential and log notations is somewhat like translating between English and Spanish. Practice helps.

15. Devise and solve four examples of your own similar to those in Exercise 14. Compare your work with that of other students.

16. Translate into exponential notation:

(a) $\log_5 25 = 2$ (b) $\log_6 1 = 0$ (c) $\log_3 x = 2$

(d) $\log_n 3 = 2$ (e) $\log_7 7 = m$ (f) $\log_3 c = 5$

(g) $\log_{10}(0.01) = -2$ (h) $\log_2 8 = n$ (i) $\log_7 1 = 0$

17. Devise and solve four examples of your own similar to those in Exercise 16. Compare your work with that of other students.

18. Solve for x:

(a) $\log_3 81 = x$ (b) $\log_{10} 100 = x$ (c) $\log_3 1 = x$

(d) $\log_7(\frac{1}{7}) = x$ (e) $\log_{10}(0.001) = x$ (f) $\log_{27} 3 = x$

16 (a) $25 = 5^2$
(c) $x = 3^2$
(e) $7 = 7^m$
(g) $0.01 = 10^{-2}$
(i) $1 = 7^0$

(a) 4
(c) 0
(e) -3

THE GRAPH OF $y = \log_2 x$

When you write $y = \log_2 x$ in exponential notation, you obtain the equation $2^y = x$. Hence the graph of $y = \log_2 x$ is identical with that of $2^y = x$, which you graphed in Exercise 8. The next two exercises are concerned with the graph of $y = \log_2 x$.

19. (a) Copy the table and fill in the values for $y = \log_2 x$.

x	$\frac{1}{16}$	$\frac{1}{8}$	$\frac{1}{4}$	$\frac{1}{2}$	1	2	4	8
$y = \log_2 x$								

Observe which numbers have a negative logarithm.

(b) Use the eight points in (a) to help graph

$$y = \log_2 x.$$

(Note. Keep equal scales for both axes.)

Compare this graph with that of Exercise 3 and that of Exercise 7.

20. In the graph of $y = \log_2 x$, what happens to y when

(a) x gets larger?   (b) x gets near 1?   (c) x gets near 0?

(d) x is negative?   (e) Does the graph cross the x-axis? If so, where?   (f) Does the graph cross the y-axis? Explain.

(g) Describe the steepness of the graph, moving from left to right.

(h) Is the graph ever horizontal? vertical? Explain.   (i) Is there a vertical line the graph does not cross? Explain.   (j) Is there a horizontal line the graph does not cross? Explain.

This is a good Exercise for discussion to check your understanding of logarithms.

21. Write the following six equations as three pairs of equations such that the two equations in a pair have the same meaning.

If someone asked you what a logarithm is, could you answer correctly? What would you say? [Hint: Read the definition of log notation again.]

(a) $y = 2^x$ (b) $x = \frac{y}{3}$ (c) $y = x^3$

(d) $y = 3x$ (e) $x = \log_2 y$ (f) $x = y^{\frac{1}{3}}$

THE GRAPH OF $y = \log_{10} x$

So far we have concentrated on the graph of $y = \log_2 x$. Now let us consider another base and graph

$$y = \log_{10} x.$$

"the log of x to the base 10."

22. (a) Copy and fill in the table:

You may use different scales for the x and y axes.

x	$\frac{1}{100}$	$\frac{1}{10}$	1	10	100
$\log_{10} x$					

(b) Use the table in (a) to graph $y = \log_{10} x$. Label the points on the graph where $x = \frac{1}{100}, \frac{1}{10}, 1, 10, 100$.

The graph of $y = \log_{10} x$ in Exercise 22 resembles that of $y = \log_2 x$. The next exercise contrasts the two graphs.

23. (a) Fill in the tables

x	$\frac{1}{8}$	$\frac{1}{4}$	$\frac{1}{2}$	1	2	4	8
$y = \log_2 x$							

x	$\frac{1}{100}$	$\frac{1}{10}$	1	10	100
$y = \log_{10} x$					

(b) Use different scales for the x and y axes.

(b) On the same coordinate system, sketch $y = \log_2 x$ and $y = \log_{10} x$, and label the curves.
(c) At what point do the graphs cross?
(d) Which graph has a larger y-value for a given x when $x > 1$?
(e) Which graph has a larger y-value for a given x when $x < 1$?

APPROXIMATIONS OF LOGARITHMS

The following example introduces a technique for making a rough estimate of a logarithm.

It is easy to compute

$$\log_2 16 \quad \text{and} \quad \log_2 32,$$

but what can be said about $\log_2 21$? Since

$$16 < 21 < 32$$

and $\log_2 x$ increases as x increases, it follows that

$$\log_2 16 < \log_2 21 < \log_2 32.$$

Thus

$$4 < \log_2 21 < 5,$$

and $\log_2 21$ is between the consecutive integers 4 and 5.

The numbers 16 and 32 are chosen because $2^4 = 16$ and $2^5 = 32$.

   Example 2. (a) Find integers x and x + 1 such that

$$x < \log_2 43 < x + 1.$$

   (b) Find integers x and x + 1 such that

$$x < \log_{10} 75 < x + 1.$$

   (c) Find integers x and x + 1 such that

$$x < \log_{10}(0.13) < x + 1.$$

Read this aloud as "the log of 43 to the base 2 is between x and x + 1."

   Solution. (a) From the inequalities

$$32 < 43 < 64$$

it follows that

$$\log_2 32 < \log_2 43 < \log_2 64$$

or

$$5 < \log_2 43 < 6.$$

   (b) The number 75 lies between 10 and 100,

$$10 < 75 < 100.$$

Thus

$$\log_{10} 10 < \log_{10} 75 < \log_{10} 100$$

or

$$1 < \log_{10} 75 < 2.$$

   (c) From the inequalities

$$0.1 < 0.13 < 1$$

it follows that

$$\log_{10}(0.1) < \log_{10}(0.13) < \log_{10} 1$$

or

$$-1 < \log_{10}(0.13) < 0.$$

24. Find an integer x such that:

(a) $x < \log_2 22 < x + 1$ (b) $x < \log_3 20 < x + 1$

(c) $x < \log_5 327 < x + 1$ (d) $x < \log_{10}(0.7) < x + 1$

(a) x = 4
(c) x = 3
[Hint: (d) find $\log_{10}(0.1)$ and $\log_{10} 1$.]

25. Devise and solve two problems of your own similar to Exercise 24.

26. Solve for the letter in each of these equations:

Hint: change to exponential notation.

(a) 25
(c) -4
(e) 100,000
(g) $\frac{1}{6}$

(a) $\log_5 x = 2$ (b) $\log_6 36 = M$ (c) $\log_3\left(\frac{1}{81}\right) = n$

(d) $\log_3\left(\frac{1}{3}\right) = k$ (e) $\log_{100} H = \frac{5}{2}$ (f) $\log_9 9 = c$

(g) $\log_x 6 = -1$ (h) $\log_2 32 = z$

27. Solve for x:

(a) 0.01
(c) 3
(e) $\frac{1}{5}$
(g) 1

(a) $-2 = \log_{10} x$ (b) $1 = \log_3 x$ (c) $x = \log_3 27$

(d) $\frac{1}{2} = \log_x 100$ (e) $-1 = \log_5 x$ (f) $\frac{2}{3} = \log_8 x$

(g) $0 = \log_{10} x$ (h) $x = \log_{27} 9$

ANOTHER PERSPECTIVE ON LOGARITHMS

The next example may seem strange to the eye, for a <u>logarithm appears as an exponent</u>!

Example 3.   (a) Find x if $2^{\log_2 4} = x$.

(b) Find x if $3^{\log_3 81} = x$.

Solution.   (a) $x = 2^{\log_2 4}$ (Putting the x on the left helps some students.)

$\log_2 x = \log_2 4$ (changing from exponential to logarithmic notation)

Thus x = 4.

(b) Similarly,

$x = 3^{\log_3 81}$ (exponential notation)

$\log_3 x = \log_3 81$ (logarithmic notation)

Thus x = 81

In general, if

IMPORTANT! Make sure at this point you understand log notation and that you can comfortably go back and forth between log notation and exponential notation.

Note: Writing A = B as B = A often helps students who have a LEFT-RIGHT problem.

If you forget the general rule, you can solve the specific case by going from exponential notation to logarithmic.

This particular identity, though really just the definition of a logarithm, gives trouble to many calculus students.

$$b^{\log_b N} = x$$

or

$$x = b^{\log_b N},$$

then

$$\log_b x = \log_b N$$

Hence

$$\boxed{b^{\log_b N} = N}$$

Example 4. (a) Find x if $x = 2^{\log_2 5}$

(b) Find x if $x = 10^{\log_{10} 29}$.

Solution. (a) Change to log notation:

$$\log_2 x = \log_2 5.$$

Thus

$$x = 5.$$

(b)

$$x = 10^{\log_{10} 29} \text{ means that}$$

$$\log_{10} x = \log_{10} 29$$

Thus x = 29.

Note that you do not divide by the word "log" here. Rather since $\log_{10}A = \log_{10}B$, then you conclude $A = B$.

28. Find x:

(a) $x = 10^{\log_{10} 100}$ (b) $x = 10^{\log_{10} 1}$ (c) $x = 10^{\log_{10} 10}$

(d) $x = 2^{\log_2 29}$ (e) $5^{\log_5 MN} = x$ (f) $10^{\log_{10} A^4} = x$

(g) $7^{\log_7 48} = x$ (h) $b^{\log_b A}$

(a) 100
(c) 10
(e) MN
(g) 48

Were you able to solve these problems easily? If not, go back and re-view the definition of logarithmic notation and Exercises 14 and 16 again.

LOGARITHM RULES

Now that you are familiar with logarithmic notation, some important properties of logarithms will be presented.

Chapter 2 presented several rules of exponents. Since a logarithm is an exponent, each exponent rule leads to a corresponding logarithm rule as the rest of this chapter will show.

LOG OF A PRODUCT

29. (a) Copy and fill in the table:

A	B	$\log_2 A$	$\log_2 B$	$\log_2 AB$
2	4			
8	32			
$\frac{1}{4}$	16			
$\frac{1}{8}$	$\frac{1}{4}$			

(b) Study the table in (a). What relation is there among $\log_2 A$, $\log_2 B$, and $\log_2 AB$?

Say the relation aloud in your own words.

30. (a) Copy and fill in the table:

A	B	$\log_{10}A$	$\log_{10}B$	$\log_{10}AB$
10	100			
100	1000			
$\dfrac{1}{100}$	10,000			
$\dfrac{1}{1000}$	100,000			

(b) Study the table in (a).   What relation is there among $\log_{10}A$, $\log_{10}B$, and $\log_{10}AB$?

31. (a)   This exercise develops the rule for the logarithm of a product. Assuming $b > 0$, <u>copy the following and explain each step.</u>

Let $A = b^c$ and $B = b^d$

   Step 1.   Then $\log_b A = c$ and $\log_b B = d$

   Step 2. $AB = b^{c+d}$

   Step 3.   From Step 2, $\log_b AB = c + d$

   Step 4.   Using Steps 1 and 3, show how to obtain the rule for the log of a product; that is, $\log_b AB$ in terms of $\log_b A$ and $\log_b B$:

> LOG OF A PRODUCT
>
> $\log_b AB = \log_b A + \log_b B$

MEMORIZE!

(b) Complete the sentence, using "log of a factor";
   "To obtain the log of a product _____ "
(c) Which rule of exponents did you use in Step 2 of part (a) in order to prove the log of a product rule?

32. Assuming $\log_{10}2 = 0.3010$, $\log_{10}3 = 0.4771$, $\log_{10}5 = 0.6990$ and $\log_{10}7 = 0.8451$, find:

(a) 0.7781
(c) 1.1761
(e) 1.3222

(a) $\log_{10}6$ (b) $\log_{10}35$ (c) $\log_{10}15$

(d) $\log_{10}14$ (e) $\log_{10}21$ (f) $\log_{10}10$

33. Using the information given in Exercise 32, find each of the following:

(a) 0.9542
(c) 1.0791

(a) $\log_{10}9$ (b) $\log_{10}8$ (c) $\log_{10}12$ (d) $\log_{10}49$

The next two rules of logarithms concern the log of a reciprocal and the log of a quotient.

LOG OF A RECIPROCAL

34. (a) Explain why the following is valid for all b > 0.

$$\boxed{\log_b 1 = 0}$$ MEMORIZE!

(b) If A > 0, then $\log_b(A \cdot \frac{1}{A}) = \log_b 1 = ?$

(c) Use the log of a product rule to rewrite $\log_b(A \cdot \frac{1}{A})$ in terms of $\log_b A$ and $\log_b(\frac{1}{A})$.

(d) Combining (b) and (c) and solving for $\log_b(\frac{1}{A})$ in terms of $\log_b A$, show how to get the rule for the log of a reciprocal:

> **LOG OF A RECIPROCAL**
>
> $\log_b(\frac{1}{A}) = -\log_b A$ MEMORIZE!

Say the rule in words, "The log of the reciprocal of a number is the negative of the log of the number."

35. Use Exercise 34 and the data given in Exercise 32 to find:

(a) $\log_{10}(\frac{1}{2})$ (b) $\log_{10}(\frac{1}{3})$ (c) $\log_{10}(\frac{1}{5})$

(d) $\log_{10}(\frac{1}{7})$ (e) $\log_{10}(\frac{1}{10})$ (f) $\log_{10}(\frac{1}{6})$

*(a) −0.3010
(c) −0.6990
(e) −1*

LOG OF A QUOTIENT

The next rule of logarithms shows how to find $\log_b(\frac{A}{B})$ in terms of $\log_b A$ and $\log_b B$. It follows easily from the first two rules of logarithms as the next exercise shows.

36. Let A and B be positive numbers.

(a) Write $\frac{A}{B}$ in terms of A and $\frac{1}{B}$.

(b) Assuming b > 0, write $\log_b(\frac{A}{B})$ in terms of $\log_b A$ and $\log_b \frac{1}{B}$.

Which rule of logarithms did you use?

(c) Use (b) and the rule for the log of a reciprocal to write the logarithm rule for the log of a quotient:

> **LOG OF A QUOTIENT**
>
> $\log_b(\frac{A}{B}) = \log_b A - \log_b B$ MEMORIZE!

Say the rule aloud "the log of a quotient is the difference of the logs."

(d) Using "log of the numerator" and "log of the denominator", complete the sentence "The log of a quotient equals _____."

* * *

The next examples will improve your skill in using the logarithm rules.

Example 5.   If $\log_{10}3 = 0.4471$ and $\log_{10}7 = 0.8451$, find

(a) $\log_{10}(\frac{1}{3})$ (b) $\log_{10}21$ (c) $\log_{10}(\frac{3}{7})$

Solution.

When you work problems like these, show all steps, as you see here. Doing so will help avoid errors.

$$\text{(a)}\ \log_{10}\frac{1}{3} = -\log_{10}3$$
$$= -0.4771$$

(b) $\log_{10}21 = \log_{10}(3 \cdot 7)$ (c) $\log_{10}(\frac{3}{7}) = \log_{10}3 - \log_{10}7$

FOR NOW, USE ONLY THE ARITHMETIC KEYS OF YOUR CALCULATOR; DO NOT USE LOG KEY YET.

$\qquad\qquad = \log_{10}3 + \log_{10}7 \qquad\qquad = 0.4771 - 0.8451$

$\qquad\qquad = 0.4771 + 0.8451 \qquad\qquad = -0.3690$

$\qquad\qquad = 1.3222$

37. If $\log_{10}2 = 0.3010$, $\log_{10}3 = 0.4771$, $\log_{10}5 = 0.6990$, and

$\log_{10}7 = 0.8451$, find each of the following:

(a) 0.3680
(c) -0.3980
(e) 0.3980
Hint: (e)
$2.5 = \frac{5}{2}$

(a) $\log_{10}(\frac{7}{3})$ (b) $\log_{10}(\frac{3}{2})$ (c) $\log_{10}(\frac{2}{5})$

(d) $\log_{10}(\frac{5}{7})$ (e) $\log_{10}(2.5)$ (f) $\log_{10}(\frac{2}{3})$

38. Use the data given in Exercise 37 to find:

(a) 1.5229
(c) 0.4771 - 1
or -0.5229
(e) -0.8751

(a) $\log_{10}(\frac{100}{3})$ (b) $\log_{10}(0.75)$ (c) $\log_{10}(0.3)$

(d) $\log_{10}(\frac{6}{7})$ (e) $\log_{10}(\frac{2}{15})$ (f) $\log_{10}(\frac{4}{9})$

LOG OF A POWER

There is one more important rule of logarithms. It concerns the "log of a power." For instance, it will relate $\log_{10}5^3$ to $\log_{10}5$. Example 6 introduces this rule, which is related to the "log of a product" rule.

Example 6.   Assume $\log_{10}3 = 0.4771$. Find (a) $\log_{10}(3^2)$

(b) $\log_{10}(3^3)$

Solution. (a) $\log_{10}(3^2) = \log_{10}3 + \log_{10}3$
$$= 2[\log_{10}3]$$
$$= 2(0.4771)$$
$$= 0.9542$$

(b) $\log_{10}(3^3) = \log_{10}3 + \log_{10}3 + \log_{10}3$

$= 3[\log_{10}3]$

$= 3(0.4771)$

$= 1.4313$

Exercise 39 is similar to Example 6.

39. Using $\log_{10}2 = 0.3010$ and $\log_{10}5 = 0.6990$, find:

(a) $\log_{10}(2^2)$ (b) $\log_{10}(2^3)$ (c) $\log_{10}(5^{-1})$ (d) $\log_{10}(5^4)$

(e) What would you guess is the relation between $\log_b(A^n)$ and $\log_b A$?

(a) 0.6020
(c)=0.6990
(e) You guess!

* * *

The next exercise proves the rule for the log of a power.

The "power of a power" rule for exponents is described by the equation,

$$(b^x)^n = b^{xn}.$$

The next exercise translates this law into the language of logarithms.

40. Copy and explain each of the following steps in the proof that $\log_b(A^n) = n \log_b A$. Let $A = b^x$, where A and b are positive.

Step 1. $\log_b A = x$

Step 2. $A^n = b^{xn}$

Step 3. $\log_b(A^n) = nx$

Step 4. $\log_b(A^n) = n \log_b A$

$$\boxed{\begin{array}{c} \text{LOG OF A POWER} \\ \log_b(A^n) = n \log_b A \end{array}}$$ MEMORIZE!

Describe the "log of a power" rule in your own words.

Example 7.   (a) Given that $\log_{10}5 = 0.6990$, find $\log_{10}(5^9)$

(b) Given that $\log_{10}7 = 0.8451$, find $\log_{10}\sqrt[3]{7}$

Solution.   (a) $\log_{10}(5^9) = 9 \log_{10}5$

$= 9(0.6990)$

$= 6.2910$

$$\text{(b) } \log_{10} \sqrt[3]{7} = \log_{10}(7^{\frac{1}{3}})$$

$$= \frac{1}{3} \log_{10} 7$$

$$= \frac{1}{3}(0.8451)$$

$$= 0.2817$$

41. Given that $\log_{10} 2 = 0.3010$, $\log_{10} 3 = 0.4771$, $\log_{10} 5 = 0.6990$ and $\log_{10} 7 = 0.8451$, find:

(a) $\log_{10} 49$ (b) $\log_{10} \sqrt{3}$ (c) $\log_{10} \sqrt[4]{2}$

(d) $\log_{10} 81$ (e) $\log_{10} \sqrt[5]{\frac{27}{4}}$ (f) $\log_{10}(2^{100})$

(a) 1.6902
(c) 0.0753

42. Find fractions x and y such that $\log_{10} \sqrt[5]{\frac{9}{8}} = x\log_{10} 2 + y\log_{10} 3$.

$y = \frac{2}{5}, x = -\frac{3}{5}$

43. (a) Write the four rules of logarithms in formula form.

   (b) State the rules in (a) in words.

44. Use the data given in Exercise 41 together with the four rules of logarithms to find each of the following to four decimal places. (Show your reasoning by recording all steps.)

As you solve each part, check that the answer is reasonable. For instance the answer to (b) could not be $\log_{10} 32 = 2.050$ because $\log_{10} 32$ is between 1 and 2.

(a) $\log_{10}(\frac{2}{3})$ (b) $\log_{10} 32$ (c) $\log_{10} \sqrt{5}$ (d) $\log_{10}(0.6)$

(e) $\log_{10} 50$ (f) $\log_{10} 60$ (g) $\log_{10}(0.25)$ (h) $\log_{10}(0.5)$

(i) $\log_{10}(0.1)$ (j) $\log_{10}(0.2)$ (k) $\log_{10}(\frac{1}{20})$ (l) $\log_{10} \sqrt[10]{3}$

(a) −0.1761
(c) 0.3495
(e) 1.6990
(g) −0.6020
(i) −1.0000
(k) −1.3010

45. Given that $\log_{10} 3 = 0.4771$ and $\log_{10} 5 = 0.6990$, together with the four rules of logarithms, evaluate each of the following:

(a) $\log_{10}(5 - 3)$ (b) $\frac{\log_{10} 5}{\log_{10} 3}$ (c) $(\log_{10} 5)(\log_{10} 3)$ (d) $\log_{10}(5 + 3)$

(a) 0.3010
(b) 1.4651
(c) 0.3335
(d) 0.9030

COMMON STUDENT ERRORS

Exercise 45 presented opportunities for making certain <u>common errors.</u> Observe carefully that

$$\log_{10}(5 - 3) \text{ is NOT equal to } \log_{10} 5 - \log_{10} 3.$$

In fact, by the "log of a quotient rule,"

These are common errors; study them carefully!

$$\log_{10}(\frac{5}{3}) = \log_{10} 5 - \log_{10} 3.$$

Quite directly,

$$\log_{10}(5 - 3) = \log_{10}2$$

$\dfrac{\log_{10}5}{\log_{10}3}$ is NOT equal to $\log_{10}(\frac{5}{3}) = \log_{10}5 - \log_{10}3$

It is computed directly by division:

$$\frac{\log_{10}5}{\log_{10}3} = \frac{0.6990}{0.4771} = 1.4651$$

Similarly the product,

$$(\log_{10}5)(\log_{10}3) \text{ is NOT equal to } \log_{10}(5 \cdot 3) = \log_{10}5 + \log_{10}3.$$

Rather,

$$(\log_{10}5)(\log_{10}3) = (0.6990)(0.4771)$$
$$\doteq 0.335$$

Finally,

$$\log_{10}(5 + 3) = \log_{10}8$$

$$\log_{10}(5 + 3) \text{ is NOT equal to } \log_{10}5 + \log_{10}3 = \log_{10}15$$

In particular, there is NO rule concerning $\log_b(x + y)$ or $\log_b(x - y)$.

EQUATIONS INVOLVING LOGARITHMS

The next example concerns an equation involving logarithms. When solving for x in such an equation, first use any or all of the log rules to simplify the expression.

Example 8.   Solve for x in the equation,

$$\log_{10}3 + 2 \log_{10}x = \log_{10}27$$

Solution. $\log_{10}3 + 2 \log_{10}x = \log_{10}27$

$$\log_{10}3 + \log_{10}x^2 = \log_{10}27$$
$$\log_{10}3x^2 = \log_{10}27$$
$$3x^2 = 27$$
$$x^2 = 9$$
$$x = 3$$

NOTE: Since x must be positive, we reject the solution x = -3 for the equation $x^2 = 9$. Also, note that we did <u>not</u> divide each side of the equation by the word "log" in going from

$$\log_{10} 3x^2 = \log_{10} 27$$

$$3x^2 = 27.$$

The basis for this step is the fact that if two numbers, A and B, have the same logarithm,

$$\log_{10} A = \log_{10} B,$$

then the numbers, A and B, are equal,

$$A = B$$

Exercises 46 and 47 present problems of this type.

46. Find x for each equation in the manner of Example 8.

(a) $\log_{10} 10 + \log_{10} 100 = \log_{10} x$ (b) $\log_{10} 20 + \log_{10} 50 = \log_{10} x$

(c) $1 + 2\log_{10} 5 + 2\log_{10} 2 = \log_{10} x$ (d) $3\log_{10} 5 + \frac{3}{2}\log_{10} 4 = \log_{10} x$

(e) $\log_{10} x = 4\log_{10} 2$ (f) $\log_{10} x + \log_{10} 9 = \log_{10} 144$

(a) 1000
(c) 1000
(e) 16

COMMON (BASE 10) LOGARITHMS

Base 10 logarithms are sometimes called "common logarithms". Henceforth the following notation will be used:

In Projects another important base is presented, base e: e ≐ 2.7182818284... It is used in calculus.

> NOTATION. The number $\log_{10} x$ will be written simply:
>
> log x.
>
> In other words, when no particular base is indicated, the base is 10.

47. Find x in each equation.

(a) $2\log 6 - \log x = \log 4$ (b) $\log(x - 2) + \log 5 = 2$

(c) $\frac{1}{2}\log 25 + \log x = \log 50$ (d) $\log(x+1) - \log(x-1) = \log 8$

(e) $\log x - \log(1 - x) = \log 1$ (f) $\log x - \log(1 - x) = \log 24$

(a) 9
(c) 10
(e) $\frac{1}{2}$

* * *

While at first they may appear strange to the eye, exponents in such expressions as $10^{0.2345}$ and $10^{-0.3503}$ are most valuable for computational purposes. The next exercise gives some practice with such exponents.

48. If you are given $2 = 10^{0.3010}$, $3 = 10^{0.4771}$, $5 = 10^{0.6990}$ and $7 = 10^{0.8451}$, find x:

(a) $6 = 10^x$

(b) $15 = 10^x$

(c) $\frac{2}{3} = 10^x$

(d) $\frac{7}{3} = 10^x$

(e) $25 = 10^x$

(f) $343 = 10^x$

(g) $\sqrt{3} = 10^x$

(h) $\sqrt[9]{7} = 10^x$

(i) $0.4 = 10^x$

(a) 0.7781
(c) -0.1761
(e) 1.3980
(g) 0.2386
(i) -0.3980

USING LOGARITHMS TO SOLVE EQUATIONS

At the beginning of this chapter, the opening problem raised questions about a rope that required solutions to three equations:

(a) $5(2^d) = 40$

(b) $5(2^d) = 80$

(c) $5(2^d) = 120$

The first two are easy! For instance,

$$5(2^d) = 40$$

reduces to

$$2^d = 8$$

or

$$d = 3.$$

(The second, in a similar manner, gives d = 4.)

But the equation

$$5(2^d) = 120$$

which reduces to

$$2^d = 24,$$

is a more difficult problem. How can d be found?

First of all,
since

$$2^4 < 24 < 2^5$$

or

$$2^4 < 2^d < 2^5$$

then

$$4 < d < 5$$

The next example shows how to find d.

Example 9. If log 2 = 0.3010 and log 3 = 0.4771, find d such that

$$2^d = 24.$$

Solution. Take the log (base 10) of both sides of the equation

$$\log 2^d = \log 24$$

$$d \log 2 = \log 24$$

then

$$d = \frac{\log 24}{\log 2}$$

$$= \frac{\log (2^3 \cdot 3)}{\log 2}$$

A calculator is most helpful in problems such as Example 9, but you still have to know logarithms well to use the calculator correctly.

$$= \frac{\log 2^3 + \log 3}{\log 2}$$

$$= \frac{3 \log 2 + \log 3}{\log 2}$$

$$= \frac{3(0.3010) + 0.4771}{0.3010}$$

$$= \frac{1.3801}{0.3010}$$

$$= 4.5850$$

Here we write
log 2 = 0.3010
and log 3=0.4771
as rounded off
values. Check
this by computing
log 2 and log 3
on your calculator.

49. Use the given logarithm values in Exercise 41 to solve for x:

    (a) $3^x = 5$ (b) $8(3^x) = 72$

    (c) $5(2^x) = 15$ (d) $3(2^x) = 24$

(a) 1.4651
(c) 1.5850

SCIENTIFIC NOTATION (OPTIONAL)

Any positive number can be expressed as the product of one number between 1 and 10 and another number of the form 10^n, where n is an integer. This is the <u>scientific notation for numbers.</u> For example,

$$141 = (1.41)10^2$$
$$0.0141 = (1.41)10^{-2}$$
$$535 = (5.35)10^2$$
$$1000 = (1)10^3$$

50. Write the following numbers in the scientific notation:

(a) $(5.23)(10^{-2})$

(c) $(2.05)(10^{-4})$

(e) $(4.67)(10^{-1})$

(g) $(1.23)(10^{2})$

    (a) 0.0523   (b) 96.7   (c) 0.000205   (d) 247,000
    (e) 0.467   (f) 6.59   (g) 123   (h) 0.00784

COMPUTATIONS WITH LOGARITHMS

Before the advent of the computer, logarithms were used to estimate the solutions of the equations such as the following:

$$x^{10} = 2, \qquad\qquad (4.027)^{25} = x, \qquad\qquad 7 = 8^x,$$
$$x = \sqrt[5]{68.2}, \qquad\qquad 10^{1.734} = x.$$

Along with the logarithmic rules, tables were used involving what were often quite tedious arithmetic calculations.

However, all classes should have access to a calculator with log and exponential keys, which make the use of a log table unnecessary. For all the remaining exercises in this main section, it is expected that the calculator will be used to compute logarithms. All given answers are rounded-off to four significant digits, or four decimal places.

The following examples show how to solve some typical problems.

*STUDY THESE
EXAMPLES
CAREFULLY!*

Example 10. Estimate x to four digits if $x = (4.027)^{25}$.

Solution.

$$x = (4.027)^{25} \quad \text{(Use calculator } y^x \text{ key;}$$
$$\text{round off answer.)}$$

$$\doteq (1.332)(10^{15})$$

No one in his or her right mind would do this example by pure arithmetic!

Example 11. Estimate x to four digits if log x = 1.7302.

Solution.

$$\log x = 1.7302$$

$$x = 10^{1.7302}$$

(calculator) $\doteq 53.73$ (rounded off)

Finding a number when only its log is given is sometimes called finding the "antilog." We will not use that term.

Example 12. Estimate x to four digits if $8^x = 7$.

Solution.

$$8^x = 7$$

$$\log 8^x = \log 7 \quad \text{(Take the log of both sides)}$$

$$x \log 8 = \log 7$$

$$x = \frac{\log 7}{\log 8}$$

(calculator) $x \doteq \dfrac{0.8451}{0.9031}$ (rounded off)

$$x \doteq 0.9358$$

*IMPORTANT!
Even if the calculator does all the "dirty work," always show the logic and steps of the problem on your paper.*

Example 13. Estimate $\sqrt[5]{68.23}$ to four digits.

Solution.

$$\sqrt[5]{68.23} = (68.23)^{\frac{1}{5}}$$

(calculator) $\doteq 2.327$ (Use either the y^x or $\sqrt[x]{y}$ key.)

Example 14. Estimate x to four digits if $x = \log_2 3$.

Solution.

$$x = \log_2 3$$

$$2^x = 3$$

$$\log 2^x = \log 3$$

$$x \log 2 = \log 3$$

$$x = \frac{\log 3}{\log 2}$$

(calculator) $\doteq \dfrac{0.4771}{0.3010}$

$$\doteq 1.585 \quad \text{(rounded off)}$$

Study each step of Examples 14 and 15 carefully. The calculator has only $\log_{10}$ or ln (base e) keys.

Make sure you understand Examples 10-15

As you solve Exercises 51-59, you may wish to refer to these examples again.

Example 15. Estimate x to four digits if $x^{0.34} = 2$.

Solution.

$$x^{0.34} = 2$$
$$\log x^{0.34} = \log 2$$
$$0.34 \log x = \log 2$$

$$\log x = \frac{\log 2}{0.34}$$

(calculator)
$$= \frac{0.3010}{0.34}$$

$$\doteq 0.8853 \quad \text{(rounded off)}$$

$$x = 10^{0.8853}$$

(calculator)

$$x \doteq 7.679$$

Show all steps as you solve the following exercises.

51. Solve for x:

    (a) $2 = 5^x$ (b) $x = 3^{2.95}$ (c) $5 = x^{1.34}$

    (d) $x = \sqrt[7]{4}$ (e) $x = (1.45)^{50}$

*51. Hint: (a)
write
$\log 2 = \log 5^x$
(a) 0.4306
(c) 3.324
(e) $1.171(10^8)$*

52. Solve for x:

    (a) $\log_5 2 = x$ (b) $\log_2 5 = x$ (c) $\log_2 10 = x$

*52. Hint: Write each in exponential notation, then take the log of each side
(a) $\frac{3010}{6990}$ or
0.4306
(c) $\frac{1}{0.3010}$ or
3.322*

53. Show that

$$\log_5 2 = \frac{\log 2}{\log 5}$$

*53. Hint: Let
$\log_5 2 = x$*

54. Use logs to estimate x if

    (a) $\log_3 2 = x$ (b) $x = (37.5)^{50}$ (c) $x = \sqrt[3]{0.3127}$ (d) $2 = x^{10}$

*54. (a) $\frac{3010}{4771}$
or 0.6309
(c) 0.6787
(d) 1.072*

55. Solve for x:

    (a) $\log x + \log 3x = 2$ (b) $\log_6 x + \log_2 8 = \log_2 64$

    (c) $\log 25 + \log x = \log 100$ (d) $\log_4 x + \log_3 27 = \log_5 25$

*55. Neither table nor calculator is needed here!
(a) $\sqrt{\frac{100}{3}}$
(c) 4*

56. Suppose that you are the only heir of a $100 bequest left in a bank by an obscure relative fifty years ago. The rate of interest has remained at 3%, compounded annually.

    (a) Write an expression showing how much your account was worth after

    (i) 1 year   (ii) 2 years   (iii) t years

    (b) How much would your account be worth now?

    (c) Suppose the $100 had been left one hundred and fifty years ago; how much would your account be worth now?

*56. CONTINUE TO SHOW ALL STEPS IN YOUR WORK!
(a)(i) 100(1.03)
(iii) $100(1.03)^t$
(c) about $8425*

57. Suppose that you leave $1,000 on deposit at 5% interest compounded annually.
    (a) Explain why $1000(1.05)^t$ describes the value of the account at time t (in years).

    (b) At what time will the account be worth $10,000?

*(b) about 47.19
    years*

58. The human population of the world is now 5.0 billion and is growing at about 2% per year. If this growth rate remains constant, the population t years from now will be

$$(5.0)(1.02)^t \text{ billion.}$$

You should find Project 5 interesting now.

(a) Do you think the population will take less than 50 years to double? more than 50 years?

(a) guess!

(b) When will the population be double what it is now? That is, what number t (time in years) satisfies this equation:

$$(5.0)(1.02)^t = 10.0?$$

(c) When do you think the population will reach 40 billion? (Some experts say that the earth can support no more than this number of people.)

(c) about 105 years

59. In Exercise 58, in how many years will the human population of the earth reach 13 billion?

about 48.25 years

* * * SUMMARY * * *

This chapter introduced the concept of a logarithm of a number. If b > 0 and

$$N = b^x,$$

then x is called "the logarithm of N to the base b" and is written

$$\log_b N = x.$$

(Particular attention was given to the base 2 and the base 10.) The equation

$$b^{\log_b N} = N$$

records the basic property of logarithms.

As you change notations back and forth, observe again the LEFT-RIGHT sense. Even now, some students may be confused when we write:
$$\log_b N = x$$
means
$$b^x = N.$$

The most important rule of logarithms asserts that

$$\log_b AB = \log_b A + \log_b B.$$

Other key rules are:

$$\log_b 1 = 0$$

$$\log_b \left(\frac{1}{A}\right) = -\log_b A$$

$$\log_b \frac{A}{B} = \log_b A - \log_b B$$

$$\log_b (A^n) = n \log_b A$$

The chapter concluded with applications of logarithms to the solution of equations, such as solving for d in

$$2^d = 24.$$

Other types of equations that can be solved by the use of logarithms are:

$$x^{0.34} = 2, \qquad x = \sqrt[5]{68.2}, \qquad \text{and } x = (4.027)^{28}$$

* * *   WORDS AND SYMBOLS   * * *

logarithm   log x means $\log_{10} x$ (common logarithm)

base b   scientific notation

$\log_b N$ base 10 logarithmic table

log rules

* * *   CONCLUDING REMARKS   * * *

Logarithms, which were introduced in the seventeenth century to simplify calculations, have had many applications since that time. One significant application involves the Richter Scale of measuring the intensity of an earthquake; reference is made to it in practically every news report of a major earthquake. Let us examine what Richter accomplished.

A scale formerly used assigned ten numbers, 1 through 10, to varying intensities of earthquake shock. For instance, 2 described "an extremely feeble shock, felt by a small number of persons at rest." The number 6 meant "a fairly strong shock, generally awakening those asleep, stopping of clocks, visible agitation of trees." A shock of extreme intensity, 10, was a "great disaster; ruins; cracks in the ground; rocks fall from mountains."

Even with reliable information, this scale was far from objective. Influence on the rating scale of the earthquake's depth and the type of ground could not easily be taken into account. In a region such as Southern California, where many earthquakes occur in sparsely populated areas and still others originate under the ocean and spread to observable areas, a measurement procedure depending upon observation is out of the question.

In 1935, Charles F. Richter introduced the scale that is now universally used to measure earthquake. As he put it:

> Despite the evident difficulties, the requirements
> of research, as well as the public interest, call for some
> estimate of the magnitude ... of each important shock.
> What was looked for was a method ... freed from the
> uncertainties of personal estimates or the accidental
> circumstances of reported effects.

After extensive study Richter proposed to define the intensity as

> the logarithm to the base 10 of the amplitude that a
> standard seismometer should register in the case of an
> earthquake of standard magnitude, which is one tenth
> of a centimeter at a distance of 100 kilometers from
> the center of the earthquake.

On this scale, "the smallest shocks are of magnitude 0; the smallest reported felt are of magnitude 1.5. ...Those of magnitude 6 are destructive over a restricted area; those of magnitude 7.5 are at the lower limit of major earthquakes."

It turns out that the Richter scale is closely related to the total energy E (measured in ergs) of the quake. The energy and the magnitude M are linked by the formula

$$\log_{10} E = 11.3 + 1.8\ M.$$

Thus if one quake has 100 times the energy of another, its Richter number is about 2 units larger than the Richter number of the smaller quake. (Why?)

Richter's logarithmic scale provides scientists and laymen with a precise and objective means of describing earthquake intensity.

* * * CENTRAL SELF QUIZ * * *

1. If $\log 2 = 0.3010$, $\log 3 = 0.4771$, and $\log 5 = 0.6990$, find
   (a) $\log 9$ (b) $\log \sqrt{2}$ (c) $\log (0.6)$
   (d) $\log (0.04)$ (e) $\log (0.2)$ (f) $\log (\frac{6}{5})$
   (g) $\log (\frac{1}{3})$ (h) $\log 64$ (i) $\log \sqrt[10]{2}$

2. Let $b^n = A$ and $b^m = B$. Express each of the following in terms of m and n.
   (a) $\log_b A$ (b) $\log_b B$ (c) $\log_b (A \cdot B)$
   (d) $\log_b (\frac{1}{A})$ (e) $\log_b (\frac{A}{B})$ (f) $\log_b A^7$

3. Let $\log A = x$, $\log B = y$, and $\log C = z$. Express each of the following in terms of x, y, and z.
   (a) $\log (\frac{A}{C})$ (b) $\log (B \cdot C)$ (c) $\log A^x$
   (d) $\log \sqrt{A}$ (e) $\log \sqrt[3]{\frac{B}{C}}$ (f) $\log (A\sqrt{BC})$
   (g) Can you find $\log (A + B)$? Explain.
   (h) Can you find $\log (A - B)$? Explain.

4. Solve for x:
   (a) $\log 11 + \log (x-2) = \log 3$ (b) $\log (5x+7) - \log (x-2) = 1$

5. Write the product in scientific notation:
   (a) $(5,000,000)(9,000,000)$ (b) $(0.00003)(0.002)$

6. Find x if   (a) $\log x = -0.2933$ (b) $\log x = 1.2933$

7. (a) If a certain population has a growth rate of 2% per year, in how many years will it be 50% more than it is now?
   (b) Explain why (a) is the same as solving the following equation for t:
   $$(1.02)^t = 1.5$$

ANSWERS TO LOGARITHMS CENTRAL SELF QUIZ

1. (a) 0.9542 (b) 0.1505 (c) -0.2219 (d) -1.3980 (e) -0.6990 (f) 0.0791
   (g) -0.4771 (h) 1.8060 (i) 0.0301   2. (a) n (b) m (c) m+n (d) -n
   (e) n-m (f) 7n   3. (a) x-z (b) y+z (c) x^2 (d) $\frac{x}{2}$ (e) $\frac{1}{3}(y-z)$
   (f) $x+\frac{1}{2}(y+z)$ (g) no (h) no   4. (a) $\frac{25}{11}$ (b) $\frac{27}{5}$ 5. (a) $(4.5)(10^{13})$
   (b) $(6)(10^{-8})$ 6. (a) 0.509 (b) 19.65   7. (a) If P is current popula-

tion, then the population in t years will be $P(1.02)^t$. Then, $P(1.02)^t = (1.5)P$ and $t \stackrel{\cdot}{=} 20.5$ years.

(b) Divide each side of the equation in (a) by P.

* * * LOGARITHMS REVIEW * * *

1. Explain each statement:

    (a) "A logarithm is an exponent."
    (b) "Logarithms give us a method for solving for an unknown exponent: for example,
    $$2 = 5^x.$$

2. Match each exponent rule with the corresponding logarithm rule:

    (a) $b^{m+n} = b^m b^n$ (1) $\log_b 1 = 0$

    (b) $b^{-n} = \dfrac{1}{b^n}$ (2) $\log_b(\frac{A}{B}) = \log_b A - \log_b B$

    (c) $b^{m-n} = \dfrac{b^m}{b^n}$ (3) $\log_b(\frac{1}{A}) = - \log_b A$

    (d) $b° = 1$ (4) $\log_b AB = \log_b A + \log_b B$

    (e) $(b^m)^n = b^{mn}$ (5) $\log_b(A^n) = n \log_b A$

3. What log rule corresponds to each of the following exponent rules?

    (a) $10^m \cdot 10^n = 10^{m+n}$ (b) $10^{-1} = \dfrac{1}{10}$ (c) $\dfrac{10^m}{10^n} = 10^{m-n}$

    (d) $(10^m)^n = 10^{mn}$

4. On the same coordinates graph $y = 2^x$ and $y = 3^x$. Which graph is higher when   (a) $x > 0$ (b) $x = 0$ (c) $x < 0$

    (d) Guess how the graphs of $y = 3^x$ and $y = 10^x$ would compare.

5. (a) Since $(-3, \frac{1}{8})$ is on the graph of $y = 2^x$, then $(?,?)$ is on the graph of $2^y = x$.

    (b) If each of the following points is on the graph of $y = 2^x$, what corresponding points will be on the graph of $2^y = x$?

        (i) $(-1, 0.5)$ (ii) $(4,16)$ (iii) $(-5, \frac{1}{32})$ (iv) (m,n)

6. If $N = 2^a$, then $\log_2 N = ?$

7. Change to logarithmic notation:

    (a) $5 = 2^x$ (b) $N = 3^4$ (c) $10 = x^2$

    (d) $0.6325 = 10^x$ (e) $M = 10^{0.1234}$ (f) $8 = b^2$

8. Change to exponential notation:

    (a) $\log_5 7 = x$ (b) $\log_3 N = 4$ (c) $\log_b 6 = 2$

    (d) $\log_7 6 = N$ (e) $\log_b N = x$ (f) $4 = \log_x 7$

9. Change to logarithmic notation:

(a) $2^x = 7$ (b) $3 = 5^n$ (c) $z^n = 2$

(d) $N = b^x$ (e) $N = \sqrt{2}$ (f) $M = 27$

10. Solve for the letter:

(a) $\log_4 N = \frac{1}{2}$ (b) $\log_6 216 = x$ (c) $\log_{100} N = \frac{5}{2}$

(d) $\log_{10}(0.0001) = x$ (e) $\log_b 8 = -\frac{3}{2}$ (f) $\log_{10}(0.1) = x$

(g) $\log_b 3 = \frac{1}{3}$ (h) $\log_b 5 = -1$ (i) $\log_b 10 = \frac{1}{4}$

(j) $x \log_2 16 = 1$ (k) $\log_2 (\frac{1}{8}) = m$ (l) $C \log_5 125 = 2$

11. (a) Compare the graphs of

$$y = 2^x \text{ and } y = \log_2 x$$

with respect to the line $y = x$.

(b) Compare the graphs of

$$y = 2^x \text{ and } y = (\tfrac{1}{2})^x$$

with respect to the y-axis.

12. Find an integer x such that

(a) $x < \log_{10} 300 < x + 1$.

(b) $x < \log_{10}(.05) < x + 1$.

(c) $x < \log_{10} 6000 < x + 1$.

13. Give an example of a number x that makes the statement true:

(a) $1 < \log_{10} x < 2$ (b) $3 < \log_{10} x < 4$ (c) $-1 < \log_{10} x < 0$

14. For each, find an integer x so that

(a) $x < \log_{10}(1250) < x + 1$ (b) $x < \log_{10}(6) < x + 1$

(c) $x < \log_{10}(0.2) < x + 1$

15. Find x:

(a) $(10^3)(10^{0.2947}) = 10^x$ (b) $\dfrac{10^{0.2349}}{10^{0.5674}} = 10^x$ (c) $(10^{2.6549})^3 = 10^x$

16. If $\log 2 = 0.3010$, $\log 3 = 0.4771$, $\log 5 = 0.6990$, and $\log 13 = 1.1139$, find

(a) $\log 26$ (b) $\log (0.6)$ (c) $\log(\frac{13}{2})$ (d) $\log (13 - 3)$

(e) $\log \sqrt[3]{5}$ (f) $\log (0.2)$ (g) $\log (\frac{1}{3})$ (h) $\log (0.005)$

17. Solve for x:

    (a) log x + log 8 = log 24
    (c) x log 100 = 2
    (e) log 3 + 2 log x - log 5 = log 6

    (b) log 13 - log x = 1
    (d) log x - log 13 = 3
    (f) $\frac{1}{4}$ log x = log 12 - 2 log 2

18. Compute x:

    (a) $2^x = 13$
    (b) $2(3^x) = 8$
    (c) $5(13^x) = 2$

19. Write as $10^{(\text{exponent})}$

    (a) $\dfrac{10^x}{10^y}$
    (b) $(10^x)^y$
    (c) $(10^x)(10^y)$
    (d) $(\sqrt[x]{10})(\sqrt[y]{10})$

20. If log X = 1, log Y = 2, and log Z = -3, use this information only to find the <u>numerical value</u> of the following <u>if possible.</u>

    Exercise 20 is a good check of your understanding of the log rules.

    (a) log (XY)
    (b) log $(\frac{Y}{Z})$
    (c) X
    (d) Y

    (e) Z
    (f) log (Y^4)
    (g) $10^{\log X}$
    (h) $\dfrac{\log X}{\log Z}$

    (i) (log X)(log Y)
    (j) log (X + Y)
    (k) $Y^{\log Z}$
    (l) log (Y - X)

21. Compute x:

    (a) $\sqrt[4]{9.324} = x$
    (b) $(1.035)^{16} = x$
    (c) $\log_6 320 = x$
    (d) $10^{1.2345} = x$

22. Solve for x:

    (a) $64.8 = (3.2)^x$
    (b) $3^{x+1} = 240$
    (c) $8^{2x} = 320$
    (d) $(4.3)^{x-1} = 85.6$

23. Solve for x:

    (a) $x^{2.3} = 5.29$
    (b) $17^x = 11$
    (c) $(2.47)^{29} = x$
    (d) $(2.3)^{-x} = 7$
    (e) $\log_2 7 = x$
    (f) $10^{\log 9} = x$

24. (a) Find

    $$\log (\tfrac{1}{2}) \text{ and } \log 2$$
    $$\log (\tfrac{3}{4}) \text{ and } \log (\tfrac{4}{3})$$

    (b) What is the relation between

    $$\log (\frac{1}{x}) \text{ and } \log x?$$

    (c) How is the logarithm of the reciprocal of a number related to the logarithm of that number?

25. Write each of the following as an expression free of the word "log".

    (a) log A = log π + 2 log r
    (b) log C = log 2 + log π + log r

26. Write as a single logarithm:

(a) $\log 2 + \log x + 3[\log x - \log y]$ (b) $\log (x + 1) + \log (x - 1)$

(c) $\log (x^2 - 1) - \log (x + 1)$ (d) $\log 4 - \log 3 + 3 \log r + \log \pi.$

27. (a) If $1 is left on deposit at a compound interest rate of 4% per year, explain to a classmate why the value of the account is equal to:

    (i) 1.04 dollars after one year,
    (ii) $(1.04)^2$ dollars after two years,
    (iii) $(1.04)^3$ dollars after three years,
    (iv) $(1.04)^n$ dollars after n years.

(b) If $1 is left on deposit at an interest rate of 4% per year, compounded annually, compute the value of the account at the end of 100 years.

28. A biologist observes that under favorable conditions a certain kind of bacteria doubles in number every day. If A is the number of bacteria present at the beginning of a given day, then the number of bacteria present after n days will be

You may wish to do Project 8 after this problem.

$$A(2^n).$$

In how many days will the number of bacteria be

(a) 4A?   (b) 16A?   (c) 7A?   (d) (2.5)A?

29. Translate each of these into decimal notation:

(a) $(9.1066)10^{-28}$ grams (the mass of an electron)

(b) $(5.976)10^{24}$ kilograms   (the mass of the earth)

(c) $(5.8803)10^{12}$ miles   (the distance light travels in one year)

30. Given that (i) X, Y, Z, and b are positive,
    (ii) $\log_b X = c$, $\log_b Y = d$, and $\log_b Z = e$,

find the following in terms of c, d, and e (if possible):

(a) $\log_b(X \cdot Y)$ (b) $\log_b \left(\dfrac{Y}{Z}\right)$ (c) $\log_b(X^n)$ (d) $\log_b(X - Y)$

(e) $\log_b \sqrt[4]{Y}$ (f) $\log_b \sqrt{\dfrac{X \cdot Y}{Z}}$ (g) $\log_b(X + Y)$ (h) $\log_b\left(\dfrac{X \cdot Y}{Z}\right)$

31. (a) Suppose that a bank pays 5% compound interest per year and that you deposit $1. In how many years will your account be worth

    (i) $1.05   (ii) $(1.05)^4$?   (iii) $(1.05)^n$?

(b) How many years will it take your deposit of $1 to be worth $2? (That is, find n such that $(1.05)^n = 2.$)

32. Suppose that you deposit A dollars at 5% compound interest per year. In how many years will your account be worth

(a) $A(1.05)$?   (b) $A(1.05)^2$?   (c) $A(1.05)^3$

(d) $A(1.05)^{29}$?   (e) $A(1.05)^n$?

33. If the population of the world is now about 5.0 billion and the population at time t (in years) is

$$(5.0)(1.02)^t$$

(a) when will the population triple?
(b) What was the population 50 years ago?
(c) Discuss your answers to (a) and (b) with regard to the "population explosion."

Hint: (b) 50 years ago is a negative number.

34. Solve for x:

Hint: (a) Recall log N means N>0.

(a) $\log x^2 + \log 5 = \log 50$ (b) $\log (x + 1) + \log (x - 1) = \log 8$

(c) $\log (x - 1)^{\frac{1}{3}} + \log 5 = \log 15$

ANSWERS TO LOGARITHMS REVIEW

1. (a) $\log_b x = n$ means $b^n = x$ (b) $2 = 5^x$ via logs and algebra becomes $\frac{\log 2}{\log 5} = x$.

2. (a) 4 (b) 3 (c) 2 (d) 1 (e) 5   3. (a) $\log AB = \log A + \log B$

(b) $\log (\frac{1}{A}) = -\log A$ (c) $\log (\frac{A}{B}) = \log A - \log B$ (d) $\log A^n = n \log A$

4. (a) $y = 3^x$ is higher than $y = 2^x$.   (b) They share the same point; (0,1) is on both graphs.   (c) $y = 2^x$ is higher than $y = 3^x$.

(d) $y = 10^x$ is higher than $y = 3^x$ for $x > 0$ and lower than $y = 3^x$ for $x < 0$.   5. (a) $(\frac{1}{8}, -3)$ (b) (i) (0.5,-1)   (ii) (16,4)   (iii) $(\frac{1}{32}, -5)$

(iv) (n,m)   6. <u>a</u>

7. (a) $\log_2 5 = x$ (b) $\log_3 N = 4$ (c) $\log_x 10 = 2$

(d) $\log 0.6325 = x$ (e) $\log M = 0.1234$ (f) $\log_b 8 = 2$

8. (a) $5^x = 7$ (b) $3^4 = N$ (c) $b^2 = 6$
(d) $7^N = 6$ (e) $b^x = N$ (f) $x^4 = 7$

9. (a) $\log_2 7 = x$ (b) $\log_5 3 = n$ (c) $\log_Z 2 = n$

(d) $\log_b N = x$ (e) $\log_{\sqrt{2}} N = 1$ or $\log_2 N = \frac{1}{2}$

(f) $\log_{27} M = 1$ or $\log_3 M = 3$

10. (a) $N = 2$ (b) $x = 3$ (c) $N = 10^5$ (d) $x = -4$ (e) $b = \frac{1}{4}$

(f) $x = -1$ (g) $b = 27$ (h) $b = \frac{1}{5}$ (i) $b = 10^4$ (j) $x = \frac{1}{4}$

(k) $m = -3$ (l) $c = \frac{2}{3}$

11. (a) mirror reflections with respect to $y = x$
(b) mirror reflections with respect to the y-axis

12. (a) $x = 2$ (b) $x = -2$ (c) $x = 3$

13. Student finds an x such that
(a) $10 < x < 100$ (b) $1000 < x < 10000$ (c) $0.1 < x < 1$

14. (a) 3   (b) 0   (c) -1   15. (a) 3.2947 (b) -0.3325   (c) 7.9377

16. (a) 1.4149   (b) -0.2219   (c) 0.8129   (d) 1   (e) 0.2330
    (f) -0.6990   (g) -0.4771   (h) -2.3010

17. (a) x = 3   (b) x = 1.3   (c) x = 1   (d) x = 13000   (e) x = $\sqrt{10}$ (f) x = 81

18. (a) 3.7006   (b) 1.2618   (c) -0.3572

19. (a) 10^{x-y} (b) 10^{xy} (c) 10^{x+y} (d) $10^{\left(\frac{x+y}{xy}\right)}$

20. (a) 3   (b) 5   (c) 10   (d) 100   (e) 0.001   (f) 8   (g) x = 10   (h) $-\frac{1}{3}$
    (i) 2   (j) impossible   (k) 0.000001   (l) impossible

21. (a) 1.747   (b) 1.734   (c) 3.219   (d) 17.16

22. (a) 3.587   (b) 3.989   (c) 1.387   (d) 4.051

23. (a) 2.063   (b) 0.8464   (c) $(2.445)(10^{11})$ (d) -2.336   (e) 2.807   (f) 9

24. (a) log 2 = 0.3010, log $(\frac{1}{2})$ = -0.3010 log$(\frac{3}{4})$ = -0.1249, log$(\frac{4}{3})$ = 0.1249

    (b) log$(\frac{1}{x})$ = - log x   25. (a) A = πr^2 (b) C = $2\pi r$

26. (a) log $\left(\frac{2x^4}{y^3}\right)$ (b) log $(x^2 - 1)$ (c) log $(x - 1)$ (d) log $(\frac{4}{3}\pi r^3)$

27. (a) Explain (b) $50.50   28. (a) 2 days (b) 4 days (c) 2.8 days (d) 1.3 days

29. (a) 0.00000000000000000000000000091066
    (b) 5,975,000,000,000,000,000,000,000   (c) 5,880,300,000,000

30. (a) c + d   (b) d - e   (c) no   (d) impossible   (e) $\frac{d}{4}$ (f) $\frac{1}{2}$(c + d - e)
    (g) impossible   (h) c + d - e

31. (a) (i) one year   (ii) four years   (iii) n years   (b) about 14.21 years.

32. (a) 1 year   (b) 2 years   (c) 3 years   (d) 29 years   (e) n years

33. (a) about 55.48 years   (b) 1.858 billion   (c) Compare the rates of
                                                      increase.

34. (a) $\sqrt{10}$ (b) 3   (c) 28

* * * LOGARITHMS REVIEW SELF TEST * * *

1. (a) Graph on the same axes

$$y = 3^x \text{ and } y = \log_3 x$$

   (b) With reference to what line is each graph in (a) the "reflection" of
       the other?

2. (a) Graph on the same axes
$$y = 3^x \text{ and } y = (\tfrac{1}{3})^x$$

   (b) With reference to what line is each a "reflection" of the other?

3. Solve for x:

   (a) $\log_4 x = 1.5$ (b) $\log_x 9 = -1$ (c) $\log_{125} 25 = x$

Hint: Since (2,9) is on the graph of $y = 3^x$, then (?,?) is on the graph of $y = \log_3 x$.

4. Solve for x:

   (a) $10^{\log_{10}14} = x$ (b) $10^{\log_{10}N} = x$ (c) $10^{\log_{10}2001} = x$

5. Given that $\log_{10}2 = 0.3010$, $\log_{10}6 = 0.7782$, and $\log_{10}7 = 0.8451$, find (if possible)

   (a) $\log_{10}42$ (b) $\log_{10}3$ (c) $\log_{10}5$ (d) $\log_{10}64$

6. Solve for x:

   (a) $2^x = 7$ (b) $(7)(8^x) = 9$

7. Explain scientific notation to a classmate.

8. Solve for x:

   (a) $\sqrt[9]{0.1275} = x$ (b) $x = \log_{11}7$ (c) $(5.648)^{25} = x$ (d) $2 = (3.14)^x$

ANSWERS TO REVIEW SELF TEST

1. (a) Graph (b) $y = x$ 2. (a) Graph (b) $x = 0$ 3. (a) 8 (b) $\frac{1}{9}$ (c) $\frac{2}{3}$

4. (a) 14   (b) N   (c) 2001   5. (a) 1.6233 (b) 0.4772   (c) 0.6990 (d) 1.806

6. (a) 2.8076   (b) 0.1209   7. Explain   8. (a) 0.7954 (b) 0.8115

   (c) $(6.271)(10^{18})$ (d) 0.6058

* * * LOGARITHMS PROJECTS * * *

1. When Robinson Crusoe was stranded on a desert island, he amused himself by estimating logarithms to the base 10. He remembered that $\log_{10}2 = 0.3010$. Then he quickly figured out $\log_{10}4$, $\log_{10}8$ and $\log_{10}5$. How did he do it?

2. (See Central opening problem). Robinson Crusoe was really curious about $\log_{10}3$. He wrote down the inequalities

   $$2^3 < 3^2$$

   and

   $$3^2 < 10.$$

   From these inequalities and a little arithmetic in the sand, he deduced that

   $$0.4515 < \log_{10}3 < 0.5000.$$

   How did he do it?

3. (See Exercise 2). Robinson wanted to pin down $\log_{10}3$ even more accurately. He observed that $3^5 < 2^8$. What did that tell him about $\log_{10}3$?

4. (See Exercise 3). Robinson was rescued before he could go further in his estimates of $\log_{10}3$.

   (a) Use his general approach to get better estimates of $\log_{10}3$.

   (b) Try to estimate $\log_{10}7$ using the same general approach.

5. At present the "underdeveloped" nations have twice the population of the "overdeveloped" nations. Assume that the population of the "underdeveloped" nations increases as the rate of 3% per year and the population of the "overdeveloped" nations increases at the rate of 1% per year. In how many years will the "underdeveloped" nations be ten times as populous as the "overdeveloped" nations?

6. Compute:

(a) $(4^{\log_4 9})^{\frac{1}{2}}$ (b) $4^{\frac{1}{2}\log_4 9}$ (c) $(4^{\frac{1}{2}})^{\log_4 9}$

(d) $4^{\log_2 9}$ (e) $8^{5\ \log_{32} 17}$

7. (a) Fill in the boxes:

(i) $10^{\log_{10}(A \cdot B)}$ = $\boxed{?}$

(ii) $10^{\log_{10} A}$ = $\boxed{?}$

(iii) $10^{\log_{10} B}$ = $\boxed{?}$

(b) Use (a) to show that

$$\log_{10}(A \cdot B) = \log_{10}A + \log_{10}B$$

(c) Use

$$10^{\log_{10}(\frac{A}{B})}   =   \boxed{?}$$

and parts (ii) and (iii) of (a) to show that

$$\log_{10} \frac{A}{B} = \log_{10}A - \log_{10}B$$

8. Radioactive substances disintegrate over a period of time by emitting particles from some of the atoms, thus changing those atoms to another substance and making the weight of the unchanged material decrease with time. If A is the original weight, in time t the weight of the unchanged material will be

$$Ab^{-t}, (b > 1).$$

This Project is an excellent problem to test your understanding of logs.

The half-life is the time it takes for half the substance to disintegrate. If the half-life of the substance is 3.85 days, what fraction of the original amount remains after

(a) 7.7 days?   (b) 30.8 days?   (c) 10 days?   (d) 1 day?

9. Where does the graph of

$$y = \log_2 x$$

cross the line y = 7?

10. Obtain up-to-date information on the size of the human population of the world. (a) What is the present value?

(b) At what rate is it growing?

(c) Sketch a graph showing its predicted values.

11.   Suppose a new car sells for $15,000 and depreciates 20% per year.

(a) What is the value of the car after 10 years?

(b) At what approximate time is the car worth $7,500? $1,000? $100?

12. Find x if $x^x = 10$.   (Experiment!)

13.
(a) Graph $y = \dfrac{\log x}{x}$ for $x \geq 1$.

(b) Estimate the value of x that yields the largest value of y. (Experiment!)

14. (a) Compute $(\log_2 8)(\log_8 2)$.

(b) Compute $(\log_{10} 100)(\log_{100} 10)$.

(c) Evaluate $(\log_a b)(\log_b a)$ for your own (convenient) choices of a and b. What do you conclude?

15. Start with the equation
$$a^{\log_a b} = b.$$
Take $\log_b$ of both sides. Explain why the result is related to Project 14.

16. (a) Write x, y, and z in terms of logarithms where
$$b^x = a, \quad a^y = c, \quad \text{and } b^z = c.$$

(b) How is z in part (a) related to x and y?

(c) Show that
$$(\log_b a)(\log_a c) = \log_b c$$

(d) Why is Project 14 (c) a special case of part (c) above?

17. Use a calculator to estimate $(1 + \frac{1}{n})^n$ for

(a) $n = 10$ (b) $n = 10^3$ (c) $n = 10^6$ (d) $n = 10^9$

(In calculus it is shown that when n is large $(1 + \frac{1}{n})^n$ gets arbitrarily close to a number whose decimal expansion begins 2.718281828...
This irrational number is denoted "e" in honor of Euler.)

18. When money is left in the bank at simple interest, interest is not paid on the interest. For instance, one dollar in an account paying 5% simple interest earns 5 cents each year that it remains in the bank. It will take 20 years to double in value. If it is left, instead, at compound interest, then interest is paid on interest. At the same rate, 5%, in 20 years one dollar earns more at compound interest, and the account grows to $(1.05)^{20}$ dollars.

(a) Evaluate $(1.05)^{20}$.

(b) How many years does it take for one dollar to grow to ten dollars at compound interest, if the interest rate is 5%?

19. This exercise describes a relation between areas and logarithms known as early as the 16th century.

   (a) Sketch the curve $y = \dfrac{1}{x}$ for $x \geq 1$ on graph paper. There should be at least five squares to the unit and x should go from 1 to 12.

   (b) Let $A(x)$ equal the area under the curve, above the x-axis, from 1 to x. By counting squares estimate the values in this table:

x	2	3	4	5	6	7	8	9	10	12
A(x)										

   (c) Does $A(2 \cdot 3)$ seem to equal $A(2) + A(3)$?
   (d) Does $A(3 \cdot 4)$ seem to equal $A(3) + A(4)$?
   (e) Does $A(3 \cdot 3)$ seem to equal $A(3) + A(3)$?
   (f) Does $A(2 \cdot 4)$ seem to equal $A(2) + A(4)$?
   (g) What rule of logarithms does the area, $A(x)$, seem to satisfy?
   (h) Estimate x if $A(x) = 1$.
   (i) If $A(x) = \log_b x$, what is the base b?

20. Logarithms are important in the study of prime numbers. Let P_n be the nth prime number. Thus

$$P_1 = 2,\ P_2 = 3,\ P_3 = 5,\ P_4 = 7,\ P_5 = 11,\ \text{and so on.}$$

While there is no exact formula for P_n, there is one that provides a good estimate when n is large. It involves logarithms with a certain base b, which is about 2.72. For simplicity, denote $\log_b x$ by the symbol $L(x)$. Thus $L(2.72) = 1$ and $L(8)$ is a little more than 2, since $(2.72)^2 < 8$. Here is a table of $L(x)$ for certain values of x.

x	1	2.72	3	8	10	20	100
L(x)	0	1	1.10	2.08	2.30	3.00	4.61

   According to advanced number theory, $n \cdot L(n)$ is a good estimate of P_n, especially when n is large.

   (a) Compare P_3 and $3 \cdot L(3)$.   (b) Compare P_8 and $8 \cdot L(8)$.

   (c) Compare P_{10} and $10 \cdot L(10)$.   (d) Compare P_{20} and $20 \cdot L(20)$.

   For large n the quotient,

$$\frac{P_n}{n\,L(n)},$$

   is near 1.

21. (See Exercise 20.) How many primes are there that are less than or equal to the number n? Advanced number theory asserts that there are approximately

$$\frac{n}{L(n)}$$

such primes.

(a) Test the assertion for n = 10.   (b) Test it for n = 20.

(c) Test it for n = 100.

* * * FIRST YEAR ALGEBRA REVIEW (OPTIONAL) * * *

Becuase algebra is essentially arithmetic using letters, algebraic fractions are added, subtracted, multiplied and divided much like arithmetic fractions. Consider addition for example:

MAKE SURE YOU UNDERSTAND HOW TO WORK WITH FRACTIONS! They are used repeatedly in later math courses.

arithmetic

$$\frac{4}{5} + \frac{6}{7} = \frac{4}{5} \cdot \frac{7}{7} + \frac{6}{7} \cdot \frac{5}{5}$$

$$= \frac{(4)(7) + (6)(5)}{(5)(7)}$$

$$= \frac{28 + 30}{35}$$

$$= \frac{58}{35}$$

algebra

$$\frac{x}{y} + \frac{z}{w} = \frac{x}{y} \cdot \frac{w}{w} + \frac{z}{w} \cdot \frac{y}{y}$$

$$= \frac{xw + zy}{yw}$$

or subtraction:

arithmetic

$$\frac{4}{9} - \frac{2}{5} = \frac{4}{9} \cdot \frac{5}{5} - \frac{2}{5} \cdot \frac{9}{9}$$

$$= \frac{(4)(5) - (2)(9)}{(9)(5)}$$

$$= \frac{20 - 18}{45}$$

$$= \frac{2}{45}$$

algebra

$$\frac{a}{b} - \frac{c}{d} = \frac{a}{b} \cdot \frac{d}{d} - \frac{c}{d} \cdot \frac{b}{b}$$

$$= \frac{ad - bc}{bd}$$

EXERCISES

1. Find the sum as a single fraction:

(a) $\frac{2}{3} + \frac{9}{11}$ (b) $\frac{m}{n} + \frac{p}{q}$ (c) $\frac{x}{3} + \frac{x-1}{4}$ (d) $\frac{x-y}{x} + \frac{x+y}{y}$

2. Find the difference as a single fraction:

(a) $\frac{7}{9} - \frac{11}{12}$ (b) $\frac{2n}{5} - \frac{n}{7}$ (c) $\frac{x}{y^2} - \frac{x}{y}$ (d) $\frac{2}{mn} - \frac{3}{np}$

Hint: (c) Write
$\frac{22}{7}$
(g) Write
$1 \cdot \frac{x}{x} + \frac{1}{x} = \frac{x+1}{x}$

3. Write the reciprocal of each of the following:

(a) 5   (b) $\frac{2}{3}$ (c) $3\frac{1}{7}$ (d) -2   (e) -1   (f) 1   (g) $1 + \frac{1}{x}$ (h) $\frac{6}{7}$

Consider multiplication:

	arithmetic		algebra

$$\frac{2}{5} \cdot \frac{5}{8} = \frac{\cancel{2}}{\cancel{5}} \cdot \frac{\cancel{5}}{(2)(2)(2)}$$

$$\frac{2}{3x^2 + x} \cdot \frac{x}{3} = \frac{2}{\cancel{x}(3x + 1)} \cdot \frac{\cancel{x}}{3}$$

$$= \frac{1}{(2)(2)}$$

$$= \frac{2}{(3x + 1)3}$$

$$= \tfrac{1}{4}$$

$$\text{or} \quad \frac{1}{9x + 3}$$

or division:

	arithmetic		algebra

$$\frac{2}{3} \div \frac{4}{7} = \frac{\dfrac{2}{3}}{\dfrac{4}{7}}$$

$$\frac{x}{y} \div \frac{z}{w} = \frac{\dfrac{x}{y}}{\dfrac{z}{w}}$$

$$= \frac{\cancel{2}}{3} \cdot \frac{7}{\cancel{4}_2}$$

$$= \frac{x}{y} \cdot \frac{w}{z}$$

$$= \frac{7}{6}$$

$$= \frac{xw}{yz}$$

To divide by a number is the same as multiplying by its reciprocal.

4. Write the product as a single fraction:

(a) $(2\tfrac{1}{3})(\tfrac{3}{7})$

(b) $(1 + \tfrac{1}{x})(\tfrac{x}{2})$

(c) $\dfrac{x^2 - 4}{x^2 - x - 2} \cdot \dfrac{x + 1}{x^2 + 5x + 6}$

(d) $\dfrac{3x - 6}{x} \cdot \dfrac{x^2 - 2x}{x^2 - 4x + 4}$

5. Write the quotient as a single fraction:

(a) $\dfrac{\dfrac{3}{10}}{\dfrac{2}{7}}$

(b) $\dfrac{\dfrac{x}{y}}{\dfrac{z}{w}}$

(c) $\dfrac{2\tfrac{3}{4}}{1\tfrac{3}{8}}$

(d) $\dfrac{x - \dfrac{1}{x}}{x + \dfrac{1}{x}}$

6. Write each of the following as a single fraction:

(a) $\dfrac{5}{11} + \dfrac{3}{7}$

(b) $\dfrac{2}{x - 1} + \dfrac{5}{x + 3}$

(c) $\dfrac{4}{2x+1} - \dfrac{1}{x + 1}$

(d) $\dfrac{3}{xy} - \dfrac{2}{xz}$

(e) $\dfrac{a}{5} - \dfrac{b}{4} + \dfrac{c}{3}$

When adding or subtracting fractions whose numerators have more than one term, enclose those numerators in parentheses to help avoid errors.

Example 1. Write as a single fraction:

$$\frac{3}{1 - x} - \frac{1 + x}{2}$$

Hint: Write
(a) $2\tfrac{1}{3} = 2 + \tfrac{1}{3}$
$= 2 \cdot \tfrac{3}{3} + \tfrac{1}{3}$
$= \tfrac{7}{3}$
(b) $1 \tfrac{1}{x} = 1 \cdot \tfrac{x}{x} + \tfrac{1}{x}$
$= \tfrac{x + 1}{x}$

Solution. Find the lowest common denominator, which is 2(1-x); then write equivalent fractions:

$$\frac{3}{1-x} - \frac{1+x}{2} = \frac{3}{1-x} \cdot \frac{2}{2} - \frac{1+x}{2} \cdot \frac{1-x}{1-x}$$

$$= \frac{(3)(2)}{(1-x)(2)} - \frac{(1+x)(1-x)}{(2)(1-x)}$$

$$= \frac{(3)(2) - (1+x)(1-x)}{2(1-x)}$$

$$= \frac{6 - (1-x^2)}{2(1-x)}$$

$$= \frac{6 - 1 + x^2}{2(1-x)}$$

$$= \frac{5 + x^2}{2 - 2x}$$

7. Write as a single fraction:

(a) $\dfrac{5}{1-x} - \dfrac{3}{x}$ (b) $\dfrac{2}{x-3} + \dfrac{3}{x+5}$ (c) $\dfrac{x+3}{6} - \dfrac{x-2}{4}$

(d) $\dfrac{1+x}{1-x} - \dfrac{1-x}{1+x}$ (e) $\dfrac{x^2}{x+3} - \dfrac{9}{x+3}$ (f) $\dfrac{3}{x-y} - \dfrac{4}{x+y}$

(g) $\dfrac{4y}{4y^2-1} - \dfrac{3}{6y-3}$ (h) $\dfrac{x}{x-y} - \dfrac{y}{x+y}$

8. Write as a single fraction:

(a) $\dfrac{\dfrac{2x+4y}{3xy}}{\dfrac{3x+6y}{6x^2y}}$ (b) $\dfrac{x-1}{1-\dfrac{1}{x}}$

(c) $\dfrac{x+\dfrac{1}{x}}{\dfrac{1}{x}-1}$ (d) $\dfrac{\dfrac{x}{y}-\dfrac{y}{x}}{\dfrac{x}{y}+\dfrac{y}{x}}$

9. The following expression occurs in deriving the quadratic formula. Write it as a single fraction.

$$\frac{b^2}{4a^2} - \frac{c}{a}$$

10. Solve for x:

(a) $7x = 0$ (b) $\dfrac{x}{3} = \dfrac{5}{9}$ (c) $\dfrac{x}{2} - \dfrac{3}{11} = 0$

Hint: (d)
Multiply each side
by abx.

(d) $\dfrac{1}{x} = \dfrac{1}{a} + \dfrac{1}{b}$ (e) $\dfrac{1}{x} - n = 2$

11. Solve for x:

(a) $\dfrac{2x}{3} + \dfrac{5x}{8} = 4$

(b) $\dfrac{7}{x - 3} = 1$

(c) $\dfrac{5}{x} - 2 = \dfrac{7 - x}{3x}$

(d) $\dfrac{x}{4} = \dfrac{x - 9}{2}$

(e) $\dfrac{6}{x - 1} = \dfrac{10}{1 - 3x}$

(f) $\dfrac{3}{x^2 - 4} - \dfrac{7}{x + 2} = \dfrac{9}{x - 2}$

Hint: Clear the equation of fractions by multiplying each side by the lowest common denominator.

12. Solve for x:

(a) $\dfrac{a}{x} + \dfrac{b}{x} + \dfrac{c}{x} = 1$

(b) $\dfrac{x}{a} = n$

(c) $\dfrac{1}{x} = \dfrac{1}{n}$

(d) $\dfrac{1}{x} - n = q$

(e) $\dfrac{x}{n} + m = m$

(f) $\dfrac{x - n}{x + n} = \dfrac{1}{7}$

13. Solve for x:

(a) $\dfrac{x - 2}{x + 3} = \dfrac{2x + 5}{2x - 1}$

(b) $\dfrac{5}{x} - \dfrac{3}{x} = 9$

(c) $\dfrac{2x - 7}{3x + 5} = \dfrac{4}{11}$

(d) $\dfrac{x - 5}{3} = \dfrac{x}{2}$

14. Solve for x:

(a) $x + \dfrac{1}{x} = 2$

(b) $x + 3 + \dfrac{2}{x} = 0$

(c) $2x + 1 = \dfrac{1}{x}$

(d) $6x = 5 + \dfrac{6}{x}$

Hint: Clear the equation of fractions. Then write: polynomial = 0.

ANSWERS TO ALGEBRA REVIEW

1. (a) $\dfrac{49}{33}$ (b) $\dfrac{mq + pn}{nq}$ (c) $\dfrac{7x - 3}{12}$ (d) $\dfrac{x^2 + 2xy - y^2}{xy}$

2. (a) $-\dfrac{5}{36}$ (b) $\dfrac{9n}{35}$ (c) $\dfrac{x - xy}{y^2}$ (d) $\dfrac{2p - 3m}{mnp}$

3. (a) $\dfrac{1}{5}$ (b) $\dfrac{3}{2}$ (c) $\dfrac{7}{22}$ (d) $-\tfrac{1}{2}$ (e) -1 (f) 1 (g) $\dfrac{x}{x + 1}$ (h) $\dfrac{7}{6}$

4. (a) 1 (b) $\dfrac{x + 1}{2}$ (c) $\dfrac{1}{x + 3}$ (d) 3 5. (a) $\dfrac{21}{20}$ (b) $\dfrac{xw}{yz}$

   (c) 2 (d) $\dfrac{x^2 - 1}{x^2 + 1}$ 6. (a) $\dfrac{68}{77}$ (b) $\dfrac{7x + 1}{x^2 + 2x - 3}$

   (c) $\dfrac{2x + 3}{2x^2 + 3x + 1}$ (d) $\dfrac{3z - 2y}{xyz}$ (e) $\dfrac{12a - 15b + 20c}{60}$

7. (a) $\dfrac{8x - 3}{x - x^2}$ (b) $\dfrac{5x + 1}{x^2 + 2x - 15}$ (c) $\dfrac{12 - x}{12}$ (d) $\dfrac{4x}{1 - x^2}$

   (e) $x - 3$ (f) $\dfrac{7y - x}{x^2 - y^2}$ (g) $\dfrac{1}{2y + 1}$ (h) $\dfrac{x^2 + y^2}{x^2 - y^2}$

8. (a) $\dfrac{4x}{3}$ (b) x (c) $\dfrac{x^2 + 1}{1 - x}$ (d) $\dfrac{x^2 - y^2}{x^2 + y^2}$

9. $\dfrac{b^2 - 4ac}{4a^2}$ 10. (a) $x = 0$ (b) $x = \dfrac{5}{3}$ (c) $x = \dfrac{6}{11}$ (d) $x = \dfrac{ab}{b + a}$

(e) $x = \dfrac{1}{2 + n}$ 11. (a) $x = \dfrac{96}{31}$ (b) $x = 10$ (c) $x = \dfrac{8}{5}$

(d) $x = 18$ (e) $= -\frac{1}{2}$ (f) $= -\dfrac{1}{16}$

12. (a) $x = a + b + c$ (b) $x = na$ (c) $x = n$

(d) $x = \dfrac{1}{n + q}$ (e) $x = 0$ (f) $x = \dfrac{4n}{3}$

13. (a) $x = -\dfrac{13}{16}$ (b) $x = \dfrac{2}{9}$ (c) $x = 9.7$ (d) $x = -10$

14. (a) $x = 1$ (b) $x = -1$ or $x = -2$ (c) $x = \frac{1}{2}$ or $x = -1$

(d) $x = \dfrac{3}{2}$ or $x = -\dfrac{2}{3}$

<table>
<tr><td>**5**
Algebra II</td><td># Functions</td></tr>
</table>

Suppose that a ball is thrown upward with a velocity of 50 feet per second.

How high will it go? At what time will it reach its maximum height? At what time will it hit the ground? The next few exercises will answer these questions and also introduce the concept of a function.

EXERCISES

1. Assume that the flight of the ball starts at ground level, which is height 0.

    (a) If there were no gravity, the ball would continue to rise at a constant velocity. In that case, after t seconds its height would be _____ feet.

    (b) A ball, when dropped from a position of rest, falls $16t^2$ feet in t seconds, due to the pull of gravity. Combine this fact with part (a) to show that the height of the thrown ball at time t is given by the formula

    $$h(t) = 50t - 16t^2.$$

    (c) According to the formula in (b), how high is the ball after 1 second? After 2 seconds? After 3 seconds?

(a) 50t feet

(b) We must accept this step as a law from physics.

(c) 1 second: 34 feet
3 seconds: 6 feet

* * *

EXAMPLE 1. A ball is thrown upward from the ground with a velocity of 96 feet per second. How high will it be after t seconds? After 2 seconds? At what time will it hit the ground?

Solution. The height, in feet, of the ball at time t is given by the formula

$$h(t) = 96t - 16t^2$$

The first term, 96t, is the height it would be if there were no gravity; the second term records the effect of gravity. After 2 seconds, that is, when t = 2, the ball is at the height

The original force (throwing) and gravity work in opposite directions.

$$h(2) = 96(2) - 16(2)^2$$
$$= 192 - 64$$
$$= 128 \text{ feet.}$$

When the ball hits the ground, h(t) = 0. To find when the ball hits the ground it is necessary to solve for t in the equation,

$$0 = 96t - 16t^2.$$

This can be solved by factoring:

$$0 = 16t(6 - t).$$

Thus,   16t = 0   or   6 - t = 0

                  t = 0   t = 6

The height is 0 at the moment the ball is thrown (t = 0) and 6 seconds later (t = 6). So the ball strikes the ground 6 seconds after it is thrown.

2. (a) Using the formula, h(t) = 50t - 16t², from Exercise 1 (b), copy and fill in the following table:

t	0.5	1	1.5	2	2.5	3
h(t)						

On the basis of the table, estimate when the ball will be highest.

(b) Copy and fill in the table.

t	1.4	1.5	1.6	1.7
h(t)				

On the basis of the table, estimate when the ball is highest.

(c) At what time will the ball strike the ground?

(d) From the data in (a) and (b) make a graph that shows how the height, h(t), depends on t. Use the horizontal axis to show time, t; use the vertical axis to show height, h(t).

"INPUT-OUTPUT" MACHINES

In the tables in Exercise 2, for each number (the time) in the first row, there is another number (the height) in the second row. The height depends on the time. Think of this as a <u>machine</u>. Given an <u>input</u> (the time), the machine produces an <u>output</u> (the height). For example, when you feed in 1, the machine tells the height at 1 second.

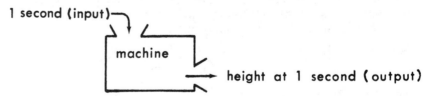

Call the machine "the height machine", and give it the symbol h. Then, rather than say that the output is "the height of the ball at 1 second", call it "the <u>value of h</u> when t is 1". The value of h when t = 1 is written h(1), which is read "h of 1". The picture above becomes

height machine

The use of a calculator saves time here.
(a) When t = 0.5, h(t) = 21; when t = 2.5, h(t) = 25.

(b) When t = 1.4, h(t) = 38.64; when t = 1.6, h(t) = 39.04. Highest sometime between t = 1.4 and t=1.6

(c) When t = 3.125 seconds

(d) Use different scales for the horizontal and vertical axes.

Note: h(1) is not multiplication; rather it is the output of the machine h when the input is 1.

3. (a) From your tables in Exercise 2, find: h(0.5), h(1.4), and h(2.5).

(b) If the machine is called m instead of h, what is m(1.4)?   m(2)?   m(3)?

(c) If the machine is called f, what is f(1.6)?

(d) If the machine is called g, what is g(1.4)?

(a) h(0.5) = 21
h(1.4) = 38.64
h(2.5) = 25

(c) f(1.6) = 39.04

Example 2. Let f be a machine defined by the formula $f(x) = x^4 + 3$.
Find f(0), f(1), f(2), and f(a).

Solution. $f(0) = (0)^4 + 3 = 3$,

$f(1) = (1)^4 + 3 = 4$,

$f(2) = (2)^4 + 3 = 19$,

$f(a) = (a)^4 + 3$.

4.   Machine f is defined by $f(x) = x^2 + 2x + 3$.   Find: (a)   f(0),   (b) f(2),

(c) $f(\frac{1}{3})$,   (d) f(-3),   (e) f(a),   (f) f(c).

5.   If the machine f is defined by $f(x) = 10^x$, find:

(a) f(2)   (b) f(0.3010)   (c) f(2.3010)   (d) the output
when the input is -1.

(a) 3

(c) $\frac{34}{9}$ or $3\frac{7}{9}$

(a) $a^2 + 2a + 3$

DOMAIN

The machines discussed so far accept any number on the number line as
an input. But consider the machine f defined by

$$f(x) = \frac{1}{x - 1} .$$

Clearly this machine does not work when the input is 1. The denominator
is 0. The number 1 is not an acceptable input for this particular machine.
However, all other numbers are acceptable.

DEFINITION. The set of all possible inputs for which the machine f
works (that is, provides an output) is called the <u>domain</u> of the machine.

For instance, the domain of the machine described by the formula

$$f(x) = \frac{1}{x - 1}$$

consists of all real numbers other than 1.

Example 3. When is the domain of the machine g defined by $g(x) = 1 + \frac{1}{x}$?

Solution. There is an output for every x except x = 0, since x = 0 leads
to division by zero. Thus, the domain consists of all numbers except 0.

6.   Find the domain of each of the following machines:

(a) f defined by $f(x) = \frac{x - 1}{x^2 + 1}$ (b) g defined by $g(x) = \sqrt{x}$

(c) h defined by $h(x) = \frac{1}{x^2 - 1}$ (d) k defined by k(x) = log x

*(c) all x except
1 and -1*

6. (e) f defined by $f(x) = \dfrac{1}{x^2 + 1}$ (f) h defined by $h(x) = \dfrac{x}{x + 2}$

(g) The machine that takes an input, cubes it, adds 3, and then divides the result by 5.

(e) all x

FUNCTION

It is customary to call a particular machine a "function". The following definition describes the basic concept.

MEMORIZE!
A function is a machine such that for each acceptable input, there is exactly one output.

> **DEFINITION.** A <u>function</u> is a method of assigning to each input from some set exactly one output.

If the input is x, the output is denoted

$$f(x) \quad \text{("the value of f at x" or "f of x").}$$

Example 4. Which of the following defines a function?

(a) $g(x) = 5$ (b) $f(x) = \dfrac{x - 1}{x + 1}$ (c) $h(x) = \pm\sqrt{x}$

Solution. (a) and (b) are functions, but (c) is not. A function must assign only one value to an input; however, h(4), for instance, leads to two outputs, 2 and -2. Thus, h is not a function.

7. Which of the following is <u>not</u> a function? Explain.

(a) $f(x) = \sqrt{x^2}$ (b) $f(x) = -\sqrt{x^2}$ (c) $f(x) = \pm\sqrt{x^2}$

(d) $f(x) = 4$ (e) $f(x) = x^2 - (x + 2)(x - 2)$ (f) $f(x) = 3^x$

The concept of a function was already used in previous chapters. For instance, the functions

$$f(x) = 10^x,$$
$$g(x) = \log x ,$$
$$\text{and } m(x) = P(x),$$

where P is a polynomial, have been examined extensively.

MEMORIZE!
Later in the chapter it will sometimes be useful to restrict the domain of certain functions.

> **DEFINITION.** The <u>domain</u> of a function is the set of inputs for which the function provides an output.

Example 5. What is the domain of each of the following functions?

(a) $f(x) = 13$ (b) $f(x) = \log_2 x$ (c) $f(x) = \dfrac{1}{2x - 1}$

Solutions.

(a) all numbers x   (b) all x > 0   (c) all x except $x = \frac{1}{2}$

8. What is the domain of each of the following functions?

(a) all except ±1
(c) all x > 0
(e) all x
(g) all x except x=1 and x=2

(a) $f(x) = \dfrac{x^2 + 1}{x^2 - 1}$ (b) $f(x) = \dfrac{x^2 - 1}{x^2 + 1}$ (c) $f(x) = \log x$

(d) $f(x) = \dfrac{1}{1 - \log x}$ (e) $f(x) = 3$ (f) $f(x) = \dfrac{3x - 3}{x - 1}$

(g) $f(x) = \dfrac{1}{x^2 - 3x + 2}$ (h) $f(x) = \sqrt{1 - x^2}$

GRAPH OF A FUNCTION

In earlier chapters, graphs were drawn of $y = 2^x$,

$$y = \log_{10} x,$$

$$\text{and } y = P(x), \text{ where } P(x) \text{ is a polynomial.}$$

Any function f involving numbers has a graph, although it may be difficult, or impossible, to sketch.

> DEFINITION. The graph of the function f consists of all points
> $$(x \, , \, y)$$
> where x is in the domain of f and
> $$y = f(x).$$

Chapter 3 included the graphing of quadratic and cubic polynomials. The next example presents the graph of a function whose formula is given by a first degree polynomial.

Example 6. Graph $f(x) = 2x - 1$.

Solution. First make a table to obtain a few points on the graph.

x	f(x)	points on the graph
-2	2(-2) - 1 = -5	(-2, -5)
-1	2(-1) - 1 = -3	(-1, -3)
0	2(0) - 1 = -1	(0, -1)
1	2(1) - 1 = 1	(1, 1)
2	2(2) - 1 = 3	(2, 3)

The graph passes through the five points computed, as illustrated.

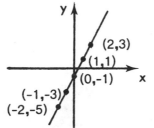

Graph of
$f(x) = 2x - 1$

It can be shown that for any constants a and b, the graph of $f(x) = ax + b$ is a straight line.

9. Graph each of these functions. In each case the graph will be a line. Though only two points are needed to determine a line, plot three points, using one as a check.

(a) $f(x) = x + 2$ (b) $f(x) = 1 - x$ (c) $f(x) = 2x + 3$

(d) $f(x) = -3x$ (e) $f(x) = \dfrac{x}{2} - 1$ (f) $f(x) = -x$

* * *

To provide another illustration, the next example discusses a function which has different formulas at different parts of its domain.

Example 7. Let

$$f(x) = \begin{cases} -x, & \text{if } x < 0 \\ 1 + x, & \text{if } 0 \le x \le 1 \\ 3, & \text{if } x > 1 \end{cases}$$

Graph the function f.

Solution. For x < 0, f(x) = -x. This is part of the line y = -x. For 0 ≤ x ≤ 1, the graph is part of the line y = 1 + x. For x > 1, the graph is part of the horizontal line y = 3. This is the graph of f:

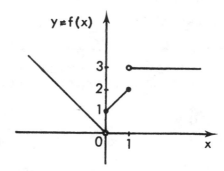

The points marked • (the filled-in dots) are present in the graph. The points marked ° (the hollow dot) are not points of the graph.

ABSOLUTE VALUE FUNCTION

The <u>absolute value</u> of a number is the distance (a nonnegative number) that the number is from zero on the number line.

> ABSOLUTE VALUE OF A NUMBER
> |x| = x, if x ≥ 0
> but |x| = -x, if x < 0

For instance, |-2| = 2 and |5| = 5.

10. Let f(x) = |x|

(a) Find f(0), f(1), f(-1), f(2), f(-2), f(-0.5), f(3.1)

(b) Sketch the graph of y = f(x).

(b) v-shaped

The function in Exercise 10 is usually called the "absolute value function."

FUNCTIONS DEFINED BY A TABLE

If the domain of a function has only a few elements, it is often more convenient to describe the function by means of a table. The first row is the input and the second row is the output. (In Exercise 2, such a table was shown.)

Functions defined by tables are common in daily life. Cashiers may use a printed table for the sales tax corresponding to the amount of purchase. The telephone directory lists various cities (the inputs) along with their area codes (the outputs). It also lists names of people (the inputs) and their telephone numbers (the outputs).

11. Give three more everyday examples of functions defined by tables, stating the input and output of each function. Compare your examples with those of your classmates.

12. It costs 44¢ per half ounce, or fraction thereof, to send a letter airmail to Europe. Let P(x) be the postage required to send a letter weighing x ounces by airmail to Europe.

    (a) What is the domain of the function P?
    (b) Sketch the graph y = P(x) for 0 < x ≤ 4.
    (c) Make a table for x ≤ 4 that a post office clerk might find useful.

This is sometimes called a "step function" because of the shape of the graph. Where will the hollow dots be? Take care that your graph shows clearly that there is exactly one output for each input.

A table defines a function if for each input there is <u>exactly</u> one corresponding output. Thus these two tables both define functions.

input	1	2	3
output	1	2	3

input	1	2	1	3
output	4	1	4	2

However, the following table does not define a function:

input	1	2	1	3
output	4	1	2	2

NOT A FUNCTION

There is more than one output when the input is 1. By the definition of function, there can be only one output for a single input.

FUNCTIONS DEFINED BY A GRAPH

As stated earlier, when graphing a function the y-coordinate is used to record f(x). Thus

$$f(x) = 2x - 3,$$

is expressed as

$$y = 2x - 3.$$

The graph is then

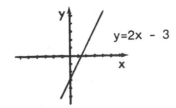

y=2x - 3

13. Graph each function on a separate coordinate system.
    (a) $f(x) = -3x + 4$ (b) $f(x) = x^2$

    (b) $f(x) = 2^x$ (d) $f(x) = \log_2 x$ (where x > 0)

Use the x-axis for inputs and the y-axis for outputs.

14. (a) How many times does a vertical line cut each graph in Exercise 13?

    (b) How can you decide whether a curve drawn in the xy-plane describes a function?

    (c) How do (a) and (b) relate to the definition, "for each input there is exactly one output?"

15. Which of these are graphs of functions?   Which are not?

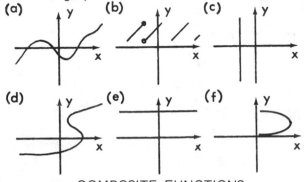

COMPOSITE FUNCTIONS

Functions can be connected to form an assembly line of two or more functions.   In the following diagram, the output of f becomes the input of g:

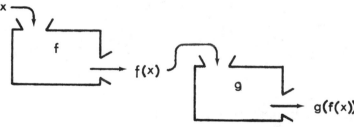

Note: If we call the function (machine) f the output is f(x). Similarly for g. Do not confuse the name of the machine with the output.

The total effect is to make a new function that takes an input x and produces an output g(f(x)) in two steps.   First f(x) is computed.   Then f(x) is fed into the g-machine as an input.

Example 8.   In the above diagram let f(x) = x - 3 and g(x) = x². Find

(a) f(5)   (b) g(f(5))   (c) g(f(2))

Solution.

(a) f(5) = 5 - 3   (b) g(f(5)) = g(2)   (c) g(f(2)) = g(2 - 3)

         = 2   = 2²   = g(-1)

                                     = 4   = (-1)²

                                                                   = 1

16. Let the functions f and g be defined by f(x) = 2x + 1 and g(x) = x².

    (a) What is f(2)?   What is g(f(2))?   (The input of the function g is f(2).)

    (b) What is g(f(3))?   g(f(-2))?   g(f(1))?   g(f(-1))?   g(f(x))?

    (c) Of course we could reverse the roles, and put x into g first and

a) g[f(2)] = 25

c) f[g(2)] = 9
  g[f(2)] = 25

then the output of g into f:

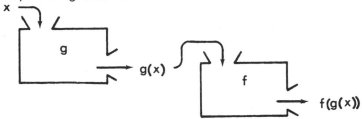

Will the answers change?   What is f(g(2))?   Does it equal g(f(2))?

(d)   Compare: f(g(-2)) and g(f(-2)); f(g(x)) and g(f(x)).   What do you conclude?

(d) $g[f(x)] =$
$4x^2 + 4x + 1$
$f[g(x)] =$
$2x^2 + 1$

* * *

A function obtained by connecting two functions, f and g, is called a
"composite function."   It is "composed" from other functions.   Of special
interest is the case where g reverses the action of f.   This case, of great
importance in trigonometry, will now be examined.

17.   Let the functions f and g be defined by: $f(x) = 2x$ and $g(x) = \dfrac{x}{2}$.

(a)   What is g(f(2))?   g(f(3))?   g(f(-7))?

(b)   Consider

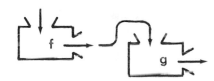

If the input of f is x, what is the output of g?

(c)   Suppose that the order of g and f is reversed:

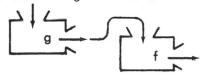

What is f(g(2))?   f(g(3))?   f(g(-7))?

(d)   Compute g(f(2)), g(f(3)), and g(f(-1)).

(e)   Compute f(g(4)), f(g(7)), and f(g(-5)).

(f)   Fill in the blanks: For all x,
        g(f(x)) = _____   and   f(g(x)) = _____

(f) Do not guess
here. Show
your work.

The functions f and g in Exercise 17 undo the effect of each other.
Two functions that are related this way are given a special name.

DEFINITION.   The functions f and g are said to be <u>inverses</u> of each
other if
        (i) f(g(x)) = x for all x in the domain of g,
    and
        (ii) g(f(x)) = x for all x in the domain of f.

Observe there are
two conditions for
f and g to be
inverses of each
other.

Thus, if f and g are a pair of inverse functions, f is called the <u>inverse</u> of g and g is called the <u>inverse</u> of f.

Example 9.   Consider the functions

$$f(x) = 3x \quad \text{(the "multiply by 3" function)}$$

and

$$g(x) = \frac{x}{3} \quad \text{(the "divide by 3" function)}.$$

Show that f and g are inverses of each other.

Solution.
$$f(g(x)) = 3(\frac{x}{3})$$
$$= x$$

and

$$g(f(x)) = \frac{1}{3}(3x)$$
$$= x$$

Thus, in Example 9, <u>the two functions are inverses; each undoes the effect of the other.   The value of the final output equals that of the original input</u>, <u>x</u>.

Example 10.   Consider the functions $h(x) = 3x - 4$ and $k(x) = \frac{x + 4}{3}$

Show that h and k are inverses of each other.

Solution.
$$h(k(x)) = 3(\frac{x + 4}{3}) - 4$$
$$= x + 4 - 4$$
$$= x$$

and

$$k(h(x)) = \frac{(3x - 4) + 4}{3}$$
$$= \frac{3x - 4 + 4}{3}$$
$$= x$$

Thus, h and k are inverses of each other.

The following is an everyday example of inverse functions. Define the function f by

f(person) = that person's Social Security number.

(The domain of f is the set of people who have Social Security numbers.)

Define a function g, whose domain is the set of all Social Security numbers, by

g(Social Security number) = the person with that Social Security number.

Note that the input of f is the person, not the name of the person. (Several persons have the same name.)

For instance, hook up the functions so that g is applied first and f afterwards.   Note what happens to the input 052-14-2677, which is the Social Security number of Alex Smart, a particular person.

052-14-2677

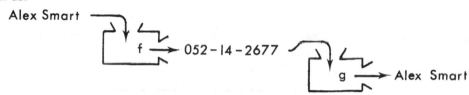

The original input for g is then the output of f.

Next, hook up the functions, g and f, so that f is first and g is applied afterwards.

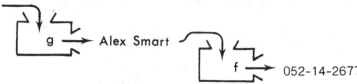

Note that the original input of f, Alex Smart, becomes the final output of g. Each of the two functions reverses the effect of the other. Alex Smart, with social security number 052-14-2677 is typical:

$$f(g(052\text{-}14\text{-}2677)) = f(\text{Alex Smart}) = 052\text{-}14\text{-}2677$$
and
$$g(f(\text{Alex Smart})) = g(052\text{-}14\text{-}2677) = \text{Alex Smart}.$$

The next two exercises explore further the notion of inverse function.

18. Use the definition of inverse functions to show that f and g are inverses of each other.

   (a) If $f(x) = 2x - 1$ and $g(x) = \dfrac{x + 1}{2}$,

   (b) If $f(x) = \dfrac{1}{x - 1}$ and $g(x) = 1 + \dfrac{1}{x}$.

19. (a) Use the definition of inverse functions to show that
$$f(x) = 2x - 3$$
   and
$$g(x) = -3x + 2$$
   are <u>not</u> inverses of each other.

   (b) Are
$$h(x) = x + 1$$
   and
$$g(x) = x - 1$$
   inverses of each other?   Explain.

20. (a) Show that the function $f(x) = x^3$ has an inverse function. What is the formula for the inverse function?

   (b) Show that the function $f(x) = x^2$ does not have an inverse.

   (c) Show that the function $f(x) = x^2$ <u>does</u> have an inverse if the domain is restricted to $x \geq 0$.   What is the formula for the inverse function?

THE GRAPH OF INVERSES AND REFLECTION OF POINTS ACROSS THE LINE y = x

If f and g are inverses, their graphs are related in a special way. The next few exercises explore this relation.

21. (a) Copy and fill in the table for $f(x) = x^3$:

x	-2	1	2	3
f(x)				

(b) Hint: See Exercise 20(a), the formula for g, the inverse of f.

(b) Use (a) to find four points on the graph of the inverse of $f(x) = x^3$.

22. As shown in Exercise 18, the functions $f(x) = 2x - 1$ and $g(x) = \dfrac{x + 1}{2}$ are inverses of each other.

(a) Fill in this table for f(x):

x	-3	0	1	5
f(x)				

(b) Fill in this table for g(x):

x	-7	-1	1	9
g(x)				

(c) Plot the points in the tables in (a) and (b) on the same coordinate axes.

(d) How are the four points in (a) related to the four points in (b)?

23. Graph the line y = x. On the same coordinate axes plot the following points:

(a) (2,1) and (1,2)   (b) (2,5) and (5,2)
(c) (-2,3) and (3,-2)   (d) (-4,-7) and (-7,-4)

(e) If you are not sure, choose some other points.

(e) From what you observe in parts (a) through (d), convince yourself that the points (a,b) and (b,a) are <u>mirror reflections</u> with respect to the line y = x.

Think of folding the graph along the line y = x. Convince yourself that the following points determine a square.

What is true about the diagonals of a square?   * * *

Consider the following "function machine" diagram where f and g are inverses:

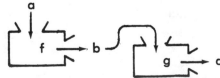

and

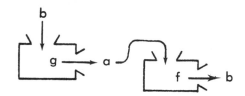

Since:
f(a) = b and
g(b) = a, then
g[f(a)] = a and
f[g(b)] = b.

This above diagram shows that if (a,b) is a point on the graph of f, then (b,a) is a point on the graph of g. The two graphs are mirror images of each other with respect to the line y = x. (See Exercise 23.)

STUDY CAREFULLY THIS DIAGRAM AND EXERCISE 23. ASK QUESTIONS!

INVERSE OF AN EXPONENTIAL FUNCTION

The next exercise returns to the exponential function of Chapter 2.

24. (a) Copy and fill in the table for $f(x) = 2^x$:

x	-3	-2	-1	0	1	2	3
f(x)							

IMPORTANT! This exercise shows the need for logarithmic notation; we need to be able to write the inverse of the exponential function in g(x) notation.

(b) Copy and fill in the table for g, the inverse of f.

x	$\frac{1}{8}$	$\frac{1}{4}$	$\frac{1}{2}$	1	2	4	8
g(x)							

(c) Show that the formula for g (in terms of x) is

$$g(x) = \log_2 x.$$

(d) Use the definition of inverse functions to show that your answer in (c) is the inverse of $f(x) = 2^x$.

(e) If given a point on the graph of f, how could you immediately plot a point on the graph of g, and vice versa?

(f) If the graph $y = 2^x$ were printed on a page in <u>wet ink</u>, how could you quickly sketch the graph of $y = \log_2 x$?

The important idea of Exercise 24 is that the inverse of an exponential is a logarithm function. The following list summarizes some pairs of functions that are inverses of each other:

Function f	Inverse Function g
"doubling"	"halving"
"cubing"	"cube root"
exponential (base 2)	log (base 2)
exponential (base 10)	log (base 10)

25. Find in g(x) notation the inverse of each of the following functions:

   (a) $f(x) = \log_2 x$ (b) $f(x) = \log_{10} x$ (c) $f(x) = 3x$

   (d) $f(x) = x - 3$ (e) $f(x) = 10^x$ (f) $f(x) = x^2;\ x \geq 0$

THE FORMULA FOR THE INVERSE OF A FUNCTION

Recall: f and g are merely the names of the functions whose outputs are f(x) and g(x) respectively. We write f(x) = y because the output f(x) is graphed on the y-axis. Similarly, g(x) = y, because the output g(x) is graphed on the y-axis when we graph the function g.

Consider two functions f and g that are inverses of each other. The next example shows how to obtain the formula for g from f and express it in terms of x and y.

Example 11. Let f be the function defined by the formula

$$f(x) = 3x - 1.$$

Find the formula for the function g that is inverse to f.

Solution. Since f(x) = y, the formula for f is

$$f(x) = 3x - 1$$

$$\text{or } y = 3x - 1.$$

The input for f is x and output is y. The function g starts with an output of f and determines what the input was. In other words, given the number x, it solves for y in the equation

$$x = 3y - 1. \quad \text{(The roles of x and y are reversed.)}$$

A little algebra finds y in terms of x:

$$x = 3y - 1$$

$$x + 1 = 3y$$

$$\frac{x + 1}{3} = y$$

$$y = \frac{x + 1}{3}$$

Thus g is given by the formula,

$$g(x) = \frac{x + 1}{3}$$

In general, to find the formula for g, starting with the formula for f, consider the following steps:

STUDY THESE STEPS CAREFULLY! MEMORIZE! The key steps are 2 and 3 for they apply to functions that are not in f and g notation.

HOW TO FIND THE FORMULA FOR AN INVERSE

1. Write the formula for f in the form y = f(x).

2. <u>Switch</u> x and y in the formula for f.

3. <u>Solve</u> for the new y in terms of x.

4. Write the formula for g in the form new y = g(x).

NOTE: In some texts, f^{-1} is used to denote the inverse of f, the function that we prefer to denote by g. (If f(a) = b, then f^{-1}(b) = a.) In this usage, do not treat "-1" as an exponent; it has the meaning of "inverse".

We prefer to avoid f^{-1} notation in this text.

26. In each case find the formula for the inverse function y = g(x) if

   (a) f(x) = 4x   (b) $f(x) = \dfrac{3x - 2}{5}$

   (c) f(x) = 2x - 3   (d) f(x) = x^3

(a) $g(x) = \frac{x}{4}$

(c) $g(x) = \frac{x + 3}{2}$

27. Express the inverse of each of the following functions in the form:

$$y = \text{a formula involving } x$$

   (a) f(x) = 10^x (b) f(x) = $\log_{10}x$

   (c) f(x) = x^5 (d) f(x) = $x^3 + 4$

(a) $y = \log_{10}x$

HORIZONTAL AND VERTICAL LINES

When we graph a function f(x) = y, a vertical line cuts the graph in at most one point (i.e., for each input x, there is exactly one output, f(x) - y). Now, if (a,b) is a point on the graph of f, then (b,a) is a point on the graph of g, the inverse of f. Hence, we may make the following observations:

   (i) A <u>vertical line</u> cuts the graph of a function in at most one point.

STUDY CAREFULLY!

   (ii) If a function has an inverse (not all functions do!), then a <u>horizontal line</u> cuts the graph of the function in at most one point.

See Exercise 23 again.

28. Consider the graph of f(x) = x^2.

   (a) Is f a function? Explain.

   (b) How many times does any vertical line cut the graph? How is this related to (a)?

   (c) How many times may a horizontal line cut the graph?

   (d) Explain why f(x) = x^2 does (or does not) have an inverse function.

The preceding discussion and exercise provide a way of deciding by a glance at a graph whether the function has an inverse.

29. The opening exercise concerns the function given by the formula

$$h(t) = 50t - 16t^2.$$

   (a) Sketch the graph and show that h does <u>not</u> have an inverse.

   (b) If the domain of h is restricted to those t such that $0 \le t \le \dfrac{25}{16}$ show from the graph that h has an inverse. Find the formula for the inverse in the form

$$t = \text{an expression in } y.$$

   Note. The inverse function tells at what time t the ball reaches a specified height y.

* * * SUMMARY * * *

This chapter introduced the idea of a function, a tool of great importance in trigonometry and calculus. It may be thought of as a machine in which for each possible input there is produced a single output. In practice, a function is usually given by a formula, such as $f(x) = 10^x$. Associated with a numerical function f is a table that displays the actual output for each input. From such a table a graph of the function can be sketched using the x-axis for the inputs and the y-axis for the outputs.

The notion of composite functions was introduced. Of special importance was the concept of a function f having an inverse g. Not every function has an inverse. In fact, a function has an inverse only if any horizontal line crosses the graph at most once. The inverse of the function $y = 10^x$ is $y = \log x$. The function x^2 has an inverse if x is restricted to being a non-negative number; then its inverse is $\sqrt{x}$.

The graphs of a function and its inverse function have simple geometric relation. Each is the reflection of the other across the line $y = x$. This relation comes from the fact that if the point (a,b) is on the graph of a function, then the point (b,a) is on the graph of its inverse.

* * * WORDS AND SYMBOLS * * *

input	output
function	domain
f(x)	inverse
the value of f at x	composite function
f of x	graph of a function

* * * CONCLUDING REMARKS * * *

One encounters functions every day. Even the daily newspaper is full of functions. One such function is the table that lists as inputs the cities of the United States and as outputs the temperature at each city.

The study of arithmetic in elementary school is in fact the study of certain numerical functions. For instance, addition is a certain function f that assigns to each pair of numbers their sum. Instead of writing

$$f(3,7) = 10$$

we write

$$3 + 7 = 10.$$

Multiplication is another function that assigns to each pair of numbers their product. Thus

$$f(3,7) = 21.$$

This is usually written,

$$3 \cdot 7 = 21.$$

It is rather odd that arithmetic is devoted to more complicated functions than the functions examined in earlier chapters, $y = b^x$ and $y = \log_b x$. Here b is fixed; the "inputs" are only single numbers for x.

Trigonometry is the study of certain functions connected with a circle. These functions have single numbers as inputs and outputs. In a certain sense trigonometry is easier than arithmetic. But in any case, the study of mathematics, either elementary or advanced, is really the study of certain functions.

* * * FUNCTIONS SELF-QUIZ * * *

1. State the domain of the following functions:

(a) $f(x) = \dfrac{2x - 2}{x - 1}$ (b) $f(x) = \dfrac{1}{1 - 2^x}$

(c) $f(x) = -\sqrt{x^2}$ (d) $f(x) = 3 + \log x$.

2. Given $f(x) = x + 2$ and $g(x) = x^2 - 1$, find

(a) $f(1)$, $g(3)$, $f(8)$,   and $g(-4)$

(b) $f(g(1))$, $g(f(1))$, $f(g(2))$,   and $g(f(0))$

(c) $f(g(x))$, $g(f(x))$.

3. Show that the functions f and g are inverses of each other.

(a) $f(x) = 2 + 3x$, $g(x) = \dfrac{x - 2}{3}$ (b) $f(x) = 2^{x-1}$, $g(x) = 1 + \log_2 x$

4. Which of the following are graphs of functions? Which of these are graphs of a function with an inverse? How do you decide?

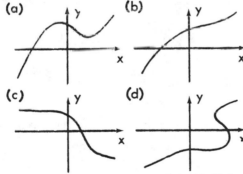

5. Find the formula for g, the inverse of f, if $f(x) = 5x + 3$.

ANSWERS TO SELF QUIZ

1. (a) all x except x = 1, (b) all x except x = 0,   (c) all x   (d) x > 0.
2. (a) $f(1) = 3$, $g(3) = 8$, $f(8) = 10$, $g(-4) = 15$
    (b) $f(g(1)) = 2$, $g(f(1)) = 8$, $f(g(2)) = 5$, $g(f(0)) = 3$
    (c) $f(g(x)) = x^2 + 1$, $g(f(x)) = x^2 + 4x + 3$
3. Show $f(g(x)) = x$ and $g(f(x)) = x$.   4. (a), (b), and (c) are functions,
    (d) is not.   (a) has no inverse, (b) and (c) do.

5. $g(x) = \dfrac{x - 3}{5}$

* * * FUNCTIONS REVIEW * * *

1. (a) Graph h   such that

$$h(x) = \begin{cases} x^2 \text{ if } x \geq 1 \\ 2 - x \text{ if } x < 1 \end{cases}$$

   (b) Is h a function?   Explain.

   (c) Does h have an inverse?   If not, why not?   If so, what is the formula defining g, the inverse of h?

2. Let f be the function defined by $f(x) = \frac{x + 1}{x^2 - 1}$.   Find

   (a) f(3)   (b) f(0)   (c) f(-3)   (d) What is the domain of f?

3. Let g be the function defined by $g(x) = \begin{cases} x^2 - 1 \text{ if } x \geq 0 \\ 2 + x \text{ if } x < 0. \end{cases}$

   (a) What is the domain of g?
   (b) Find: g(0), g(- ½),   g(g(-1)).
   (c) For what x is g(x) = 0?
   (d) Sketch y = g(x).

4. Find the domain of each of the following functions:
   (a) $f(x) = \sqrt{x + 2}$ (b) $g(x) = 1 + \log x$
   (c) $h(x) = \frac{1}{\sqrt{x}}$ (d) $k(x) = \frac{1}{2 - \log_2 x}$

5. Show that f and g are inverses of each other if
   (a) $f(x) = 2x + 7$ and $g(x) = \frac{x - 7}{2}$
   (b) $f(x) = 1 + 2^x$ and $g(x) = \log_2(x - 1)$ and x > 1
   (c) $f(x) = x^4$ and $g(x) = x^{\frac{1}{4}}$ for $x \geq 0$.

6. In this exercise, all curves and graphs are in the xy-plane.

   (a) If a vertical line cuts a curve more than once, then that curve is
   not the graph of a function.   Explain.

   (b) Fill in the blanks: a graph is that of a function if any _____
   line intersects the graph in at most _____.

   (c) If a graph is that of a function, then the function has an inverse
   if each _____ line intersects the graph in at most _____.

7. (a) Draw a graph of a function which does not have an inverse.

   (b) Find a horizontal line which intersects the graph of the function
   in (a) more than once.

8. Which of the following are graphs of functions?   Explain.

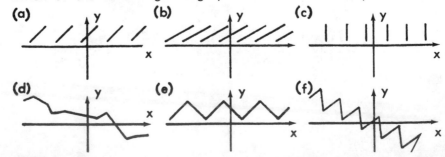

(g)

(h)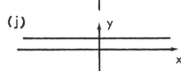

(i)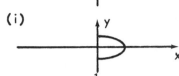

(j)

9. Let $f(x) = \dfrac{1}{1 - x}$.

    (a) Find $f(2)$, $f(f(2))$, $f(f(f(2)))$.

    (b) $f(f(2))$ can be sketched

    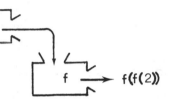

    Sketch $f(f(f(2)))$.

    (c) Find $f(3)$, $f(f(3))$, $f(f(f(3)))$.

    (d) Find $f(f(x))$.

    (e) Show that the function g defined by $g(x) = \dfrac{x - 1}{x}$ is the inverse of $f(x) = \dfrac{1}{1 - x}$.

10. (a) Let b be a positive number. Use the definition of the inverse to show that f and g are inverses of each other, if $f(x) = b^x$ and $g(x) = \log_b x$.

    (b). Use (a) to explain why
    $$b^{\log_b x} = x \qquad \text{and} \qquad \log_b b^x = x.$$

11. Find the formula for g, the inverse of f, if f is defined by

    (a) $f(x) = 7x - 9$ (b) $f(x) = \dfrac{x}{10}$ (c) $f(x) = \log_5 x$

12. If the point (m,n) is on the graph of
    $$y = 2^x,$$
    then $(?,?)$ is on the graph of
    $$y = \log_2 x.$$

13. Match each curve with the appropriate statement.

    (a)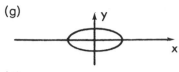

    (b)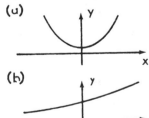

    (i) graph of a function that does not have an inverse

    (ii) graph of a function that has an inverse

(c)

(iii) not the graph of a function

ANSWERS TO REVIEW

1.   (a)

$h(x) = x^2$ if $x \geq 1$

$h(x) = 2 - x$ if $x \leq 1$

(b)   Yes, there is only one value of h(x) for each x.

(c)   There is no inverse, for a horizontal line cuts the curve twice.

2.   (a) $\frac{1}{2}$,   (b) -1,   (c) $-\frac{1}{4}$,   (d)   All real numbers except 1 and -1.

3.   (a) all x,   (b) -1, $1\frac{1}{2}$, 0   (d)

(c) x = -2, 1

$(0,2)$ $(0,-1)$

$g(x) = x^2 - 1$ if $x \geq 0$

$g(x) = 2 + x$ if $x < 0$

4.   (a) $x \geq -2$ (b) x > 0   (c) x > 0   (d) x > 0 except x = 4

5.   (a), (b), (c) show f(g(x)) = g(f(x)) = x   6. (a) There is only one output
for each input.   (b) vertical, one point   (c) horizontal, one point.

7.   (a) student example (e.g., f(x) = x²),   (b) student work

8.   (a), (d), (e), (h), (j) are functions, the rest are not.

9.   (a) -1, $\frac{1}{2}$, 2   (b) 2

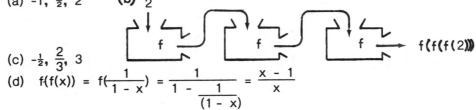

(c) $-\frac{1}{2}$, $\frac{2}{3}$, 3

(d) $f(f(x)) = f(\frac{1}{1-x}) = \dfrac{1}{1 - \dfrac{1}{(1-x)}} = \dfrac{x-1}{x}$

(e)   Show f(g(x)) = x = g(f(x))   [Note: g(x) is not defined at x = 0,
f(x) is not defined at x = 1.]

10. (a)   Since $f(x) = b^x$ and $g(x) = \log_b x$, then $f(g(x)) = b^{\log_b x} = x$ and
$g(f(x)) = \log_b b^x = x$.

(b)   The log and exponential functions are inverses.

11. (a) $g(x) = \dfrac{x+9}{7}$ (b) g(x) = 10x   (c) $g(x) = 5^x$.   12. (n,m)

13. (a) - (i)   (b) - (ii)   (c) - (iii)

* * * FUNCTIONS REVIEW SELF TEST * * *

1.   (a)   Define the domain of a function.   (b) Define a function.

2. If f and g are functions which are inverses of each other,
then f(g(x)) = _____ for all x in the domain of g.
and g(f(x)) = _____ for all x in the domain of f.

3. (a) If h is a function, a vertical line cuts the graph in at most how
many points?

(b) If h in part (a) has an inverse, a horizontal line cuts the graph in
at most how many points?

(c) How are the graphs of h and g related with respect to the line y = x?

4. (a) Draw a graph that is not the graph of a function.

(b) Draw the graph of a function that does not have an inverse.

(c) Draw the graph of a function that does have an inverse.

5. (a) If f and g are inverses of each other and f(x) = 4x + 3, then
g(x) = ____?____

(b) Show how to check your answer in (a).

6. If k(x) = 10^x and ℓ(x) = log x, then find

(a) k(2)   (b) ℓ(0.1)   (c) ℓ(k(-4))   (d) If (a,b) is on the graph of
ℓ then (? , ?) is on the graph of ____?____ .

7. If f and g are functions, and f(x) = x^2 - 1, g(x) = 2x - 5 find:

(a) g(f(x))   (b) f(g(x))   (c) g(f(2))   (d) f(g(-1)).

8. Show that f and g, defined by

$$f(x) = x^2 \quad \text{and} \quad g(x) = \sqrt{x},$$

are inverses of each other if x ≥ 0.

ANSWERS

1. See main section.   2. x, x.   3. (a) one, (b) one,
(c) They are (mirror) reflections across the line y = x.
4. Compare your work with classmates.
(a) It must violate Exercise 3(a). Compare.   (b) It must not violate
Exercise 3(a).   (c) As in (b), but at least one horizontal line must intersect the graph more than once.

5. (a) g(x) = $\frac{x - 3}{4}$. (b) Check that f(g(x)) = x and g(f(x)) = x.

6. (a) 100, (b) -1, (c) -4, (d) (b,a) is on the graph of k.
7. (a) $2x^2$ - 7, (b) $4x^2$ - 20x + 24, (c) 1, (d) 48.
8. Use the definition of inverses.

* * * FUNCTION PROJECTS * * *

1. (a) Let f(x) = 10^x. Show that f(m + n) = f(m)•f(n).

(b) What other functions satisfy the equation in (a)?

2. (a) Let f(x) = 10x. Show that f(m + n) = f(m) + f(n).

(b) What other functions satisfy the equations in (a)?

3. (a) Let $f(x) = \log_{10}x$. Show that $f(mn) = f(m) + f(n)$.

    (b) What other functions satisfy the equation in (a)?

4. (a) Let $f(x) = x^{10}$. Show that $f(mn) = f(m) \cdot f(n)$.

    (b) What other functions satisfy the equation in (a)?

5. Let $f(x) = \dfrac{1}{x}$

    (a) Show that f has an inverse function.

    (b) What is the inverse function?

6. Let $f(x) = 1 - x$.

    (a) Show that f has an inverse function.

    (b) What is the inverse function?

7. Let $f(x) = x^3 + x$.

    (a) Show that f has an inverse function, g.

    (b) Find $g(2)$ and $g(10)$.

    (c) Estimate $g(3)$.

8. Let $f(x) = \dfrac{1}{1 - x}$.

    (a) Find $f(2)$, $f(-1)$, and $f(\tfrac{1}{2})$.   (b) Find $f(3)$, $f(-\tfrac{1}{2})$, and $f(\tfrac{2}{3})$.

    (c) Find $f(f(x))$.   (d)   Find $f(f(f(x)))$.

9. (a) The function $f(x) = x^2$ has the property that $f(-x) = f(x)$. (Check
         this.) For which exponents n will the function $g(x) = x^n$ have the
         property that $g(-x) = g(x)$?

    (b) Show that if n is odd, the function $g(x) = x^n$ has the property that
         $g(-x) = -g(x)$.

    (c) A function, g whose domain is the real numbers, is called an even
         function if $g(-x) = g(x)$; it is called an odd function if $g(-x) = -\overline{g(x)}$.
         Why do you think that these terms are used?

    (d) Show that for any function f, whose domain is the set of real num-
         bers, the function
         $$g(x) = \frac{f(x) + f(-x)}{2}$$

         is even and the function
         $$h(x) = \frac{f(x) - f(-x)}{2}$$
         is odd.

    (e) Prove that any function f given in (d) is the sum of an even func-
         tion and an odd function.

10. Let f be a numerical function whose domain is the real numbers.
     Assume that $f(1) = 1$ and that for all pairs of number, m and n,
     $f(m + n) = f(m) + f(n)$.

     (a) Find $f(2)$, $f(3)$.   (b) Find $f(\tfrac{1}{2})$.

     (c) Find $f(0)$.   (d) Show that if x is rational $f(x) = x$.

* * * FIRST YEAR ALGEBRA REVIEW (OPTIONAL) * * *

INEQUALITIES

Let $\underline{a}$ and b be two numbers on the number line.

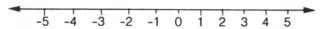

Assume that $\underline{a}$ is the number to the left of b. We say "a is less than b" and write

$$a < b.$$

Or we may say "b is greater than a" and write b > a.

Recall that when a < b, and c is any number, then

$$a + c < b + c \text{ and } a - c < b - c.$$

When a < b and c is positive, then

$$ac < bc \text{ and } \frac{a}{c} < \frac{b}{c}.$$

When a < b and c is negative, then

$$ac > bc \text{ and } \frac{a}{c} > \frac{b}{c}.$$

Multiplying or dividing by a negative number changes the order of the inequality. "<" becomes ">" and vice versa.

(Note that "<" is replaced by ">".)

1. If x < 1, which of the following is true?

    (a) 3x < 3

    (b) x – 1 < 0

    (c) x + 9 < 10

    (d) $\frac{x}{7} < \frac{1}{7}$

    (e) -2x < -2

    (f) $^{-}\frac{x}{4} < ^{-}\frac{1}{4}$

Inequalities may be solved by techniques like those for solving equations. However, there are two exceptions; multiplying or dividing by a negative number changes the order of the inequality.

Example 1. Solve for x:

$$\frac{-x}{3} + 5 < 4$$

Solution.

$$\frac{-x}{3} + 5 < 4$$

$$\frac{-x}{3} < -1 \qquad \text{(subtract 5)}$$

$$x > 3 \qquad \text{(multiply by -3)}$$

(Any number x larger than 3 satisfies the original inequality.)

2. Solve for x:

(a) $6x < 1$

(b) $x - 2 > 14$

(c) $\dfrac{x}{3} < 5$

(d) $\dfrac{x}{2} + 7 < -3$

(e) $-6x + 5 < 8$

(f) $9x - 5 < 2x + 4$

(g) $-\dfrac{x}{7} + 5x + 2 < 0$

(h) $-6x + \dfrac{2}{5} < 3$

3. If $1 < N$, compare

(a) N and N^2

(b) N^2 and N^3.

4. If $0 < N < 1$, compare

(a) N and N^2

(b) N^2 and N^3.

5. If $-1 < N < 0$, compare

(a) N and N^2

(b) N^2 and N^3.

SIMULTANEOUS LINEAR EQUATIONS

6. Graph each pair of equations. Estimate where each pair of lines cross each other. (Use a different coordinate system for each pair of equations.)

(a) $\begin{cases} x + 2y = 7 \\ x - y = 1 \end{cases}$

(b) $\begin{cases} x + 2y = -3 \\ 2x - y = 4 \end{cases}$

(c) $\begin{cases} x + 3y = 14 \\ 2x - y = 0 \end{cases}$

(d) $\begin{cases} 2x - y = 2 \\ 3x + 2y = 17 \end{cases}$

The following examples illustrate an easier way to find the point of intersection. No lines have to be graphed. Instead algebraic techniques determine precisely the point where they cross.

Example 2.   Solve for x and y:

$$\begin{cases} 2x - y = 2 \\ 3x + 2y = 17 \end{cases}$$

Use the curly brackets every time the two euqations are written together.

Solution.

$$\begin{cases} 2(2x - y) = 2(2) \qquad \text{(multiply by 2)} \\ 3x + 2y = 17 \end{cases}$$

This approach is called the "addition-subtraction" method.

$$\begin{cases} 4x - 2y = 4 \\ \underline{3x + 2y = 17} \qquad \text{(add)} \end{cases}$$

$$7x \qquad\quad = 21$$

$$x = 3$$

$$2(3) - y \ = 2 \qquad \text{(Solve for y by substituting}$$
$$6 - y = 2 \qquad x = 3 \text{ in one of the original}$$
$$-y = -4 \qquad \text{equations.)}$$
$$y \ = 4$$

The point of intersection is (3,4).

Check: $2(3) - 4 \overset{?}{=} 2$ $\qquad\qquad$ $3(3) + 2(4) \overset{?}{=} 17$

$\qquad\qquad$ $6 - 4 \overset{?}{=} 2$ $\qquad\qquad\qquad$ $9 + 8 \overset{?}{=} 17$

$\qquad\qquad\quad$ $2 = 2$ $\qquad\qquad\qquad\qquad$ $17 = 17$

$\qquad\qquad\quad$ yes $\qquad\qquad\qquad\qquad\quad$ yes

$\qquad$ The next example illustrates the "substitution" method for solving Example 2.

Example 3.   Solve for x and y:

$$\begin{cases} 2x - y = 2 \\ 3x - 2y = 17 \end{cases}$$

Solution.

$$\begin{cases} 2x - 2 = y \\ 3x + 2y = 17 \end{cases}$$ (Solve for y in one of the equations.)

Or you may solve for x and replace that value in the original equation.

$\qquad$ $3x + 2(2x - 2) = 17$ (Replace y with 2x - 2 in the lower

$\qquad\quad$ $3x + 4x - 4 = 17$ equation.)

$\qquad\qquad\qquad$ $7x = 21$

$\qquad\qquad\qquad$ $x = 3$

$\qquad$ Then substitute x = 3 in one of the original equations to solve for y. Check as in Example 2.

7. Solve Exercise 6 (a) and 6 (b) using the addition-subtraction method.

8. Solve Exercises 6 (c) and 6 (d) using the substitution method.

9. Solve for x and y:

(a) $\begin{cases} 3x + 3y = 3 \\ 3x - 2y = 13 \end{cases}$ $\qquad$ (b) $\begin{cases} 3x + 2y = 13 \\ 2x + 5y = 5 \end{cases}$

(c) $\begin{cases} 4x - 3y = -3 \\ 3x + 5y = -24 \end{cases}$ $\qquad$ (d) $\begin{cases} 5x + y = 11 \\ 3x + 2y = 8 \end{cases}$

10. Solve for x and y:

(a) $\begin{cases} \dfrac{x}{3} + \dfrac{y}{4} = 6 \\ 3x - 3y = 4 \end{cases}$ $\qquad$ (b) $\begin{cases} 3(y - 3) = 2(x + 2) \\ 3(x - 2) - 2(y + 1) = 0 \end{cases}$

(c) $\begin{cases} \dfrac{5x}{2} + \dfrac{7y}{4} + 1 = 0 \\ 3x + 5y + 1 = 0 \end{cases}$ $\qquad$ (d) $\begin{cases} 1 + \dfrac{x}{3} = \dfrac{y}{2} \\ 1 + \dfrac{x}{9} = \dfrac{y}{4} \end{cases}$

ANSWERS TO ALGEBRA REVIEW

1. (a), (b), (c), (d) $\qquad$ 2. (a) $x < \dfrac{1}{6}$ $\quad$ (b) $x > 16$ $\quad$ (c) $x < 15$ $\quad$ (d) $x < -20$

$\quad$ (e) $x > -\frac{1}{2}$ $\quad$ (f) $x < \dfrac{9}{7}$ $\quad$ (g) $x < -\dfrac{7}{17}$ $\quad$ (h) $x > -\dfrac{13}{30}$ $\quad$ 3. (a) $N^2 > N$

3. (b) $N^3 > N^2$ 4. (a) $N > N^2$ (b) $N^2 > N^3$ 5. (a) $N < N^2$

(b) $N^2 > N^3$

6. (a) (3,2)   (b) (1,-2)

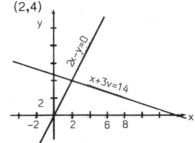

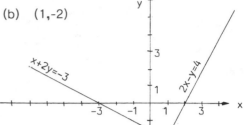

(c) (2,4)   (d) (3,4)

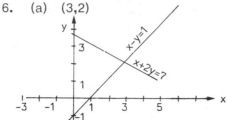

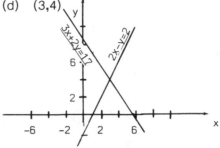

7. (a) $\begin{cases} x + 2y = 7 \\ \underline{x - y = 1} \\ 3y = 6 \end{cases}$ (b) $\begin{cases} x + 2y = -3 \\ 2x - y = 4 \end{cases}$

$y = 2$ $2x + 4y = -6$

$x + 4 = 7$ $\underline{2x - y = 4}$

$x = 3$ $5y = -10$

point (3,2) $y = -2$

$x - 4 = -3$

$x = 1$

point (1, -2)

8. $\begin{cases} x + 3y = 14 \\ 2x - y = 0 \end{cases} = \begin{cases} x + 3y = 14 \\ 2x + 0 = y \end{cases}$

substitute 2x for y: $x + 3(2x) = 14$

$x + 6x = 14$

$7x = 14$

$x = 2$

Substitute 2 for x: $2 + 3y = 14$

$3y = 12$

$y = 4$

point (2,4)

9. (a) (3,-2)   (b) (5,-1)   (c) (-3,-3)   (d) (2,1)   10. (a) $(\frac{76}{7}, \frac{200}{21})$

(b) (10,11)   (c) $(-\frac{13}{29}, \frac{2}{29})$ (d) (9,8)

* * *

* * * CENTRAL SECTION * * *

You are familiar with the fact that points on the number line, the x-axis, can be added (or subtracted) and multiplied (or divided, except for division by zero). Can we manage to do the same thing for points in the (x,y)-plane? That is, can we find a way of adding (or subtracting) and multiplying (or dividing) points in the plane? [The addition and multiplication must obey the usual rules of arithmetic, such as a(b+c) = ab + ac.] This chapter is devoted to answering this question.

* * *

A NEW NUMBER SYSTEM

The numbers that are used in arithmetic and algebra can be represented on a line, the so-called number line. The following diagram shows a few points and their corresponding numbers:

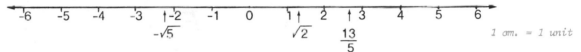

1 cm. = 1 unit

The number line is rich, for it represents all the integers,

$$. . ., -2, -1, 0, 1, 2, . . .$$

all the other rational numbers, such as

$$\frac{13}{5}, \frac{5}{7}, \text{ and } \frac{3}{4},$$

and all the irrational numbers, such as

$$\sqrt{2}, \sqrt{5}, \text{ and } \sqrt{11}.$$

You will need a Centimeter ruler and protractor for this chapter.

But the number line is only a small part of a larger number system. This chapter will show that every point in the (x,y)-coordinate plane can be thought of as a number. In other words, it will introduce a way of "adding" any two points in the plane, and a way of "multiplying" them. However, to do this we will need to consider a new plane, for the familiar one does not have the properties we will need.

GEOMETRIC ADDITION OF TWO POINTS

Let P and Q be points in the (x,y)-plane.

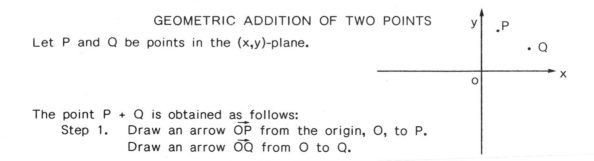

The point P + Q is obtained as follows:
   Step 1.   Draw an arrow $\overrightarrow{OP}$ from the origin, O, to P.
             Draw an arrow $\overrightarrow{OQ}$ from O to Q.

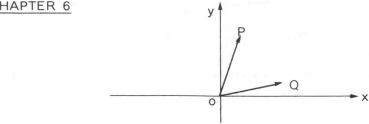

Step 2.   Move the arrow $\overrightarrow{OP}$ parallel to itself until its "tail" is at Q. The head of the arrow is now at the point called P + Q.

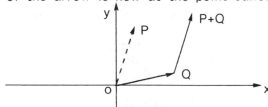

On the other hand, we could have chosen to move $\overrightarrow{OQ}$ parallel to itself until the tail is at P, and obtain Q + P. Both approaches give the same point because we arrive at the same vertex of the parallelogram.

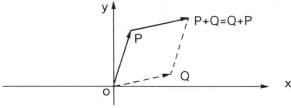

These arrows are often called "vectors", each has length and direction.

The new point, P + Q, can be represented by an arrow, which is the diagonal of a parallelogram formed by the arrows representing the points P and Q.

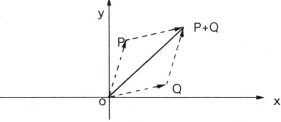

The same definition of resultant appears in the study of forces in elementary physics.

The arrow from O to P + Q is sometimes called the "resultant" of the arrows from O to P and from O to Q.

1.   Draw P + Q after tracing these figures on your paper.

(a) 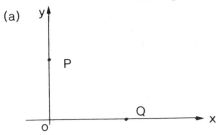   (b)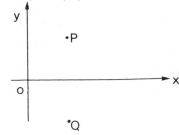

2. Draw P + Q if P and Q are the points shown in these diagrams. (Show the arrows used.)

(a)
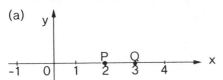

(b)
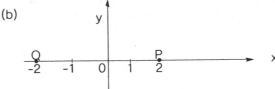

(c)
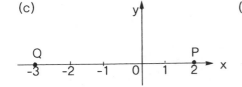

(d)
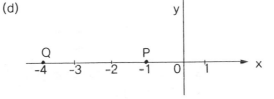

Note. In Exercise 2 the points P and Q happen to be on the x-axis, or what is called "the real number line." The definition of addition using arrows gives the same result as obtained by ordinary arithmetic addition.

3. Show using arrows for the points P and Q in the diagrams below that P + Q = 0.

(a)
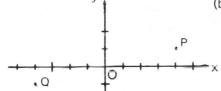

(b)
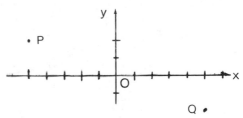

4. (a) Give two more examples of points P and Q such that P + Q = 0.

(b) What is the geometric relation of P and Q with respect to the origin?

(b) reflections across the origin.

Exercises 3 and 4 suggest the following definition.

OPPOSITE

DEFINITION. Let P be a point in the plane. If

P + Q = 0

then point Q is called the <u>opposite</u> of P. (Q is that point such that 0 is the midpoint of the line segment PQ). Q is also denoted "-P", and is also sometimes called the negative of P.

Adding opposites in the plane gives the same result as in arithmetic on the number line.

Also, observe in the above definition that points P and Q are <u>reflections</u> of each other with respect to the origin, the point 0.

On the number line, opposites are the same distance from zero and are reflections across zero. In this new plane, opposites are the same distance from the origin and are reflections across the origin.

QUADRANTS

A point that is neither on the x-axis nor the y-axis is in one of four quadrants, as shown in this diagram:

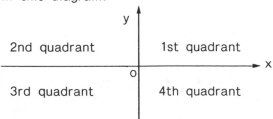

5. (a) If P = -2, find -P.
   (b) If Q = 7, find -Q.
   (c) If P is on the y-axis, where is -P?
   (d) If P is in the second quadrant, where is -P?

6. (a) If both P and Q are on the y-axis, where is P + Q?
   (b) If P is on the positive y-axis and Q is on the positive x-axis, where is P + Q?
   (c) If P is on the negative x-axis and Q is on the negative y-axis, where is P + Q?
   (d) If P is on the positive x-axis and Q is on the negative y-axis, where is P + Q?

7. Where could P + Q be
   (a) if P and Q are in the first quadrant?
   (b) if P and Q are in the second quadrant?
   (c) if P is in the first quadrant and Q is in the third quadrant?
   (d) if P is in the first quadrant and Q is in the second quadrant?

There may be more than one possibility for each of these cases.

* * *

Now that addition of points in the plane has been described, the multiplication of points will be considered. The next section presents another way of describing points in the plane. This description will be needed in the definition of multiplication.

DESCRIBING A POINT USING ANGLE AND DISTANCE

Let P be any point in the plane. It can be identified by a two-fold description:

(1) The distance it is from the origin, and

(2) The angle it determines with respect to the positive x-axis.

The following diagram illustrates this:

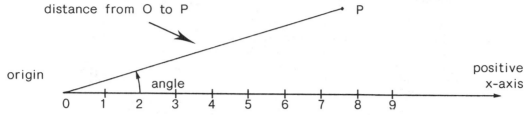

distance from O to P

P

origin

angle

positive
x-axis

0 1 2 3 4 5 6 7 8 9

Observe that in the above figure the moving arm of the angle θ is rotated <u>counterclockwise</u> with reference to the fixed arm along the positive x-axis. This is defined as a <u>positive rotation</u>. If the moving arm is rotated <u>clockwise</u>, the rotation is <u>negative</u>. If the point P is the origin, no angle is assigned.

8. Using a centimeter ruler and a protractor, redraw the following diagrams and estimate the distance from the origin to P and the angle θ for each point P shown.

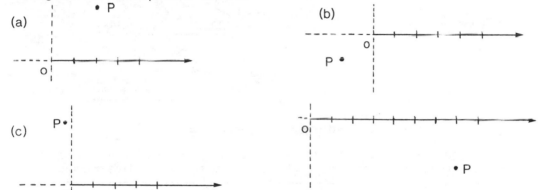

(a)

• P

(b)

P •

(c)

P •

• P

Conversely, when you are given an angle and a distance from the origin of a certain points, you can locate that point.

Example 1. Draw the point P that has an angle of 30° and is 4 units from the origin, O.

Solution.   We think of ourselves as standing at the origin, O, and facing in the direction of the positive x-axis. We turn 30° (counterclockwise) and then go ahead 4 units.

4   P

30°

x

Exercise 9 concerns the location of a point when given an angle and the distance from the origin.

9.   Draw the point P that has

Remember that a negative angle describes a clockwise rotation.

(a)   angle 180° and is 3 units from 0.
(b)   angle 0° and is 4 units from 0.
(c)   angle 540° and is 2 units from 0.
(d)   angle 90° and is 1 unit from 0.
(e)   angle -60° and is 5 units from 0.

* * *

Note that the point with an angle of 30° and a distance 4 units from 0 is the same as the point with an angle of 390° and a distance 4 units from 0. In fact, a point may be described by an infinite number of angles. If θ is one angle describing the angle P, then you may add (or subtract) 360° from θ as many times as you wish and still describe the same point P. Thus the points P, Q, R, and T described below are the same:

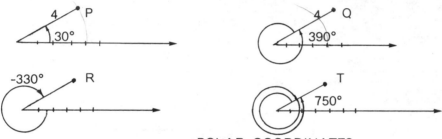

All of these pictures describe the same point.

POLAR COORDINATES

The next definition is useful in describing a point in terms of its distance from the origin and the angle formed with the positive x-axis.

We use the Greek letter, θ ("theta"), to represent the angle here. Also, observe that r (the radius) is non-negative.

DEFINITION. (Radius and angle of P) The distance, r, from the origin, 0, to point P is called the radius of P. An angle θ that the ray from 0 to P makes with the positive x-axis is called an angle of P. (The origin has radius zero; it is not assigned an angle.)

In short, a point P is described by its radius r (a non-negative number) and its angle θ:

$$P = (r, \theta).$$

As you will soon see, some problems are easier to solve using "polar coordinates" than with "rectangular," or (x,y)-coordinates.

We call (r,θ) the polar coordinates of the point. The angle is positive if the rotation is counterclockwise. The angle is negative if the rotation is clockwise. The radius is never negative.

Example 2. Draw:

(a)   The point (2, 135°).   (b) The point (2,-30°).

Solution.

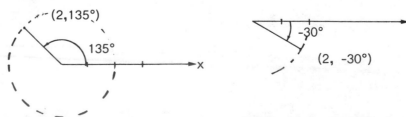

10. (a)   Sketch three points P that have an angle of 60° and describe them with polar coordinates.
    (b)   Sketch the set of all points P that have angle 60°.

11. What is the angle of a point P that is situated
    (a)   on the positive x-axis?   (b) on the negative x-axis?   *(a) 0° (or 360°, 720°, or −360° etc.)*

12. (a)   Sketch three points P that have radius 2.
    (b)   Sketch the set of all points P that have radius 2.

13. (a)   If points P and Q both have angle 90°, what can be said about the angle of P + Q?
    (b)   If P has angle 90° and Q has angle 0°, what can be said about the angle of P + Q?   *(a) 90°*
    (c)   If P has angle 90° and Q has angle 180°, what can be said about the angle of P + Q?
    (d)   If P has radius 5 and Q has radius 2, what can be said about the radius of P + Q?   *(c) between 90° and 180°*

14. Draw the point P = (r,θ) in each of these cases:
    (a)   (3, 135°)   (b) (1, 300°)   (c) (4, 210°)
    (d)   (2, −45°)   (e) (3, 420°)   (f) (5, 315°)

15. If the angle of P is 60°, what is the angle of
    (a)   P + P?   (b) −P?   *(a) 60°*

16. Find r and θ for each of the points P_1, P_2, P_3, P_4, P_5 shown in this diagram.

$P_1 = (3, 45°)$

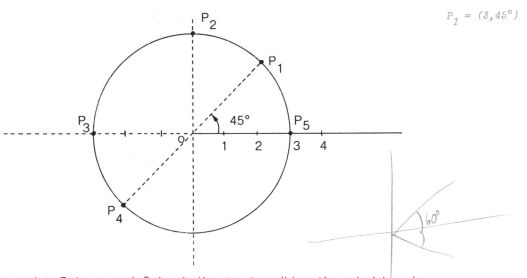

Describing point P by r and θ is similar to describing the wind by giving its speed and the direction it is going. The polar description is needed to describe how to multiply points P and Q. The rectangular description of P, in terms of its x- and y-coordinates, will be used later in the chapter.

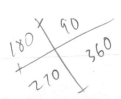

GEOMETRIC MULTIPLICATION OF TWO POINTS

DEFINITION.   The <u>product</u>, P • Q, of points P and Q is defined as follows:

1.   The <u>angle</u> of P • Q is the <u>sum</u> of the angles of P and Q.

2.   The <u>radius</u> of P • Q is the (ordinary) <u>product</u> of the radius of P and the radius of Q.

STUDY CAREFULLY THIS VERY IMPORTANT DEFINITION

Say this definition in your own words. Verify it with another student.

Summarizing this, we have the following definitions of multiplication of points:

$$\text{If } P = (r_1, \theta_1) \text{ and } Q = (r_2, \theta_2)$$

$$\text{then } P \cdot Q = (r_1 r_2, \theta_1 + \theta_2).$$

That is, the <u>radius</u> of the product is the <u>product</u> of the radii.   The <u>angle</u> of the product is the <u>sum</u> of the angles.

An example will show how this definition works in practice.

Example 3.   Draw the product P • Q of the points P and Q shown in the following diagram if Q = (2,45°) and P = (3,30°).

Q_•   •P

0 1 2 3 4 5

Solution.   First determine the radii and angles of P and Q by drawing the appropriate rays and arcs that P and Q determine.

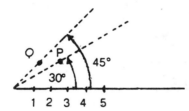

As the diagram shows,

P has angle 30° and radius 3,
Q has angle 45° and radius 2,

The product P • Q therefore has

angle (30° + 45°) = 75°

and

radius (2)(3) = 6.

With this information we sketch P • Q.
P • Q = (6, 75°)

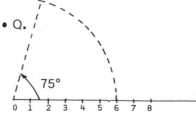

17. Compute and sketch P • Q if P and Q are the points given in the following diagrams:

(a)

(b)

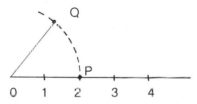

(c)

(d)

Exercises 17(d)-19 illustrate an important principle: If P and Q lie on the x-axis, their product P • Q, as defined in this chapter, coincides with their ordinary product as real numbers. The next exercise goes further into this observation.

18. (a) Using the definition of multiplication given in this chapter, compute the product P • Q of the points shown in this diagram.

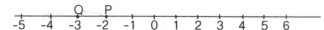

(b) Using the approach in (a) explain why the product of two negative real numbers is a positive real number.

19. (a) Using the definition of multiplication in this chapter, compute and draw the product P • Q, where P and Q are points in the following diagram.

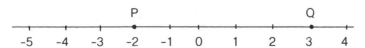

(b) Using the approach in (a), explain why the product of a negative real number and a positive real number is a negative real number.

* * *

The points in the plane can now be both added and multiplied. The next section describes another important resemblance to the arithmetic of the number line.

THE RECIPROCAL

In ordinary arithmetic, the <u>reciprocal</u> of a nonzero real number n is that number x which is the solution of the equation

$$nx = 1.$$

The reciprocal of n is denoted $\frac{1}{n}$. Every real number other than 0 has a reciprocal. The next few exercises show that the idea of reciprocal applies to points in the plane.

Note: A negative angle denotes a negative rotation; that is, a clockwise direction of rotation.

20. Point P has radius 3 and angle 30°, point Q has radius $\frac{1}{3}$ and angle -30°. Find P • Q in (r,θ) form.

21. Let P have radius 4 and angle 60°. Find the radius and angle of Q such that P • Q = 1.

22. Show that for each point P in the plane, other than 0, there is a point Q such that P • Q = 1.

20. 1

21. $r = \frac{1}{4}$

* $\theta = -60°$*

* * *

As the preceding exercises show, each point on the number line, other than zero, has a reciprocal. We denote this reciprocal by $\frac{1}{P}$.

COMPLEX NUMBERS

The points in the plane can be added and multiplied. With this notion of addition and multiplication, the points in the plane form a new number system. We call this system the <u>complex numbers</u>. Because the real numbers are graphed on the x-axis (the real number line), observe the real numbers are contained in the complex numbers. That is, <u>a real number is also a complex number.</u> In general, if x is a real number it is represented by the point (x,0) in rectangular coordinates.

The negative numbers were introduced in the Renaissance to solve an equation such as

$$x + 3 = 0.$$

The rational numbers were introduced more than 3000 years ago to solve an equation such as

$$2x = 3.$$

The irrational numbers were introduced about 2500 years ago to solve an equation such as

$$x^2 = 2.$$

But the real numbers do <u>not</u> provide a solution for the equation

$$x^2 = -1.$$

The next section shows that the complex numbers provide a solution, in fact two solutions, for that equation.

WHY -1 HAS A SQUARE ROOT

The real number 1 is represented in the complex plane by the point with rectangular coordinates (1,0). Similarly, (-1,0) represents -1. The point (1,0) has polar coordinates (1, 0°) while the point (-1,0) has polar coordinates (1, 180°). In general we identify the point (x,0) with the real number x.

The point with rectangular coordinates (0, 1) is called i. Hence, the point with rectangular coordinates (0, -1) is $-i$.

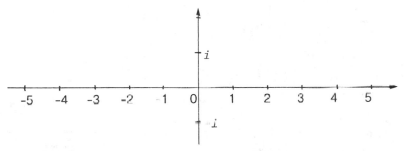

Consider arrows here. A number added to its opposite is zero (the origin).

23. (a) What are the radius and angle of i?
    (b) What is the radius of $i \cdot i$?
    (c) What is the angle of $i \cdot i$?
    (d) Show that $(-i)^2 = -1$.

Hint: (d) Show i^2 and -1 have the same radius and angle.

NOTE. Exercise 23 shows that -1 has a square root. That square root is not a real number (which would lie on the x-axis). Rather, it is the number i and lies on the y-axis. The next exercise shows that there is another square of -1.

24. (a) What are the radius and angle of $-i$?
    (b) What is the radius of $(-i)(-i)$?
    (c) What is the angle of $(-i)(-i)$?
    (d) Show tha $(-i)^2 = -1$.

NOTE. Exercises 23 and 24 show that -1 has two square roots, neither of which lies on the x-axis. One is i and the other is $-i$. (You may wish to carefully think through Exercises 23 and 24 again.)

* * *

This is an important exercise. Make sure that you understand it.
(a) $-i$ (c) i
(e) i (g) $-i$

25. Use the fact that $i^2 = -1$ to express each of the following more simply as 1, -1, i, or $-i$.

    (a) i^3 (b) i^4 (c) i^5 (d) i^6

    (e) i^9 (f) i^{30} (g) i^{43} (h) i^{100}

    (i) Could you compute i^{2001} easily? Explain.

Note. In general, where we multiply by i, the effect is a rotation counterclockwise of 90°. For example

$$i = 1 \cdot i = (1, 0°) \cdot (1, 90°) = (1, 90°) = i$$
$$i^2 = i \cdot i = (1, 90°) \cdot (1, 90°) = (1, 180°) = -1$$
$$i^3 = i^2 \cdot i = (1, 180°) \cdot (1, 90°) = (1, 270°) = -i$$
$$i^4 = i^3 \cdot i = (1, 270°) \cdot (1, 90°) = (1, 360°) = 1$$

and so on ....

* * *

The following section will show that all complex numbers may be written by using the real numbers and the single complex number i.

THE a + bi FORM OF A COMPLEX NUMBER

The next few exercises develop a simple notation for complex numbers.

26. Consider the complex number i and the complex number 3.

    (a) What are the radii of the above numbers?
    (b) What are the angles of the above numbers?

The results in (c) and (d) should be the same

    (c) Use (a) and (b) to draw the product $3 \cdot i$, or $3i$.
    (d) Draw the sum $i + i + i$.

> NOTATION. If b is a real number it is customary to write the product b $\cdot$ i simply as
> $$bi.$$

Make a good guess for the location of these points.

27. Sketch each of the points:

    (a) $-3i$ (b) $\sqrt{2}i$ (c) $4.1i$ (d) $-6.5i$ (e) $7i$

28. If b lies on the x-axis (that is, b is a real number) where does bi lie?

29. Both 2 and $3i$ are complex numbers. Draw the point that represents the complex number $2 + 3i$.

Use arrows as at the beginning of this chapter.

30. Draw the point that represents each of these complex numbers:
    (a) $1 + i$ (b) $-2 + 3i$ (c) $-1 - 4i$ (d) $6 + 0i$

Exercises 29 and 30 illustrate an important idea: the complex number a + bi is represented by the point whose rectangular coordinates are (a, b).

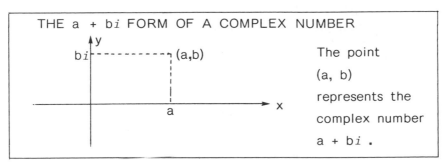

THE a + bi FORM OF A COMPLEX NUMBER

The point (a, b) represents the complex number a + bi.

Numbers of the form bi where b ≠ 0 are sometimes called "imaginary numbers." Keep in mind that they, as well as the real numbers, are complex numbers!

"Imaginary numbers" are of the form 0 + bi where b≠0. Real numbers are of the form a + 0·i

Example 4.   Express the point (-2, 3) in the form a + bi for appropriate numbers <u>a</u> and b.

Solution.   The point (-2, 3) is shown in this diagram.

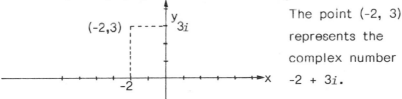

The point (-2, 3) represents the complex number -2 + 3i.

It is the same as the point obtained by adding the points -2 and 3i by the earlier parallelogram approach to addition. The point (-2, 3) is -2 + 3i .   This observation is emphasized in the following definition, which depends on the definition of addition of complex numbers.

RECTANGULAR FORM OF A COMPLEX NUMBER

Every complex number can be written uniquely in the form

a + bi,

where <u>a</u> and <u>b</u> are real numbers. This description is called the <u>rectangular</u>, or a + bi, form. Occasionally it is called the <u>cartesian</u> form.

It is easy to add and subtract complex numbers when they are written in the "rectangular" form a + bi . For instance,

$$(2 + 3i) + (4 + 5i) = 6 + 8i.$$

Also, we have

$$3 + 5i + (-3 - 5i) = 0 + 0i$$

which is the origin. This shows that $-3-5i$ is the "opposite" of $3 + 5i$, for it is the complex number we add to $3 + 5i$ to get 0. Observe that if P is a complex number, then its opposite, which we call $-P$, is <u>opposite</u> it across the origin, as shown in this diagram.

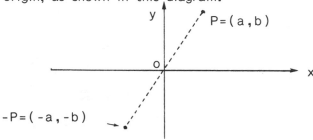

In general, if $P = (a, b)$ then $-P = (-a, -b)$.

CHANGING FROM (r, θ) FORM TO $a + bi$ FORM

The next example shows how to change a complex number from <u>polar coordinate</u> form (r, θ) to <u>rectangular coordinate</u> form $a + bi$.

Example 5.   If $P = (2, 150°)$ find the $a + bi$ (rectangular) form of P.

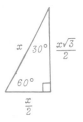

Recall from Geometry: These triangles will be used many times in the work ahead. Know well the relation among the sides.

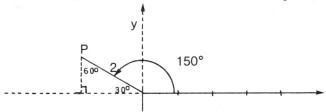

Solution.   Note the 30°-60°-right triangle (the reference triangle shown dotted); thus the rectangular coordinates of P are $(-\sqrt{3}, 1)$. The $a + bi$ form of the complex number is $-\sqrt{3} + i$.

Example 6.   If $Q = (3, 225°)$ in polar form, write the $a + bi$ (rectangular) form of Q.

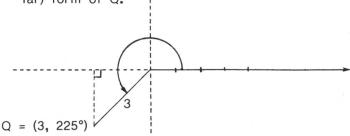

Solution.   Observe that a 45°-45° right triangle as a reference is involved, and that the rectangular coordinates of Q are

$(- \dfrac{3\sqrt{2}}{2} , - \dfrac{3\sqrt{2}}{2})$. Thus, the a + b$i$ (rectangular) form of the complex number is $\dfrac{-3\sqrt{2}}{2} - \dfrac{3\sqrt{2}i}{2}$.

Watch here that you do not write

$\dfrac{-3\sqrt{2}}{2} - \dfrac{3\sqrt{2i}}{2}$.

The i is outside the radical; for this reason it may be better to write

$\dfrac{-3\sqrt{2}}{2} - \dfrac{3i\sqrt{2}}{2}$.

31. Write the point P in a + bi (rectangular) form if the polar coordinates of P are the following: (Draw a figure for each.)

    (a) (4, 60°)   (b) (7, -45°)   (c) ($\sqrt{18}$, 120°)

    (d) (10, 270°)   (e) (1, 420°)   (f) (1, 630°)

(a) $2 + 2i\sqrt{3}$

(c) $-\dfrac{3}{2}\sqrt{2} + \dfrac{3}{2}i\sqrt{6}$

(e) $\dfrac{1}{2} + i\dfrac{\sqrt{3}}{2}$

Hint: (31) First find the coordinates of P in rectangular form.

CHANGING FROM a + bi FORM TO (r, θ) FORM

In changing from the a + bi form to the polar form of a complex number P, first find the rectangular coordinates of P. Then compute r and θ. (The following exercises depend only on angles that are conveniently related to 30° or 45°.)

Example 7. Find the radius and angle of P = -2 + 2i.

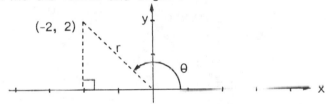

Solution.   Note that the rectangular coordinates of the point are (-2, 2) and that a 45°-45° right reference triangle is involved. Thus, r = 2$\sqrt{2}$ and θ = 135°.

32. Give the radius and angle of each of the following numbers:

    (a) 3 + 3i (b) $\sqrt{2} - \sqrt{2}i$ (c) $\dfrac{-\sqrt{3}}{2} - \dfrac{i}{2}$ (d) 0 - 5i

33. Show that the complex number a + bi has radius $\sqrt{a^2 + b^2}$.

(a) 45°
(c) 210°

* * *

ARITHMETIC OF COMPLEX NUMBERS

The addition and multiplication of complex numbers satisfy the basic rules of arithmetic:

These rules are basic to ordinary arithmetic.

Addition	Multiplication	
0 + P = P	1 • P = P	(0 and 1 are identity elements)
P + Q = Q + P	P • Q = Q • P	(commutative rules for + and •)
(P + Q) + R = P + (Q + R)	(P • Q) • R = P • (Q • R)	(associative rules for + and •)

Addition (con't) | Multiplication (con't.)

For each P there is a point, -P, such that P + (-P) = 0

For each P ≠ 0 there is a point $\frac{1}{P}$, such that (opposite and reciprocal)

$$P \cdot \left(\frac{1}{P}\right) = 1$$

Finally, the distributive rule holds:

$$P \cdot (Q + R) = P \cdot Q + P \cdot R,$$

and

$$(Q + R) \cdot P = Q \cdot P + R \cdot P.$$

These laws will not be proved here; their proof is left to the projects. Instead, they will be used to show how to carry out the arithmetic of complex numbers that are expressed in a + bi (rectangular) form.

Example 8.   Compute the product

$$(2 + 3i)(4 + 5i)$$

Solution.   We will justify the key steps as follows:

$$(2 + 3i)(4 + 5i) = (2 + 3i)4 + (2 + 3i)5i \qquad \text{distributive rule}$$

$$= 2 \cdot 4 + 3i \cdot 4 + 2 \cdot 5i + 3i \cdot 5i \qquad \text{distributive rule}$$

$$= 2 \cdot 4 + (4 \cdot 3)i + (2 \cdot 5)i + (3 \cdot 5)i^2 \qquad \begin{array}{l}\text{rearrangement} \\ \text{(using commu-} \\ \text{tative \& asso-} \\ \text{ciative rules)}\end{array}$$

Always let $i^2 = -1$ immediately to help avoid errors!

$$= 8 + 12i + 10i + 15i^2 \qquad \text{(computing)}$$

$$= 8 + 22i + 15i^2 \qquad \text{distributive rule}$$

$$= 8 + 22i + 15(-1) \qquad (i^2 = -1)$$

$$= -7 + 22i$$

34. (a) Plot accurately 2 + 3i and 4 + 5i from Example 8 and also their product, -7 + 22i.
    (b) Use a centimeter ruler to measure the radii of the three points in (a).   Check that the radius of -7 + 22i is approximately equal to the product of the radii of 2 + 3i and 4 + 5i .
    (c) Use the Pythagorean **Theorem** to check your answers in (b).
    (d) With a protractor measure the angles of 2 + 3i and 4 + 5i , and also of -7 + 22i.   Check that the sum of the first two angles is equal to the angle of -7 + 22i.

* * *

Stop here. Reflect on what you have learned in this chapter thus far.

Example 8 shows how to compute with complex numbers in rectangular form: Just treat complex numbers like real numbers, but whenever you have i^2 replace it by -1.

In particular, we have two ways of finding the product of complex numbers given in rectangular form:

REVIEW THE
CHAPTER.

    (1)   multiply algebraically (using the basic algebra laws),

or (2)   change to polar form and multiply geometrically (using the angles and radii).

35. Compute $(1 + i)(1 - i)$

(a) and (b)   2

    (a) algebraically.
    (b) geometrically, after determining their angles and radii.

36. Compute $(\sqrt{3} + i)(\sqrt{3} + i)$

    (a) algebraically.
    (b) geometrically, using angles and radii.

*(a) and (b)
$2 + 2i\sqrt{3}$
Suggestion:
Generally you may
wish to write
$i\sqrt{3}$ rather than
$\sqrt{3}i$, because
careless writing
may change $\sqrt{3}i$
to $\sqrt{3i}$. Radicals
must be written
carefully.*

37. Compute:

    (a) $(5 - 3i)(2 + 4i)$ (b) $(-2 + 4i)(3i)$

    (c) $(2 + 5i)(2 - 5i)$ (d) $\left(\dfrac{\sqrt{2}}{2} + \dfrac{\sqrt{2}i}{2}\right)^2$

* * *

Adding complex numbers in $a + bi$ form is a simple algebraic addition as the next example illustrates.

*37.   (a) $22 + 14i$
    (c) 29*

    Example 9. Compute $(-3 + 4i) + (5 - 7i)$

    Solution. $(-3 + 4i) + (5 - 7i) = -3 + 4i + 5 - 7i$

$$= -3 + 5 + 4i - 7i$$

$$= 2 - 3i$$

    These steps can be justified by the commutative, associative, and distributive rules.

    Note, it is just as easy to subtract two complex numbers.   For example,

$$(2 + 3i) - (1 + 4i) = 2 + 3i - 1 - 4i$$

$$= 1 - i$$

If $a + bi$ and $c + di$ are complex numbers, then

$$(a + bi) + (c + di) = (a + c) + (b + d)i$$

and

$$(a + bi) - (c + di) = (a - c) + (b - d)i$$

*Simply collect
"like terms"*

38. Compute:

(a) $(-3i)(-2i)$

(b) $(2 + i)(2 - i)$

(c) $(2 + 3i) + (4 - 7i)$

(d) $(1 - 6i) + (1 + 6i)$

(e) $(3 - i) - (2 - 5i)$

(f) $(-2 - 7i)(-4i)$

HOW TO DIVIDE COMPLEX NUMBERS

Finding the quotient of two complex numbers involves a little mathematical trick, which depends on the idea of Exercise 39.

39. Show that

(a) $(5 - 3i)(5 + 3i) = 34$

(b) $(1 + i)(1 - i) = 2$

(c) $(2 + 7i)(2 - 7i) = 53$

(d) $(a + bi)(a - bi) = a^2 + b^2$ (a real number)

* * *

The complex numbers a + bi and a = bi are <u>conjugates</u> of each other.

Division of complex numbers depends on the fact that the product

$$(a + bi)(a - bi) = a^2 + b^2,$$

is a real number.

The number a - bi is called the <u>conjugate</u> of the number a + bi. The conjugate of P is denoted $\overline{P}$. (Thus, $\overline{a + bi}$ = a - bi .)

* * *

40. Draw 3 + 4i and its conjugate 3 - 4i ; show that their product is 25,

Hint:
(b) If P has angle θ what is the angle of P̄?

(d) Consider a reflection.

(a) algebraically.

(b) geometrically, using the definition of multiplication.

(c) Why is the product of a complex number P and its conjugate $\overline{P}$ always a real number? Explain in detail using a picture.

(d) What is the geometric relation between these conjugates?

* * *

The next example is typical of all division problems involving complex numbers expressed in rectangular form.

Example 10.   Compute $\dfrac{2 + 3i}{4 + i}$

Solution.   First <u>multiply</u> numerator and denominator <u>by the conjugate of the denominator.</u>

$$\frac{2 + 3i}{4 + i} = \frac{(2 + 3i)(4 - i)}{(4 + i)(4 - i)}$$

$$= \frac{(2 + 3i)(4 - i)}{4^2 - i^2}$$

$$= \frac{(2 + 3i)(4 - i)}{17}$$

$$= \frac{8 - 2i + 12i - 3i^2}{17}$$

$$= \frac{11 + 10i}{17}$$

$$= \frac{11}{17} + \frac{10i}{17}$$

Why should the radius of $2 + 3i$ equal the product of the radius of $4 + i$ and the radius of $\frac{11}{17} + \frac{10}{17}i$? Check that it does.

The quotient is expressed in rectangular form and can be easily plotted.

* * *

41.   (a)   Compute $\dfrac{3 - 5i}{6 + 2i}$ in a + bi form.

Hint: Use conjugate as in Example 10.

(b)   As a check, multiply your answer by 6 + 2i. The result should be 3 - 5i . (Why?)

42.   Compute in a + bi form.

(a) $\dfrac{1 + i}{1 - i}$ (b) $\dfrac{6 + 3i}{2 + i}$ (c) $\dfrac{-3 + i\sqrt{2}}{4 + i\sqrt{2}}$ (d) $\dfrac{3 + 2i}{2 - 3i}$

(e) $\dfrac{4}{5 - 2i}$ (f) $\dfrac{i}{-1 + i}$

(a) i

(c) $-\dfrac{5}{9} + \dfrac{7i\sqrt{2}}{18}$

(e) $\dfrac{20}{29} + \dfrac{8}{29}i$

THE SQUARE OF A COMPLEX NUMBER

Not every real number is the square of a real number. For example, there is no real number x such that

$$x^2 = -1,$$

nor is there a real number x such that
$$x^2 = -7.$$

IMPORTANT!

In fact, only the nonnegative real numbers are the squares of real numbers. The situation is much different for complex numbers. We have already seen that -1 is the square of both i and $-i$, and as you will see (using the polar form of a complex number), <u>every complex number is the square of a complex number.</u> In other words, "every complex number has a square root." In fact, except for zero, every complex number has <u>two</u> square roots.

This may be a surprise.

43. The complex number P has radius 3 and angle 30°.

    (a) What is the radius of P²?

    (b) What is the angle of P²?

    As Exercise 43 illustrates, to square a complex number,

<u>square its radius</u> and <u>double its angle.</u>

    The next few exercises show how to find the square roots of <u>any</u> complex number in polar form.

THE SQUARE ROOTS OF A COMPLEX NUMBER

DRAW FIGURES!

Do not rush when doing these exercises! Make sure you understand how you get the answers.

(a) i   (b) i

(a) $\frac{i}{4}$ (b) $\frac{i}{4}$

*(a) 3
(b) 30° or 210°
(or these plus n360° for any integer n)*

44. Compute and draw P² if P has

    (a) radius 1 and angle 45°.   (b) radius 1 and angle 225°.

45. (Review Exercise 44.) Draw the two square roots of i.

46. Compute P² if P has

    (a) radius ½ and angle 45°.   (b) radius ½ and angle 225°.

47. (Review Exercise 46.) Draw the two square roots of $\frac{i}{4}$.

48. Compute P² if P has

    (a) radius 2 and angle 60°.   (b) radius 2 and angle 240°.

49. (a) (Review Exercise 48.) Draw the <u>two</u> square roots of $-2 + 2\sqrt{3}\,i$.

    (b) Write one of the square roots in (a) in $a + bi$ form.

    (c) Square the number in (b) algebraically. The result should be $-2 + 2\sqrt{3}\,i$.

50. Let P be the complex number with radius 9 and angle 60°. If Q² = P,

    (a) what must the radius of Q be?

    (b) what must the angle of Q be?

51. Use Exercise 50 to compute the two square roots of P in polar form if P = (9, 60°).

52. (a) Write the complex number P of Exercise 51 and its <u>two</u> square roots in rectangular form.

    (b) Check each of the square roots in rectangular form by squaring them algebraically.

* * *

    Recall that in Exercises 23 and 24 it was shown that -1 has two square roots, which were called i and $-i$. The next example shows how to find the square roots of other complex numbers.

Example 11.   Find the square roots of P = -16 in a + bi form.

Solution.   Observe that the radius of -16 is 16 and the angle is 180°.
Then,
$$Q^2 = P = (16, 180°).$$
One possible value of Q is
$$Q_1 = (4, \frac{180°}{2})$$
$$= (4, 90°), \text{ or simply } 4i.$$
Another value for Q is

$$Q_2 = (4, \frac{180° + 360°}{2})$$

Remember that P is also (16, 180° + 360°)

$$= (4, 270°) \text{ or } -4i.$$

Thus the two square roots of -16 are 4i and -4i.

53. (a) Find the two square roots of i in polar form.

   (b) Draw the answers in (a). State them in a + bi form.

Hint: First change to polar form.

54. Find and draw the two square roots of $-1 + \sqrt{3}i$ in a + bi form.

55. If P has radius r and angle θ, show that the square roots of P have radius $\sqrt{r}$ and angle $\frac{\theta}{2}$ or $\frac{\theta + 360°}{2}$.

56. Let P have radius r and angle θ.

   (a) What is the radius of the circle on which the points representing the square roots of P lie? Explain.

(a) $\sqrt{r}$

   (b) How are the points representing the square roots of P arranged on the circle in (a)? Explain.

(b) ends of a diameter.

57. Draw any point in the plane. Using a protractor, centimeter ruler, and calculator, estimate and plot its two square roots.

58. Compute the two square roots of P = (2, 80°).

One square root is ($\sqrt{2}$, 40°); what is the other?

* * *

The polar coordinates of a point are not unique. For example,
$$(8, 30°) = (8, 390°) = (8, 750°) = \ldots$$
and
$$(8, 30°) = (8, -330°) = (8, -690°) = \ldots$$

You can add or subtract multiples of 360° to the angle θ and still describe the same point. This observation is quite useful when calculating cube roots (or 4th roots, etc.) of a complex number.

THE CUBE ROOTS OF A COMPLEX NUMBER

The ideas in the preceding section can be generalized to find the cube, the cube roots, or higher powers and roots of complex numbers. The next exercise introduces the fact that every complex number (other than 0) has three cube roots.

59. (a) If P has radius 1 and angle 120°, show that $P^3 = 1$.
    (b) If Q has radius 1 and angle 240°, show that $Q^3 = 1$.
    (c) If R has radius 1 and angle 0°, show that $R^3 = 1$.
    (d) Plot and làbel in polar notation the points P, Q and R, the three cube roots of 1.
    (e) Write in a + bi form notation the three cube roots of 1.

(e) One is
$-\frac{1}{2} + \frac{\sqrt{3}}{2}i$

If you are curious cube
$-\frac{1}{2} + \frac{\sqrt{3}}{2}i$

algebraically. Does the result equal 1?

NOTE.   For emphasis, we plot the three cube roots of 1 and show their polar coordinates.

STUDY CAREFULLY

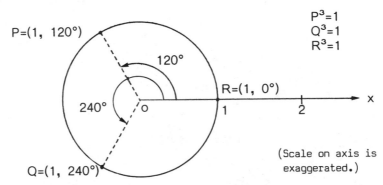

Of course R is just our old friend, the real number 1. (1^3 does equal 1 after all.) <u>The cube roots of 1 are equally spaced on the circle and are</u> <u>the vertices of an equilateral triangle.</u> As will be illustrated in the next few exercises, the three cube roots of any complex number are the vertices of an equilateral triangle inscribed in a circle with the center at the origin.

(a) 2
(b) $\frac{0°}{3} = 0°$

$\frac{0° + 360°}{3} = 120°$

$\frac{0° + 2(360°)}{3} = 240°$

one root is

$-1 - i\sqrt{3}$

Hint: what will be the radius of the circle?

* * *

60. (a) If $P^3 = 8$, what must the radius of P be?
    (b) The angle of 8 is 0°, or 360°, or 720°, and so on. What possibilities are there for the angle of P, a cube root of 8?
    (c) Plot the three cube roots of 8.

61. Write the three cube roots of 8 in a + bi (rectangular) form. (See Exercise 60.)

62. Describe the three cube roots of 8 in terms of an appropriate circle. (See Exercises 60 and 61.)

63. (a) Explain why the three cube roots of the complex number P = (r, θ). are $Q_1 = (\sqrt[3]{r}, \frac{\theta}{3})$

$$Q_2 = (\sqrt[3]{r}, \frac{\theta + 360°}{3})$$

$$Q_3 = (\sqrt[3]{r}, \frac{\theta + 2(360°)}{3})$$

    (b) Describe the circle on which the cube roots lie. How are the roots spaced on that circle?

64. (a) If $P^3 = -1$, what must be the radius of P?

(b) The angle of -1 is 180°, or 540°, or 900°, or 1260°, and so on. What are the possibilities for the angle of P, a cube root of -1?

(c) Plot in polar form the three cube roots of -1.

65. (a) Write the three cube roots of -1 in a + bi form. (See Exercise 64. First use polar form, then change to a + bi form.)

(b) Cube each root in (a) algebraically to check that it is a cube root of -1.

66. (a) Find and plot in polar form the three cube roots of -8.

(b) Write the three cube roots of -8 in rectangular form.

67. (a) If $P^3 = i$, what must be the radius of P?

(b) What are the possibilities for the angle of P, a cube root of i ?

(c) Plot in polar form the three cube roots of i .

(d) Write in a + bi form the three cube roots of i.

68. Write the <u>polar form</u> of the points representing the three cube roots of $4\sqrt{2} - 4\sqrt{2}\,i$. (State all angles θ in a form such that 0° ≤ θ < 360°.)

Remember the radius is always a positive number.
(a) 1
(b) 60°, 180° 300° and so on, 60° + n120°

(b) one root is 1 − i√3

(a) 1
(c) The angles are 30°, 150°, 270°

One root is (2, 105°)

THE FOURTH ROOTS OF A COMPLEX NUMBER

We have considered square roots and cube roots. It turns out (by similar reasoning) that there are four fourth roots of a complex number (other than 0); they are equally spaced around a circle and thus are the vertices of a square that is inscribed in that circle with the origin as a center. In other words, for any complex number P ≠ 0, the equation

$$x^4 - P = 0$$

has four complex roots.

69. In order to find the four fourth roots of 1, answer the following questions.

(a) If $P^4 = 1$, what must be the radius of P?

(b) The angle of 1 is 0°, or 360°, or 720°, or 1080°, and so on. What are the possibilities for the angle of P?

(c) Sketch the fourth roots of 1 in polar form.

(d) Express the four fourth roots of 1 in a + bi form.

70. Show that each of these numbers is a fourth root of P = (r, θ).

(a) $\left(\sqrt[4]{r}, \frac{\theta}{4}\right)$

(b) $\left(\sqrt[4]{r}, \frac{\theta + 360°}{4}\right)$

(c) $\left(\sqrt[4]{r}, \frac{\theta + 720°}{4}\right)$

(d) $\left(\sqrt[4]{r}, \frac{\theta + 1080°}{4}\right)$

(e) Say in your own words how to find the four fourth roots of (r, θ) in polar form.

two roots are
$\frac{\sqrt{2}}{2} + i\frac{\sqrt{2}}{2}$ and
$\frac{\sqrt{2}}{2} - i\frac{\sqrt{2}}{2}$

71. Express the four fourth roots of -1 in a + bi form.

COMPLEX NUMBERS AND THE QUADRATIC FORMULA

So far we have considered equations such as

$$Q^2 = P, \quad Q^3 = P, \quad Q^4 = P, \text{ etc.}$$

where we may write in polar form P = (r, θ). Now our attention turns to the typical quadratic equation

$$ax^2 + bx + c = 0,$$

where a, b, and c are real numbers.

For instance, consider the equation

$$x^2 + 7 = 0$$

or

$$x^2 = -7.$$

Note that $i\sqrt{7}$ is a solution, since

$$(i\sqrt{7})(i\sqrt{7}) = i^2(\sqrt{7})^2 = (-1)(7) = -7$$

Thus

$$i\sqrt{7} = \sqrt{-7}$$

Also, $-i\sqrt{7}$ is a solution, for

$$(-i\sqrt{7})(-i\sqrt{7}) = i^2(\sqrt{7})^2 = -1(7) = -7.$$

Thus -7 has two square roots, $i\sqrt{7}$ and $-i\sqrt{7}$. Let us denote the first one as $\sqrt{-7}$. Thus, $\sqrt{-7} = i\sqrt{7}$.

Generally, if n is a positive real number, then $\sqrt{-n}$ is written $i\sqrt{n}$. For instance,

$$\sqrt{-15} = i\sqrt{15},$$

and

$$\sqrt{-9} = i\sqrt{9}$$
$$= 3i$$

and

$$\sqrt{-12} = i\sqrt{12}$$
$$= 2i\sqrt{3}.$$

IMPORTANT! As a first step, change to i notation when simplifying the square root of a negative number.

CAUTION! The rules for radicals,

$$\sqrt{ab} = \sqrt{a}\sqrt{b}$$

and

$$\sqrt{\frac{a}{b}} = \frac{\sqrt{a}}{\sqrt{b}}$$

apply <u>only</u> for real numbers. For instance,

$$\sqrt{-2}\sqrt{-2} = (i\sqrt{2})(i\sqrt{2})$$
$$= i^2\sqrt{4}$$
$$= -2. \text{ (Correct)}$$

It is NOT valid to write:

$$\sqrt{-2}\sqrt{-2} = \sqrt{(-2)(-2)}$$
$$= \sqrt{4}$$
$$= 2. \qquad \text{(Incorrect)}$$

The next exercise will provide practice in simplifying radicals.

72. Simplify by writing each number in terms of i and reducing each radical.

    (a) $\sqrt{-2}$ (b) $\sqrt{-8}$ (c) $\sqrt{-125}$ (d) $\sqrt{-98}$ (e) $\sqrt{-72}$ (f) $\sqrt{-175}$

As a first step in simplifying expressions such as $\sqrt{-8}$, write $\sqrt{-8} = i\sqrt{8}$

(a) $i\sqrt{2}$
(c) $5i\sqrt{5}$
(f) $5i\sqrt{7}$

* * *

Until now, we have been unable to solve a quadratic equation,

$$ax^2 + bx + c = 0$$

where $b^2 - 4ac < 0$. The next example demonstrates how to solve any such equation with the aid of complex numbers.

Example 12. Solve for x:
$$x^2 + x + 1 = 0.$$

Solution. By the quadratic formula;

$$x = \frac{-b \pm \sqrt{b^2 - 4ac}}{2a}$$

$$= \frac{-(1) \pm \sqrt{(1)^2 - 4(1)(1)}}{2(1)}$$

$$= \frac{-1 \pm \sqrt{-3}}{2}$$

$$= \frac{-1 + i\sqrt{3}}{2} \quad \text{(Change to } i \text{ notation.)}$$

IMPORTANT! When you meet the square root of a negative number, change to i notation in the next step.

73. (a) Solve for x: $x^2 + x + 1 = 0$.   (b) Check each of the solutions in (a) by substituting them in the original equation.

74. Solve for x:
    (a) $x^2 - 4x + 6 = 0$ (b) $3x^2 + 2x + 2 = 0$
    (c) $3x^2 = x - 3$ (d) $4x^2 + 2 = 5x$
    (e) $2x^2 - 3x + 5 = 0$ (f) $x^2 + x + 1 = 0$

(a) $2 \pm i\sqrt{2}$

(c) $\frac{1 \pm i\sqrt{35}}{6}$

(e) $\frac{3 \pm i\sqrt{31}}{4}$

75. In the quadratic formula,
$$x = \frac{-b \pm \sqrt{b^2 - 4ac}}{2a}$$
the expression $b^2 - 4ac$ is called the <u>discriminant</u>. For what <u>value</u> of the discriminant will the equation $ax^2 + bx + c = 0$ have

    (a) exactly one (double) solution?   (b) two real solutions?
    (c) two complex solutions that are not real numbers?

The discriminant allows us to describe the solutions of a quadratic.

SOME TERMINOLOGY

If P is a complex number

the <u>radius</u> of P is often called the <u>magnitude</u> of P,

the <u>angle</u> of P is often called the <u>amplitude</u> of P,

the radius of P is often denoted $|P|$, which is the distance from the origin to point P.

Thus, in $a + bi$ form,
$$|a + bi| = \sqrt{a^2 + b^2}.$$

To show this, apply the Phythagorean **Theorem** to the right triangle in the following diagram.

Observe that $|a + bi|$ is the distance from $a + bi$ to the origin, just as $|x|$ is the distance the number x is from the origin on the real number line. If $P = a + bi$ and $(a,b) = (r,\theta)$ then

$$r = |P| = \sqrt{a^2 + b^2}$$

Example 13.   Compute $|2 - 3i|$.

Solution. $|2 - 3i| = \sqrt{(2)^2 + (-3)^2}$

$$= \sqrt{4 + 9}$$

$$= \sqrt{13}$$

76. Compute:

(a) $|-3 + 4i|$ (b) $|5 - 12i|$ (c) $\left|\frac{1}{2} + \frac{i}{2}\right|$

(d) $|-1 - i\sqrt{3}|$ (e) $|-8 + 15i|$ (f) $\left|-\frac{\sqrt{3}}{2} - \frac{i}{2}\right|$

(a) 5

(c) $\frac{1}{\sqrt{2}}$ or $\frac{\sqrt{2}}{2}$

(e) 17

* * * SUMMARY * * *

The number line is the geometric picture for the set of real numbers. These real numbers include the integers,

$$. . . , -3, -2, -1, 0, 1, 2, 3, . . .$$

the rational numbers, such as $\frac{3}{7}$, and the irrational numbers, such as $\sqrt{2}$.

This number line appears in daily life as thermometers and measuring sticks.

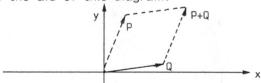

Every point on the number line represents a real number and every real number has a point on the number line representing it.

This chapter showed that the <u>line of real numbers</u> is just a small part of the <u>plane of complex numbers.</u> Taking a geometric viewpoint, we can call any point in the (x,y)-plane a complex number. Two such points, P and Q, are <u>added</u> with the aid of this diagram:

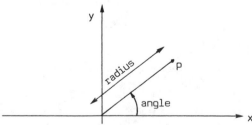

Each point P other than the origin has an angle and a radius, as shown in this diagram:

Each point P other than the origin has an angle and a radius, as shown in this diagram:

To <u>multiply</u> two points, multiply their radii and add their angles. Of greatest importance was the fact that the equation

$$P \bullet P = -1$$

has a solution. In other words, the complex numbers provide a square root of -1. In fact, it was shown that the complex numbers provide solutions of any quadratic equation

$$ax^2 + bx + c = 0,$$

where a, b, and c are real, even when the discriminant, $b^2 - 4ac$, is negative.

It was shown that any complex number can be expressed uniquely in the form

$$P = a + bi,$$

where a and b are real numbers.

Cube roots and fourth roots of a complex number were also investigated. The solutions of the equation

$$x^3 = P$$

were shown to be the vertices of an equilateral triangle. The solutions of

$$x^4 = P$$

are vertices of a square.

* * * WORDS AND SYMBOLS * * *

real number

complex number (point in the plane)

addition of complex numbers (by arrows or by parallelogram)

opposite

radius (magnitude) of a complex number

angle (amplitude) of a complex number

polar form (r, θ) of a complex number

multiplication of complex numbers (multiply their radii and add their angles)

reciprocal

i

rectangular form, $a + bi$, of a complex number

imaginary

conjugate

quadrant

$|P|$ (radius of complex number P)

* * * CONCLUDING REMARKS * * *

Renaissance mathematicians had frequently met equations whose solutions required the square root of -1, but they were reluctant to get involved with such a strange concept. Nevertheless, Gauss, by the year 1815, had put the complex numbers on a solid foundation, giving them their name, introducing the symbol "i" for $\sqrt{-1}$.

In 1893, Steinmetz, an electrical engineer with General Electric, found that the complex numbers were exactly what he needed to describe the mathematics of alternating current. Alternating current had become of great importance then for it permitted electrical power to be transmitted long distances at low cost.

Today every electrical engineer uses $\sqrt{-1}$, though he (or she) gives it the symbol j . (The symbol i is reserved in electrical engineering texts for current.)

For more applications of complex numbers see S.K. Stein, Mathematics: The Man-Made Universe, Freeman, New York.

This application is typical of the history of mathematics. Often the concepts that a mathematician creates are applied years later to a field that the mathematician never had in mind.   Alternating current had not even been discovered when Gauss's work was published.   Seven years later, in 1822, Faraday wrote in his notebook, "convert magnetism into electricity," a goal that he achieved in 1831 when he created the first alternating current.

* * * COMPLEX NUMBERS SELF QUIZ * * *

1. Let P = (3, 60°) and Q = (2, 30°).   Find

   (a) P + Q in rectangular form.

   (b) P • Q in rectangular form,

      (i)   geometrically (first finding the radius and angle of P • Q).

      (ii)   algebraically (first writing P and Q in rectangular form).

   (c) Check that the two results in (b) agree.

2. Which of these are equal?

   (a) i^{35} (b) $-i$ (c) i^{69} (d) i^{18}

   (e) i^4 (f) i^3 (g) 1   (h) -1

3. (a) Express the two square roots of -3 in polar form.

   (b) Draw the two square roots in (a).

   (c) Express the two square roots in (a) in rectangular form.

   (d) Square the two roots in (c) algebraically.   The results should be -3.

4. (a) Plot the three cube roots of $27i$.

   (b) Express each root in (a) in polar form.

   (c) Express each root in (a) in rectangular form.

5. Solve for x:

   (a) $x^2 - 2x - 1 = 0$ (b) $x^2 - 2x + 3 = 0$

   (c) Check your answers by substitution in the original equations.

6. Find the radius and angle (between 0° and 360°) of these complex numbers:

   (a) $2 - 2i$ (b) $-4i$ (c) -8

   (d) 17 (e) $-\sqrt{3} + i$ (f) $-1 - i$

7. Compute:

   (a) $|8 - 15i|$ (b) $|\sqrt{2} - i\sqrt{3}|$

8. Let $P = 2 + 2i$. Plot

   (a) the conjugate of P.   (b) the reciprocal of P.

   (c) the square of P.   (d) -P.

9. If P has angle 20° and Q has angle 30°, what can be said about the angle of

   (a) P • Q ?   (b) $\frac{1}{P}$?   (c) P^2?   (d) Q^3?

   (e) $\overline{Q}$?   (f) P + Q?   (g) $P^2 • Q^2$?

10. Express in rectangular form, $\dfrac{2 - 3i}{4 + i}$.

ANSWERS TO SELF QUIZ

1. (a) $(\frac{3}{2} + \sqrt{3}) + i(\frac{3\sqrt{3}}{2} + 1)$ (b) $6i$ 2. (a), (b), (f) all equal $-i$

   (d), (h) equal -1.   (e), (g) equal 1   (c) equals i .

3. (a) $(\sqrt{3}, 90°)$, $(\sqrt{3}, 270°)$ (c) $\sqrt{3}, -i\sqrt{3}$

4. (b) $(3, 30°)$, $(3, 150°)$, $(3, 270°)$ (c) $\frac{3\sqrt{3}}{2} + \frac{3i}{2}$, $\frac{-3\sqrt{3}}{2} + \frac{3i}{2}$, $-3i$

5. (a) $1 \pm \sqrt{2}$ (b) $1 \pm i\sqrt{2}$ 6. (a) $(\sqrt{8}, 315°)$ (b) $(4, 270°)$

   (c) $(8, 180°)$ (d) $(17, 0°)$ (e) $(2, 150°)$ (f) $(\sqrt{2}, 225°)$

7. (a) 17 (b) $\sqrt{5}$ 8. The points to be plotted are

   (a) $2 - 2i$ (b) $(\frac{1}{\sqrt{8}}, -45°)$ (c) $8i$ (d) $-2 - 2i$

9. (a) 50°   (b) -20°   (c) 40°   (d) 90°   (e) -30°

   (f) between 20° and 30°   (g) 100°   10. $\frac{5}{17} - \frac{14}{17}i$

* * * COMPLEX NUMBERS REVIEW * * *

1. If $|P| = 2$ and $|Q| = 3$, what can be said about

   (a) $|P • Q|$?   (b) $|P + Q|$?   (c) $|5P|$?   (d) $|P^3|$?

2. Express in the $a + bi$ form:

   (a) $(2 + i)(1 - 3i)$

   (b) $\dfrac{2 + i}{1 - 3i}$

3. Express in polar form:

   (a) $4i$　　　　(b) $3 + 3i$　　　(c) -5　　　(d) $\dfrac{-5}{2} - \dfrac{5\sqrt{3}}{2}i$

4. There are four complex solutions of the equation
$$z^4 = -8 - 8\sqrt{3}\,i$$

   (a) Express each in polar form.

   (b) Plot them.

5. P, Q, and R are three complex numbers such that $P \cdot R = Q$. If $P = (4, 205°)$ and $Q = (8, 315°)$, express R in polar form.

6. Express in rectangular form:

Use the polar form to compute.

   (a) $(\dfrac{-1}{2} + \dfrac{\sqrt{3}}{2}i)^{10}$

   (b) $(\dfrac{\sqrt{2}}{2} + \dfrac{i\sqrt{2}}{2})^9$

   (c) P^3 if $P = (\sqrt{2},\ 135°)$

   (d) $(-1 + i)^7$

7. Compute and plot the two square roots of

   (a) -2

   (b) $-1 + i\sqrt{3}$

8. One of the four fourth roots of a certain complex number is $(2, 70°)$. Plot all four of the fourth roots of that number.

9. Solve for x:

   (a) $x^2 - 4x + 13 = 0$　　　(b) $x^2 + 16 = 0$　　　(c) $x^3 + 8 = 0$

10. Let a, b, and c be real numbers, $a \neq 0$. How many of the solutions of of the equation
$$ax^2 + bx + c = 0$$

   (a) are real if $b^2 - 4ac > 0$?　　　(b) are real if $b^2 - 4ac < 0$?

11. Express in rectangular form:

   (a) $\dfrac{1}{-i}$　　　(b) $\dfrac{2i}{i - 3}$　　　(c) $\dfrac{2i - 3}{3i + 2}$　　　(d) $(1 + 2i)^3$

12. Express in rectangular form:

   (a) $\dfrac{1}{\sqrt{-2}}$　　　(b) $\sqrt{\dfrac{-1}{2}}$　　　(c) $\sqrt{\dfrac{-5}{8}}$　　　(d) $\dfrac{6}{\sqrt{-3}}$

13. Express in rectangular form:

   (a) i^3　　　(b) $\sqrt{-48}$　　　(c) $\sqrt{9}\sqrt{-3}$　　　(d) $\sqrt{12}\sqrt{-9}$

   (e) $\sqrt{-35}\sqrt{-7}$　　　　(f) $(2 - \sqrt{-3})(2 + \sqrt{-3})$

14. Show that $\sqrt{-9}\sqrt{-16}$ is <u>not</u> $\sqrt{(-9)(-16)}$.

15. What happens to any complex number (in terms of radius and **angle**) when you multiply it by

    (a) i ?   (b) i^3?   (c) i^2?   (d) i^4 ?   (e) $-i$?

16. Write in a + bi form:

$$i^{100} + i^{101} + i^{102} + i^{103}$$

17. For any complex number, there are

    (a) 2 square roots, 180° apart,   (b) 3 cube roots, ___° apart,

    (c) 4 fourth roots, ___° apart,   (d) 5 fifth roots, ___° apart,

    (e) in general, ___nth roots, ___° apart.

ANSWERS TO REVIEW

1.   (a) 6   (b) between 1 and 5   (c) 10   (d) 8.   2. (a) $5-5i$ (b) $\frac{-1}{10} + \frac{7}{10}i$

3.   (a) (4, 90°)   (b) $(3\sqrt{2}, 45°)$ (c) (5, 180°)   (d) (5, 240°)

4.   (a) (2, 60°), (2, 150°), (2, 240°), (2, 330°)   (b) They form a square inscribed in the circle of radius 2 and center at the origin.   5. (2,110°)

6.   (a) $\frac{-1}{2} + \frac{\sqrt{3}}{2}i$ (b) $\frac{\sqrt{2}}{2} + \frac{i\sqrt{2}}{2}$ (c) $2 + 2i$ (d) $-8 - 8i$

7.   (a) $i\sqrt{2}, -i\sqrt{2}$ (b) $\frac{\sqrt{2}}{2} + \frac{\sqrt{6}}{2}i$ and $\frac{-\sqrt{2}}{2} - \frac{\sqrt{6}}{2}i$ 8. The other three roots are (2, 160°), (2, 250°), (2, 340°). The four are the vertices of a square.

9.   (a) $2 \pm 3i$ (b) $\pm 4i$ (c) $-2, 1 + \sqrt{3}i, 1 - \sqrt{3}i$ 10. (a) two (b) none

11.   (a) i (b) $\frac{1}{5} - \frac{3}{5}i$ (c) i (d) $-11 - 2i$

12.   (a) $\frac{-i}{\sqrt{2}}$ (b) $\frac{i}{\sqrt{2}}$ (c) $\frac{\sqrt{10}}{4}i$ (d) $-2\sqrt{3}i$

13.   (a) $-i$ (b) $4\sqrt{3}i$ (c) $3\sqrt{3}i$ (d) $6\sqrt{3}i$ (e) $-7\sqrt{5}$ (f) 7

14. $-12 \neq 12$ 15. (a) rotates + 90°   (b) rotates + 270°

    (c) rotates + 180°   (d) rotates + 360°   (e) rotates - 90°

16.   (a) $0 + 0i$, or simply 0   17. (b) 120° (c) 90° (d) 72° (e) n. $\left(\frac{360}{n}\right)°$

* * * COMPLEX NUMBERS REVIEW SELF TEST * * *

1. Express in the form a + bi:

    (a) $2 + 400i^3 - 300i^8 + 10i^{103}$ (b) $\frac{2-i}{5+7i}$

2. Express in polar form, if A = (4, 165°) and B = (3, 135°)

    (a) A · B   (b) B^3 (c) $\overline{B}$ (the conjugate of B)

    (d) $-B$ (the opposite of B)

3. Solve for z:

(a) $z^2 + 6z + 25 = 0$ (b) $z^2 + 6z - 25 = 0$

4. Plot the five fifth roots of

(a) 1   (b) i (c) –1

5. If P is a positive real number and Q is a complex number of magnitude 1 and angle θ, what is

(a) the angle of P•Q?   (b) the magnitude of P • Q?

ANSWERS

1. (a) $-298 - 410i$ (b) $\frac{3}{74} - \frac{19}{74}i$

2. (a) (12, 300°)   (b) (27, 45°)   (c) (3, 135°)   (d) (3, –45°)

3. (a) $-3 \pm 4i$ (b) $-3 \pm \sqrt{34}$

4. In each case they form the vertices of a regular 5-sided figure inscribed in the circle of radius 1, center at origin. One of the vertices is

(a) 1   (b) (1, 18°)   (c) (1, 36°)   5. (a) θ   (b) P

* * *   COMPLEX NUMBERS PROJECTS   * * *

Hint: Change to polar form

1. Compute $(\frac{\sqrt{2}}{2} + i\frac{\sqrt{2}}{2})^{1003}$ 2. Compute $(\frac{-1}{2} + i\frac{\sqrt{3}}{2})^{2001}$

3. Let P have radius 0.9 and angle 60°.

(a) Plot P^n for n = 1, 2, 3, 4, 5, 6, 7, 8.

(b) What happens to P^n as n gets large?

(c) What happens in (b) if P has a radius greater than one?

4. Let P and Q be complex numbers. What relation is there between the radii of P, Q, and

(a) P • Q?   (b) P + Q?

Recall that $\overline{P}$ is the conjugate of P.

5. Let $P = a + bi$ and $Q = c + di$.

(a) Prove that $\overline{P+Q} = \overline{P} + \overline{Q}$.   (b) Prove that $\overline{P \cdot Q} = \overline{P} \cdot \overline{Q}$.

6. (a) If P is a real number, prove that $\overline{P} = P$.

(b) If $\overline{P} = P$, must P be real?

Hint: (c) Then a = 0 in a + bi there is no real part.

(c) If P is imaginary, prove that $\overline{P} = -P \cdot \overline{P} = -P$.

(d) If $\overline{P} = -P$, must P be imaginary?

7. Find all complex numbers P such that $\frac{1}{P} = \overline{P}$.

8. Let P have radius r. Prove that $P\overline{P} = r^2$.

9. Let a, b, and c be complex numbers, with a ≠ 0. Prove that the equation
$$ax^2 + bx + c = 0$$
has a complex solution.

10. Let a, b, and c be real numbers, with a ≠ 1. Prove that the roots of the equation

$$ax^2 + bx + c = 0$$

are conjugates, that is, if one root is P, the other is $\overline{P}$.

11. Let P be a complex number. Let $\overline{P}$ be the conjugate of P.

   (a) Using the geometric definitions of addition and multiplication, show that $P + \overline{P}$ and $P \cdot \overline{P}$ are real numbers.

   (b) Write P as $a + bi$. Using the algebraic approach, show that $P + \overline{P}$ and $P \cdot \overline{P}$ are real numbers.

12. (See Exercise 11.) Let P be a complex number.
   (a) The Product of $(x - P)(x - \overline{P})$ is a polynomial of degree two. Show that it has real coefficients.

   (b) Prove that any complex number is the root of polynomial of degree two with real coefficients.

   (c) Find a polynomial of degree two with real coefficients of which $3 + 4i$ is a root.

13. (a) Show that $x^3 - 1 = (x - 1)(x^2 + x + 1)$.

   (b) Use (a) to show that $x^3 - 1 = 0$ has three solutions, one real and two involving the complex number i.

While $x^3 -1 = 0$ or $x^3 - 1$, has three complex number solutions, $x^3 = 1$ has only one real solution. That is, any real number has only one cube root that is a real number.

14. Let P be a complex number of radius 1 and angle θ. Show that $P \cdot Q$ is obtained from Q by increasing its angle by θ. In other words, to draw $P \cdot Q$, just rotate Q around the origin by the angle θ.

15. (This exercise concerns the distributive law for complex numbers. See Project 14.)

   Assume that $|P| = 1$. Let Q and R be typical complex numbers. Show by diagram that $P \cdot (Q + R) = P \cdot Q + P \cdot R$. (You may assume that when you rotate a parallelogram, the result is still a parallelogram.)

16. (a) If $P^5 = 1$, what is the radius of P, a fifth root of 1?

   (b) What are possibilities for the angle of P?

17. Plot the five fifth roots of -2.

18. Can you devise a way to add points in 3-dimensional space? The addition should satisfy all the rules of arithmetic, such as $P + Q = Q + P$ and $(P + Q) + R = P + (Q + R)$ for all points P, Q, and R and for points P and Q there is a unique point X such that $P + X = Q$. [Note. It has been proved impossible to introduce addition and multiplication for the points in 3-dimensional space such that they obey the usual rules of arithmetic and algebra as found at the end of the First Year Algebra Review in Chapter 2.]

* * *   FIRST YEAR ALGEBRA REVIEW (OPTIONAL)   * * *

Recall the product and quotient rules for square roots:

NOTE. These are the <u>only</u> rules for square roots of real numbers.

$$\boxed{\begin{array}{c} \text{SIMPLIFYING RADICALS} \\[4pt] \sqrt{a}\,\sqrt{b} = \sqrt{ab} \quad \text{(Product rule)} \\[4pt] \dfrac{\sqrt{a}}{\sqrt{b}} = \sqrt{\dfrac{a}{b}} \quad \text{(Quotient rule)} \end{array}}$$

For instance,

$$\sqrt{3}\,\sqrt{27} = \sqrt{81}$$
$$= 9$$

and

$$\frac{\sqrt{8}}{\sqrt{2}} = \sqrt{\frac{8}{2}}$$
$$= \sqrt{4}$$
$$= 2$$

Example 1.   Simplify:

    (a) $\sqrt{45}$ (b) $\sqrt{108} - \sqrt{75}$ (c) $\sqrt{6}(\sqrt{8} + \sqrt{27})$

Solution.

    (a) $\sqrt{45} = \sqrt{3^2 \cdot 5}$ (b) $\sqrt{108} - \sqrt{75} = \sqrt{36 \cdot 3} - \sqrt{25 \cdot 3}$
$$= 3\sqrt{5} \qquad\qquad\qquad\qquad\quad = 6\sqrt{3} - 5\sqrt{3}$$
$$= \sqrt{3}$$

    (c) $\sqrt{6}(\sqrt{8} + \sqrt{27}) = \sqrt{6}(\sqrt{4 \cdot 2} + \sqrt{9 \cdot 3})$
$$= \sqrt{6}(2\sqrt{2} + 3\sqrt{3})$$
$$= 2\sqrt{12} + 3\sqrt{18}$$
$$= 2\sqrt{4 \cdot 3} + 3\sqrt{9 \cdot 2}$$
$$= 2 \cdot 2\sqrt{3} + 3 \cdot 3\sqrt{2}$$
$$= 4\sqrt{3} + 9\sqrt{2}$$

* * *

1. Simplify (assume no letters are negative numbers):

    (a) $\sqrt{112}$ (b) $\sqrt{128} - \sqrt{8}$ (c) $\sqrt{45} - \sqrt{20}$ (d) $\sqrt{6}(\sqrt{6} - \sqrt{24} + \sqrt{3})$

    (e) $(\sqrt{8} + \sqrt{27})(\sqrt{24} - \sqrt{54})$ (f) $(\sqrt{7} - \sqrt{3})(\sqrt{7} + \sqrt{3})$

    (g) $\sqrt{80} - \sqrt{20} + \sqrt{5}$ (h) $\sqrt{28} - 2\sqrt{112} + 3\sqrt{63}$

    (i) $5\sqrt{8x} - \sqrt{18x} - 3\sqrt{32x}$ (j) $3\sqrt{20n^3} - 2\sqrt{45n^3} + 7\sqrt{5n^3}$

    (k) $\sqrt{24} + \sqrt{54} - 5\sqrt{6}$

RATIONALIZING THE DENOMINATOR

Expressions such as $\dfrac{1}{\sqrt{x}}$ and $\dfrac{1}{1-\sqrt{5}}$ can be written in an equivalent form free of radicals in the denominator. This is called "rationalizing the denominator."

Example 2. Simplify:

(a) $\dfrac{1}{\sqrt{x}}$ (b) $\dfrac{1}{1+\sqrt{5}}$

Solution. (a) $\dfrac{1}{\sqrt{x}} = \dfrac{1}{\sqrt{x}} \cdot \dfrac{\sqrt{x}}{\sqrt{x}}$ (b) $\dfrac{1}{1+\sqrt{5}} \cdot \dfrac{1-\sqrt{5}}{1-\sqrt{5}} = \dfrac{1-\sqrt{5}}{1-5}$

$= \dfrac{\sqrt{x}}{x}$ $= \dfrac{1-\sqrt{5}}{-4}$

or $\dfrac{\sqrt{5}-1}{4}$

Observe in each case the expression was multiplied by 1 in a form that got rid of the radical in the denominator.

2. Simplify. In each case, rationalize the denominator. (Assume x > 0.)

(a) $\dfrac{x}{\sqrt{x}}$ (b) $\dfrac{1}{1-\sqrt{x}}$ (c) $\dfrac{2}{\sqrt{2}-1}$ (d) $\sqrt{\dfrac{1}{x}}$

(e) $\sqrt{3\tfrac{1}{2}}$ (f) $\sqrt{x+\dfrac{1}{x}}$ (g) $\sqrt{2x+\dfrac{x}{4}}$

(h) $\dfrac{1}{\sqrt{x-1}}$ (i) $\sqrt{\dfrac{b^2}{4a^2}-\dfrac{c}{a}}$

SOLVING QUADRATIC EQUATIONS

Sometimes a quadratic equation can be solved by factoring.

Example 3. Solve for x:
$$3x^2 + 2x - 5 = 0$$
Solution.
$$(3x + 5)(x - 1) = 0$$

then $3x + 5 = 0$ or $x - 1 = 0$

$3x = -5$ $x = 1$

$x = \dfrac{-5}{3}$

Check that $-\dfrac{5}{3}$ and 1 are solutions by substituting them in the original equation,

$3x^2 + 2x - 5 = 0$

3. Solve for x by factoring.

(a) $x^2 + 6x - 7 = 0$ (b) $3x^2 + 2 = 5x$

(c) $6x^2 + 7x = 20$ (d) $7x^2 - x = 0$

Example 4. Solve for x:
$$x^2 - 5 = 0$$

Solution.

$$x^2 - 5 = 0$$

$$(x - \sqrt{5})(x + \sqrt{5}) = 0$$

$$\text{then } x - \sqrt{5} = 0 \quad \text{or} \quad x + \sqrt{5} = 0$$

$$x = \sqrt{5} \qquad x = -\sqrt{5}$$

thus $x = \sqrt{5}$ or $-\sqrt{5}$.

4. Solve for x:

(a) $x^2 - 2 = 0$ (b) $x^2 - 17 = 0$

(c) $2x^2 - 3 = 0$ (d) $3x^2 - 11 = 0$

* * *

Recall the identities,

$$(a + b)^2 = a^2 + 2ab + b^2$$

$$(a - b)^2 = a^2 - 2ab + b^2$$

Both are useful in the following example:

Example 5. Fill in the blanks so that each statement is true.

(a) $(x + \_)^2 = x^2 + 4x + \_$ (b) $(x - \_)^2 = x^2 - 14x + \_$

Solution.

(a) $(x + 2)^2 = x^2 + 4x + 4$ (b) $(x - 7)^2 = x^2 - 14x + 49$

5. Fill in the blanks so that each statement is true.

(a) $(x + \_)^2 = x^2 + 10x + \_$ (b) $(x - 6)^2 = \_ - \_ + \_$

(c) $(x - \_)^2 = x^2 - x + \_$ (d) $(x + \_)^2 = x^2 + 3x + \_$

(e) $(x - \_)^2 = x^2 - 5x + \_$ (f) Describe the method in your own words.

* * *

An equation such as $x^2 + 4x - 1 = 0$ cannot be solved by factoring. However, it can be solved by "completing the square," as shown in the next example.

Example 6. Solve for x by completing the square:

$$x^2 + 4x - 1 = 0$$

Solution.

$$x^2 + 4x \qquad = 1 \qquad \text{(Prepare to complete the square).}$$

$$x^2 + 4x + 4 = 1 + 4 \qquad \text{(Complete the square)}$$

$$(x + 2)^2 \quad = 5$$

$$\text{then } x + 2 = \pm \sqrt{5}$$

$$x = -2 \pm \sqrt{5}$$

6. Solve for x by completing the square:

(a) $x^2 - 2x - 1 = 0$ (b) $x^2 - 3x - 5 = 0$

Exercise 6 contained only instances where the coefficient of x^2 is 1. The next example illustrates a case where the coefficient of x^2 is not 1.

Example 7. Solve by completing the square:

$$3x^2 + 2x - 7 = 0$$

Solution.

$$x^2 + \frac{2}{3}x - \frac{7}{3} = 0 \qquad \text{(Divide by 3.)}$$

$$x^2 + \frac{2}{3}x = \frac{7}{3} \qquad \text{(Prepare to complete the square.)}$$

$$x^2 + \frac{2}{3}x + \frac{1}{9} = \frac{7}{3} + \frac{1}{9} \qquad \text{(Complete the square.)}$$

$$\left(x + \frac{1}{3}\right)^2 = \frac{7}{3} \cdot \frac{3}{3} + \frac{1}{9}$$

$$\left(x + \frac{1}{3}\right)^2 = \frac{22}{9} \qquad \text{(Add fractions.)}$$

$$x + \frac{1}{3} = \pm \sqrt{\frac{22}{9}} \qquad \text{(Find square root.)}$$

$$x + \frac{1}{3} = \pm \frac{\sqrt{22}}{3} \qquad \text{(Simplifying radical)}$$

$$x = \frac{-1}{3} \pm \frac{\sqrt{22}}{3}$$

$$\text{or} \quad x = \frac{-1 \pm \sqrt{22}}{3}$$

That is, $x = \dfrac{-1 + \sqrt{22}}{3}$ or $x = \dfrac{-1 - \sqrt{22}}{3}$; both numbers are solutions.

7. Solve for x by completing the square:

(a) $x^2 - 3x - 5 = 0$ (b) $2x^2 - 5x - 1 = 0$

(c) $4x^2 - 5x - 3 = 0$ (d) $3x^2 - 7x - 1 = 0$

THE QUADRATIC FORMULA

Exercise 8 develops a formula for solving equations such as

$$ax^2 + bx + c = 0,$$

where a, b, and c are fixed numbers and $a \neq 0$. It is called the quadratic formula,

$$x = \frac{-b \pm \sqrt{b^2 - 4ac}}{2a}$$

and will solve any quadratic equation if solutions exist. Completing the square is used to derive the formula.

8. Explain each of the following steps:

Solve for x: $ax^2 + bx + c = 0$ (where $a \neq 0$)

Step 1. $x^2 + \dfrac{b}{a}x + \dfrac{c}{a} = 0$

Step 2. $x^2 + \dfrac{b}{a}x \qquad = -\dfrac{c}{a}$

Step 3. $x^2 + \dfrac{b}{a}x + \left(\dfrac{b}{2a}\right)^2 = \left(\dfrac{b}{2a}\right)^2 - \dfrac{c}{a}$

Step 4. $\left(x + \dfrac{b}{2a}\right)^2 \qquad = \dfrac{b^2}{4a^2} - \dfrac{c}{a}$

Step 5. $\left(x + \dfrac{b}{2a}\right)^2 \qquad = \dfrac{b^2}{4a^2} - \dfrac{4ac}{4a^2}$

Step 6. $\left(x + \dfrac{b}{2a}\right)^2 \qquad = \dfrac{b^2 - 4ac}{4a^2}$

Step 7. $x + \dfrac{b}{2a} \qquad = \pm \sqrt{\dfrac{b^2 - 4ac}{4a^2}}$

Step 8. $x + \dfrac{b}{2a} \qquad = \pm \dfrac{\sqrt{b^2 - 4ac}}{2a}$

Step 9. $x \; = \dfrac{-b}{2a} \pm \dfrac{\sqrt{b^2 - 4ac}}{2a}$

Step 10.

$$\boxed{\; x = \dfrac{-b \;\pm\; \sqrt{b^2 - 4ac}}{2a} \qquad \text{The quadratic formula} \;}$$

Example 8.   Use the quadratic formula to solve for x:
$$2x^2 + 1 = 5x$$

Solution.
$$2x^2 - 5x + 1 = 0 \qquad \text{(Change to } ax^2 + bx + c = 0 \text{ form.)}$$

Observe that $a = 2$, $b = -5$, and $c = 1$.

$$x = \dfrac{-b \pm \sqrt{b^2 - 4ac}}{2a}$$

$$x = \dfrac{-(-5) \pm \sqrt{(-5)^2 - 4(2)(1)}}{2(2)}$$

Use parentheses in place of the letters when substituting.

$$x = \dfrac{5 \pm \sqrt{25 - 8}}{4}$$

$$x = \dfrac{5 \pm \sqrt{17}}{4}$$

9. Solve for x:

(a) $x^2 - 4x + 1 = 0$

(b) $2x^2 + 3x = 1$

(c) $2x^2 + x = 6$

(d) $3x^2 + 4x - 3 = 0$

(e) $3x + 1 = 2x^2$

(f) $3x^2 = 11x + 4$

Hint: Write (b)(c)(e)(f) in the form

$ax^2 + bx + c = 0.$

10. Solve for x by any method:

(a) $x^2 + 2x - 9 = 0$

(b) $3x^2 + 5x + 2 = 0$

(c) $2x^2 + 3x - 2 = 0$

(d) $5x^2 + 2x = 0$

(e) $3x^2 + 3x = 1$

(f) $4x^2 - 6x + 1 = 0$

ANSWERS TO ALGEBRA REVIEW

1. (a) $4\sqrt{7}$ (b) $6\sqrt{2}$ (c) $\sqrt{5}$
(d) $3\sqrt{2} - 6$ (e) $-4\sqrt{3} - 9\sqrt{2}$ (f) 4
(g) $3\sqrt{5}$ (h) $3\sqrt{7}$ (i) $-5\sqrt{2x}$
(j) $7n\sqrt{5n}$ (k) 0

2. (a) $\sqrt{x}$ (b) $\dfrac{1 + \sqrt{x}}{1 - x}$ (c) $2\sqrt{2} + 2$

(d) $\dfrac{\sqrt{x}}{x}$ (e) $\dfrac{\sqrt{14}}{2}$ (f) $\dfrac{\sqrt{x^3 + x}}{x}$

(y) $\dfrac{3\sqrt{x}}{2}$ (h) $\dfrac{\sqrt{x - 1}}{x - 1}$ (i) $\dfrac{\sqrt{b^2 \quad 4ac}}{2a}$

3. (a) $1, -7$ (b) $\dfrac{2}{3}, 1$ (c) $\dfrac{4}{3}, \dfrac{-5}{2}$ (d) $0, \dfrac{1}{7}$

4. (a) $\pm\sqrt{2}$ (b) $\pm\sqrt{17}$ (c) $\pm\sqrt{\dfrac{3}{2}} = +\dfrac{\sqrt{6}}{2}$ (d) $\pm\sqrt{\dfrac{11}{3}} = \pm\dfrac{\sqrt{33}}{3}$

5. (a) $5, 25$ (b) $x^2 - 12x + 36$ (c) $\dfrac{1}{2}, \dfrac{1}{4}$ (d) $\dfrac{3}{2}, \dfrac{9}{4}$ (e) $\dfrac{5}{2}, \dfrac{25}{4}$

6. (a) $1 \pm \sqrt{2}$ (b) $\dfrac{-1 \pm \sqrt{5}}{2}$

7. (a) $\dfrac{3 \pm \sqrt{29}}{2}$ (b) $\dfrac{5 \pm \sqrt{33}}{4}$ (c) $\dfrac{5 \pm \sqrt{73}}{8}$ (d) $\dfrac{7 \pm \sqrt{61}}{6}$

8. 1. Divide 2. Prepare to complete square 3. Complete square
4. Factor 5. Common denominator 6. Add fractions
7. Square root 8. Simplify 9. Solve for x 10. Add fractions

9. (a) $2 \pm \sqrt{3}$ (b) $\dfrac{-3 \pm \sqrt{17}}{4}$ (c) $\dfrac{3}{2}, -2$

(d) $\dfrac{-2 \pm \sqrt{13}}{3}$ (e) $\dfrac{3 \pm \sqrt{17}}{4}$ (f) $4, -\dfrac{1}{3}$

10. (a) $-1 \pm \sqrt{10}$ (b) $-1, -\dfrac{2}{3}$ (c) $\dfrac{1}{2}, -2$

(d) $0, -\dfrac{2}{5}$ (e) $\dfrac{-3 \pm \sqrt{21}}{6}$ (f) $\dfrac{3 \pm \sqrt{5}}{4}$

* * * CUMULATIVE REVIEW, CHAPTER 4 - 6 * * *

1. Write in logarithmic notation:

   (a) $x = 2^7$ (b) $3^4 = y$ (c) $7^2 = 49$ (d) $b^x = N$

2. Write in exponential notation:

   (a) $2 = \log_4 16$ (b) $6 = \log_2 64$ (c) $\log_b 4 = 3$ (d) $x = \log_b N$

3. Solve for x:

   (a) $\log_5 x = 3$ (b) $x\log 100 = 2$ (c) $\log_x 4 = -1$

   (d) $\log_9 3 = x$ (e) $10^{\log 2} = x$ (f) $\log(10)^{3.4} = x$

4. Solve for x:

   (a) $x = (3.259)^{14}$ (b) $x = \log_7 5$ (c) $x^7 = 3$

   (d) $\log_2 x = 1.34$ (e) $\sqrt[4]{6.759} = x$ (f) $(3)(7^x) = 10$

   (g) $10^{\log 2.7} = x$ (h) $x^{1.3} = 4.8$ (i) $(5.9)^x = 3.4$

5. How much will $100 be worth after seven years if it is left on deposit at a compound interest rate of 5%?

6. Which of the following are functions?

   (a) $x = y$ (b) $x^2 = y^2$ (c) $x = 2y$

   (d) $y = 2x$ (e) $x^2 = y$ (f) $x = y^2$

   (g) $y = \log x$ (h) $y = 10^x$ (i) $y = x^2 - x + 1$

7. Which of the following functions have inverses?

   (a) $y = 2^x$ (b) $y = 2x$ (c) $y = \log_2 x$

   (d) $y = x^2$ (e) $y = 2x - 1$ (f) $y = x^3$

8. In terms of g(x), write the formula of the inverse, g, if the function, f, is defined by

   (a) $f(x) = 7x$ (b) $f(x) = 10^x$ (c) $f(x) = \log_7 x$

   (d) $f(x) = 5x + 2$ (e) $f(x) = x - 3$ (f) $f(x) = \frac{2}{3}x + 4$

9. How can you tell by looking at a graph

   (a) whether it is a function?
   (b) If a function whether it has an inverse?

10. (a) If (2, -7) is on the graph of a function, f, then what point is on g, the inverse of f?

    (b) What is the geometric relation between the points in (a)?

11. If $P = (5, 40°)$ and $Q = (3, 20°)$ find in $a + bi$ form:

    (a) $P \cdot Q$ (b) P^2 (c) Q^3 (d) $\frac{1}{P}$

12. If $P = \sqrt{3} + i$ and $Q = \sqrt{3} - i$, find in a + bi form:

    (a) P + Q   (b) P • Q   (c) P^5 (d) $\dfrac{P}{Q}$ (e) -Q

Hint: (c)
Multiply in (r, θ)
form

13. Find all P in a + bi form such that

    (a) $P^4 = -1$ (b) $P^3 = i$ (c) $P^2 = -i$

14. Describe geometrically all P such that $P^{10} = 1$.

15. Solve for x:

    (a) $2x^2 + x + 5 = 0$ (b) $x^2 - 2x + 9 = 0$

* * *

ANSWERS TO CUMULATIVE REVIEW

1. (a) $\log_2 x = 7$ (b) $4 = \log_3 y$ (c) $2 = \log_7 49$ (d) $x = \log_b N$

2. (a) $4^2 = 16$ (b) $2^6 = 64$ (c) $4 = b^3$ (d) $b^x = N$

3. (a) 125   (b) 1   (c) $\frac{1}{4}$ (d) $\frac{1}{2}$ (e) 2   (f) 3.4

4. (a) $(1.525)(10^7)$ (b) 0.8271   (c) 1.170
    (d) 2.532   (e) 1.612   (f) 0.6187
    (g) 2.7   (h) 3.342   (i) 0.6895

5. $x = 100(1.05)^7 = \$140.70$ (about)   6. (a), (c), (d), (e), (g), (h), (i)

7. (a), (b), (c), (e), (f)

8. (a) $g(x) = \dfrac{x}{7}$ (b) $g(x) = \log x$ (c) $g(x) = 7^x$

    (d) $g(x) = \dfrac{x - 2}{5}$ (e) $g(x) = x + 3$ (f) $g(x) = \dfrac{3}{2}x - 6$

9. (a) Any vertical line intersects, at most, one point.
    (b) Any horizontal line intersects, at most, one point.

10. (a) (-7, 2)   (b) reflections across y = x

11. (a) (15, 60°)   (b) (25, 80°)   (c) (27, 60°)   (d) $(\frac{1}{5}, -40°)$

12. (a) $2\sqrt{3}$ (b) 4   (c) $-16\sqrt{3} + 16i$ (d) $\dfrac{1 + i\sqrt{3}}{2}$ (e) $-\sqrt{3} + i$

13. (a) $\dfrac{\sqrt{2}}{2} + \dfrac{\sqrt{2}}{2}i$, $\dfrac{-\sqrt{2}}{2} + \dfrac{\sqrt{2}}{2}i$, $\dfrac{-\sqrt{2}}{2} - \dfrac{\sqrt{2}}{2}i$, $\dfrac{\sqrt{2}}{2} - \dfrac{\sqrt{2}}{2}i$

    (b) $\dfrac{\sqrt{3}}{2} + \dfrac{1}{2}i$, $\dfrac{-\sqrt{3}}{2} + \dfrac{1}{2}i$, $-i$ (c) $\dfrac{-\sqrt{2}}{2} + \dfrac{\sqrt{2}}{2}i$, $\dfrac{\sqrt{2}}{2} - \dfrac{\sqrt{2}}{2}i$

14. Ten points equally spaced on a circle of radius 1 with center at the origin. One point is at (1, 0), the others are 36° apart.

15. (a) $\dfrac{-1 \pm i\sqrt{39}}{4}$ (b) $1 \pm 2i\sqrt{2}$

* * * CENTRAL SECTION * * *

Twenty students live in a certain dormitory. Each night three are chosen to prepare dinner. Can a month pass without the same three persons being chosen as cooks? Can a year pass? For how many nights can dinner be prepared without having to use the same three cooks twice?

EXERCISES

1. What would you guess are the answers to the preceding questions?

The next exercise concerns a smaller dormitory where the questions can be answered easily by the making of a list.

2. Assume that the dormitory has only six residents, named

A, B, C, D, E, and F.

(a) List all ways of selecting three cooks. For example, A, C, F, is one possibility and A, B, C is another.

(b) How many consecutive nights can the cooks be chosen in a different way each time?

Later we will return to these opening questions.

SETS AND SUBSETS

A <u>set</u> is any collection of objects, whether they are numbers, letters, people or what not. Thus,

{a, b, c, d, e }

is a set of five letters. The individual objects in a set are called <u>elements</u> or <u>members</u>. For instance, c is an element of the set { a, b, c, d, e }. The set {1, 2, A, B } has four elements.

> DEFINITION: A set that has n elements is called an n-set.

Example 1.   {a, b, c, d, e } is a 5-set

{A, B, C, D, E, F } is a 6-set

{c } is a 1-set

The residents of the dormitory described in the first paragraph of this chapter form a 20-set.

3. Give an example of a

(a) 3-set of numbers   (b) 7-set of letters   (c) 2-set of cities

* * *

The <u>order that the elements are listed in the set is not important.</u> Thus {a,b,x } and {x,a,b } are the same set. On the other hand, {a,b,x } and {a,b,c } are different sets.

In short, <u>two sets are equal only when they have the same elements</u>.

The dormitory problem involves the number of ways we can choose a 3-set from a 20-set. The next definition will be useful in talking about the dormitory problem later.

When we write A⊂B then every member of A is also a member of B.

> DEFINITION: If each element of the set A is an element of the set B, then A is a <u>subset</u> of B. We also say, "A is contained in B", and write
>
> $$A \subset B$$

Example 2. { a,c} is a subset of { a,b,c,d,e } . We write

$$\{a,c\} \subset \{a,b,c,d,e\}$$

Also

$$\{c,d,e\} \subset \{a,b,c,d,e\}$$

Note that a set is a subset of itself. For instance,

$$\{a,b,c,d,e\} \subset \{a,b,c,d,e\}$$

4.   Let
A = set of all quadrilaterals
B = set of all squares
C = set of all parallelograms
D = set of all rectangles

(a) no
(c) yes

(a) Is A ⊂ B?   (b) Is B ⊂ A?   (c) Is C ⊂ A   (d) Is D ⊂ C?

Hint: There are ten such subsets

5.   List all the 2-sets that are subsets of the 5-set {a,b,c,d,e }

6.   List all 1-sets that are subsets of the 3-set { a,b,c } .

Hint: There are three such subsets.

THE COMPLEMENT OF A SUBSET

Let B be a subset of the set A,

Pictorially: B ⊂ A.

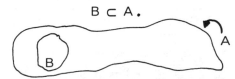

The part of A that is not in B is called the <u>complement of B</u> (in A).

Pictorially:

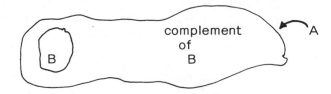

More precisely, the definition is as follows:

> DEFINITION: Let B be a subset of A. The <u>complement of B</u> (in A) is the subset of A that consists of the elements of A that are not in B.

The complement of set B is sometimes denoted: B' or B̄, or -B, or ~B. However, you will not need such notations for this course.

Example 3. Let A = {a,b,c,d,e} and B = { b,e } .

The complement of B is { a,c,d } .

The "complement" provides a shortcut for dealing with some subsets, as the next several exercises will show.

7. Let A = { a,b,c,d,e } :

(a) Make a list of the ten 2-sets contained in A. With each such 2-set list its complement.

(b) Use (a) to show that the number of 2-sets contained in a 5-set is the same as the number of 3-sets contained in a 5-set.

The number of ways you can choose a 2-set is the same as the number of ways you can choose the complement of a 2-set from a 5-set.

8. (a) List the 1-sets contained in { a,b,c,d,e,f } .

(b) For each 1-set in (a) list its complement.

(c) Using (a) and (b), determine the number of 5-sets contained in a 6-set.

9. Let A be a 15-set.

(a) How many 1-sets are contained in A?

(b) How many 14-sets are contained in A?

Do not make a list.

THE NUMBER $\binom{n}{k}$

The opening problem concerns the number of 3-sets contained in a 20-set. To help deal with this problem efficiently, we introduce the following definition:

> DEFINITION: Let n and k be integers such that $0 < k \leq n$. From a given n-set it is possible to choose various k-sets. The number of ways of choosing from a fixed n-set all possible k-sets is denoted
>
> $$\binom{n}{k}$$
>
> The symbol $\binom{n}{k}$ is read "n choose k", or "n over k".

The condition $0 < k \leq n$ is very important here.

Note that $\binom{n}{k}$ is not the quotient n divided by k.

Observe that $\binom{n}{k}$ is a <u>number,</u> as Example 4 illustrates.

In terms of sets, recite to yourself what each of these numbers means.

Example 4. Exercise 2 shows that
$$\binom{6}{3} = 20.$$

Exercise 5 shows that

$$\binom{5}{2} = 10.$$

The opening problem concerns

$$\binom{20}{3}.$$

This chapter is devoted to the study of the numbers $\binom{n}{k}$ and some of their applications. In particular <u>it will develop ways of computing</u> $\binom{n}{k}$ <u>without having to make a list of all the k-sets contained in an n-set.</u>

10. Review Exercise 7. What is the relation between $\binom{5}{2}$ and $\binom{5}{3}$?

11. Review Exercise 8. What is the relation between $\binom{6}{1}$ and $\binom{6}{5}$?

Hint: Compare the number of ways you can choose a 3-set from a 15-set with the number of ways you can not choose the remaining 12-set.

12. Explain, in a short paragraph, why

$$\binom{n}{1} = \binom{n}{n-1}$$

14. Explain why

$$\binom{n}{k} = \binom{n}{n-k}$$

See the hint for Exercise 12.

for $0 < k < n$.

15. (a) Show that $\{a,b,c,d,e\} \subset \{a,b,c,d,e\}$.

(b) Evaluate $\binom{5}{5}$, that is, how many ways can you choose a subset of five elements from a set of five elements.

16. (a) Evaluate $\binom{1}{1}$.   (b) Evaluate $\binom{2}{2}$.

(c) For any positive integer n, evaluate $\binom{n}{n}$.

THE EMPTY SET

Exercise 14 concerns the equation

$$\binom{n}{k} = \binom{n}{n-k}$$

for $k = 1, 2, 3, ..., n - 1$. If we let $k = n$, the equation becomes

$$\binom{n}{n} = \binom{n}{n-n}$$

that is,

$$\binom{n}{n} = \binom{n}{0} .$$

Now, $\binom{n}{n}$ makes sense; in fact, according to Exercise 16,

$$\binom{n}{n} = 1.$$

But up to this point $\binom{n}{0}$ has not been defined. Let us pause to see what meaning the symbol $\binom{n}{0}$ should have.

The symbol $\binom{n}{0}$ should stand for the number of ways that we can, from a given n-set, choose a 0-set. But what is a 0-set? The following definition turns out to be useful.

> DEFINITION: A 0-set is the set that has no elements whatsoever. There is only one such set, called the <u>empty set</u>. It is denoted ∅. The empty set is considered to be a subset of any set.

It is important to remember that the empty set is a subset of every set.

Example 5.   Compute $\binom{4}{0}$.

Solution.   Let $\{a,b,c,d\}$ be the 4-set in question. How many ways can we choose from $\{a,b,c,d\}$ <u>none</u> of the elements a,b,c,d? There is only one way: Choose none of them. Thus,

$$\binom{4}{0} = 1.$$

17. Evaluate:

(a) $\binom{1}{0}$ 　　　　(b) $\binom{2}{0}$ 　　　　(c) $\binom{17}{0}$ 　　　　(d) $\binom{20}{0}$

18. Explain why

(a) $\binom{3}{0} = \binom{3}{3}$ 　　　　(b) $\binom{5}{0} = \binom{5}{5}$ 　　　　(c) $\binom{1}{0} = \binom{1}{1}$

(d) $\binom{n}{k} = \binom{n}{n-k}$ when $k = n$.

19. Evaluate $\binom{10}{k}$ for $k = 0, 1, 9, 10$.

20. Evaluate $\binom{n}{0}$, $\binom{n}{1}, \binom{n}{n-1}, \binom{n}{n}$.

$*\;*\;*$

We turn now to a shortcut for computing $\binom{n}{k}$. As Exercise 20 shows, if $k = 0, 1, n-1$, or n, it is not too difficult to find $\binom{n}{k}$. However, finding $\binom{n}{2}$ is troublesome. (How many ways can you choose a 2-set from an n-set?)

This is not an easy question to answer. The problem of the dormitory cooks concerns $\binom{20}{3}$. Who wants to list all of the 3-sets contained in a 20-set?

And if there were such a person, how would he know that his list was complete? So we must turn to ways of finding $\binom{n}{k}$ efficiently.

PASCAL'S TRIANGLE

Keep in mind that $\binom{n}{k}$ is a number.

We first develop a rapid way of computing $\binom{n}{k}$ when n is a fairly small number.

21. (a) Evaluate $\binom{4}{2},\binom{4}{3}$, and $\binom{5}{3}$ by making lists of subsets.

   (b) Check that
$$\binom{5}{3} = \binom{4}{2} + \binom{4}{3}.$$

EXERCISES 21-25 REQUIRE CAREFUL CONSIDERATION!

The next exercise shows <u>why</u> the equation in Exercise 21 (b) is true without making a list of the elements.

22. Let $\{a,b,c,d,e\}$ be the 5-set to be considered. The number $\binom{5}{3}$ records the number of 3-sets that can be chosen from $\{a,b,c,d,e\}$. There are two types of such 3-sets that can be chosen:

   "Those in which e is an element"

and "Those in which e is <u>not</u> an element".

Suppose e is to be a member of the chosen 3-set. Then there are $\binom{4}{2}$ ways to choose the other two members. On the other hand, if e is not a member of the chosen 3-set then there are $\binom{4}{3}$ ways to choose a 3-set from the remaining members. Study the results of this exercise carefully.

   (a) Explain why $\binom{4}{2}$ is the number of 3-sets of which e is an element.

   (b) Explain why $\binom{4}{3}$ is the number of 3-sets of which e is <u>not</u> an element.

   (c) Use (a) and (b) to show why
$$\binom{5}{3} = \binom{4}{2} + \binom{4}{3}$$

23. Let A = $\{a,b,c,d,e,f,g,h\}$ be an 8-set. There are $\binom{8}{4}$ ways to choose a subset with four elements.

   (a) Show that there are $\binom{7}{3}$ ways to choose a 4-set from set A if that 4-set <u>contains</u> the element h.

   (b) Show that there are $\binom{7}{4}$ ways to choose a 4-set if that 4-set does <u>not contain</u> the element h.

   (c) Explain why
$$\binom{8}{4} = \binom{7}{3} + \binom{7}{4}$$

24. Explain why
$$\binom{7}{4} = \binom{6}{3} + \binom{6}{4}$$

25. Let n be larger than 6. Explain why

$$\binom{n}{6} = \binom{n-1}{5} + \binom{n-1}{6}$$

The idea of Exercises 21-25 is expressed in the following useful
Theorem.

THEOREM 1. For any positive integer n and any integer k, $1 \le k \le n-1$,

$$\binom{n}{k} = \binom{n-1}{k-1} + \binom{n-1}{k}$$

This diagram records Theorem 1:

You may find it convenient to give Theorem 1 the name "Pascal's Formula". Study this discussion carefully.

Add these two numbers:

You get the number

For instance,

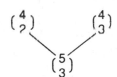

records the equation

$$\binom{5}{3} = \binom{4}{2} + \binom{4}{3}$$

This is the general pattern:

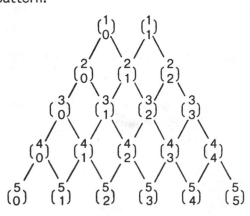

and so on.

We already know the numbers in the two outside diagonals; each of them equals 1. That is, $\binom{n}{0} = 1 = \binom{n}{n}$.

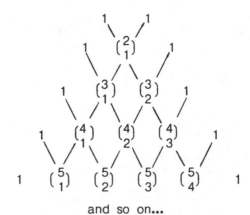

and so on...

Theorem 1 ("Pascal's Theorem") enables one to fill in any row if you know the row above it. Thus, with the aid of Theorem 1 you can compute

$$\binom{5}{1} \qquad \binom{5}{2} \qquad \binom{5}{3} \qquad \binom{5}{4}$$

if you know the values in the row above:

$$\binom{4}{0} = 1 \qquad \binom{4}{1} = 4 \qquad \binom{4}{2} = 6 \qquad \binom{4}{3} = 4 \qquad \binom{4}{4} = 1$$

Place the known values in the row when n = 4:

$$1 \qquad 4 \qquad 6 \qquad 4 \qquad 1 \qquad (n = 4)$$

Then write the <u>sums</u> of the adjacent numbers below:

$$1 \qquad 4 \qquad 6 \qquad 4 \qquad 1 \qquad (n = 4)$$
$$5 \qquad 10 \qquad 10 \qquad 5 \qquad (n = 5)$$

26. Use the preceding technique to fill in the next row (n = 6). In other words, use the technique to find

$$\binom{6}{1} \qquad \binom{6}{2} \qquad \binom{6}{3} \qquad \binom{6}{4} \qquad \binom{6}{5} \qquad \binom{6}{6} \ .$$

27. Using the result of Exercise 26, evaluate

   (a) $\binom{6}{2}$ (b) $\binom{6}{3}$

(a) 15

28. Fill in the row where n = 7.

29. Using Exercise 28, find

   (a) $\binom{7}{2}$ (b) $\binom{7}{3}$

(b) 35

* * *

The "n = 0" row consists of one number, $\binom{0}{0}$. Now, there is exactly one 0-set in a 0-set.   Thus

$$\binom{0}{0} = 1.$$

The rows, starting with n = 0, form a triangle in

It is important to remember there is a zero row. You will see later why this is convenient.

$$\binom{0}{0} \qquad\qquad\qquad (n = 0)$$

$$\binom{1}{0} \quad \binom{1}{1} \qquad\qquad\qquad (n = 1)$$

$$\binom{2}{0} \quad \binom{2}{1} \quad \binom{2}{2} \qquad\qquad\qquad (n = 2)$$

$$\binom{3}{0} \quad \binom{3}{1} \quad \binom{3}{2} \quad \binom{3}{3} \qquad\qquad\qquad (n = 3)$$

$$\binom{4}{0} \quad \binom{4}{1} \quad \binom{4}{2} \quad \binom{4}{3} \quad \binom{4}{4} \qquad\qquad\qquad (n - 4)$$

and so on.

When evaluated as integers, this triangle becomes

```
                1                    (n = 0)
             1     1                 (n = 1)
          1     2     1              (n = 2)
       1     3     3     1           (n = 3)
    1     4     6     4     1        (n = 4)
```

and so on.

These are just the first few rows of a triangle that continues downward without end.   This endless triangle is called <u>Pascal's triangle.</u>

30. (a) Fill in Pascal's triangle all the way through the row for n = 8.

(b) Make two triangles, one in $\binom{n}{k}$ notation, the other using integers.

* * *

Pascal's triangle provides a way to compute $\binom{n}{k}$ without making a list of subsets.   However, the larger n is the longer it takes to find $\binom{n}{k}$.   The computation goes row by row until the nth row is reached.

15

31. Use Pascal's triangle to find how many cooking crews of 4 persons can be chosen from 6 people.

35

32. Use Pascal's triangle to find how many ways you can choose a 4-set from a 7-set.

33. Use Pascal's triangle to evaluate

(a) 70
(b) 84

(a) $\binom{8}{4}$ (b) $\binom{9}{3}$.

Hint: Compare $\binom{n}{k}$

and $\binom{n}{n-k}$.

34. In Exercise 30, you filled in Pascal's triangle through n = 8. Each row reads the same backwards as forwards. Is that a coincidence or is there a reason why this happens? Explain.

* * *

The problem that introduced this chapter concerns the number $\binom{20}{3}$. This

Remember that the top number in Pascal's triangle is the "zero row."

number could be computed with the aid of Pascal's triangle by filling in the values of each row until the "n = 20" row is reached. While this could take only a few minutes, there is a faster way. It is possible to compute $\binom{20}{3}$ directly, by a very simple formula which will now be described.
NOTE. The reason the formula works will be reserved for the next chapter which concerns counting.

FACTORIAL NOTATION

The short formula for calculating $\binom{n}{k}$ will involve the product of all integers from 1 through the positive integer n,

$$1 \cdot 2 \cdot 3 \cdots (n-2) \cdot (n-1) \cdot n$$

This product will be denoted

$$n!$$

and called "n factorial".

For instance,

$$5! = 1 \cdot 2 \cdot 3 \cdot 4 \cdot 5 = 120$$

If you have a calculator compute n! for larger and larger n. Note how quickly it grows.

35. Evaluate   (a) 2!   (b) 3!   (c) 4!   (d) 6!

In particular,

$$1! = 1$$

(a) 2
(d) 720

As you will see, it is very convenient if the value of 0! = 1. Thus we make the following special definition:

> DEFINITION: Zero factorial is defined to be 1,
> $$0! = 1.$$

The next example illustrates a shortcut in working with factorials.

Example 6. Evaluate $\frac{7!}{6!}$

Solution. The hard way is to compute 7! and 6!, then divide. The easy way is to note a lot of cancellations:

$$\frac{7!}{6!} = \frac{7 \quad \cancel{6} \quad \cancel{5} \quad \cancel{4} \quad \cancel{3} \quad \cancel{2} \quad \cancel{1}}{\cancel{6} \quad \cancel{5} \quad \cancel{4} \quad \cancel{3} \quad \cancel{2} \quad \cancel{1}} = 7$$

36. Use the shortcut of Example 6 to evaluate

(a) $\dfrac{10!}{9!}$ (b) $\dfrac{100!}{99!}$ (c) $\dfrac{n!}{(n-1)!}$

(a) 10

37. (a) Show that $\dfrac{10!}{8!} = 10 \cdot 9$ (Use cancellations)

(b) Show that

$$\frac{n!}{(n-2)!} = n(n-1)$$

38. Evaluate

(a) $\dfrac{100!}{99!}$ (b) $\dfrac{100!}{98!}$ (c) $\dfrac{100!}{97!}$

(b) 9900

39. Evaluate

(a) $\dfrac{100!}{98! \, 2!}$ (b) $\dfrac{20!}{3! \, 17!}$

(a) 4950

SHORT FORMULA FOR $\binom{n}{k}$

Now comes the main theorem of the chapter. It provides a short formula for $\binom{n}{k}$ in terms of factorials.

THEOREM 2. (Number of k-sets contained in an n-set)

$$\binom{n}{k} = \frac{n!}{k! \, (n-k)!}$$

Accept on faith, memorize, and use this formula. Later, we will prove it.

<u>Memorize this formula.</u> It tells how many k-sets are contained in an n-set.

The next exercise already displays the power of Theorem 2 by answering the questions that opened the chapter.

40. (a) Use the formula in Theorem 2 to show that

$$\binom{20}{3} = 1140$$

(b) How many nights can three cooks be chosen from a set of twenty people without duplicating the same set of three cooks?

As Exercise 40 shows, more than three years will pass before the dormitory must use the same three cooks twice!

* * *

41. Evaluate $\binom{6}{2}$

(a) by making a Pascal's triangle,

(b) by using the formula of Theorem 2.

42. Use Theorem 2 to show that

(a) $\binom{n}{1} = n$ (b) $\binom{n}{n-1} = n$

330

43. From an 11-set, how many ways can you choose a 4-set?

(a) 36
(c) 15,504
(e) 499,500

44. Compute:

(a) $\binom{9}{2}$ (b) $\binom{12}{4}$ (c) $\binom{20}{5}$ (d) $\binom{50}{3}$ (e) $\binom{1000}{998}$

286

45. How many 3-sets are contained in a 13-set?

46. Show that $\binom{100}{3} = \dfrac{100 \cdot 99 \cdot 98}{1 \cdot 2 \cdot 3}$

47. Show that $\binom{45}{4} = \dfrac{45 \cdot 44 \cdot 43 \cdot 42}{1 \cdot 2 \cdot 3 \cdot 4}$

* * *

$\binom{100}{5} = \dfrac{100!}{5!\,95!}$

$= \dfrac{100 \cdot 99 \cdot 98 \cdot 97 \cdot 96 \cdot 95!}{5! \quad 95!}$

$= \dfrac{100 \cdot 99 \cdot 98 \cdot 97 \cdot 96}{1 \cdot 2 \cdot 3 \cdot 4 \cdot 5}$

Exercises 46 and 47 suggest a short cut for evaluating

$$\frac{n!}{k!\,(n-k)!}$$

when k is small: (1)   write out $1 \cdot 2 \cdots k$ in the denominator,

(2)   then write n above 1, n – 1 above 2, and so on, to form the numerator, which is
$n(n-1)(n-2) \cdots (n-k+1)$

For instance $\binom{100}{5} = \dfrac{100 \cdot 99 \cdot 98 \cdot 97 \cdot 96}{1 \cdot 2 \cdot 3 \cdot 4 \cdot 5}$

48. Evaluate (a) $\binom{88}{2}$ (b) $\binom{100}{3}$

(a) 3828

49. Use Theorem 2 to evaluate (a) $\binom{8}{0}$ (b) $\binom{8}{8}$ (c) $\binom{n}{0}$ (d) $\binom{n}{n}$

* * *

Most of this chapter has been devoted to defining the numbers $\binom{n}{k}$ and and then showing how to compute them by using Pascal's triangle and by the formula

$$\binom{n}{k} = \frac{n!}{k!\,(n-k)!}$$

The rest of the chapter concerns the use of the numbers $\binom{n}{k}$ in algebra.

THE BINOMIAL THEOREM

In programming and other applications of mathematics it is sometimes necessary to multiply out an expression of the form $(a + b)^n$, where a and b are numbers and n is a positive integer.

For example,

$$(a + b)^2 = a^2 + 2ab + b^2$$

50. By multiplying, show that
$$(a + b)^3 = a^3 + 3a^2b + 3ab^2 + b^3.$$

51. By multiplying, show that
$$(a + b)^4 = a^4 + 4a^3b + 6a^2b^2 + 4ab^3 + b^4$$

52. As in Exercise 50 and 51, find $(a + b)^5$.

The labor of multiplying $(a + b)^n$ increases rapidly as n gets larger, as Exercises 50-52 may already suggest. But look back at the first few rows of Pascal's triangle. Compare the row for n = 4 to the expansion of $(a + b)^4$ as in Exercise 51. The numerical coefficients of the terms in

$$1a^4 + 4a^3b + 6a^2b^2 + 4ab^3 + 1b^4$$

are   1   4   6   4   1,

exactly the "n = 4" row of Pascal's triangle.

CAREFUL!

Again, the top row in Pascal's triangle corresponds to n = 0. Thus do not merely count rows from the top to find the row where n = 4.

53. (a) Do the numerical coefficients in the expansion of $(a + b)^3$ match the n = 3 row of Pascal's triangle?

(b) Does the n = 5 row of Pascal's triangle match the coefficients in the expansion of $(a + b)^5$?

54. How do you think the expression $(a + b)^1$ relates to Pascal's triangle?

* * *

Why should Pascal's triangle be related to the expansion of $(a + b)^n$? Or, to ask a more specific question, when we expand $(a + b)^5$ why is the <u>coefficient</u> of a^2b^3 equal to $\binom{5}{3}$?

The answer is not hard to find. To compute $(a + b)^5$ we multiply out the five factors

$$(a + b)(a + b)(a + b)(a + b)(a + b).$$

How do we get products of the form a^2b^3? We simply choose <u>b</u> from three of the five $(a + b)$ factors and <u>a</u> from the other two. One way is to choose <u>b</u> from each of the first three factors and <u>a</u> from the fourth and fifth factors. (This is shown by the underlined b's.)

$$(a + \underline{b})(a + \underline{b})(a + \underline{b})(a + b)(a + b)$$

Another way is to choose b from the second, fourth, and fifth factors and and a from the other factors.

$$(a + b)(a + \underline{b})(a + b)(a + \underline{b})(a + \underline{b})$$

Every time we choose b from three of the five factors and a from the other two we get a product of the form a^2b^3. The total number of ways of doing this is $\binom{5}{3}$, the number of 3-sets contained in a 5-set. This observation may be generalized to explain why the expansion of $(a + b)^n$ is related to the numbers $\binom{n}{k}$ and hence to Pascal's triangle. For this reason the numbers $\binom{n}{k}$ are called <u>binomial coefficients</u>. (The expression a + b is called a "binomial.") The preceding argument is the basic of Theorem 3.

THEOREM 3. (The Binomial Theorem.) When $(a + b)^n$ is expanded, the coefficient of the term involving

$$a^{n-k}b^k \text{ is } \binom{n}{k}.$$

That is,

Note that k is the power of $\underline{b}$ in the expansion not the power of $\underline{a}$.

$$(a + b)^n = \binom{n}{0}a^nb^0 + \binom{n}{1}a^{n-1}b^1 + \binom{n}{2}a^{n-2}b^2 + \ldots + \binom{n}{n-1}ab^{n-1} + \binom{n}{n}a^0b^n$$

Example 7. Use the binomial theorem to compute $(a + b)^4$ directly.

Solution. The binomial theorem for n = 4 asserts that

OBSERVE THE PATTERN HERE.

$$(a + b)^4 = \binom{4}{0}a^4b^0 + \binom{4}{1}a^3b + \binom{4}{2}a^2b^2 + \binom{4}{3}a\,b^3 + \binom{4}{4}a^0b^4,$$

which is simply

$$a^4 + 4a^3b + 6a^2b^2 + 4ab^3 + b^4.$$

NOTE. The <u>sum</u> of the two exponents in any term equals 4, which is the exponent of the binomial $(a + b)^4$. This is true for any n. That is, in the expansion of $(a + b)^n$ the sum of the exponents will be n for any term of the expansion.

55. Use the binomial theorem to expand $(a + b)^3$. Compare your answer with that in Exercise 50.

56. Use the binomial theorem to expand $(a + b)^5$. Compare the result with that in Exercise 52.

(a) 35
(c) 21

57. In the expansion of $(a + b)^7$ use the binomial theorem to find the coefficient of each of the following terms:

(a) a^3b^4 (b) a^4b^3 (c) a^2b^5 (d) $a\,b^6$

(a) 1
(c) 220
(e) 792

58. In the expansion of $(a + b)^{12}$ find the coefficient of

(a) a^{12} (b) $a^{11}b$ (c) a^9b^3 (d) a^6b^6 (e) a^5b^7

* * *

The letters $\underline{a}$ and $\underline{b}$ in $(a + b)^n$ can be any numbers. For instance, if $a = 1$, $b = x$, and $n = 4$, the binomial theorem asserts that

$$(1 + x)^4 = \binom{4}{0}1^4x^0 + \binom{4}{1}1^3x^1 + \binom{4}{2}1^2x^2 + \binom{4}{3}1^1x^3 + \binom{4}{4}1^0x^4,$$

which reduces to

$$1 + 4x + 6x^2 + 4x^3 + x^4.$$

The next few exercises illustrate the form of the expansion of $(a + b)^n$ when a and b are replaced by various expressions.

59. Use the binomial theorem to expand $(x - y)^5$.

60. Use the binomial theorem to expand $(1 - x)^4$.

61. Use the binomial theorem to expand

(a) $(1 + 2x)^4$ (b) $(1 + x)^6$ (c) $(x + y)^3$ (d) $(1 - 2x)^5$

* * *

SIGMA NOTATION

In the previous exercises, sums of a particular form were presented. Since such sums occur often, we now introduce a convenient notation for them, called $\underline{sigma\ notation}$, named after the Greek letter Σ, which corresponds to the S of Sum. Sigma notation is sometimes called "summation notation".

DEFINITION OF SIGMA NOTATION
The sum $a_1 + a_2 + a_3 + \dots + a_n$ may be denoted by the symbol

$$\sum_{i=1}^{n} a_i\ ,$$

which is read "the sum of a sub i as i goes from 1 to n."

In the above definition i is called the $\underline{summation\ index}$. Any other letter could be used instead. such as j or k .

Example 8. Write the sum $1^2 + 2^2 + 3^2 + 4^2$ in sigma notation.

Solution. Since the ith summand is the square of i and the summand extends from $i = 1$ to $i = 4$, we have

$$1^2 + 2^2 + 3^2 + 4^2 = \sum_{i=1}^{4} i^2$$

Example 9. Compute $\sum_{i=1}^{5} i$

Solution. This is short for $1 + 2 + 3 + 4 + 5$, or 15.

59. Hint: first write it as $[x + (-y)]^5$.

60. Hint: first write it as $[1 + (-x)]^4$.

61. (a) Hint: write $(1 + 2x)^4$ as $[1 + (2x)]^4$
(a) $1 + 8x + 24x^2 + 32x^3 + 16x^4$
(c) $x^3 + 3x^2y + 3xy^2 + y^3$.

Example 10. Compute $\sum_{j=3}^{5} (2j-1)$

Solution. This is short for $[2(3) - 1] + [2(4) - 1] + [2(5) - 1]$ which is

$5 + 7 + 9$, or 21.

Note.   As Example 10 shows, the lower or beginning index is not necessarily 1; it could be any number. Also, a sigma notation of a sum is not unique. That is, there may be more than one way to write a sum in sigma notation. This occurs because we may use different lower and upper numbers for the index along with a different algebraic formula to give the same results.

62. Compute

(a) $\sum_{i=1}^{6} 2i$

(b) $\sum_{i=3}^{8} (i+1)$

(c) $\sum_{i=1}^{3} \left(i^2 - 1\right)$

(d) $\sum_{i=0}^{4} \dfrac{1}{2^i}$

(e) $\sum_{i=0}^{5} (-1)^i$

(f) $\sum_{i=1}^{4} i^3$

(g) $\sum_{i=1}^{5} (3i-1)$

(h) $\sum_{i=0}^{4} \dfrac{i}{i+1}$

(i) $\sum_{i=1}^{3} i(i+2)$

63. Write in sigma notation.

Note. Two students may have different answers here and both may be correct.

(a) $1 + 2 + 3 + 4 + 5$ (b) $3 + 4 + 5 + 6$ (c) $3 + 5 + 7 + 9$

(d) $\dfrac{1}{2} + \dfrac{1}{3} + \dfrac{1}{4} + \dfrac{1}{5}$ (e) $\dfrac{1}{8} + \dfrac{1}{16} + \dfrac{1}{32}$ (f) $2 + 5 + 8 + 11$

(g) $1 + 2 + 2^2 + 2^3 + \ldots + 2^{50}$ (h) $x^3 + x^4 + x^5 + x^6$

64. Explain why we may write the Binomial Theorem as

$$(a + b)^n = \sum_{k=0}^{n} \binom{n}{k} a^{n-k} b^{k}.$$

65. Write each of the following in sigma notation.

(a) $(a + b)^3$ (b) $(x + y)^7$ (c) $(z + w)^5$

* * * SUMMARY * * *

A problem in choosing cooks in a dormitory led to the study of the number of k-sets contained as subsets of an n-set. This number is denoted $\binom{n}{k}$. The relation

$$\binom{n}{k} = \binom{n-1}{k-1} + \binom{n-1}{k}$$

provides the basis for Pascal's triangle, which is of use in computing $\binom{n}{k}$ when n is small.

Using the definition of n factorial,

$$n! = 1 \cdot 2 \cdot 3 \cdots (n-2)(n-1)n,$$

we stated without proof that

$$\binom{n}{k} = \frac{n!}{k!\,(n-k)!}.$$

Why this formula is correct will be shown in the next chapter.

We also pointed out that the numbers $\binom{n}{k}$ also appear as coefficients of the expansion of $(a+b)^n$. The term with $a^{n-k}b^k$ has as a coefficient the number $\binom{n}{k}$. For this reason the number $\binom{n}{k}$ is usually called the "binomial coefficient $\binom{n}{k}$."

The binomial theorem gives the coefficient of $a^{n-k}b^k$ in the expansion of $(a+b)^n$,

$$\binom{n}{0}a^n b^0 + \binom{n}{1}a^{n-1}b + \binom{n}{2}a^{n-2}b^2 + \ldots + \binom{n}{n-1}ab^{n-1} + \binom{n}{n}a^0 b^n.$$

In sigma (or summation) notation, the Binomial Theorem may be written,

$$(a+b)^n = \sum_{k=0}^{n} \binom{n}{k}a^{n-k}b^k.$$

In many algebra and probability texts, $\binom{n}{k}$ is called "the number of com-binations of n things taken k at a time." It is sometimes denoted by the symbol

$$C_k^n \qquad \text{or} \qquad {}_nC_k$$

The equation

$$\binom{n}{k} = \binom{n-1}{k-1} + \binom{n-1}{k}$$

is the basis of Pascal's triangle, which starts out

Be sure to study the summary and main sections carefully. Ask questions.

$\sum_{k=0}^{n}$ *may also be written as* $\Sigma_{k=0}^{n}$

Only the $\binom{n}{k}$ notation will be used in this text.

"Combination" and "set" have the same meaning, as you will see in Chapter 8.

```
                        1
                   1         1
              1         2         1
         1         3         3         1
    1         4         6         4         1
```

Each number (other than the 1's) is the sum of the two numbers just above it to the right and left.

In the next chapter it will be proved that

$$\binom{n}{k} = \frac{n!}{k! \ (n - k)!}.$$

The binomial theorem asserts that the coefficient of the term
$$a^{n-k}b^k$$
in the expansion of $(a + b)^n$ is $\binom{n}{k}$.

A <u>set</u> is any collection of objects, called <u>elements</u> or <u>members.</u>

A set is a <u>subset</u> of another set if each of its elements is an element of the second set.

A set with n elements is called an <u>n-set.</u>

There is only one 0-set, called the <u>empty set,</u> also denoted ∅.

Let the set B be a subset of the set A. Then the set consisting of the elements of A not in B is called the <u>complement</u> of B (in A). If A is an n-set and B is a k-set, the complement of B is a set containing n - k elements.

* * * WORDS AND SYMBOLS * * *

set   Pascal's triangle

subset   n factorial   n!

element, member   sigma notation

A ⊂ B (A is a subset of B)   Binomial Theorem

$\binom{n}{k} = \frac{n!}{k! \ (n - k)!}$ $(a + b)^n = \sum_{k=0}^{n} \binom{n}{k} a^{n-k}b^k$

* * * CONCLUDING REMARKS * * *

The formula for the coefficient of x^k in the expansion of $(1 + x)^n$ was known to the Indian mathematician Bhaskara in 1150 and the French mathematician Levi ben Gerson in 1321. Pascal's triangle had been studied long before Pascal was born, but Pascal in the 17th century was the first to relate the coefficient of x^k in the expansion of $(1 + x)^n$ to the number of k-sets in an n-set.

Isaac Newton in 1665, before he was twenty-three, had noticed that the binomial theorem holds even when the exponent n is not an integer.   He claimed that if -1 < x < 1 then

$$(1 + x)^n = 1 + \frac{n}{1}x + \frac{n(n - 1)}{1 \cdot 2}x^2 + \frac{n(n - 1)(n - 2)}{1 \cdot 2 \cdot 3}x^3 + \cdots$$

(where ... means that further terms are formed in the same pattern without stopping.)

For instance, Newton showed that

$$(1 + x)^{\frac{1}{2}} = 1 + \frac{\frac{1}{2}}{1}x + \frac{\frac{1}{2}(\frac{1}{2} - 1)}{1 \cdot 2}x^2 + \frac{\frac{1}{2}(\frac{1}{2} - 1)(\frac{1}{2} - 2)}{1 \cdot 2 \cdot 3}x^3 + \cdots$$

an equation which asserts that

$$\sqrt{1 + x} = 1 + \frac{x}{2} - \frac{x^2}{8} + \frac{x^3}{16} - \cdots$$

* * * BINOMIAL THEOREM SELF QUIZ * * *

1. Mike has ten good friends. He wishes to invite them five at a time to a series of parties at his house. How many parties may he give and still not invite the same five people to his house twice?

2. Use factorial notation to prove that $\binom{8}{4} = \binom{7}{3} + \binom{7}{4}$.

3. In the expansion of $(x + y)^{10}$ find the coefficient of

    (a) $x^9 y$ (b) $x^5 y^5$ (c) $x^3 y^7$ (d) $x^7 y^3$

4. Expand and write the coefficients as integers when possible:

    (a) $(x - y)^5$ (b) $(2x + 1)^4$ (c) $(2x - \frac{1}{2})^6$ (d) $(3x - 2y)^3$

5. If each student in a class is to select 6 problems from a set of 20 problems, how many students must be in the class before there will certainly be a repetition of a 6-problem set?

6. This row appears as a line in Pascal's triangle:

    1   9   36   84   126   126   84   36   9   1

    (a) To which n does it correspond?

    (b) Use it to fill in the next row of Pascal's triangle.

7. Compute.

    (a)   (b)

    $$\sum_{i=5}^{8} (4i - 1)$$ $$\sum_{i=0}^{4} i^3$$

ANSWERS:

1.   252

2. $\binom{8}{4} = \dfrac{8 \cdot 7 \cdot 6 \cdot 5}{1 \cdot 2 \cdot 3 \cdot 4} = 70$, while $\binom{7}{3} + \binom{7}{4} = \dfrac{7 \cdot 6 \cdot 5}{1 \cdot 2 \cdot 3} + \dfrac{7 \cdot 6 \cdot 5 \cdot 4}{1 \cdot 2 \cdot 3 \cdot 4} = 35 + 35 = 70$

3.   (a) 10   (b) 252   (c) 120   (d) 120

4.   (a) $x^5 - 5x^4 y + 10x^3 y^2 - 10x^2 y^3 + 5xy^4 - y^5$

     (b) $16x^4 + 32x^3 + 24x^2 + 8x + 1$

     (c) $64x^6 - 96x^5 + 60x^4 - 20x^3 + \dfrac{15}{4}x^2 - \dfrac{3}{8}x + \dfrac{1}{64}$

     (d) $27x^3 - 54x^2 y + 36xy^2 - 8y^3$

5.   38761   6. (a) n = 9

     (b) 1   10   45   120   210   252   210   120   45   10   1

7.   (a) 100   (b) 100

* * *   BINOMIAL THEOREM REVIEW   * * *

1.   This exercise shows that a single mathematical idea can come in many
     disguises.   The map below shows Bill's house and his school, which is
     located at 4th Avenue and 6th Street.   Point D indicates a drug store
     at 2nd Avenue and 2nd Street.   Point B indicates a bakery at 3rd Ave-
     nue and 1st Street.

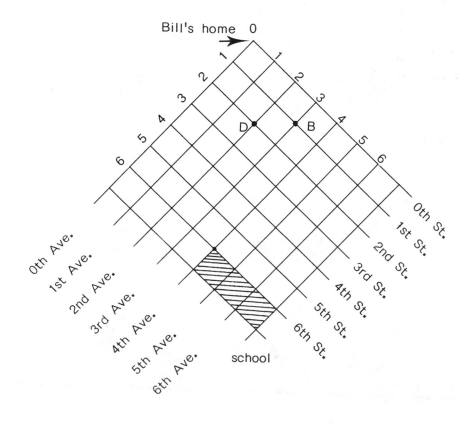

How many routes does Bill have from his home to

(a) the bakery at B?
(b) the drugstore at D?
(c) the corner of 1st Avenue and 2nd Street?
(d) the corner of 0th Avenue and 5th Street?
(e) the school at 4th Avenue and 6th Street?
(f) Do you see a familiar pattern?

Bill always takes one of the shortest possible routes.

2. Evaluate each of the following in the simplest way.

(a) the number of 3-sets contained in a 7-set.
(b) the number of 4-sets contained in a 7-set.
(c) the number of 5-sets contained in an 8-set.
(d) the number of 5-sets contained in an 11-set.

3. Write as an integer:

(a) $\binom{5}{2}$ (b) $\binom{4}{1}$ (c) $\binom{6}{3}$

4. Use the factorial notation for $\binom{n}{r}$ to show that $\binom{5}{2} = \binom{5}{3}$.

5. Expand in the easiest manner possible (write the coefficients as integers):

(a) $(x + 1)^4$ (b) $(x + 1)^7$ (c) $(2x + 1)^5$ (d) $(x - 2y)^6$

6. Find the sum:

$$\binom{5}{5} + \binom{5}{4} + \binom{5}{3} + \binom{5}{2} + \binom{5}{1} + \binom{5}{0}$$

7. Fill in the blanks:

$(H + T)^0 = \qquad\qquad 1$

$(H + T)^1 = \qquad\qquad H \quad + \quad T$

$(H + T)^2 = \qquad\qquad H^2 \quad + \quad 2HT \quad + \quad T^2$

$(H + T)^3 = \qquad\qquad H^3 \quad + \quad \underline{?}H^2T \quad + \quad \underline{?}HT^2 \quad + \quad T^3$

$(H + T)^4 = \qquad \underline{?}H^4 \quad + \quad \underline{?}H^3T \quad + \quad \underline{?}H^2T^2 \quad + \quad \underline{?}HT^3 \quad + \quad T^4$

$(H + T)^5 = \underline{?}H^5 \quad + \quad \underline{?}H^4T \quad + \quad \underline{?}H^3T^2 \quad + \quad \underline{?}H^2T^3 \quad + \quad \underline{?}HT^4 \quad + \quad \underline{?}T^5$

8. Compute:

(a) $\binom{100}{0}$ (b) $\binom{100}{1}$ (c) $\binom{100}{2}$ (d) $\binom{100}{99}$ (e) $\binom{100}{98}$

9. Use factorial notation to prove that

$$\binom{8}{5} = \binom{7}{5} + \binom{7}{4}$$

10. Write the expansion of $(1 + 2x)^5$ with integer coefficients.

11. Write the integer that is the coefficient of the term

    (a) HT^2 in the expansion of $(H + T)^3$,
    (b) H^3T^2 in the expansion of $(H + T)^5$,
    (c) $H^5 T^8$ in the expansion of $(H + T)^{13}$.

12. Expand:

$$\sum_{k=0}^{6} \binom{n}{k} a^{n-k} b^k$$

13. Compute:

    (a) $\displaystyle\sum_{i=0}^{4} 3i^2$ (b) $\displaystyle\sum_{i=3}^{5} i^3$ (c) $\displaystyle\sum_{j=1}^{3} (2j-3)$

14. Write in sigma notation.

    (a) $\dfrac{1}{2^2} + \dfrac{1}{3^2} + \dfrac{1}{4^2} + \dfrac{1}{5^2}$ (b) $2 + 4 + 8 + 16 + 32$

ANSWERS.

1. (a) 4   (b) 6   (c) 3   (d) 1   (e) 210   (f) Pascal's triangle.

2. (a) 35   (b) 35   (c) 56   (d) 462

3. (a) 10 (b) 4 (c) 20   4. $\dfrac{5!}{2! \ 3!} = \dfrac{5!}{3! \ 2!}$

5. (a) $1 + 4x + 6x^2 + 4x^3 + x^4$

    (b) $1 + 7x + 21x^2 + 35x^3 + 35x^4 + 21x^5 + 7x^6 + x^7$

    (c) $32x^5 + 80x^4 + 80x^3 + 40x^2 + 10x + 1$

    (d) $x^6 - 12x^5 y + 60x^4 y^2 - 160x^3 y^3 + 240x^2 y^4 - 192xy^5 + 64y^6$ 6. 32

7. $(H + T)^3 = H^3 + 3H^2T + 3HT^2 + T^3$

    $(H + T)^4 = H^4 + 4H^3T + 6H^2T^2 + 4HT^3 + T^4$

    $(H + T)^5 = H^5 + 5H^4T + 10H^3T^2 + 10H^2T^3 + 5HT^4 + T^5$

8. (a) 1   (b) 100   (c) 4950   (d) 100   (e) 4950

9. Show that $\binom{8}{5} = 56$, $\binom{7}{5} = 21$, $\binom{7}{4} = 35$

10. $1 + 10x + 40x^2 + 80x^3 + 80x^4 + 32x^5$ 11. (a) 3   (b) 10   (c) 1287

12. $a^6 + 6a^5 b + 15a^4 b^2 + 20a^3 b^3 + 15a^2 b^4 + 6ab^5 + b^6$

13. (a) 90   (b) 216   (c) 3

14. (a) $\displaystyle\sum_{i=2}^{5} \dfrac{1}{i^2}$ (b) $\displaystyle\sum_{i=1}^{5} 2^i$

* * * BINOMIAL THEOREM REVIEW SELF TEST * * *

1. This picture represents 20 lines, no two of which are parallel and no three meet in a single point. Compute the total number of points of intersection.

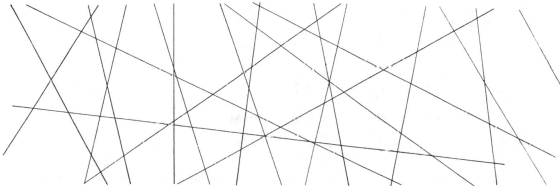

2. Compute

(a) 6!   (b) 0!   (c) $\binom{4}{2}$ (d) $\binom{9}{3}$ (e) $\binom{50}{48}$

3. Write as a single binomial coefficient $\binom{n}{k}$:

(a) $\binom{3}{2} + \binom{3}{3}$ (b) $\binom{9}{4} + \binom{9}{5}$ (c) $\binom{n-1}{k-1} + \binom{n-1}{k}$

4. In how many ways may a sub-committee of 3 people be selected from a committee of 8 people?

5. Use the binomial theorem to expand each of the following.

(a) $(x + y)^3$ (b) $(x - y)^6$ (c) $(x - 2y)^4$

ANSWERS.

1.   190   2. (a) 720 (b) 1   (c) 6   (d) 84   (e) 1225

3.   (a) $\binom{4}{3}$ (b) $\binom{10}{5}$ (c) $\binom{n}{k}$ 4.   56

5.   (a) $x^3 + 3x^2y + 3xy^2 + y^3$

(b) $x^6 - 6x^5y + 15x^4y^2 - 20x^3y^3 + 15x^2y^4 - 6xy^5 + y^6$

(c) $x^4 - 8x^3y + 24x^2y^2 - 32xy^3 + 16y^4$

* * *BINOMIAL THEOREM PROJECTS * * *

1. Consider the following pattern:

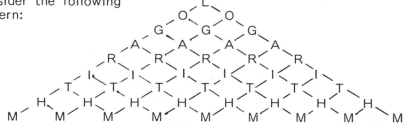

These projects should increase your understanding and appreciation of the binomial theorem.

If you follow a pattern from top down, with how many different paths can you spell out each of the following "words"?

(a) LO   (b) LOG   (c) LOGA   (d) LOGAR

(e) LOGARITHM ?

2. Consider Project 1. Find the sum of the integers in the row of Pascal's triangle that corresponds to

(a) n = 3   (b) n = 5   (c) n = 8   (d) any n

3. (a) Compute $\binom{5}{0} + \binom{5}{1} + \binom{5}{2} + \binom{5}{3} + \binom{5}{4} + \binom{5}{5}$

(b) Evaluate the sum in (a) by showing that it equals $(1 + 1)^5$.

4. (a) Compute $\binom{5}{0} - \binom{5}{1} + \binom{5}{2} - \binom{5}{3} + \binom{5}{4} - \binom{5}{5}$

(b) Evaluate the quantity in (a) by showing that it equals $(1 - 1)^5$.

5. Assuming n is large, write each of the following numbers in order from largest to smallest.

A calculator would be useful here.

$$n^2, \quad 2^n, \quad n!, \quad (1.01)^n$$

6. "Stirling's formula" is important in statistics. It provides a very good estimate of n! when n is large. This estimate asserts that

$$\sqrt{2\pi n} \left(\frac{n}{e}\right)^n$$

is "close" to n! for any large n. (This means that the ratio

$$\frac{\sqrt{2\pi n} \left(\frac{n}{e}\right)^n}{n!}$$

is near 1). The number e is approximately 2.718; π is approximately 3.142.

(a) Use Stirling's formula to estimate 5!

(b) Is the estimate too small or too large?

7. Using appropriate subsets of the n-set $\{1,2,3, \dots n\}$, explain why each of the following equations is valid.

(a) $\binom{n}{k} = \binom{n}{n - k}$ (b) $\binom{n}{k} = \binom{n - 1}{k - 1} + \binom{n - 1}{k}$

8. Using the formula

$$\binom{n}{k} = \frac{n!}{k! \, (n - k)!}$$

prove that

$$\binom{n}{k} = \binom{n - 1}{k - 1} + \binom{n - 1}{k}.$$

9. (a) From the equation $(1 + x)^n(1 + x)^n = (1 + x)^{2n}$ deduce that for each integer m, $0 \le m \le n$

$$\binom{2n}{m} = \binom{n}{0}\binom{n}{m} + \binom{n}{1}\binom{n}{m-1} + \binom{n}{2}\binom{n}{m-2} + \cdots + \binom{n}{m-1}\binom{n}{1} + \binom{n}{m}\binom{n}{0} .$$

(b) Check it for n = 5 and m = 3.

10. This exercise concerns the behavior of $(1 + \frac{1}{n})^n$ for increasing values of the positive integer n.

(a) Express $(1 + \frac{1}{n})^n$ as a decimal for n = 1,2,3, and 4.

(b) The four numbers in (a) increase as n increases from 1 to 4.

Show that
$$(1 + \frac{1}{4})^4 < (1 + \frac{1}{5})^5$$

by expanding each with the aid of the binomial theorem. (Compare the five summands in the expansion of $(1 + \frac{1}{4})^4$ to the first five summands in the expansion of $(1 + \frac{1}{5})^5$. __Do not__ express any as decimals.)

(c) Using the techniques in (b), show that
$$(1 + \frac{1}{n})^n < (1 + \frac{1}{n+1})^{n+1} .$$

(d) As (c) shows, $(1 + \frac{1}{n})^n$ increases as n increases. However, it does not get arbitrarily large. In fact, again using the binomial theorem, show that $(1 + \frac{1}{n})^n$ is less than

$$1 + 1 + \frac{1}{2!} + \frac{1}{3!} + \cdots + \frac{1}{(n-1)!} + \frac{1}{n!}$$

(e) Show that the sum in (d) is less than 3.

(f) Putting parts (c) and (d) together, show that as n gets large, $(1 + \frac{1}{n})^n$ approaches a number that is not larger than 3. (In fact it approaches the number known as e, whose decimal expansion begins 2.71828.... This number is of great importance in calculus.)

11. How many subsets of $\{1, 2, 3, \ldots, 100\}$ contain the set $\{1,2,3\}$?

12. Write down a row of 1's, such as

$$1\ 1\ 1\ 1\ 1\ 1\ 1$$

Fill in the row above according to this rule: above a given number x place the sum of all the numbers to the left of x in the row together with x. In this case the next row is

$$1\ \ 2\ \ 3\ \ 4\ \ 5\ \ 6\ \ 7.$$

The first three rows are

$$1 \quad 3 \quad 6 \quad 10 \quad 15 \quad 21 \quad 28$$
$$1 \quad 2 \quad 3 \quad 4 \quad 5 \quad 6 \quad 7$$
$$1 \quad 1 \quad 1 \quad 1 \quad 1 \quad 1 \quad 1$$

(a) Complete enough rows to obtain a square arrangement.

(b) How is the resulting array related to the Pascal triangle?

13. Let r be a number larger than 1. Thus, $r = 1 + h$, where $h > 0$.

(a) Using the binomial theorem, show that
$$r^n \geq 1 + nh \qquad \text{for } n = 1, 2, \dots$$

(b) From (a) deduce that as n gets large, r^n gets arbitrarily large.

What happens to $(\frac{1}{8})^n$?

14. (See Project 13) Show that if $0 < s < 1$, then as n gets large s^n approaches 0.

15. (See Projects 13 and 14)

(a) What happens to 1.0001^n as n increases?

(b) What happens to 0.9999^n as n increases?

16. (a) Explain how we can think of $(a + b)^2 = a^2 + 2ab + b^2$, when a and b are positive, using the following diagram in the plane.

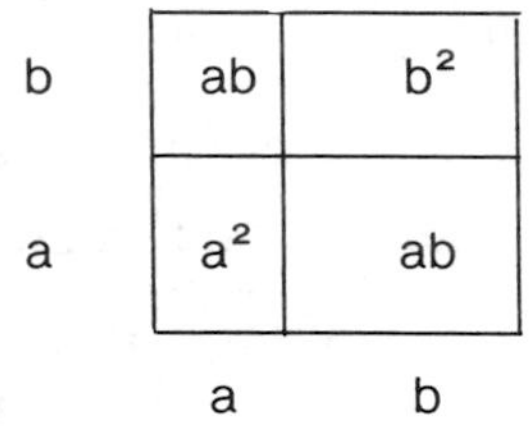

(b) Similar thinking can lead us to "cut up a cube" into rectangular solids if we consider a and b as positive numbers. Explain how $(a + b)^3 = a^3 + 3a^2b + 3ab^2 + b^3$ is related to this idea.

(c) Sketch a picture of (b). It will involve the sketches of several rectangular boxes.

(d) Explain (a) - (c) to a classmate.

* * * CENTRAL SECTION * * *

In a certain state a license plate consists of three letters followed by three digits. Letters, as well as digits, can appear more than once. Some examples:

| ZGR 351 |

| KGB 007 |

| LOG 123 |

Another pattern is three digits followed by three letters. How many more license plates does this provide?

How many different license plates are possible with this system?

* * *

1. What would you guess is the answer to the opening question?

Note: We will return to this problem after considering some simpler problems.

2. A warehouse uses several fork lifts. Each has a license consisting of one of the three letters

A, B, or C

   followed by one of the four digits

0, 1, 2, 3

   (a) Make a complete list of all possible licenses.

   (b) How many are there?

   (b) 12

3. The warehouse in Exercise 2 purchases more forklifts and requires more licenses. If a license now consists of one of the letters,

A, B, C, D, E

   followed by one of the five digits,

0, 1, 2, 3, 4

   (a) list all the possible licenses.

   (b) How many are there?

   (b) 25

4. How many licenses can be made by choosing any one of the 26 letters

A, B, C, . . . , Z

   followed by any one of the ten digits

0, 1, 2, . . . , 9 ?

   (Do not make a list)

* * *

260

The counting principle illustrated in the preceding exercises is expressed in the following general rule.

*This is a very
powerful idea.
The previous
problems should
make it plausible.*

THE MN COUNTING PRINCIPLE

If there are M ways to choose one thing and for each such choice there are N ways to choose another, then the two choices can be made consecutively in

$$MN$$

different ways.

---

5. How many license plates are possible if a license consists of any two letters of the English alphabet? (For example, CC, AC, CA, PQ are four cases. A letter may be repeated.)

676

6. A man owns 6 shirts and 8 pants. How many different "costumes" can he wear?

7. A man has 6 shirts, 8 pants and 5 hats. How many different "costumes" can he wear?

240

* * *

As Exercise 7 shows, the MN principle applies to more than two choices. Exercise 8 continues this idea.

8. A trucker has a choice of 5 routes from San Francisco to Denver, 4 routes from Denver to Chicago and 6 routes from Chicago to New York.

San Francisco ⟮⟯ Denver ⟮⟯ Chicago ⟮⟯ New York

How many different ways can he go from San Francisco to New York?

9. Let us say that there are 200 possible first names, 1000 possible middle names and 1000 possible family names. How many different three-part names are possible? Would there be enough to give each person in the USA a different name?

*Is the number of
names reasonable?
Are more possible?*

10. Return to the opening question. How many different license plates can be made consisting of three letters followed by three digits? (Assume that any letter or digit may be repeated.)

COUNTING WHEN DUPLICATION IS PROHIBITED

The number of three-letter license plates is

$$(26)(26)(26) = 17,576$$

This includes licenses that have duplicated letters, such as

| CDC | MMM | SKS |

Here is a different type of question:

How many three-letter license plates do you think can be made if <u>no duplication</u> of a letter is permitted? (Assume that letters are chosen from the usual 26-letter alphabet.)

Before answering this question, we present some simpler exercises.

11. (a) List all 2-letter license plates that can be made using the letters A, B, or C <u>without</u> <u>duplication</u>.

(b) How many are there?

12. How many 2-letter license plates without duplication can be made using the letters A, B, C, D, or E?

13. How many 3-letter license plates can be made using the letters A, B, C, or D without duplication?

14. A person owns 5 books, A, B, C, D, and E. In how many different orders can be put them on a shelf? (Do not make a list.)

* * *

Example 1. How many three-digit numbers use no digit more than once? (Assume that 0 can be a left digit.)

Solution. The possible digits are 0, 1, 2, . . . , 9. They are to be placed in the three blanks

____   ____   ____

to form a three digit number.

There are ten choices for the first blank, 0, 1, 2, . . . , 9. Since <u>duplication</u> <u>is</u> <u>not</u> <u>permitted</u>, for each choice of the first blank, there are nine choices for the second blank. So far, there are

$$(10)(9)$$

ways to fill in the first two blanks. For each of these $(10)(9)$ ways of filling in the first two blanks there are eight choices for the third blank (not more, because no duplication is permitted). Thus there are

$$(10)(9)(8) = 720$$

three-digit numbers without duplicated digits.

15. How many three-letter license plates are possible if duplication is not permitted?

* * *

Recall that a <u>set</u> is a collection of things. <u>The</u> <u>order,</u> <u>or</u> <u>arrange-</u> <u>ment</u> <u>of</u> <u>the</u> <u>members</u> <u>of</u> <u>a</u> <u>set</u> <u>is</u> <u>not</u> <u>important.</u> However, in an <u>arrange-</u> <u>ment</u>, the <u>order</u> <u>is</u> <u>important.</u>

(b) 6
Hint: How many choices are there for the first letter? Second letter?

13. 24
Does the number change if you think left-to-right or right-to-left?

14. 120

15,600

STUDY THESE IDEAS CAREFULLY!

Sometimes an arrangement is called a permutation.

In the next section we alternately use the idea of <u>set</u> (the order of listing is <u>not</u> important) and <u>arrangement</u> (the order of listing <u>is</u> important) to obtain the formula

$$\binom{n}{k} = \frac{n!}{k!(n-k)!}$$

which was stated without proof in the preceding chapter.

ARRANGEMENTS OF k THINGS

Let us now turn to the problem of counting the number of possible arrangements of objects in a line.

There are six arrangements.

16. A man owns three books, called A, B, and C. List all the ways he can arrange them on a shelf. For instance, one arrangement is BCA.

17. Every day four people line up at the bus stop. In how many different orders can they form this line? For convenience call the people A, B, C, and D. (Make a list.)

18. Using the MN counting principle explain without making a list why the answer in Exercise 17 is 24.

120

19. In how many ways can a person arrange five books on a shelf? (Do not list the possibilities.)

20. In your own words explain the principle:

Recall it is customary to denote the product of all the integers from k down to 1 simply as k!, called k factorial (see the Binomial Theorem Chapter.)

```
┌─────────────────────────────────────────────────┐
│        NUMBER OF ARRANGEMENTS OF k OBJECTS        │
│  It is possible to arrange k different objects in a line in │
│        k · (k - 1) · (k - 2) ··· 3 · 2 · 1        │
│  different ways.                                  │
└─────────────────────────────────────────────────┘
```

21. Express the answer to Exercise 17 as a factorial.

22. Express the answer to Exercise 18 as a factorial.

23. Express the answer to Exercise 19 as a factorial.

24. How many "words" 26 letters long, using no letter twice, can be made with the 26 letters of the alphabet. (Express your answer as a factorial.)

WHY $\binom{20}{3}$ IS EQUAL TO $\frac{20!}{3!17!}$

The next few exercises lead to an explanation of the formula $\binom{n}{k} = \frac{n!}{k!(n-k)!}$

The problem that started off the Binomial Theorem Chapter concerns a dormitory of 20 students that selects a different set of three cooks every night. But now the dormitory changes its routine and assigns each of the three cooks a specific place at the long kitchen work table.

First place	Second place	Third place

At "first place" the salad is prepared, at "second place" the entrée, and at "third place" the dessert. This raises a new problem: in how many different ways can the assignment of three cooks be made now? Remember, each is assigned a specific task and there are 20 residents in the dormitory.

The next few exercises outline two approaches to the problem. A comparison of the two approaches will lead to the formula

$$\binom{20}{3} = \frac{20!}{3!17!}$$

25. Why is it that the number of ways to assign the cooks to the three tasks is equal to $20 \cdot 19 \cdot 18$?

26. Another approach is to choose the arrangements of three cooks in two steps. <u>First</u>, choose a <u>set</u> of three cooks. Then <u>for each</u> such choice of three cooks compute how many ways we can arrange them at the work table. Using this approach, explain why there are

Hint: Recall the MN counting principle.

$$\binom{20}{3}(3!)$$

ways to arrange three cooks at the table in a dormitory of 20 students.

27. (a) Using Exercises 25 and 26, explain why

$$\binom{20}{3}(3!) = 20 \cdot 19 \cdot 18$$

(b) From (a) deduce that

$$\binom{20}{3} = \frac{20 \cdot 19 \cdot 18}{3!}$$

Hint: $\binom{20}{3} \cdot 3! = 20 \cdot 19 \cdot 18$

(c) From (b) deduce that

$$\binom{20}{3} = \frac{20!}{3!17!}$$

* * *

The reasoning in Exercises 25 - 27 explains <u>why</u> $\binom{20}{3}$ is given by the formula

$$\frac{20!}{3!17!}$$

A similar argument in Exercises 29-30 shows <u>why</u>

$$\binom{n}{k} = \frac{n!}{k!(n-k)!}$$

The next exercise illustrates one more specific case.

28. Suppose another dormitory has 25 residents. Each cook crew consists of four cooks assigned to four different tasks at a long table shown in the diagram below:

| salad | entree | dessert | vegetable |

Using the approach in Exercises 25 and 26, explain why

(a) $\binom{25}{4} 4! = 25 \cdot 24 \cdot 23 \cdot 22$

and therefore

(b) $\binom{25}{4} = \frac{25!}{4! \, 21!}$

29. Suppose that there are n students in the dormitory and four work areas in the kitchen. Use this situation to answer the following questions,

(a) Explain why $\binom{n}{4}(4!) = n(n-1)(n-2)(n-3)$

(b) Deduce from (a) that $\binom{n}{4} = \frac{n(n-1)(n-2)(n-3)}{4!}$

(c) Deduce from (b) that $\binom{n}{4} = \frac{n!}{4!(n-4)!}$

30. What size dormitory and what number of work tables would you use to explain why

$$\binom{n}{k} = \frac{n!}{k!(n-k)!}$$

(Do not write out the general explanation.)

* * *

Exercises 25 - 30 have outlined <u>why</u> formula $\frac{n!}{k!(n-k)!}$ gives the number of k-sets contained as subsets of an n-set.

Sometimes the notation $_nP_k$ is used instead of P^n_k.

Incidently, the number of ways to assign to specific work tables the three cooks out of the dormitory of twenty residents is sometimes called "the number of <u>permutations</u>, or arrangements, of 3 objects from a set of 20 objects." This number is denoted P^{20}_3. As observed in Exercises 20 - 30,

$$P^n_k = \binom{n}{k} k!$$

Note how a <u>permutation</u> differs from a set.

In a <u>permutation</u>, the arrangement, or order of listing is important. In a <u>set</u> (sometimes called a "combination"), the arrangement or order is not important. Exercise 31 illustrates both of these ideas.

31. (a) From a class of 30 students a committee of four students is to
          be formed. How many ways can the committee be formed?
          Explain your answer.

     (a) 27,405

     (b) The same class is to elect officers, a president, vice-
          president, secretary and treasurer. How many ways can this be
          done? Explain your answer.

     (b) 657,720
     Are you surprised
     by the answers
     here?

32. There are five people entered in a contest.

     (a) Suppose that the first prize is $15, the second prize is $10,
          and the third prize is $5. In how many different ways may the
          prize money be awarded?

     (a) 60

     (b) Suppose instead that there are three equal prizes of $10 each.
          How many different ways may the prize money be awarded?

     (b) 10

                            * * *

     We now turn to a completely different type of counting problem. It
will not use the MN principle.

COUNTING SEATS IN AN AUDITORIUM

     If an auditorium has 10 rows and each row has 15 seats, then there
are

                    (10)(15) = 150 seats

But a typical auditorium is narrower at the front than at the back.
Assume in each example or exercise that from the front to the back the
successive rows lengthen by a fixed number of seats. This diagram shows
an auditorium where each row (other than the front row) has 2 more seats
than the row in front of it.

```
               ┌───────┐
               │ Stage │
               └───────┘

        . . . . . . . . .
       . . . . . . . . . .
      . . . . . . . . . . .          (Auditorium
     . . . . . . . . . . . .              seats)
    . . . . . . . . . . . . .
```

The front row has 10 seats, the next 12 seats, the next 14 and so on.

33. How many seats are there in the auditorium shown? (Show your work.)   *70*

In Exercise 33 you probably found the total number of seats by adding the number of seats in each row. For auditoriums with just a few rows this method is adequate. But for a large auditorium this approach would be laborious. We shall develop a shortcut for finding the total number of seats in any auditorium. First, we put the problem in a mathematical setting.

Consider an auditorium in which the front row has a seats. The next row has, say d more seats; thus the second row has

$$a + d \text{ seats.}$$

The third row has d more seats than the second row, hence

$$a + 2d \text{ seats.}$$

The fourth row has

$$a + 3d \text{ seats.}$$

And so on until we reach the last row. If there are n rows in the auditorium, there are

$$a + (n - 1)d \text{ seats}$$

in the back, or last row. Pictorially,

Why does this formula have "n - 1" rather than n?

Could you find the number of seats in the 7th row? 10th row? or 15th row? How?

Row 1		a seats
Row 2		a + d seats
Row 3		a + 2d seats
etc.	etc.	etc.
.	.	.
.	.	.
Row n		a + (n - 1)d seats

The numbers, a, d, and n, completely describe the design of the auditorium seating.

34. What are a, d, and n in the auditorium sketched before Exercise 33?

35. Sketch the design of a small auditorium for which a = 5, d = 3, and n = 4.

36. What are a, d, and n for the auditorium sketched below?

```
          . . . . . .
        . . . . . . . .
      . . . . . . . . . .
    . . . . . . . . . . . .
```

The general question is:

>How many seats are there in an auditorium in which
>>the front row has $\underline{a}$ seats
>>
>>there are n rows;
>>
>>each row (other than the front row) has d more seats
>>than the row in front of it?

Schematically, the auditorium looks like this. For convenience let $\underline{b}$
denote the number of seats in the back row.

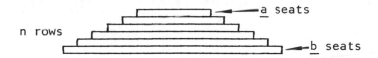

There are n rows. The front row has $\underline{a}$ seats. The back row has $\underline{b}$
seats. Let S be the total number of seats.

Now copy the diagram, turn it upside down, and place it next to the
given diagram:

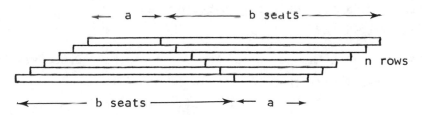

Two copies of the auditorium produces a figure made up of

>n rows

each row having

>a + b seats.

If S is the sum of the seats, then

>$$2S = n(a + b)$$

or

$$S = \frac{n(a + b)}{2}$$

A trick like this was used in Chapter 1 to find the formula for the sum of a geometric progression.

To understand the formula, study the above diagram carefully.

The boxed formula quickly tells how many seats there are in the auditorium.
It says, "find the average of the number of seats in the front and back
rows, then multiply by the number of rows."

To express the answer in terms of a, n, and d, replace b by a + (n - 1)d. Thus the total number of seats is also given by the formula

$$S = n\left[\frac{a + (a + (n - 1)d)}{2}\right]$$

or

$$S = \frac{n[2a + (n - 1)d]}{2}$$

However the formula for S in terms of a, n, and b is easier to remember. The two longer formulas are unnecessary if you recall that

$$b = a + (n - 1)d$$

and that

$$S = n\left(\frac{a + b}{2}\right)$$

37. An auditorium has 20 rows. The front row has 25 seats. The back row has 44 seats.

(a) 690

(b) 1

(a) What is S, the total number of seats?

(b) What is d?

38. An auditorium has 30 rows. The front row has 40 seats. Each row has three seats more than the row in front of it.

(a) What is a?   (b) What is d?

(c) What is b?   (d) What is S?

39. Find the sum of all integers from 1 to 10. (Think of this as describing an auditorium.)

55

40. Find the sum of all integers from 1 to 100.

5050

41. At your first birthday your cake had one candle. At your second birthday it had 2 candles. At your third it had 3 candles. And so on. How many candles are used on all the birthday cakes through your 18th birthday?

*   *   *

ARITHMETIC PROGRESSIONS

The auditorium problem, like the candle problem of Exercise 41, involves only integers.   The ideas extend to numbers that need not be integers.   We now make this generalization.

DEFINITION. Let <u>a</u> and <u>d</u> be any numbers and let n be a positive integer. The sequence of n numbers formed by starting at <u>a</u> and adding d repeatedly,

$$a, a + d, a + 2d, a + 3d, \ldots , a + (n - 1)d$$

is called an <u>arithmetic</u> <u>progression</u>. The <u>first</u> <u>term</u> is <u>a</u>. The difference is <u>d</u>. The <u>last</u> <u>term</u> is

$$a + (n - 1)d,$$

which is also denoted <u>b</u>.

Compare an arithmetic progression with a geometric progression.

Example 2. Write out the arithmetic progression with
$$a = 4.5 \qquad d = 0.5 \qquad n = 3$$
Solution. The first term is 4.5 and there are three terms,
$$4.5, \ 4.5 + 0.5, \ \text{and} \ 4.5 + 0.5 + 0.5,$$
that is,
$$4.5, \ 5, \ \text{and} \ 5.5$$

42. Write all the terms of the arithmetic progression in which
  (a) $a = 5 \qquad d = -\frac{1}{3} \qquad n = 4$

  (b) $a = \frac{1}{2} \qquad d = 1\frac{1}{2} \qquad n = 5$

  (c) $a = 6 \qquad d = -1 \qquad n = 7$

43. The first term of an arithmetic progression is <u>a</u> and the difference is <u>d</u>. What is
  (a) the second term?
  (b) the third term?
  (c) the tenth term?
  (d) the nth term?

(a) $a + d$

(c) $a + 9d$

* * *

THE SUM OF AN ARITHMETIC PROGRESSION

The formula

$$S = n\left(\frac{a + b}{2}\right)$$

was developed for the case where <u>a</u> and d, hence b, are integers. In fact, the formula is valid for all arithmetic progressions as will now be shown.

To show that the total number of seats in an auditorium is

$$n\left(\frac{a + b}{2}\right)$$

we made a second copy of the auditorium, turned it upside down, and found how many seats there were in two copies. We will do something like that now, algebraically, not geometrically.

Let

$$S = a + (a + d) + (a + 2d) + \ldots + b$$

Now write the progression backwards. The first term is now b. The last term is now <u>a</u>. Thus

$$S = b + (b - d) + (b - 2d) + \ldots + a$$

Add the two forms for S:

$$S = a + (a + d) + (a + 2d) + \ldots + b$$
$$S = b + (b - d) + (b - 2d) + \ldots + a$$

sum: $2S = \underbrace{(a + b) + (a + b) + (a + b) + \ldots + (a + b)}_{n \text{ times}}$

In short,

$$2S = n(a + b),$$

hence

$$\boxed{S = \frac{n(a + b)}{2}}$$ ← Sum of any arithmetic progression on n terms, first term <u>a</u>, last term <u>b</u>.

Say: "The sum
of an arithmetic
progression is the
average of the
first and last
terms times the
number of terms."

(a) 4
(b) $\frac{1 - 2\sqrt{2}}{7}$

44. The first term of an arithmetic progression of 8 terms is $\sqrt{2}$ and the last term is $1 - \sqrt{2}$.

   (a) What is the sum?

   (b) What is the common difference?

45. An arithmetic progression of eight terms has first term $\sqrt{3}$, and fixed difference equal to $-\frac{1}{2}$.

(a) $\sqrt{3} - \frac{7}{2}$

(b) $4(2\sqrt{3} - \frac{7}{2})$

   (a) What is its first term? (b) What is its sum?

46. An arithmetic progression has five terms. The first is -6 and the last is 6.

   (a) What is its sum?   (b) What is the common difference?

(b) 3

   (c) What are the other terms?

47. A staircase has wider steps at the bottom than at the top. The lowest step is 5 feet wide and the top step is 3 feet wide. It has 13 steps. The steps narrow by the same fixed amount as you walk up.

   (a) How wide is the next to the bottom step?

   (b) What is the total of the widths of all the 13 steps?

* * * SUMMARY * * *

A problem concerning license plates introduced the first problem in counting. It was solved by the MN principle. We then considered a similar problem in which duplications are not permitted.

Next we turned to the number of arrangements of k things in a line (there were k!). Indirectly the concept provided an explanation why

$$\binom{n}{k} = \frac{n!}{k!(n-k)!} \cdot$$

The final counting problem concerned seats in an auditorium. This problem generalized, finding the sum of any arithmetic progression. The formula for the sum is

$$S = \frac{n(a+b)}{2} ,$$

where n is the number of terms, a is the first term, and b is the last term. Thus, b = a + (n - 1)d where d is the common difference.

Moreover, the main results in the chapter are as follows:

If there are M ways to choose one thing and for each such choice there are N ways to make a second choice, then there are MN ways to make both choices together. (The MN principle).

The MN principle can be used to count the number of choices when duplications are not permitted.

The number of arrangements of k different objects on a line is

$$k \cdot (k - 1) \cdot (k - 2) \cdot \ldots \cdot 3 \cdot 2 \cdot 1 = k!$$

To justify the formula

$$\binom{20}{3} = \frac{20!}{3!17!} ,$$

count certain arrangements in two ways. Consider a set of 20 objects and count how many different ways there are to arrange three of them in a line. The answer is equal to

$$\binom{20}{3} \cdot 3!$$

and this is also equal to

$$20 \cdot 19 \cdot 18$$

Hence

$$\binom{20}{3} = \frac{20 \cdot 19 \cdot 18}{3!} = \frac{20!}{3!17!}$$

Compare the way we form an arithmetic progression with that of a geometric progression (Ch. 1). Also, compare the formulas for the last term for each progression.

Let n be a positive integer and a and d be numbers. The sequence of n numbers

$$a, a + d, a + 2d, \ldots, a + (n - 1)d$$

is called an <u>arithmetic</u> progression. The <u>first term</u> is <u>a</u>. The <u>differ-ence</u> is <u>d</u>. The <u>last term</u> is a + (n - 1)d, which is also denoted <u>b</u>. The sum of the terms in an arithmetic progression whose first term is <u>a</u>, whose last term is <u>b</u>, and which has n terms, is

$$S = \frac{n(a + b)}{2}.$$

In words, "The sum is equal to the number of terms times the average of the first term and last term ."

* * * WORDS AND SYMBOLS * * *

MN counting principle	<u>b</u> last term
arrangement (permutation)	number of terms
set (combination)	<u>d</u> common (fixed) difference
arithmetic progression	S sum
<u>a</u> first term	

* * * CONCLUDING REMARKS * * *

An arithmetic progression is formed by repeatedly <u>adding</u> a fixed number (called the difference). A geometric progression is formed by repeatedly <u>multiplying</u> by a fixed number (called the ratio). The terms of a geometric progression, such as

$$1, 2, 4, 8, 16, 32, 64, \ldots$$

can increase by larger and larger amounts, whereas the terms of an arithmetic progression increase by a fixed amount, hence more slowly. This contrast was used by Malthus in his book, <u>An Essay on the Principle of Population</u>, published in 1798.

In the recent past, the ideas of Malthus were some-times ridiculed. Do you think he was a prophet ahead of his time?

He remarks, "I say, that the power of population is indefinitely greater than the power in the earth to produce subsistence for man.

"Population, when unchecked, increases in a geometrical ratio. Sub-sistence increases only in an arithmetical ratio. A slight acquaintance with numbers will show the immensity of the first power in comparison of the second.

"I see no way by which man can escape from the weight of this law which pervades all animated nature . . . no agrarian regulations . . .

could remove the pressure of it even for a single century."

The pressure exists today. Population increases every year at about the rate of two percent. Millions of people are still undernourished in spite of the major innovations in farming practice in the almost 2 centuries since Malthus made his prediction. The pressure of which Malthus wrote may someday soon lead to a food crisis that may seem as sudden as the energy crisis, all because of the contrast of the growth rates of geometric and arithmetic progressions.

* * * COUNTING SELF QUIZ * * *

1. Vehicles in a certain country have license plates consisting of four of the letters

    A, B, C, D, . . . , Z

    while those in another country consist of six of the digits

    0, 1, 2, 3, . . . , 9.

    Which country can have the greater number of license plates without duplicating a license number? Letters or digits may be repeated.

2. A tennis team has nine members. How many different singles practice matches can be held?

3. A certain theater has 10 seats in the first row, 13 in the second row, 16 in the third, 19 in the fourth, and so on for 10 rows.

    (a) How many seats are in the last row?

    (b) How many seats does the theater have?

4. A football team of eleven students is waiting in line at the box office to see a movie. As the students begin arguing about who should be first in line, the coach decides to order each student where to stand in line. In how many different ways can the coach arrange the eleven students?

5. Ann goes into the 47 Flavors Ice Cream Store. How many different triple decker cones can she buy if

    (a) the three scoops have different flavors and their order does not matter?

    (b) the three scoops have different flavors and their order does matter?

    (c) flavors may be repeated and their order does matter?

6. Explain <u>why</u>

$$\binom{15}{4} = \frac{15!}{4!11!}$$

7. Texas automobile license plates have three letters followed by three digits. How many licenses may be issued if any digit and any letter may be used?

ANSWERS TO COUNTING SELF QUIZ

1. Country B; 1,000,000 is larger than 456,976   2. 36
3. (a) 37   (b) 235   4. 11!
5. (a) 16,215   (b) 97,290   (c) 103,823
6. See explanation in main section for the reason why

$$\binom{20}{3} = \frac{20!}{3!17!}$$

7. 17,576,000

* * * COUNTING REVIEW * * *

1. How many ways can you choose to spell ABRACADABRA using the following diagram of letters? The kth letter of the word can be any of the letters in the kth row.

Drop down a line for each letter as you spell the word.

```
                A
              B   B
            R   R   R
          A   A   A   A
        C   C   C   C   C
      A   A   A   A   A   A
        D   D   D   D   D
          A   A   A   A
            B   B   B
              R   R
                A
```

2. A basketball coach had 17 candidates come out for the team. She must cut the traveling squad to 12 players. How many ways can this be done?

3. Suppose in Exercise 2 that the basketball squad has been cut to 12 players: four forwards, six guards, and two centers. In how many ways may the coach choose a playing team of two forwards, two guards and one center?

4. Suppose we want to color the five squares below using the colors red, white, blue, green, and yellow. How many different ways can we

do this so that each square is a different color.

5. How many different ways are there to color the five squares below with the colors red and blue so that two squares are red and three are blue?

6. Suppose you wish to travel by air from city A to city B and then to city C. If five airlines have flights from A to B, and three airlines have flights from B to C, how many ways can one travel by air from A to B to C.

7. In a class of 25 students, a committee of five students is to be selected. How many different ways can this be done? (Two committees are considered the same if they have the same members.)

8. Find the number of five-element subsets of a set with seven elements.

9. Compute the coefficient of H^2T^{18} in the expansion of $(H + T)^{20}$.

10. A football team has ten offensive plays. How many different series of three plays can they have? (A play may be repeated.)

Hint: The order of plays is important in football.

11. A little theatre group has 14 actors. They wish to put on a play that has five leading parts, each to be played by a different actor. In how many ways may the leading parts be assigned? (Assume the sex of the actors makes no difference.)

12. Find the number of odd integers x such that $101 \leq x \leq 999$.

13. Find the sum of the arithmetic progression

$$101 + 103 + 105 + \ldots + 999.$$

14. Compute.

(a) $\displaystyle\sum_{i = 1}^{100} i$ (b) $\displaystyle\sum_{i = 1}^{100} (3i + 4)$

15. A classroom has 40 desks and 30 students. In how many different ways may the students be seated? (Leave your answer in factor form; do not multiply it out.)

16. Suppose you have a key ring. In how many different orders can you place on the key ring

(a) 1 key?   (b) 2 keys?   (c) 3 keys?

(d) 4 keys?   (e) 5 keys?   (f) n keys, n $\geq$ 3?

You may wish to experiment here. Turning over the key ring does not change the order of the keys.

17. New York car licenses have four digits followed by three letters. How many cars can be licensed this way?

18. Rhode Island car licenses have five digits. How many cars can be licensed this way.

19. Illinois car licenses have three letters followed by three digits. How many cars can be licensed this way?

20. There are about 100 million cars in the United States. Could they all be licensed if a license is made of six letters?

21. A composer, wanting to give the musicians some freedom, writes a symphonic suite in 5 movements that can be played in any order. How many choices are available? (Assume that the conductor, not the players, makes the choice. Otherwise, there would be chaos!)

22. A certain bachelor can cook eggs in five different ways, prepare baked, fried, mashed, or boiled potatoes, and drink coffee, tea or milk. His breakfast consists of eggs, potatoes, and a beverage.

    (a) How long can he live without duplicating a menu? (Assume that the diet provides him enough energy so that he can cook.)

    (b) Assume now that he makes his breakfast into a three-course meal, taking the eggs, potatoes, and beverage in different orders, for the sake of variety. How many different breakfasts are possible? He now considers order important.

23. A woman owns 6 cats and 6 dogs. Each day she takes 2 cats and 2 dogs for a walk. How many days can she take a walk without duplicating the same four pets?

ANSWERS TO COUNTING REVIEW

1. $\binom{10}{5} = 252$ 2. 6188   3. 180   4. 120   5. 10

6. 15   7. 53,130   8. 21   9. 190   10. 1000

11. 240,240   12. 450   13. 247,500   14. (a) 5050 (b) 15,550

15. $\dfrac{40!}{10!}$ 16. (a) 1 (b) 1 (c) 1 (d) 3 (e) 12 (f) $\dfrac{(n-1)!}{2}$

17. 175,760,000   18. 100,000   19. 17,576,000   20. Yes

21. 120   22. (a) 60 days (b) 360 days   23. 225 days

* * * COUNTING REVIEW SELF TEST * * *

1. Recall from geometry that two points determine a line. If no three
   points are in the line, how many different lines are determined by
   (a) three points?   (b) four points?
   (c) seven points?   (d) n points ($n \geq 2$)?

2. Use Exercise 1 and your knowledge of geometry to determine the
   number of diagonals in a regular
   (a) 3-gon   (b) 4-gon   (c) 5-gon
   (d) 6-gon   (e) n-gon

3. Recall from geometry that three points not in a line determine a
   plane. If no four points are in the same plane, how many planes
   are determined by
   (a) four points?   (b) five points?
   (c) seven points?   (d) n points ($n \geq 3$)?

4. How many commitees consisting of three girls and two boys can be
   selected from a class of 15 girls and 18 boys?

5. A <u>die</u> (plural, "dice") is a cube with a different number of dots
   (1, 2, 3, 4, 5, or 6) on each face. How many different results are
   there when you roll
   (a) one die?   (b) a red die and a green die?
   (c) three dice, one red, one green and one white?

6. The "zonk" (a fictitious animal) makes a single jump of two centimeters when he is three months of age. Thereafter, at each doubling
   of his age, he makes another jump, but the jump is two centimeters
   greater than the time before. How far does the zonk jump at age
   (a) 6 months?   (b) 1 year?   (c) 2 years?
   (d) 8 years?   (e) 128 years (the average life span of a zonk)?
   (f) What is the sum of all his jumps from the age of three months
        through 128 years?
   (g) The size of the jumps produces what kind of a progression?
   (h) The times of the jumps produce what kind of a progression?

ANSWERS TO COUNTING REVIEW SELF TEST

1. (a) 3   (b) 6   (c) 21   (d) $\binom{n}{2}$

2. (a) 0   (b) 2   (c) 5   (d) 9   (e) $\dfrac{n^2 - 3n}{2}$

3. (a) 4   (b) 10   (c) 35   (d) $\binom{n}{3}$

4. $69,615 \left(= \binom{15}{3}\binom{18}{2}\right)$ 5. (a) 6   (b) 36   (c) 216

6. (a) 4 cm.   (b) 6 cm.   (c) 8 cm.   (d) 12 cm. (e) 20 cm.

   (f) 110 cm. (g) arithmetic   (h) geometric

* * * COUNTING PROJECTS * * *

1. The many branches of mathematics overlap. For instance, there is a close relationship between arithmetic progressions and the number of 2-sets in an n-set, a concept introduced in the preceding chapter. This project outlines that relationship.

   (a) Let A ={1, 2, 3, . . . , n} be a set with n members, $n > 5$. Show that there are five 2-sets that can be chosen in which the larger number of the set is 6.

   (b) More generally, for any integers r, $2 \leq r \leq n$, the number of 2-sets in A that have their larger element equal to r is r - 1.

   (c) Deduce from (b) that

$$\binom{n}{2} = 1 + 2 + 3 + \ldots + (n - 1).$$

   (d) Using the formula for $\binom{n}{2}$ from the preceding chapter, show that

$$1 + 2 + 3 + \ldots + (n - 1) = \frac{n(n - 1)}{2}$$

   (e) Does the formula in (d) agree with the results in the present chapter concerning the sum of an arithmetic progression?

2. If you add more and more numbers from this sequence

$$\frac{1}{1 \cdot 2} + \frac{1}{2 \cdot 3} + \frac{1}{3 \cdot 4} + \frac{1}{4 \cdot 5} + \ldots$$

   do the resulting sums tend to get arbitrarily large or stay below some fixed number? Explain this question.

3. (a) Prove that the area of a trapezoid of height $\underline{h}$ and bases of lengths $\underline{a}$ and $\underline{b}$ is $h\left(\frac{a + b}{2}\right)$

   (b) What formula in this chapter does the formula in (a) resemble?

4. The integers 1, 2, 3, . . . , n, . . . form an endless arithmetic progression. Their reciprocals form an endless sequence of decreasing numbers,

$$\frac{1}{1}, \frac{1}{2}, \frac{1}{3}, \ldots, \frac{1}{n}, \ldots$$

   Let $S_n = \frac{1}{1} + \frac{1}{2} + \frac{1}{3} + \ldots + \frac{1}{n}.$

   (a) Using a table of reciprocals or a hand calculator, compute S_{10} and S_{20}.

(b) As n gets arbitrarily large, what do you think happens to S_n?
Does it get arbitrarily large, or does it stay less than some
fixed number?

5. Multiply the numbers in the geometric progression

$$1, \frac{1}{2}, \frac{1}{4}, \ldots , \left(\frac{1}{2}\right)^{n-1}, \ldots$$

by the numbers in the arithmetic progression

$$1, 2, 3, \ldots , n, \ldots$$

You obtain the numbers,

$$1, 2\left(\frac{1}{2}\right), 3\left(\frac{1}{2}\right)^{2}, 4\left(\frac{1}{2}\right)^{3}, \ldots n\left(\frac{1}{2}\right)^{n-1}, \ldots$$

Let S_n denote the sum of the first n numbers formed this way.
For instance,

$$S_3 = 1 + 2\left(\frac{1}{2}\right) + 3\left(\frac{1}{2}\right)^{2}$$
$$= 1 + 1 + \frac{3}{4}$$
$$= 2.75$$

(a) Compute S_4, S_5, and S_{10} to three decimal places.

(b) As n gets larger, S_n approaches 4. This is how Bernoulli,
early in the 18th century, obtained this result. He wrote out
this sequence of equations:

$$1 + \frac{1}{2} + \frac{1}{4} + \frac{1}{8} + \ldots = 2$$
$$\frac{1}{2} + \frac{1}{4} + \frac{1}{8} + \ldots = 1$$
$$\frac{1}{4} + \frac{1}{8} + \ldots = \frac{1}{2}$$
$$\frac{1}{8} + \ldots = \frac{1}{4}$$
$$\ldots$$

Adding, he claimed that

$$1 + 2\left(\frac{1}{2}\right) + 3\left(\frac{1}{4}\right) + 4\left(\frac{1}{8}\right) + \ldots = 4.$$

Explain his reasoning.

6. The numbers a, a + d, a + 2d, ... form an arithmetic progression.
What kind of progression do the numbers

$$2^{a}, 2^{a+d}, 2^{a+2d}, \ldots$$ form?

7. Sequences of numbers can be formed in many ways, not just by "adding a fixed number" or "multiplying by a fixed number."   Early in the thirteenth century the mathematician, Fibonacci, introduced the sequence

$$1, 1, 2, 3, 5, 8, 13, \ldots$$

now called "the Fibonacci sequence." Each number from the third on is the sum of the two preceding numbers. He introduced it while answering a question about rabbits: A pair of rabbits is placed in a cage and produces another pair every month. Each of the new pairs does the same once they reach the age of one month. How many pairs of rabbits will there be at the end of a year?

(a) Show that the answer is 377.

(b) Calculate the ratios between successive Fibonacci numbers. For instance

$$\frac{5}{3} \doteq 1.667$$

What happens to these ratios as you go further out in the sequence?

8. (a) By inspecting the diagram below, show that
$$2(1 + 2 + 3 + 4 + 5) = 5 \cdot 6$$

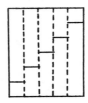

(b) Use the idea in (a) to show that
$$1 + 2 + \ldots + n = \frac{n(n + 1)}{2}$$

9. (a) By inspecting the diagram below show that
$$(1 + 2 + 3 + 4)^2 = 1^3 + 2^3 + 3^3 + 4^3$$

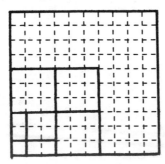

(b)   Draw the appropriate diagram to show that

$$(1 + 2 + 3 + 4 + 5)^2 = 1^3 + 2^3 + 3^3 + 4^3 + 5^3$$

10.   Consider the odd integers:

$$1, \ 3, \ 5, \ 7, \ 9, \ 13, \ . \ . \ .$$

In 1615 Galileo observed that

$$\frac{1 + 3}{5 + 7} = \frac{1 + 3 + 5}{7 + 9 + 11} = \frac{1 + 3 + 5 + 7}{9 + 11 + 13 + 15}$$

Are these equations coincidences or does the pattern continue without end?

11.   Are any arithmetic progressions also geometric progressions?

12.   The diagram below shows that $1 + 3 + 5 + 7 = 4^2$

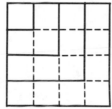

(a)   What diagram shows that

$$1 + 3 + 5 + 7 + 9 = 5^2 \ ?$$

(b)   Using the formula for the sum of an arithmetic progression, show that

$$1 + 3 + 5 + 7 + \ . \ . \ . \ + (2k - 1)$$

is a square for any positive integer k.

13.   (a)   Let A = $\left\{1, \ 2, \ 3, \ 4, \ 5, \ 6, \ 7\right\}$. Make a list of 3-sets contained in A such that each 2-set in A is a subset of exactly one of the 3-sets in your list.

(b)   Show why no such list can be found for the case when

$$A = \left\{1, \ 2, \ 3, \ . \ . \ . \ , \ 35\right\}.$$

9

Algebra II

Probability

In a certain game of chance a player rolls two dice at a time. On each roll the score is the total number of dots (called "pips") showing on top of the two dice. For instance, this diagram shows a roll with a score of 8.

Note: Chapter 9 depends on Chapters 7 and 8.

The score can be anywhere from a low of 2 (= 1 + 1) to a high of 12 (= 6 + 6). Suppose that the player rolls until he scores a 9 or a 7. If he scores the 9 first, he wins. If the 7 appears first, he loses.

If he plays this game many times, will he in the long run tend to win more often than lose, or will he win and lose about equally often?

* * *

1. What is your opinion of the player's chances in the above game?

2. Experiment using a pair of dice. Play the game twenty times and record how often you win (with a "9") or lose (with a "7").

win (9)	
lose (7)	

You may have to roll the dice many times before a game is completed.

3. Collect the data that the other students obtained in Exercise 2. What does your combined data suggest in the answer to the opening question?

4. Using the data in Exercise 3, determine the fraction, or percent, of the games that were won.

* * *

The opening question concerns "chance" or "randomness." The mathematics devoted to the study of chance is called the <u>theory of probability</u>. It was founded some 300 years ago by a gambling mathematician by the name of Cardano. Now it is the foundation of the insurance industry, and a major tool in business planning. This chapter introduces the basic ideas of probability. In particular it will provide a way to answer the opening question <u>without rolling the dice a single time</u>. Furthermore, it will tell the fraction of times that the player wins in the long run.

To develop the foundations of probability consider a simpler problem that concerns the tossing of coins. Later we will return to the dice.

TOSSING A COIN

5.   When a coin is tossed it can turn up either heads (H) or tails (T).
     Toss a coin ten times and record how many times it turns up heads
     (H) and how many times tails (T).

6.   Collect the results of Exercise 5 from other students. What is the
     total number of

     (a) H's?   (b) T's?

7.   (a) What percent of tosses turned up H?

     (b) Does the result of (a) seem reasonable? Why?

PROBABILITY

If the coin in Exercise 5 is "honest" or "fair" (that is, if heads and tails are equally likely to come up), then in the long run of tosses H and T would each occur about 50% of the time. This is expressed in the assertion:

$$\text{the probability of H is } 50\% = 0.5 = \frac{1}{2}$$

Note that this probability can be written as follows:

$$\text{probability of H} = \frac{\text{number of ways to get heads}}{\text{total number of ways the coin can fall}} = \frac{1}{2}.$$

In tossing the coin there are two <u>equally likely</u> possible outcomes. In an experiment with N equally likely outcomes, the probability of a particular outcome can be found in a similar way: count how many ways that particular outcome can occur and divide by the total number N of equally likely outcomes.

DEFINITION. If in an experiment the total number of equally likely outcomes is N, and if a particular outcome can occur in M different ways, then the <u>probability</u> of that outcome is given by the quotient:

$$\frac{\text{number of ways that the outcome can occur}}{\text{total number of equally likely possible outcomes}} = \frac{M}{N}$$

TOSSING TWO COINS

When two coins (say, a penny and a nickel) are tossed at the same time, there are four possibilities resulting from the fact that either coin may be a head or a tail:

8. In the next exercise you will toss the pair of coins 24 times. About
    how many of those 24 times do you <u>guess</u> the result will be

    (a) two heads?   (b) one head?   (c) no heads?

    (d) Compare your predictions with the predictions made by others in
         the class.

* * *

9. This exercise is done more quickly if two people work together, one
    tossing the coins, the other recording the outcomes. (a) Toss two
    coins 24 times and tally the results in this table:

2 heads HH	1 head HT	0 heads TT	Total
			24

    (b) Does the data gathered in this exercise seem to confirm your pre-
         dictions in Exercise 8?

10. Collect the data of Exercise 9 from the experiments of other students.
    What percent of the total number of tosses is

    (a) 2 heads?   (b) 1 head?   (c) 0 heads?

11. (a) How do your predictions in Exercise 8 compare with the data in
         Exercise 10?

    (b) What do you think will happen to the percents in Exercise 10 as
         more experiments are performed?

THEORY OF TWO COINS

    Now consider the probabilities theoretically rather than experiment-
ally. Let the two coins be a penny and a nickel. There are four possible
outcomes:

	penny	nickel
(1)	H	H
(2)	H	T
(3)	T	H
(4)	T	T

*We make the coins
different in
order to describe
the analysis more
easily.*

Each of these four cases is <u>equally likely</u>, since the coins are "fair."

One out of the four outcomes is "2 heads." Thus we say,

"the probability of 2 heads is $\frac{1}{4}$ or 25%."

Similarly,

"the probability of 0 heads is $\frac{1}{4}$ or 25%."

But the case of one head and one tail can occur in two different ways. Thus,

"the probability of getting exactly one head is $\frac{2}{4}$, or $\frac{1}{2}$, or 50%."

12. How do the theoretical probabilities just computed compare with

(a) your predictions in Exercise 8?

(b) the percentages found in Exercise 10?

THE THEORY OF THREE COINS

Now examine the probabilities when three coins (a penny, a nickel, and a dime) are tossed. If you wish, experiment by tossing coins. However, the exercises will move directly into the theory.

When the three coins are tossed, what is the probability of getting three heads? Two heads? One head? Zero heads?

The table below lists all (eight) possible outcomes. Since there are eight outcomes, all equally likely, each has probability $\frac{1}{8}$.

How do you know the list shows all possible outcomes?

	penny	nickel	dime	probability
(1)	H	H	H	1/8
(2)	H	H	T	1/8
(3)	H	T	H	1/8
(4)	H	T	T	1/8
(5)	T	H	H	1/8
(6)	T	H	T	1/8
(7)	T	T	H	1/8
(8)	T	T	T	1/8

The case of <u>three heads</u> occurs only once, namely in line (1). Thus, the probability of getting three heads when tossing three fair coins is $\frac{1}{8}$. This means that in the long run (on the average), $\frac{1}{8}$ of the experiments will result in three heads.

13. Show that the probability of getting precisely two heads when tossing three coins is $\frac{3}{8}$.

Which lines of the table have two H and one T?

14. What is the probability of getting precisely one head when tossing three coins?

$\frac{1}{8}$

15. What is the probability of getting no heads when tossing three coins?

Next we turn to the case of four coins. As we do so, we develop the notation and the principles that will also apply to the general case, that of tossing N coins.

THE THEORY OF FOUR COINS

Before continuing, recall the MN counting principle from Chapter 8.

THE MN COUNTING PRINCIPLE

If there are M ways to choose one thing and there are N ways to choose another, and if this second choice does not depend on the first choice, then the two choices can be made in MN different ways.

The above counting principle can be extended to include any number of independent choices, as was discussed in Chapter 8. We now use it in counting the number of ways several coins can fall when tossed.

Ignore the very remote possibility that a coin might stand on its edge after it falls.

16. (a) If you toss a penny, how many ways can it fall?

(b) If you toss a penny and a nickel, how many different ways can they fall?

(c) If you toss a penny, a nickel, and a dime, how many different ways can they fall?

(d) If you toss a penny, a nickel, a dime, and a quarter, how many different ways can they fall?

(e) In how many different ways can a set of n coins fall?

(f) Imagine tossing a penny, nickel, dime, and quarter as in (d). Copy and fill in this table:

(a) 2
(c) 8
(e) 2^n

penny	nickel	dime	quarter	probability
H	H	H	H	1/16
H				
H				
H				
H				
H				
H				
H				
T				
T				
T				
T				
T				
T				
T				
T				

The first row is already filled in.

Remembering how three coins can fall will help you fill in this table for four coins.

The following notation will be used throughout the rest of this chapter.

NOTATION.   Let n be a positive integer and k be an integer such that $0 \le k \le n$.   The following symbol denotes the probability that when you toss n coins you get exactly k heads:

$$Pr(n,k)$$

For instance

$$Pr(3,2) = \frac{3}{8} \quad \text{(See Exercise 13)}$$

and

$$Pr(2,1) = \frac{2}{4}$$

17.   With the aid of the table in Exercise 16 show that

(a) $Pr(4,4) = \frac{1}{16}$ (b) $Pr(4,3) = \frac{4}{16}$ (c) $Pr(4,2) = \frac{6}{16}$

18.   Find the following:

(a) $Pr(4,1)$ (b) $Pr(4,0)$

(a) $\frac{1}{4}$

* * *

The table below lists the various $Pr(4,k)$ of Exercises 17 and 18 in order of increasing k.

$Pr(4,0)$	$Pr(4,1)$	$Pr(4,2)$	$Pr(4,3)$	$Pr(4,4)$
$\frac{1}{16}$	$\frac{4}{16}$	$\frac{6}{16}$	$\frac{4}{16}$	$\frac{1}{16}$

PROBABILITY AND PASCAL'S TRIANGLE

Observe that the numbers of the fractions are the n = 4 row of Pascal's triangle, discussed in Chapter 7. In other words, the numerators of the fractions are the binomial coefficients:

$$\binom{4}{0} \quad \binom{4}{1} \quad \binom{4}{2} \quad \binom{4}{3} \quad \binom{4}{4}$$

Is this a coincidence? Or is there a relation between the numerators and the subsets of a 4-set?

It is <u>not</u> a coincidence. To see this, look closely at the computation of $Pr(4,2)$, for example. We have to count all rows of the table of exercise 16(f) that have exactly <u>two</u> <u>heads</u>. These are the six rows.

penny	nickel	dime	quarter
H	H	T	T
H	T	H	T
H	T	T	H
T	H	H	T
T	H	T	H
T	T	H	H

To produce such a row we <u>choose</u> <u>two</u> of the <u>four</u> <u>coins</u> to be heads. Thus there are

$$\binom{4}{2} = 6$$

choices for the particular result we want (exactly two heads). The same reasoning applies in general to show that the numerator of a probability fraction is given by a binomial coefficient. Now, there are two ways for each coin to fall and there are four coins; so, using the MN principle, there are $2 \cdot 2 \cdot 2 \cdot 2 = 2^4$ or 16 equally likely ways for the four coins to fall. Hence, the probability of tossing 4 coins and getting exactly 2 heads is

$$Pr(4,2) = \frac{\binom{4}{2}}{16} = \frac{6}{16}$$

19. Show why $Pr(4,1) = \dfrac{\binom{4}{1}}{16}$

NOTE. The <u>numerator</u> of the probability fraction $Pr(4,k)$ is $\binom{4}{k}$, where $0 \le k \le 4$. The <u>denominator</u>, found by the MN principle, is 2^4. There is no need to fill out a table listing all the ways the coins can fall.

20. Show why $Pr(3,1) = \dfrac{\binom{3}{1}}{8}$

21. Show why $Pr(3,2) = \dfrac{\binom{3}{2}}{8}$

22. The denominator of $Pr(4,k)$ is the number of rows in the table of Exercise 16. There is a row in the table for each way the four coins can fall. Explain to a classmate how to use the MN principle to show that there are 2^4 ways to fill in that table.

23. (a) Explain to a classmate why $Pr(5,k) = \dfrac{\binom{5}{k}}{2^5}$

Describe how k-sets contained in a 5-set are used as well as the MN-principle.

   (b) Compute $Pr(5,1)$.   (c) Compute $Pr(5,3)$.

24. (a) Explain why the probability of throwing 9 coins and getting

exactly 5 heads is $Pr(9,5) = \dfrac{\binom{9}{5}}{2^9}$.

(b) Why is the denominator 2^9?   (c) Why is the numerator $\binom{9}{5}$?

(d) Compute $Pr(9,1)$.   (e) Compute $Pr(9,3)$.

25. Explain the formula below.

This formula is one of the key ideas in this chapter. Make sure you understand why it is true.

> The probability when tossing n coins of getting exactly k heads:
>
> $$Pr(n,k) = \frac{\binom{n}{k}}{2^n}$$

26. Which is more likely to occur, "getting exactly one head when you toss three coins" or "getting exactly two heads when you toss six coins?" Explain.

27. Which is more likely to occur, "tossing four coins and getting exactly two tails" or "tossing six coins and getting exactly three heads?" Explain.

* * *

The opening problem concerns a gambler who rolls two dice and wins if he scores a "nine" before a "seven." Soon we will turn to solving this problem. The work with coins will provide the necessary tools.

THE THREE BASIC LAWS OF PROBABILITY

The study of the tossed coins illustrates three principles of probability.

> I.   THE "EQUALLY LIKELY" PRINCIPLE
> If an experiment can result in any of n <u>equally likely</u> outcomes, then the probability of getting a particular outcome is $\frac{1}{n}$.

The following example illustrates the "equally likely" principle.

Example 1. When a coin is tossed, there are two possible outcomes,

In this case n = 2.

heads or tails. Find the probability of tossing heads.

Solution. The first principle asserts that the probability of tossing one coin and obtaining heads is $\frac{1}{2}$.

28. When you toss a fair <u>die</u> (singular of "dice") there are six equally likely outcomes.

(a) What is the probability of scoring 3 when you toss a single die?

(b) What is the probability of scoring a 5? a 2?

* * *

┌───┐
│ 11.   THE "AND" PRINCIPLE   │
│   The probability that two events (which do not   │
│   depend on each other) occur is the <u>product</u> │
│   of their individual probabilities.   │
└───┘

We sometimes say an unfair die is "loaded."

Associate "and" with "times."

The next example illustrates the use of the "and" principle.

Example 2. What is the probability that when you toss a penny and a nickel that the penny turns up heads and the nickel turns up tails?

Solution.   The probability that the penny is heads is $\frac{1}{2}$. The probability that the nickel is tails is also $\frac{1}{2}$. Hence, by the "and" principle, the probability that "penny is heads" <u>and</u> "nickel is tails" is the product,

$$\frac{1}{2} \cdot \frac{1}{2} = \frac{1}{4}$$

As a check, look back at the discussion before Exercise 12. Example 2 can also be solved by inspection of the table of possibilities. Since there are 4 possible outcomes when the penny and nickel are tossed, all equally likely, the probability that the penny is heads <u>and</u> the nickel is tails is $\frac{1}{4}$.

29. Suppose you toss a penny and roll a die at the same time. What is the probability that the penny turns up heads and the die turns up 3? $\frac{1}{12}$

30. You roll two dice, one red and one green. What is the probability that

(a) The red die turns up 4 and the green die turns up 2?   *(a) $\frac{1}{36}$*

(b) The red die turns up 5 and the green die turns up 5?

* * *

The next exercise can be solved by making a list, as in the discussion before Exercise 13. Instead, use the "and" principle, which applies to any number of events, not just to two events.

Use the "and"
principle.

31. Whan a penny, nickel, and dime are thrown, what is the probability that

(a) all are tails?   (b) all are heads?

(a) $\frac{1}{8}$

(c) the penny and nickel are°heads, and the dime is tails?

(c) $\frac{1}{8}$

(d) the penny is heads, the nickel is tails, and the dime is heads?

32. (a) What is the probability when rolling three dice of getting all ones?

(a) $\frac{1}{216}$

(b) If you roll three dice one thousand times, about how many times would you expect to get three ones?

33. (a) What is the probability when rolling three dice of getting an odd number (that is, 1, 3, or 5) on all three dice?

(a) $\frac{1}{8}$

(b) If you rolled three dice one thousand times, about how many times would you expect to get an odd number on all three?

(c) $\frac{1}{8}$

(c) In the long run, what fraction of the time when you roll three dice would you expect to get all odd numbers?

* * *

The third shortcut for computing probabilities is known as the "at least one of," sometimes called the "either-or" principle.

Associate
"either-or" with
"add." The idea
of non-overlapping
events is very
important here.

> III.   THE "AT LEAST ONE OF" (EITHER-OR) PRINCIPLE
> The probability that at least one of several <u>nonoverlapping</u> events occurs is the <u>sum</u> of their individual probabilities.

The word "nonoverlapping" means that no two of the events can occur at the same time. For instance, the event

"getting a tail when you toss a penny"

and the event

"getting a head when you toss a penny"

are nonoverlapping. However, the event

"getting at least one head when you toss a penny and a nickel"

and the event

"getting at least one tail when you toss a penny and a nickel"

are overlapping.

Why?

The next example illustrates the use of the "at least one" principle.

Example 3. Find the probability of getting exactly one head when tossing a penny and a nickel.

Solution.   Either

(1) the penny is heads <u>and</u> the nickel is tails,

<u>or</u>

(2) the penny is tails <u>and</u> the nickel is heads.

By the "and" principle the first event has probability $\frac{1}{4}$.

$$\frac{1}{2} \cdot \frac{1}{2} = \frac{1}{4}$$

The second possibility also has probability $\frac{1}{4}$ for the same reason:

$$\frac{1}{2} \cdot \frac{1}{2} = \frac{1}{4}$$

By the "at least one" principle, the probability that at least one of the two events (one <u>or</u> the other) occurs is the <u>sum</u> of the probabilities. Thus the probability is

$$\frac{1}{2} \cdot \frac{1}{2} + \frac{1}{2} \cdot \frac{1}{2} = \frac{1}{4} + \frac{1}{4}$$

$$= \frac{1}{2}$$

Note how "and" and "or" translate into product and sum here.

Example 3 was solved just before Exercise 12 by making a table of all possible outcomes. Compare the results of each approach.

34. Suppose you roll two dice, one red and one green. What is the probability that you will roll

   (a) six on red <u>and</u> one on green?

   (b) six or two on red <u>and</u> one on green?

   (a) $\frac{1}{36}$

35. Suppose you roll a die and toss a coin. What is the probability that

   (a) the die will be three and the coin tails?

   (b) the die will be three or five and the coin will be tails.

   (a) $\frac{1}{12}$

36. If you toss a coin ten times what is the probability that

   (a) all ten tosses will be heads?

   (b) exactly one toss will be heads?

   (a) $\frac{1}{1024}$

37. What is the probability that, when tossing a red die and a green die, the red die will be a two or a five and the green die will be a one, a three, or a five?

   $\frac{1}{6}$

38. Suppose you toss a penny, a nickel, and a dime. What is the probability that

   (a) $\frac{1}{8}$

   (a) only the penny is heads?

   (b) only the nickel is heads?

   (c) $\frac{1}{8}$

   (c) only the dime is heads?

(d) Use (a), (b), and (c) and get the principles of probability to compute the probability of getting exactly one head when you toss a penny, a nickel, and a dime.

39. What is the probability of getting a four or a six when rolling a die?

$\frac{1}{3}$

* * *

We wish to return to the opening problem, which concerns tossing a "nine" before a "seven" with two dice. First, however, let us compute the probabilities of getting various scores when you toss two dice.

PROBABILITIES CONCERNING TWO DICE

When you roll two dice, say a red one and a green one, your score can be any integer from 2 ("snake eyes") up to 12 ("box cars"). For $n = 2, 3, \ldots, 12$, let

$$Pr(n)$$

be the probability of getting a score of n when you roll two dice.

Pr(n) can be estimated by rolling two dice many times. Any data would be of use in getting an understanding of the various probabilities. Some scores are fairly rare, others are much more frequent. However, the various probabilities can be computed with the aid of the three principles of probability. The following exercises show this.

40. (a) Show that when rolling two dice that $Pr(2) = \frac{1}{36}$.

(b) Which of the three principles did you use in (a)?

41. This exercise concerns Pr(3) when rolling two dice.

(a) Fill in this table, which will list all the ways of scoring a 3 with two dice.

red die	green die	probability
1	2	?
?	?	?

(b) Using (a) and the principles of probability, show that

$$Pr(3) = \frac{2}{36} \text{ or } \frac{1}{18}$$

42. (Look over Exercises 40 and 41.) About how many times more often will you score 3 with two dice compared to scoring a 2?

43. This exercise concerns Pr(4).

(a) List all the ways you can score 4 with a red die and a green die.

(b)   Use (a) and the principles of probability to show that

$$Pr(4) = \frac{1}{12} .$$

44.   Show that $Pr(5) = \frac{1}{9}$

45.   (a)   Complete the following table which shows the ways you can score when throwing two dice (red and green).

green

		1	2	3	4	5	6
	1					6	
	2			5			
red	3	4					
	4						10
	5				9		
	6						

(Record the scores in the table)

(b)   How can you compute

$$Pr(2), \; Pr(3), \; . \; . \; . \; , \; Pr(12)$$

by inspection of the table in (a).

46.   Use the principles of probability and the table in Exercise 45 to make the appropriate computations, then copy and fill in the following table.

Score (two dice)	Pr(n)	Pr(n) reduced	Score (two dice)	Pr(n)	Pr(n) reduced
2	$\frac{1}{36}$	$\frac{1}{36}$	8	$\frac{?}{36}$	?
3	$\frac{2}{36}$	$\frac{1}{18}$	9	$\frac{?}{36}$	?
4	$\frac{3}{36}$	$\frac{1}{12}$	10	$\frac{?}{36}$	?
5	$\frac{4}{36}$	$\frac{1}{9}$	11	$\frac{?}{36}$	?
6	$\frac{?}{36}$	?	12	$\frac{?}{36}$	?
7	$\frac{?}{36}$	?			

47. What score is most likely when you roll two dice?

48. If you roll two dice 36 times, about how many times on the average would you expect to get a score of

(a) 6?   (b) 8?   (c) 5?   (d) 9?   (e) 7?   (f) 11?   (g) 2?

(a) five
(c) four
(e) six
(g) one

ROLLING ONE SCORE BEFORE ANOTHER

At the beginning of the chapter, we had the situation where a gambler rolls two dice until he scores a 9 or a 7. If he scores the 9 first, he wins; if he scores the 7 first, he loses. What is the probability that he will win? The next exercise provides a background for getting the theoretical answer to this question.

49. Work with two other students. One student rolls two dice, but doesn't look at the result of his throw. The second student watches the scores and when a 9 or 7 comes up, says "stop." You then look at which case happens and record 9 or 7.

   (a)   Do the whole experiment 10 times. Record the number of wins (9's) and losses (7's).

   (b)   Which is more frequent?

* * *

The next exercise leads to a way of computing the probabilities of "win" or "loss" with regard to the 9 or 7 being rolled first.

50. Since we need only look at the final throw of the experiment, when a 9 or 7 occurs for the first time, the possibilities are restricted. Refer to the table in Exercise 45 to answer the following.

Would this be a "fair" game for the player?

   (a)   In how many different ways can two dice show the score 9?

   (b)   In how many different ways can two dice show the score 7?

   (c)   Explain why there are only ten results in the table in Exercise 45 that concern us in this problem.

   (d)   Using (a), (b), and (c), show that the probability of scoring a win is $\frac{2}{5}$ (that is, scoring a 9 before a 7).

   (e)   Were your experimental results in Exercise 49 close to the theoretical results in (d)?

Explain every answer for these problems by using the table in Exercise 49.

51. What is the probability of scoring a 7 before a 9?

52. What is the probability of scoring a 2 ("snake eyes") before a 7?

52. $\frac{1}{7}$

53. What is the probability of scoring a 7 before a 2?

54. $\frac{1}{3}$

54. What is the probability of scoring a 2 before a 3?

55. $\frac{1}{3}$

55. What is the probability of scoring a 2 or a 3 before a 7?

56. When two coins are tossed repeatedly, what is the probability of getting "two heads" before getting "one head and one tail?"

56. $\frac{1}{3}$

The next two exercises illustrate an important principle.

57. Suppose you have a die that you know is "fair." You roll it and on each of ten successive rolls you score a 2. What is the probability that you will score a 2 on the next roll?

58. Suppose that you toss 100 times a coin that you know is "fair," and score a head every time. What is the probability that you will score a head on the next toss?

Remember that each toss is independent of the others!

THE BOY-GIRL PROBLEM

Assume that any baby is as likely to be a girl as a boy, that is,

$$Pr(girl) = \frac{1}{2} = Pr(boy)$$

(Note. In a particular family this may not be true biologically.)

Remember the "principles of probability" and the coin problem.

59. What is the probability that a family with two children will have

    (a) exactly two girls?

    (b) exactly one boy and one girl?

    (c) at least one boy?

(a) $\frac{1}{4}$

(c) $\frac{3}{4}$

60. What is the probability that a family with three children will have

    (a) exactly one girl?

    (b) exactly three boys?

    (c) at least two girls?

(a) $\frac{3}{8}$ (c) $\frac{1}{2}$

Assume that the sex of each child is independent of that of the other children. There is however some evidence that parents who have produced three daughters and no sons will tend to produce another girl much more than 50% of the time.

61. What is the probability that a family of five children will have exactly

    (a) no girls?   (b) one girl?   (c) two girls?

    (d) three girls?   (e) four girls?   (f) five girls?

61. (a) $\frac{1}{32}$

(c) $\frac{5}{16}$ (e) $\frac{5}{32}$

62. A certain box contains three black balls and two white balls all the same size and made of the same material. If you are blindfolded, what is the probability that when you reach in the box and choose one ball it will be

    (a) black?   (b) white?   (c) black or white?

62. (a) $\frac{3}{5}$ (c) 1

CARDS

A bridge deck has 52 cards consisting of four suits:

spades, hearts, diamonds, and clubs.

Spades and clubs are black suits; hearts and diamonds are red. Each suit has 13 cards:

ace, king, queen, jack, 10, 9, 8, . . . , 3, 2 (deuce).

The king, queen, and jack are called "face cards" because they have pictures rather than just the "pips" that the other cards have.

63. Suppose that from a well-shuffled bridge deck of 52 cards you draw a card at random. What is the probability that it will be

(a)   the ace of spades?   (b) a king?   (c) a club?

(d)   a face card?   (e) a black card?

(f)   the king of clubs or the ten of diamonds?

(a) $\frac{1}{52}$ (c) $\frac{1}{4}$
(e) $\frac{1}{2}$

64. (a) From a bridge deck you draw a card at random. Suppose the card is the queen of spades. You draw another card (without putting the first card back). What is the probability that the second card is the king of hearts?

(a) $\frac{1}{51}$

(b) You draw two cards at random from a bridge deck. What is the probability that one is the queen of spades and the other is the king of hearts?

65. You draw two cards at random from a bridge deck. What is the probability that

(a) $\frac{1}{2652}$

(a)   they are the ten of spades and the three of diamonds?

(b)   they are a king and a queen?

(b) $\frac{4}{663}$ (c) $\frac{1}{221}$

(c)   both are aces?

(d)   neither is an ace?

(d) $\frac{188}{221}$ (e) $\frac{396}{2652}$

(e)   at least one of them is an ace?

(f)   Which principles of probability did you use in (a) through (e)?

PROBABILITY THAT AN EVENT DOES NOT HAPPEN

The probability of picking a diamond out of a deck of 52 cards is $\frac{13}{52}$ or $\frac{1}{4}$. The probability of not picking a diamond is $\frac{39}{52}$ or $\frac{3}{4}$. Note that these two probabilities add up to 1. This illustrates the following general principle:

> If the probability that an event happens is p
>
> and
>
> the probability that the event does <u>not</u> happen is q,
>
> then
>
> $$p + q = 1.$$
>
> In other words,
>
> $$q = 1 - p.$$

The probability of an event happening or not happening is 1 (a certainty).

The next three exercises concern this observation.

66. Suppose you roll a single fair die. What is the probability that you

    (a) will roll a three?   (b) will not roll a three?

67. Suppose you toss eight coins. What is the probability that

    (a) you get zero, one, or two heads?

    (b) you get more than two heads?

68. Suppose you roll two dice. What is the probability of

    (a) scoring a 7 or an 11?   (b) not scoring a 7 or an 11?

69. If the probability that an event will happen is $\frac{13}{17}$, what is the
    probability that it will not occur?

* * * SUMMARY * * *

This chapter concerns <u>probability</u>, the study of chance. It was
illustrated by the computation of the probabilities involved in tossing
coins, the boy-girl problem, rolling dice, and drawing cards.

It was shown that the probability of getting <u>exactly</u> k heads when you
toss n coins is

$$Pr(n,k) = \frac{\binom{n}{k}}{2^n}$$

Then the three basic laws of probability were stated.

   I.   The "equally likely" principle

   II. The "and" principle (multiply the probabilities)

   III. The "at least one of" or "either-or" principle (add the proba-
        bilities)

These principles were applied to coins, dice, and cards. In particular, the probability of getting a score of n when tossing two dice, Pr(n),
was computed.

The chapter also developed a method for calculating the probability that one event occurs before another event.

If the probability that an event will occur is p, then the probability that the event will not occur is 1 - p.

*   *   *   WORDS AND SYMBOLS   *   *   *

Probability $Pr(n)$, rolling dice

Laws of probability

    "equally likely" principle

    "and" principle $Pr(n,k) = \dfrac{\binom{n}{k}}{2^n}$, coins

    "at least one of" principle

*   *   *   CONCLUDING REMARKS   *   *   *

Probability has its origins in the insuring of ships' cargoes during the Renaissance and in the theory of gambling in the 17th century.

Probability plays a key role in daily life. The weather report may predict "a sixty percent chance of rain," or a Gallup poll reports a change from 27% to 30% in a public opinion. These polls rely on 1500 to 1800 interviews. How dependable are they? Is a change from 27% to 30% meaningful? What is the probability that the change of 3% is due to chance?

Probability is also being used to determine whether juries are being chosen fairly or with bias.

When buying insurance against the theft of a bicycle or a stereo set, we think in terms of probabilities. What is the fair premium to insure a $100 bicycle against theft for one year? Two dollars? Five dollars? Ten?

When we buy auto insurance should we get a "fifty dollar deductible" or a "one hundred dollar deductible?" (In a "fifty dollar deductible" the car owner pays the first fifty dollars of any damage.) The "hundred dollar deductible" should cost less than the "fifty dollar deductible," but how much less?

Many questions like these face the average citizen, who will find probability and statistics a tool for coping with the gambles and choices that must be made in daily life.

* * * PROBABILITY SELF QUIZ * * *

1. The Sooper Sluggers are an amateur softball team. From past experience they know that the probability of their beating team A is $\frac{2}{5}$, while that of beating team B is $\frac{4}{7}$. What is the probability that
   (a) they will beat both team A and team B when they next play them?
   (b) they will lose to team A but will beat team B?
   (c) they will lose to both team A and to team B?

2. What is the probability that you will score a 10 when you roll two fair dice?

3. What is the probability that a family of five children will have
   (a) exactly two boys and three girls?
   (b) at least one girl?

4. (a) How large a number can the probability of an event be?
   (b) How small a number?

5. A gambler rolls a die until it turns up 2. What is the probability that it first turns up 2 on the fourth roll?

ANSWERS TO SELF QUIZ

1. (a) $\frac{8}{35}$ (b) $\frac{12}{35}$ (c) $\frac{9}{35}$

2. $\frac{1}{12}$ 3. (a) $\frac{5}{16}$ (b) $\frac{31}{32}$

4. (a) 1   (b) 0   5. $\frac{125}{1296}$

* * * PROBABILITY REVIEW * * *

1. Explain how Pascal's triangle is involved in computing Pr(5,2), the probability that when you toss five fair coins you will get exactly two heads.

2. Suppose the probability of an event happening is $\frac{a}{a + b}$. Use algebra to show that the probability of the event not happening is $\frac{b}{a + b}$.

3. Mr. Green has taught algebra for many years. Of his 2000 students of the past, only 140 have made an "A." If you are in his algebra class with 30 students, about how many students will probably get an "A?"

4. You and your two best friends are in an algebra class of 30 students. In a moment of weakness, the teacher announces that three students will be chosen at random and that they will not have to do any home-

work that night. What is the probability that you and your friends will all be chosen?

5. What fraction of the time would you expect to get all heads when you toss

(a) one penny? (b) two pennies? (c) ten pennies? (d) n pennies?

6. What fraction of the time would you expect to score a 5 when you toss a die?

7. Using Pascal's triangle, find the probability of getting 3 heads when you toss 6 coins.

8. If one line of Pascal's triangle begins

$$1 \qquad 9 \qquad 36 \qquad 84 \ldots$$

how does the next line below begin?

9. Two equally matched teams play in the World Series. The first to win four games is the victor. If team A is behind, 1 game to 3 games for team B, what is the probability that team A wins the Series?

10. If you were to make a list of one hundred families that have exactly two children, about how many would you expect to consist of

(a) two boys?   (b) one boy and one girl?   (c) two girls?

11. In what fraction of the families with exactly three children would you guess there is one girl and two boys?

12. Out of 16,000 families with exactly five children, how many would you expect to have two girls and three boys?

13. What is the probability that a family with six children will have all girls?

14. When rolling two dice, what is the probability of

(a) rolling a 4?   (b) rolling a 6?   (c) rolling a 4 or 6?

15. When rolling two dice, what is the probability of

(a) rolling 7 or 11?

(b) rolling 5 and rolling another 5 before rolling 7 (a win)?

16. What is the probability of drawing from a deck of bridge cards

(a) an ace or king?

(b) an ace and then a king?

(c) Do (b) if the ace is returned and then the deck is shuffled.

17. In a jar there are 8 green balls, 3 white balls, and 5 red balls. What is the probability that you will draw

(a) a green or white ball?

(b) a red ball followed by a green ball, if the red ball is put back in the jar before the second ball is drawn?

(c) Do (b) if the red ball is not put back in the jar.

ANSWERS TO PROBABILITY REVIEW

1. See text for discussion of numerator $\binom{5}{2}$ and denominator 2^5.

3. About 2   4. $\frac{1}{4060}$

5. (a) $\frac{1}{2}$ (b) $\frac{1}{4}$ (c) $\frac{1}{1024}$ (d) 2^{-n}

6. $\frac{1}{9}$ 7. $\frac{5}{16}$

8. 1, 10, 45, 120, . . .

9. $\frac{1}{8}$

10. (a) 25   (b) 50   (c) 25

11. $\frac{3}{8}$ 12. 5000   13. $\frac{1}{64}$

14. (a) $\frac{1}{12}$ (b) $\frac{5}{36}$ (c) $\frac{2}{9}$

15. (a) $\frac{2}{9}$ (b) $\frac{2}{45}$

16. (a) $\frac{2}{13}$ (b) $\frac{4}{663}$ (c) $\frac{1}{169}$

17. (a) $\frac{11}{16}$ (b) $\frac{5}{32}$ (c) $\frac{1}{6}$

* * * PROBABILITY REVIEW SELF TEST * * *

1. A gambler rolls three dice. What is the probability of getting a total of 6?

2. A die is constructed in such a way that two of the faces show a 1 and four of the faces show a 2.

(a) What is the probability of rolling a 1 with a single such die?

(b) What is the probability of scoring a total of 2 with two such dice?

(c) What is the probability of scoring a total of 4 with two such dice?

(d) What is the probability of scoring a total of 3 with two such dice?

3. (Pascal's triangle might be helpful for this exercise.) What is the probability of getting at least 30% and at most 70% heads when you toss

(a) three pennies?   (b) four pennies?

(c) five pennies?   (d) nine pennies?

(e) a million pennies? (Make an intelligent estimate; don't compute it.)

4. What is the probability of getting an ace and also a 10, jack, queen, <u>or</u> king, when dealt two cards from a full deck (without jokers)?

The Projects have some interesting problems.

ANSWERS TO PROBABILITY SELF TEST

1. $\frac{5}{108}$

2. (a) $\frac{1}{3}$ (b) $\frac{1}{9}$ (c) $\frac{4}{9}$ (d) $\frac{4}{9}$

3. (a) $\frac{3}{4}$ (b) $\frac{3}{8}$ (c) $\frac{5}{16}$ (d) $\frac{105}{128}$ (e) Near 1

4. $\frac{32}{663}$

* * * PROBABILITY PROJECTS * * *
THE GAME OF CRAPS

Now we are ready to examine the game of "craps," the rules of which are simple:

A player rolls two dice. If on the first roll he scores a 7 or 111 (a "natural"), he wins. If on the first roll the player scores 2,3, or 12 ("craps"), he loses. If the first roll is any other number (4,5,6,8,9,10), the player continues to roll until he scores that number again, in which case he wins, or until a 7 appears, in which case he loses.se he loses.

1. Play 10 games of dice (or "craps") and record the number of wins and losses. Tally the total class results on the blackboard. Do you conclude from your results that a player should expect to win if he plays a great number of games? Explain.

2. (a) What is the probability of winning on the first roll? Explain.

(b) What is the probability of losing on the first roll? Explain.

(c) What is the probability that the game ends on the first roll? Explain.

(d) What is the probability the game continues beyond the first roll?

(a) $\frac{2}{9}$ *(b)* $\frac{1}{9}$

(c) $\frac{1}{3}$

3. Suppose the player scores 4 on his first roll. To go on to win, he must score a 4 before a 7.

    (a)   Roll a pair of dice until you score a 4 or a 7. Record which score, the 4 or 7, appeared first.

    (b)   Repeat (a) 10 times and combine your data with those of the class on the blackboard. What percent of the time did the 4 appear before the 7?

4. In finding the probability that the player rolls a 4 before a 7, you may disregard all throws except the last one, which is either a 4 or a 7.

    (a)   What is the total number of ways a 4 or a 7 can appear?   *(a)* 9

    (b)   What is the probability that a 4 appears on the last throw?   *(b)* $\frac{1}{3}$
(i.e. that a 4 appears before a 7.) How did this compare with Exercise 3(b)?

Example. The player can win by rolling a 4 on the first toss and then rolling a 4 before a 7. Now, in $\frac{3}{36}$ of the games the player will begin with a score of 4. (Why?) Of these games that begin with a 4, the player will win one third of the time. (Why?) Thus, the probability that the player will throw a 4 on his first roll <u>and</u> go on to win is

$$\left(\frac{3}{36}\right) \cdot \left(\frac{1}{3}\right) = \frac{1}{36}$$

(This is the "<u>and</u>" principle of probability.) Another way of saying this is, "On the average, out of every 36 games, the player will win one by opening with a 4 and then scoring a 4 before a 7."

5. What is the probability that the player wins by rolling a 9 on the first roll and then rolls another 9 before rolling a 7? Explain. $\frac{2}{45}$

6. What is the probability that the player wins by rolling an 8 on his first roll and then rolls another 8 before rolling a 7? Explain. $\frac{25}{396}$

7. (a)   What is the probability that the player will win in the game of craps? (Show all work. Which principles of probability did you use?)   *(a)* $\frac{244}{495}$

    (b)   Comment on the statement: "If the player gambles at craps long enough, he will surely lose."

ODDS

If an experiment has a probability for success of $\frac{4}{7}$, then the probability for failure is $\frac{3}{7}$. We say the <u>odds</u> in favor of the experiment are 4 to 3. A gambler, betting on success would have to wager $4 to win $3.

8. (a) What are the odds that a crap game will continue beyond the first roll?

(b) 1 to 2

(b) If the player rolls a 4 on the first roll, what are the odds that he will win?

9. (a) What is the probability that you will draw an ace from a bridge deck of cards?

(b) What are the odds against your drawing an ace in (a)?

10. (a) If the probability of your winning a certain game is $\frac{1}{2}$, and you pay $1 to play the game, how much should the winning prize be if the game is to be fair? Explain.

(b) If the probability of your winning is $\frac{2}{3}$ and you pay $1 to play, how much should the prize be if the game is fair? Explain.

(c) fifty cents

(c) If the <u>odds</u> of winning are 2 and 1 and you pay $1 to play, how much should the prize be if the game is fair? Explain.

11. Would you be willing to play the following game? You toss two fair dice. If the sum is any of the four numbers 2, 3, 4, or 5, you win. If the sum is any of the six numbers 7, 8, 9, 10, 11, or 12, your opponent wins. If the sum is 6 no one wins. Since you win on 4 numbers but lose on 6 numbers, your opponent says, "I'll pay you $6 when you win if you pay me $4 when you lose." Would you play his game? Explain.

12. (a) If the player at craps first rolls a 5, and is willing to bet $1 that he will win, how much should the opponent bet to make the betting fair?

(b) What is the relation between "odds" and the amount each person bets if the betting is to be fair?

(c) $(\frac{1-p}{p})A$

(c) If the probability of winning a game is p and it costs $A to play each game, what would the winning prize be, to make the game fair?

THE BUG

13. A bug is on a promenade, wandering on a straight line.

He goes an inch forward or an inch backward, at random. Then he

pauses, and, forgetting what he did, continues once again his promenade, going an inch backward or an inch forward. Unfortunately,
two inches from where he starts, on either side of him are bug traps.

With good luck, he may have a long walk. But with bad luck, his
walk may finish after he travels a mere 2 inches.

(a) What is the probability that he travels <u>exactly</u> 2 inches before
     falling into a trap?

(b) What is the probability that he travels <u>exactly</u> 4 inches before
     falling into a trap?

Hint: You can flip a coin to determine whether he goes right or left; what does TT mean? HH?

(c) Exactly 6 inches?

14. A and B are playing "matching coins;" each begins with two coins.
     What is the probability that B will have no coins left after two
     flips if he loses when he "matches"?

15. Let a die be weighted (or "loaded") so that the probability of a
     number appearing when the die is tossed is proportional to the
     given number. For example, 4 has twice the probability of appearing
     as 2.

(a) If p is the probability that 1 appears, what is the probability
     that 2 appears? that 3 appears?

Hint: (b) What must Pr(1) + Pr(2) + ... + Pr(6) be?

(b) Find p, the probability that 1 appears.

(c) Find the probability that an even number appears.

(c) $\frac{12}{21}$

16. We have already computed the probability that when tossing two dice
     a 4 will turn up before a 7. We will see that geometric series can
     be used to find this probability.

(a) What is Pr(4)? Pr(7)?

(b) What is the probability that neither a 4 nor a 7 turns up?

(c) What is the probability that neither 4 nor 7 turns up on the
     first two throws and 4 turns up on the 2nd throw?

(d) What is the probability that neither 4 nor 7 turns up on the
     first two throws and 4 turns up on the 3rd throw?

(e) What is the probability that neither 4 nor 7 turns up on the
     first n throws and 4 turns up on the $(n + 1)^{st}$ throw?

(d) $Pr = \frac{3}{4} \cdot \frac{1}{12}$

(e) $Pr = (\frac{3}{4})^n \cdot \frac{1}{12}$

(f) What is the probability that a 4 turns up before a 7? Express
     the answer as the sum of a geometric series, then find its sum
     by the formula in Chapter 1.

17. Three dice are designed as follows. One has six 4's. One has two 6's and four 2's. One has four 5's and two 1's. You and your opponent will each choose <u>one</u> of the dice. Each person rolls his or her own die and the one with the higher number wins.

    (a) Which die would you choose?

    (b) Or, would you be generous and let your opponent choose first?

CUMULATIVE REVIEW, CHAPTERS 7 - 9

1. Write in the form $\binom{n}{k}$:

    (a) $\binom{5}{4} + \binom{5}{5}$ (b) $\binom{7}{3} + \binom{7}{4}$

2. Compute:

    (a) $\binom{5}{3}$ (b) $\binom{7}{4}$ (c) $\binom{10}{8}$ (d) $\binom{100}{3}$

3. Expand with integer coefficients:

    (a) $(x + y)^4$ (b) $(x - 2y)^3$ (c) $(2x - 3y)^5$

4. In the expansion of $(x + y)^{10}$, write the integer coefficient of the term containing:

    (a) $x^2 y^8$ (b) $x^5 y^5$ (c) $x^7 y^3$

5. How many car license plates are possible if there are two letters followed by four digits and repetitions of letters and digits are possible?

6. Mr. Smith is a very popular chemistry teacher; 31 students want to get into one of his classes, but there are only 28 lab stations. How many different ways may he select students if he allows only 28 students to enroll?

7. In a round table seating arrangement assume there is no head of the table position. In how many ways may you seat

    (a) 1 person?   (b) 2 persons?   (c) 3 persons?

    (d) 4 persons?   (e) 5 persons?   (f) n persons?

8. Find the sum of the arithmetic progression,

$$1 + 2 + 3 + \ldots + 100$$

9. There are 30 students in class A and 27 students in class B. If class A qualifies for three representatives to student council and class B qualifies for two, in how many ways may the two classes send five representatives to student council?

10. Compute the number of diagonals in a regular decagon.

Hint: The figure has ten sides!

11. Compute the probability of tossing five coins and getting exactly three heads.

12. Compute the probability of a family with four children having exactly two girls and two boys.

13. Toss a die and a coin. What is the probability that

    (a) the coin is tails <u>and</u> the die is 6?

    (b) the coin is tails <u>or</u> the die is 6?

14. If the probability of success of a certain event is $\frac{2}{5}$, what is the probability of failure?

15. A gambler rolls two dice. What is the probability he will roll a 7

    (a) on the first roll?

    (b) for the first time on the third roll?

16. Write in sigma notation:

    (a) $2 + 4 + 6 + \ldots + 1000$ (b) $(x + y)^{10}$ expanded

17. Compute:

    (a) $\displaystyle\sum_{i=1}^{4} i^4$ (b) $\displaystyle\sum_{j=1}^{5} (2j - 1)$

ANSWERS TO CUMULATIVE REVIEW, CHAPTERS 7 - 9

1. (a) $\binom{6}{5}$ (b) $\binom{8}{4}$

2. (a) 10   (b) 35   (c) 45   (d) 161,700

3. (a) $x^4 + 4x^3y + 6x^2y^2 + 4xy^3 + y^4$

   (b) $x^3 - 6x^2y + 12xy^2 - 8y^3$

   (c) $32x^5 - 240x^4y + 720x^3y^2 - 1080x^2y^3 + 810xy^4 - 243y^5$

4. (a) 45   (b) 252

5. 6,760,000   6. 4,495

7. (a) 1   (b) 1   (c) 2   (d) 6   (e) 24   (f) $(n-1)!$

8. 5050

9. $\binom{30}{3} \cdot \binom{27}{2} = 1,425,060$

10. $\binom{10}{2} - 10 = 35$

11. $\frac{5}{16}$

12. $\frac{3}{8}$

13. (a) $\frac{1}{12}$ (b) $\frac{2}{3}$

14. $\frac{3}{5}$

15. (a) $\frac{1}{6}$ (b) $\frac{25}{216}$

16. (a) $\displaystyle\sum_{i=1}^{500} 2i$ (b) $\displaystyle\sum_{k=0}^{10} \binom{10}{k} x^{n-k} y^k$

17. (a) 354   (b) 25

* * * CENTRAL SECTION * * *

This chapter is an introduction to analytic geometry, the branch of mathematics that applies algebra to geometry. The techniques of analytic geometry will be illustrated by the development of the circle, parabola, ellipse, hyperbola, and line.

Four problems concerning a lost prospector lead to the core ideas of the chapter.

THE LOST PROSPECTOR PROBLEMS

DIRECTIONS: For each of Problems A, B, C, and D, do the following. (Assume the land is flat.)

Read these directions carefully.

(a) Make a scale drawing, with 1 centimeter representing 1 mile.

(b) Plot several points where the prospector may be.

(c) When you have enough points in (b), sketch a smooth curve (or line) through them, showing all possible places where the lost prospector may be.

The figures you will draw will be at the larger scale of one mile = one cm.

EQUIPMENT NEEDED: centimeter ruler, compass, and graph paper will be required throughout the chapter.

Problem A. "Nevada Joe", a lost prospector, radios back to his headquarters at the mine, "I'm lost and very weak . . ." He stops speaking. Just then, a dynamite charge goes off at the mine. The men at the mine hear the sound of the blast transmitted back over Joe's radio 25 seconds later. Where should they look for Joe? (Sound travels about a mile in five seconds; consider radio waves to be instantaneous.)

(How far?)

•- - - - - - - - - - - - - - - - •
mine   Joe

Carry out steps (a)-(c) of the above directions and make an accurate scale drawing. Indicate where Joe might be.

* * *

Problem B.   Suppose instead that Nevada Joe had said, "I'm just as far from the highway as from the mine . . ." His radio goes dead. Carry out directions (a)-(c) and make an accurate scale drawing. First show at least five possible locations where Nevada Joe might be. Assume that the highway is straight.

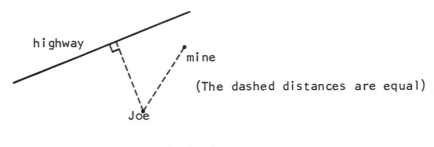

(The dashed distances are equal)

* * *

Problem C.   Suppose that Nevada Joe says, "I'm just as far from the mine as I am from Dead Man's Oasis." Then his radio goes dead.

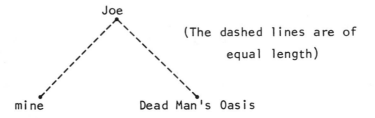

(The dashed lines are of equal length)

Carry out directions (a)-(c) and plot at least five places where Joe might be.

* * *

Problem D.   Nevada Joe says, "I'm lost. I sprained my ankle. I came in a straight line from Dead Man's Oasis. If I go in a straight line from here to the mine, my total trip will be 10 miles." The oasis and the mine are 8 miles apart.

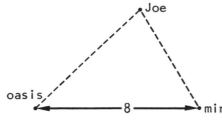

(The sum of the lengths of the dashed line segments is 10 miles)

Compare your work on Problems A, B, C, D with that of other students.

Carry out directions (a)-(c) and plot at least five points where Joe might be.

* * *

In the exercises that follow use graph paper when appropriate. From time to time the exercises will refer back to the four opening problems.

EXERCISES

1. The mine is at M = (0,0) in the (x,y)-coordinate plane in Problem A.

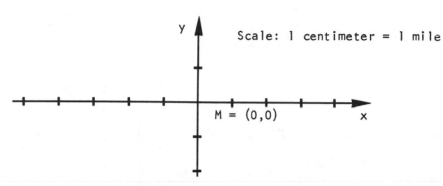

Scale: 1 centimeter = 1 mile

M = (0,0)

At which of the following points do you <u>guess</u> Joe could be? (Use a ruler to measure the distances.)

(a) (2,3)   (b) (4,1)   (c) (-3,4)   (d) (3,2)   (e) (0,-5)   (f) (4,3)

(a) no
(c) yes
(e) yes

2. In Problem B, assume that the mine is at the point M = (0,1) and the highway has the equation y = -1.

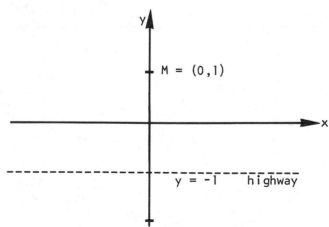

M = (0,1)

y = -1   highway

At which of the following points might Joe be? (Use a ruler to measure the distances.)

(a) (-1,1)   (b) (2,1)   (c) (0,0)   (d) (4,4)   (e) (-2,3)   (f) (-4,4)

(a) no
(c) yes
(e) no

3. In Problem C, assume that the mine is at M = (-2,0) and Dead Man's Oasis is at D = (2,2).

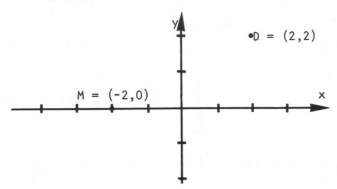

At which of the following points might Joe be? (Use a ruler to measure the distances.)

(a) (0,1)   (b) (1,-1)   (c) (0,0)   (d) (2,0)   (e) (2,-3)   (f) (-2,5)

(a) yes
(c) no
(e) yes

4. In Problem D, if Dead Man's Oasis is at D = (4,0), the mine is at M = (-4,0), and the <u>sum</u> of the distances from Joe to the mine and to the oasis is 10 miles,

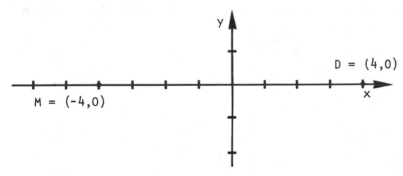

at which of the following points might Joe be? (Use a ruler to measure the distances.)

(a) (0,0)   (b) (7,3)   (c) (0,3)   (d) (0,-3)   (e) (5,5)   (f) $(3,\frac{12}{5})$

(a) no
(c) yes
(f) yes

*   *   *

All work so far in this chapter has been experimental. In each problem the places where Joe may be either form a curve or a straight line. In order to deal with curves and lines algebraically (and swiftly), certain tools of analytic geometry are needed. The next few exercises introduce them.

First, the distance between points on a line or in the plane will be expressed algebraically, in terms of coordinates.

ABSOLUTE VALUE

Recall for any <u>real</u> <u>number</u> x (a number on the number line), the <u>absolute</u> <u>value</u> of x is written

$$|x|$$

and is defined as follows:

> DEFINITION. $|x| = x$ if $x \geq 0$,
>
>   but
>
> $|x| = -x$ if $x < 0$.

The definition is stated here in case this chapter is studied before another one where the definition is needed.

Example 1. Evaluate:

(a) $|2|$ (b) $|-3|$ (c) $|0|$ (d) $|-5|$ (e) $|7.3|$

Solution.

(a) $|2| = 2$ (b) $|-3| = 3$ (c) $|0| = 0$ (d) $|-5| = 5$ (e) $|7.3| = 7.3$

The absolute value of a number x is merely the distance between x and 0 on the number line.

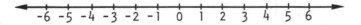

This assertion may be checked by studying the solutions to Example 1.

5. Find: (a) $|-20|$ (b) $|7|$ (c) $\left|\frac{2}{3}\right|$ (d) $|-0.5|$

(a) 20 (c) $\frac{2}{3}$

THE DISTANCE BETWEEN TWO POINTS ON THE NUMBER LINE

The absolute value of x,

$$|x|,$$

gives the distance that any number x is from 0 on the number line. The next exercise suggests a way of finding the distance between <u>any</u> two points on the number line.

6. Copy and fill in this table. Use the number line to help fill in the first empty column.

a	b	a-b	|a-b|	distance between a and b
3	5			
6	-2			
-2	6			
-7	-3			

Exercise 6 suggests this useful formula:

> The distance between two points on the number line, located at a and b, is
>
> $$|a-b|.$$
>
> (Read as "the absolute value of a minus b.")

Example 2. Find:

(a)   the distance between 4 and -2

(b)   the distance between -3 and 5

Solution.

(a) $|4 - (-2)| = |4 + 2| = 6.$

(b) $|-3 - 5| = |-8| = 8.$

7. Without drawing the points, find the distance between

(a) 5   (c) 11
(e) 5

(a)   8 and 3   (b)   3 and 8   (c)   8 and -3

(d)   5 and -5   (e)   -2 and -7   (f)   2 and 7

8. (a)   Fill in the table:

x	|x|	|-x|	$(x)^2$	$|x|^2$
2				
-3				
4				
0				
-5				

(b)   For any x how do the entries in the last two columns compare?

9. (a) Fill in the table:

a	b	$\lvert a-b\rvert$	$\lvert b-a\rvert$	$(a-b)^2$	$\lvert a-b\rvert^2$
1	5				
-2	7				
0	-4				

(b) Show that (a-b) is equal to the opposite of (b-a).

Hint:
$-(b-a) = -1(b-a)$
$\qquad = -b + a$
$\qquad = (a - b)$

(c) Use the fact that $(x)^2 = \lvert x\rvert^2$ to show that $(a-b)^2 = \lvert a-b\rvert^2$.

THE DISTANCE BETWEEN TWO POINTS IN THE PLANE

The point of Exercises 8 and 9 is that the square of a number equals the square of the absolute value of that number:
$x^2 = \lvert x\rvert^2$ *and thus* $(a-b)^2 = \lvert a-b\rvert^2.$

The distance between two points on the x-axis or on the y-axis of a coordinate system is given by the formula $\lvert a-b\rvert$. The next few exercises develop a formula for finding the distance between any two points in the (x,y)-plane.

10. Plot the points (-1,4) and (7,2) on graph paper.

(a) Estimate the distance between them by measuring with a ruler.

(b) Is your answer precise or approximate?

Observe that $\lvert a-b\rvert = \lvert b-a\rvert$.

* * *

Recall the Pythagorean Theorem from geometry:

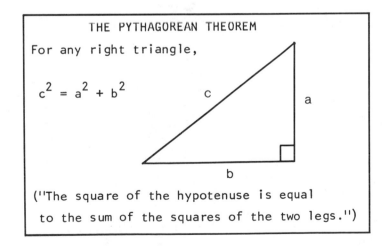

THE PYTHAGOREAN THEOREM
For any right triangle,

$c^2 = a^2 + b^2$

("The square of the hypotenuse is equal to the sum of the squares of the two legs.")

The next exercise uses the Pythagorean Theorem to obtain a formula for the distance between two points in terms of their coordinates. In particular it will allow you to determine precisely the distance between the points (-1,4) and (7,2) in Exercise 10, without any picture as a reference.

11. Let $A = (x_1, y_1)$ and $B = (x_2, y_2)$ be two points in the (x,y)-plane.

(Side BC is parallel to the x-axis; side AC is parallel to the y-axis.)

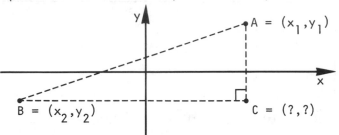

Read "x_1" as "x sub one" or "x one;" Similarly for x_2, y_1 and y_2.

(a) What are the coordinates of point C in the above diagram?

(b) Explain why the legs of triangle ABC have lengths $|x_1 - x_2|$ and $|y_1 - y_2|$.

Hint: (b) Points A and C lie on a line; points B and C lie on a line.

(c) Explain why the distance from A to B is

$$\sqrt{|x_1 - x_2|^2 + |y_1 - y_2|^2}$$

If notation bothers you here, see Exercise 9 again.

(d) Use (c) to justify the following important formula:

THE DISTANCE FORMULA

The distance, d, between the points (x_1, y_1) and (x_2, y_2) is

$$d = \sqrt{(x_1 - x_2)^2 + (y_1 - y_2)^2}$$

Example 3. Find the distance, d, between $(2,1)$ and $(3,7)$:

Solution. By the distance formula,

$$d = \sqrt{(2-3)^2 + (1-7)^2}$$
$$= \sqrt{(-1)^2 + (-6)^2}$$
$$= \sqrt{37}$$

Observe that the distance formula allows us to find the distance between the two points (x_1, y_1) and (x_2, y_2) <u>without</u> <u>drawing</u> <u>any</u> <u>pictures</u>.

12. (a) Find the distance between $(3,-6)$ and $(8,6)$ by drawing a picture and measuring the distance as accurately as you can.

(b) Solve (a) without drawing a picture.

13. Without measuring, find the distance between the two points:

    (a)   (2,3) and (6,5)   (b)   (-4,6) and (1,-3)

    (c)   (0,9) and (-7,-8)   (d)   (5,-4) and (-7,6)

    (e)   (-3,-2) and (4,-1)   (f)   (2,3) and (-1,0)

(a) $\sqrt{20} = 2\sqrt{5}$
(c) $\sqrt{338} = 13\sqrt{2}$
(e) $5\sqrt{2}$

THE CIRCLE

The next few exercises will be useful in the study of the circle and in solving the opening Problem A.

Recall the definition of a circle from geometry:

> A <u>circle</u> is a curve that consists of all points P in a plane such that the distance from P to a fixed point is a constant. The fixed point is called the <u>center</u> and the constant distance is called the <u>radius</u>.

14. Find the distance between

    (a)   (0,0) and (x,y)   (b)   (2,3) and (x,y)   (c)   (h,k) and (x,y)

15. Suppose the mine is at (0,0) and Joe is at some point (x,y) that is 4 miles from the mine.

    (a)   Show why $\sqrt{(x-0)^2 + (y-0)^2} = 4$.

    (b)   Sketch all the points where Joe might be.

    (c)   Explain why the points in (b) are described by $x^2 + y^2 = 16$.

    (d)   Could Joe be at (3,2)? At (1,2)? Explain.

    (e)   Determine if the point (2.1,3.4) is inside, on, or outside the circle by:

        (i)   drawing a graph   (ii)   using the distance formula

Suggestion:
(b) You may wish to use a larger scale. Let 1 centimeter = 1 mile.

(e) (ii) Use your calculator!

16. Consider the circle with the center at the origin and radius 10. For each of the following points, determine whether the point lies inside, on, or outside the circle. Explain your answer.

    (a) (4,9)   (b) (8,-7)   (c) (-6,-8)   (d) $(\frac{23}{3},\frac{19}{3})$ (e) $(-\sqrt{74},\sqrt{26})$

(a) inside
(c) on circle
(e) on circle

17. Consider the circle with the center at the origin and radius 13. Find a value of y such that each point lies on the circle:

    (a) (5,y)   (b) (-12,y)   (c) (10,y)   (d) $(-4\sqrt{3},y)$

    (e)   Check also by drawing the circle with a compass.

(a) $y = \pm\, 12$
(C) $y = \pm\, \sqrt{69}$

18. Consider the circle with the center at the origin and radius 5.

(a) Describe how you could test whether or not a point (x,y) lies on the circle.

(b) If the point (x,y) lies on the circle, what is the value of $x^2 + y^2$?

(c) If the point (x,y) lies inside the circle, what can be said about the value of $x^2 + y^2$?

(d) If the point (x,y) lies outside the circle, what can be said about $x^2 + y^2$?

(e) Describe how the equation $x^2 + y^2 = 25$ can be used to test whether the point P = (x,y) lies on the circle with center at the origin and radius 5.

THE EQUATION OF A CIRCLE WITH THE CENTER AT THE ORIGIN

A <u>circle</u> with the center at the origin and radius r is described by the equation,

The equation $x^2 + y^2 = r^2$ is equivalent to the equation $\sqrt{(x-0)^2 + (y-0)^2} = r$, which asserts that the distance from the origin (center) to a point (x,y) on the circle is equal to r, the radius of the circle.

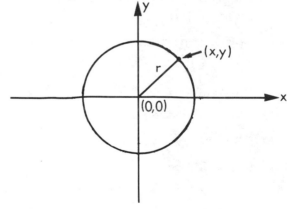

19. Write the equation of the circle in the (x,y)-coordinate plane if the center of the circle is at the origin and the radius is

(a) $x^2 + y^2 = 9$
(c) $x^2 + y^2 = 81$
(e) $x^2 + y^2 = 1$

(a) 3   (b) $\frac{4}{5}$ (c) 9   (d) r   (e) 1   (f) $\sqrt{2}$

20. Suppose the mine is at (2,3) and Joe is at some point (x,y) that is 4 units away.

Hint: 20(b) refer back to Exercise 14(b).

(a) Sketch all the points where Joe might be.

(b) Explain why the points (x,y) in (a) are described by the algebraic equation,

$$(x - 2)^2 + (y - 3)^2 = 16$$

21. Do Exercise 20 if Joe is 10 miles from the mine and the mine is at

(a) (-5,4)   (b) (6,-1)   (c) (-2,-3)

22. Explain why the equation $(x - 1)^2 + (y - 3)^2 = 49$ could be used to test whether a point (x,y) lies on the circle with the center at the point $(1,3)$ and radius 7.

* * *

The method used to solve the last few exercises also provides the following general formula for the equation of a circle with a <u>given</u> <u>center</u> and a <u>given</u> <u>radius</u>. The formula is just a restatement of the distance formula, which is really the Pythagorean Theorem.

EQUATION OF A CIRCLE IN THE PLANE

The <u>circle</u> whose center is at the point (h,k) and whose radius is r has an equation

$$(x - h)^2 + (y - k)^2 = r^2.$$

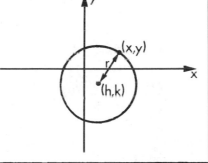

The above equation is a <u>test</u> to determine whether the point (x,y) lies on the circle or not. If x and y satisfy the equation, then the point (x,y) is on the circle; if x and y do <u>not</u> satisfy the equation, then the point (x,y) is <u>not</u> on the circle.

THERE ARE IMPORTANT IDEAS HERE!

This same idea will be used for all of the curves we study. In general, the equation of a curve is an algebraic <u>test</u> to determine whether or not a point (x,y) is or is not on the curve. The introduction of this general approach is due to Descartes, a leading French mathematician of the 17th century. In honor of this founder of analytic geometry, the coordinates in the plane are called "Cartesian coordinates".

The Latin form of Descartes (pronounced "Day-cart") is Cartesius, thus the term "Cartesian."

23. Write the equation of the circle with center C and radius r if

(a) C = (4,5), r = 3   (b) C = (-1,6), r = 9   (c) C = (-7,-2), r = 6

(a) $(x-4)^2 + (y-5)^2$ $= 9$

(c) $(x+7)^2 + (y+2)^2$ $= 36$

24. Without making a drawing, determine if the point $(5,7)$ lies inside, on, or outside the circle with the center at $(3,5)$ and radius 4.

25. Describe each of the following circles, giving the radius and the location of the center:

(a) $(x - 0)^2 + (y - 0)^2 = 3^2$ (b) $x^2 + y^2 = 49$

(c) $x^2 + y^2 - 1 = 0$ (d) $(x + 2)^2 + (y - 5)^2 = 10$

(e) $(x - 3)^2 + y^2 = 3$ (f) $x^2 + y^2 + 2y + 1 = 1$

Margin note:
(a) center (0,0)
*   radius 3*
(c) center (0,0)
*   radius 1*
(e) center (3,0)
*   radius $\sqrt{3}$*
Hint: (f)
$y^2 + 2y + 1 = (y+1)^2$

* * *

THE CIRCLE AND "COMPLETING THE SQUARE"

The quadratic formula was developed in Chapter 3 by "completing the square", that is, adding $(\frac{a}{2})^2$ to $x^2 + ax$ to obtain $(x + \frac{a}{2})^2$. Now "completing the square" will be used to analyze certain equations involving x and y. For instance, it will allow us to show that the equation

$$x^2 - 4x + y^2 + 6y = 5$$

describes a circle.

26. (a) Find the radius and the coordinates of the center of the circle described by the equation:

$$x^2 + (y - 1)^2 = 9.$$

(b) Show that

$$x^2 + y^2 - 2y + 1 = 9$$

describes a circle. What are its radius and its center?

(c) Show that

$$x^2 + y^2 - 2y - 8 = 0$$

describes a circle.

* * *

Margin note:
READ CAREFULLY!
This section shows
how to use
"completing the
square" to write
the equation of a
circle where the
center is not at
the origin.

Exercise 26 shows that

$$x^2 + y^2 - 2y - 8 = 0$$

is the equation of a circle by comparing it with the equation

$$x^2 + (y - 1)^2 = 9,$$

which is clearly the equation of a circle.

The method of "completing the square" will enable us to <u>start</u> with an equation, such as

$$x^2 + y^2 - 2y - 8 = 0,$$

and then rewrite it in a form from which the radius and center can easily
be deduced. Example 4 illustrates the technique.

Example 4. Find the center and radius of the circle:

$$x^2 - 4x + y^2 + 6y = 5$$

Solution. First, rewrite the equation to leave space for completing the
two squares:

$$(x^2 - 4x + \_\_) + (y^2 + 6y + \_\_) = 5.$$

Then complete the squares:

$$(x^2 - 4x + 4) + (y^2 + 6y + 9) = 5 + 4 + 9.$$

Thus,

$$(x - 2)^2 + (y + 3)^2 = 18.$$

It is easy to see that this last form is the equation of a circle
whose center is $(2,-3)$ and whose radius is $\sqrt{18}$, or $3\sqrt{2}$. This is not
obvious in the original equation.

The method shows that the graph of

$$x^2 + y^2 + Ax + By + C = 0$$

is a circle for any constants A, B, and C. (Or perhaps, the equation may
describe the empty set or a point. For instance, there are no points on
the graph of $x^2 + y^2 + 1 = 0$, while $x^2 + y^2 = 0$ describes a point, the
origin.)

NOTE: You may wish to refer to Project 25, where it is shown that the
graph is

a circle, if $A^2 + B^2 > 4C$

a point, if $A^2 + B^2 = 4C$

the empty set, if $A^2 + B^2 < 4C$.

27. Without plotting points, find the center and radius of the circles:

(a) $x^2 + 4x + y^2 = 5$

(b) $x^2 + y^2 + 6x = 9$

(c) $x^2 + 2x + y^2 + 6y = 0$

(d) $x^2 + 10x + y^2 + 8y = 3$

(e) $x^2 - 6x + y^2 - 10y = 15$

(f) $x^2 + 2x + y^2 + 4y = 1$

*(a) center (-2,0)
    radius 3*
*(c) center (-1,-3)
    radius $\sqrt{10}$*
*(e) center (3,5)
    radius 7*

* * *

THE PARABOLA

Recall Problem B: Joe says, "I'm just as far from the (straight) highway as from the mine . . . " Where might he be?

The next few exercises explore this problem.

28.   In Problem B, suppose Joe is at some point J = (x,y), the mine is at the point M = (0,1), and the highway is the line y = -1. Sketch the figure.

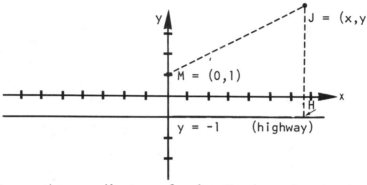

(a)   What are the coordinates of point H, the point on the highway nearest the point J? Label the coordinates in the figure.

(b)   Express the distance JH in terms of x and y.

(c)   Express the distance JM in terms of x and y.

(d)   Explain each of the following steps:

$$\text{(i)} \quad \sqrt{(x - 0)^2 + (y - 1)^2} = \sqrt{(x - x)^2 + (y - (-1))^2}$$

$$\text{(ii)} \quad \sqrt{x^2 + (y - 1)^2} = \sqrt{(y + 1)^2}$$

$$\text{(iii)} \quad x^2 + (y - 1)^2 = (y + 1)^2$$

$$\text{(iv)} \quad y = \frac{1}{4} x^2$$

Look back over the steps to see exactly how the radicals disappeared and all that was left is the simple equation
$y = \frac{1}{4} x^2.$

(e)   Using the last equation in (d), find some points and sketch a smooth curve through those points showing all possible locations where Joe may be.

29.   (This exercise continues Exercise 28.) At which of the following points may Joe be?

(a) no
(c) yes
(e) no

(a)   (2,3)   (b)   (1,1)   (c)   (2,1)

(d)   (0,0)   (e)   (-8,8)   (f)   (-4,4)

30. On the <u>same</u> <u>coordinate</u> <u>axes</u>, plot several points (x,y) satisfying each of the following equations and then sketch a smooth curve through those points. Label each curve with its equation.

   (a) $y = x^2$ (b) $y = -x^2$ (c) $y = 3x^2$

   (d) $y = -3x^2$ (e) $y = \frac{1}{3}x^2$ (f) $y = -\frac{1}{3}x^2$

* * *

The equation,

   $y = ax^2$ (<u>a</u> is a constant, or fixed, number other than 0),

The simplest form of an equation of a parabola.

describes a curve, called a <u>parabola</u>. (More will be said about parabolas after Exercise 33.)

Observe that the parabola $y = ax^2$ takes one of two positions:

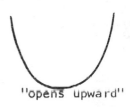   

"opens upward"   "opens downward"

31. Using Exercise 30, explain the influence of a on the parabola $y = ax^2$ where

   (a) a is positive or a is negative

   (b) $|a| < 1$ or $|a| > 1$ (compare $y = ax^2$ to $y = x^2$ in these cases).

* * *

If x and y are interchanged, the equation $y = ax^2$ becomes $x = ay^2$. As Exercise 32 illustrates, this describes a parabola "on its side", where it "opens right" or "opens left".

32. Graph all the following equations with reference to the same coordinate axes. Label each graph with its equation.

   (a) $x = y^2$ (b) $x = -y^2$ (c) $x = 3y^2$

   (d) $x = -3y^2$ (e) $x = \frac{1}{3}y^2$ (f) $x = -\frac{1}{3}y^2$

Note: Where $x \geq 0$, The graph $y = \sqrt{x}$ describes part of the parabola $y^2 = x$ where $y \geq 0$.

33. (a) Compare the graphs of $y = x^2$ and $x = y^2$.

(continued)

(b) How does the graph of $x = ay^2$ compare with the graph of $x = y^2$, if

(i) a is positive?   (ii) a is negative?

* * *

The following definition provides a precise geometric description of a parabola. Compare it with Problem B at the beginning of the chapter.

A point P on the parabola is the same distance from a line (directrix) as it is from a point (focus).

DEFINITION. A PARABOLA is a curve that consists of all points that are the same distance from a given line and a given point not on that line. The point is called the focus and the line is called the directrix. The point on the curve nearest the directrix is called the vertex.

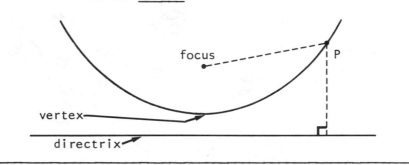

Example 5. Find the equation of the parabola whose focus is (0,2) and whose directrix is the line $y = -2$.

Solution. (Refer to the accompanying diagram:)

$$PQ = PF$$

$$\sqrt{(x-x)^2 + (y-(-2))^2} = \sqrt{(x-0)^2 + (y-2)^2}$$

$$(x-x)^2 + (y-(-2))^2 = (x-0)^2 + (y-2)^2$$

$$(y+2)^2 = x^2 + (y-2)^2$$

$$y^2 + 4y + 4 = x^2 + y^2 - 4y + 4$$

$$8y = x^2$$

$$y = \frac{1}{8}x^2$$

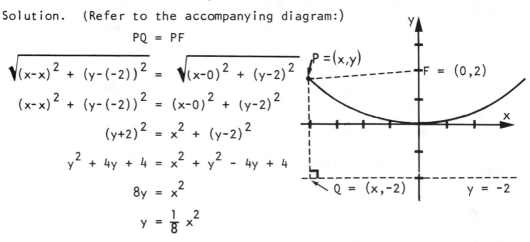

Generally, it can be shown that if the focus is at (0,c) and the directrix is the line $y = -c$, then the parabola opens upward (or downward) and has an equation

$$4cy = x^2$$
$$\text{or} \quad y = \frac{1}{4c} x^2.$$

Similarly, if the <u>focus</u> is at $(c,0)$ and the <u>directrix</u> is the line $x = -c$, then the parabola opens to the <u>left</u> (or the <u>right</u>) and has an equation,

$$4cx = y^2$$
$$x = \frac{1}{4c} y^2.$$

34. Find the simplified form of the equation of each of the following parabolas (show all work):

   (a) The focus is $(0, \frac{1}{4})$ and the directrix is the line $y = -\frac{1}{4}$.

   (a) $y = x^2$

   (b) The focus is $(-3,0)$ and the directrix is the line $x = 3$.

   (c) Draw the focus and directrix of the parabola $y^2 = 20x$.

35. This exercise compares the graphs of $y = x^2$ and $y = x^2 + 5$.

   (a) Fill in this table:

x	-2	-1	0	1	2
x^2					
$x^2 + 5$					

   (b) Use the table in (a) to graph $y = x^2$ and $y = x^2 + 5$ on the same axes.

   (c) Show that if the point (a,b) is on the graph of $y = x^2$, then the point $(a, b + 5)$ is on the graph of $y = x^2 + 5$.

   (d) Use (c) to explain how the graph of $y = x^2 + 5$ compares to the graph of $y = x^2$.

   (e) Why does $y = x^2 + 5$ describe a parabola?

   (f) Where is the focus? The directrix?

36. (a) How does the graph of $y = x^2 - 3$ compare with the graph of $y = x^2$?

   (b) Where is the focus? The directrix?

   (a) lower the graph $y = x^2$ three units.

* * *

The next example concerns a parabola whose focus is not on the y-axis.

Example 6.   Find the equation of the parabola whose focus is (1,2) and whose directrix is the line y = -2.

Solution.   Refer to the diagram:

$$PF = PQ$$

$$\sqrt{(x-1)^2 + (y-2)^2} = \sqrt{(x-x)^2 + (y-(-2))^2}$$

$$(x-1)^2 + (y-2)^2 = (x-x)^2 + (y-(-2))^2$$

$$(x-1)^2 + (y-2)^2 = (y+2)^2$$

$$x^2 - 2x + 1 + y^2 - 4y + 4 = y^2 + 4y + 4$$

$$x^2 - 2x + 1 = 8y$$

$$\frac{1}{8}(x^2 - 2x + 1) = y$$

$$\text{or} \qquad y = \frac{1}{8}x^2 - \frac{1}{4}x + \frac{1}{8}$$

37.   Draw a rough sketch of the data, then write the simplified form of the equation of the parabola, solving for y in terms of x:

(a) $y = -\frac{1}{2}(x^2 - 4x + 5)$
$= -\frac{1}{2}x^2 + 2x - \frac{5}{2}$

    (a)   the focus is (2,-1) and the directrix is the x-axis.

    (b)   the focus is (-1,3) and the directrix is y = 1.

*   *   *

As Example 6 and Exercise 37 show, an equation of a parabola may take the form, $y = ax^2 + bx + c$. We make the following observation.

An equation of a parabola in which the directrix is parallel to one of the two coordinate axes may take the form:

    (1) $y = ax^2 + bx + c$ (directrix is parallel to x-axis, parabola opens upward or downward)

or (2) $x = ay^2 + by + c$ (directrix is parallel to y-axis, parabola opens right or left)

In both cases, $a \neq 0$.

---

38.   Which of the following is an equation of a parabola?   Explain.

(a) yes
(c) yes

    (a) $y = -x^2 + 2x + 1$ (b) $y = x^2 - 4$

(Continued next page)

(c) $x = y^2 - 2x + 4$

(d) $y = 2x + 3$

(e) $y^2 = -x^2 + 9$

(f) $y^2 = -x^2 - 2x + 5$

THE LINE

In opening Problem C Joe says, "I'm just as far from the mine as from Dead Man's Oasis . . . "

39.   Suppose in Problem C that the mine is at M = (-2,0), Dead Man's Oasis is at D = (2,2), and Joe is at J = (x,y).

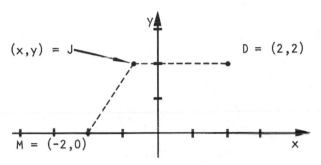

(a)   Express the distance JM in terms of x and y.

(b)   Express the distance JD in terms of x and y.

(c)   Explain why

    (i) $\sqrt{(x + 2)^2 + y^2} = \sqrt{(x - 2)^2 + (y - 2)^2}$

    (ii) $(x + 2)^2 + y^2 = (x - 2)^2 + (y - 2)^2$

    (iii) $2x + y - 1 = 0$

        or $y = -2x + 1$

(d)   Use the last equation in (c) to draw the graph showing possible locations for Joe.

$*$ $*$ $*$

As Exercise 39 shows, an equation of the form ax + by + c = 0 describes a line. This will be discussed in more detail in the next chapter.

EQUATION OF A LINE

An equation of the form,

$$ax + by + c = 0$$

describes a line. It is assumed that a, b, and c are constants and not both a and b are zero.

Sometimes it is more convenient to write an equation such as

$$3x + 4y - 7 = 0$$

This form of an equation of a line will be discussed in detail in the next chapter.

in the form

$$y = -\frac{3}{4}x + \frac{7}{4}.$$

This expresses y in terms of x and makes substitution easier when finding points on the graph.

40. Draw the graphs of these equations:

Recall from geometry that two points determine a line. However, always plot a third point as a check.

(a) $2x - 3y + 6 = 0$ (b) $y = 0$

(c) $y = x^2 + 3$ (d) $y = -2x + 5$

(e) $x = -3y + 4$ (f) $y - 2x = 4$

(g) $x = 0$ (h) $x = -y^2$

Which of the above are lines?

* * *

THE ELLIPSE

In Problem D, Joe is saying, "The <u>sum</u> of the distances from here to the mine and from here to the oasis is 10 miles." If the oasis and the mine are 8 miles apart, where might Joe be?

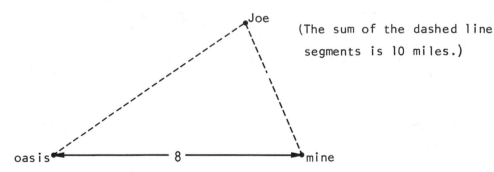

(The sum of the dashed line segments is 10 miles.)

41. "Dynamite Dan", Nevada Joe's partner back at the mine, decided to make a scale drawing of the situation before going to look for Joe. (Work together with other students and make a scale drawing similar to that of Dynamite Dan's. Let 1 inch = 1 mile.)

(a) Place a sheet of graph paper over thick cardboard (or a board) and place two thumb tacks 8 inches apart.

(b) Tie the ends of a piece of string to the tacks so that the portion of the string between the tacks is 10 inches long.

(c)   Stretch out the string with the point of your pencil and trace the figure you get when the string is taut.

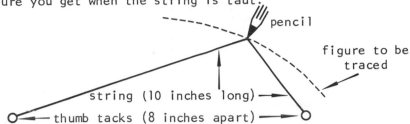

(d)   Describe the shape of the curve you drew in (c).

42.   (a)   Decrease the distance between the tacks to 6 inches. Compare the resulting figure with that in Exercise 41.

(b)   What happens to the shape of the figure as the tacks get closer and closer together? (Is the circle a special case of the ellipse? Explain.)

(c)   What happens to the shape of the figure as the tacks get farther and farther apart (but always less than 10 inches apart)?

43.   What happens to the figure in Exercise 41 if the two tacks coincide?

Compare your curve to those sketched by other students.

DEFINITION. An <u>ELLIPSE</u> is a curve that consists of all points P such that the <u>sum</u> of the distances from P to two fixed points is constant. Each of the two fixed points is call a <u>focus</u>.

PA + PB = constant

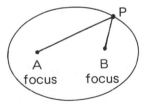

---

*"Focus" is pronounced "FO-kus".
The plural of focus is "foci"
pronounced "FO-sigh". An ellipse has two foci (except when the two foci coincide, then it is a circle).*

The next example obtains the equation of a specific ellipse.

Example 7. In Problem D, let the mine be the point M = (4,0) and the oasis be at the point D = (-4,0), and let these points be the foci of the ellipse. The <u>sum</u> of the distances from Joe's position, J = (x,y), to the mine and oasis is 10 miles. Find the equation of the ellipse that describes where Joe might be.

Solution.   Joe is at some point J = (x,y).

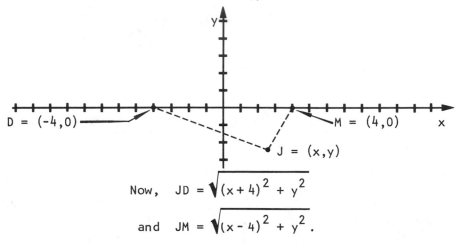

Now, $JD = \sqrt{(x+4)^2 + y^2}$

and $JM = \sqrt{(x-4)^2 + y^2}$.

It is assumed that   JD + JM = 10

or $\sqrt{(x+4)^2 + y^2} + \sqrt{(x-4)^2 + y^2} = 10$.

Write the expression so that only one radical is on a side, in order to have a less complicated expression when each side is squared.

$$\sqrt{(x+4)^2 + y^2} = 10 - \sqrt{(x-4)^2 + y^2}$$

$$(x+4)^2 + y^2 = 100 - 20\sqrt{(x-4)^2 + y^2} + (x-4)^2 + y^2$$

$$x^2 + 8x + 16 + y^2 = 100 - 20\sqrt{(x-4)^2 + y^2} + x^2 - 8x + 16 + y^2$$

$$16x - 100 = -20\sqrt{(x-4)^2 + y^2}$$

$$4x - 25 = -5\sqrt{(x-4)^2 + y^2}$$

Square again to remove the radical:

$$16x^2 - 200x + 625 = 25[(x-4)^2 + y^2]$$

$$16x^2 - 200x + 625 = 25[x^2 - 8x + 16 + y^2]$$

$$16x^2 - 200x + 625 = 25x^2 - 200x + 400 + 25y^2$$

$$225 = 9x^2 + 25y^2$$

or

$$1 = \frac{x^2}{25} + \frac{y^2}{9}$$

Stop here!
Look over the
algebraic steps
that got rid of
all the radicals.
A good exercise
would be that of
working Example 7
without looking at
the text.

This last form of the equation of the ellipse is most convenient, because it shows at a glance where the curve crosses the axes. For example, when y = 0, then x = ± 5; but when x = 0, then y = ± 3. Thus, four points are easy to find:

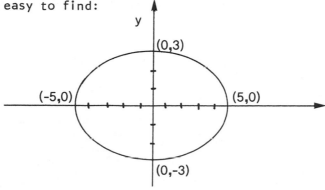

Draw a smooth curve through these points (The parts of the ellipse in each quadrant are congruent).

In Projects 2-5, the following general assertion is justified:

THE ELLIPSE

The graph of the equation,

$$\frac{x^2}{a^2} + \frac{y^2}{b^2} = 1,$$

is an ellipse that crosses the x-axis at (a,0) and (-a,0) and the y-axis at (0,b) and (0,-b).

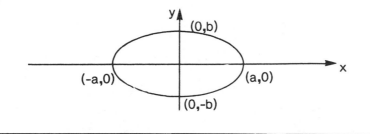

If $a > b$ it is shown in Projects 2-5 that the "length of the string" is 2a and the foci are at

$$\left(\pm\sqrt{a^2 - b^2},\ 0\right)$$

If the foci are on the y axis, $b > a$. The ellipse is then taller than it is wide.

Comment. For a rough sketch of the ellipse,

$$\frac{x^2}{a^2} + \frac{y^2}{b^2} = 1,$$

plot the points (a,0), (-a,0), (0,b), and (0,-b), then draw a smooth egg-shaped curve through these points, as in the above figure.

44. Draw a rough sketch of each ellipse:

(a) $\dfrac{x^2}{36} + \dfrac{y^2}{9} = 1$

(b) $\dfrac{x^2}{4} + \dfrac{y^2}{25} = 1$

(continued)

Hint: (d) Rewrite in the form
$$\frac{x^2}{a^2} + \frac{y^2}{b^2} = 1$$

(c) $\dfrac{x^2}{49} + \dfrac{y^2}{100} = 1$ (d) $9x^2 + 4y^2 = 36$

45.   (a)   Find an equation of the ellipse with foci at (3,0) and (-3,0) if the <u>sum</u> of the distances from a point on the ellipse to the two foci is 10.

      (b)   Simplify the equation in (a) to the form $\dfrac{x^2}{a^2} + \dfrac{y^2}{b^2} = 1$.

      (c)   Draw a rough sketch of the ellipse in the coordinate plane.

46.   Identify each of the following equations as to whether it is a circle, a parabola, a line, or an ellipse. Do <u>not</u> draw the figure.

Do not rush through Exercise 46. Be able to give reasons for each answer.

(a) parabola
(c) line
(e) circle
(g) ellipse

(a) $y = 3x^2 + 4x - 9$ (b) $(x - 3)^2 + y^2 = 16$

(c) $7x - 3y = 12$ (d) $\dfrac{x^2}{81} + \dfrac{y^2}{64} = 1$

(e) $x^2 + y^2 - 4y = 0$ (f) $x = y^2 + y - 1$

(g) $4x^2 + 25y^2 = 100$ (h) $y = \dfrac{3}{8}x - 21$

THE CONIC SECTIONS

The ellipse and parabola were defined in terms of distances to points or to a line. They have algebraic descriptions also;

$$\frac{x^2}{a^2} + \frac{y^2}{b^2} = 1 \qquad \text{(ellipse)}$$

$$y = ax^2 + bx + c \qquad \text{(parabola)}.$$

The ancient Greek mathematicians defined an ellipse and parabola in terms of the way a plane cuts a double cone. In these diagrams a plane cuts just one of the two cones to yield an ellipse or a parabola.

ellipse

parabola

Another general type of conic section is obtained by a plane that cuts both cones. This cross-section is called a "hyperbola". A hyperbola consists of two separate pieces.

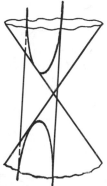

hyperbola

Rather than use this definition, we will approach the hyperbola through another Nevada Joe calamity.

THE HYPERBOLA

Problem E.   Nevada Joe says over his radio, "I'm exhausted. Help me! From where I am the __difference__ of the distances to the mine and Dead Man's Oasis is 6 miles . . . " If the mine and the oasis are 10 miles apart, where might Joe be?

A compass will be useful in working Problem E. Compare your diagram with that of other students.

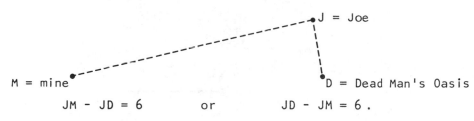

$$JM - JD = 6 \qquad \text{or} \qquad JD - JM = 6.$$

(There are two equations, since it is not known whether Joe is nearer the mine or nearer the oasis. They could be replaced by the single equation,

$$|JM - JD| = 6.)$$

* * *

47.   If the mine is at M = (-5,0) and Dead Man's Oasis is at D = (5,0), at which of the following points might Joe be?

(a) (3,0)   (b) (3,4)   (c) (4,3)

(d) (-3,0)   (e) (0,4)   (f) (-1,-4)

Hint: Use the distance formula

(a) yes
(c) no
(e) no

* * *

The next example obtains a simple equation that describes all points where Joe (of Problem E) might be.

Example 8.   Suppose in Problem E that the mine is at $M = (-5,0)$ and the oasis is at $D = (5,0)$ and that Joe is at some point $J = (x,y)$:

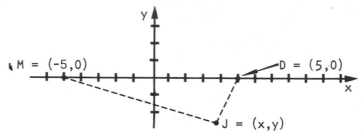

Find a simple equation (without radicals) that describes where Joe might be.

Solution.   Consider the case where Joe is nearer the oasis than the mine, that is,

$$JM - JD = 6.$$

(The other case will lead to the same final equation.)   The following steps translate this geometric statement into a simple algebraic equation.   Study them carefully.

$$JM - JD = 6$$

$$\sqrt{(x+5)^2 + y^2} - \sqrt{(x-5)^2 + y^2} = 6$$

Prepare to square $\sqrt{(x+5)^2 + y^2} = 6 + \sqrt{(x-5)^2 + y^2}$

Square both sides $(x+5)^2 + y^2 = 36 + 12\sqrt{(x-5)^2 + y^2} + (x-5)^2 + y^2$

Expand $x^2 + 10x + 25 + y^2 = 36 + 12\sqrt{(x-5)^2 + y^2} + x^2 - 10x + 25 + y^2$

Collect terms $20x - 36 = 12\sqrt{(x-5)^2 + y^2}$

Divide by 4 $5x - 9 = 3\sqrt{(x-5)^2 + y^2}$

Square $25x^2 - 90x + 81 = 9[(x-5)^2 + y^2]$

Expand $25x^2 - 90x + 81 = 9(x^2 - 10x + 25 + y^2)$

Distributive $25x^2 - 90x + 81 = 9x^2 - 90x + 225 + 9y^2$

Collect terms $16x^2 - 9y^2 = 144$

Divide by 144 $\dfrac{x^2}{9} - \dfrac{y^2}{16} = 1$

The graph in Problem E is an example of a hyperbola, which will now be defined geometrically.

DEFINITION. A <u>HYPERBOLA</u> is a curve (in two separate pieces) that consists of all points P such that the absolute value of the <u>difference</u> of the distances from P to two fixed points is constant. Each of the two fixed points is called a focus.

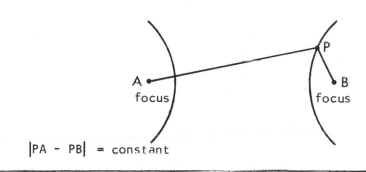

|PA - PB| = constant

In Example 8 the foci are at (5,0) and (-5,0). The "constant difference" is 6. The set of points where Joe might be is a hyperbola, and its equation, in simplest form, was found to be

$$\frac{x^2}{9} - \frac{y^2}{16} = 1.$$

Note: $\frac{x^2}{9} + \frac{y^2}{16} = 1$ is an equation of an ellipse ("sum of the distances").
$\frac{x^2}{9} - \frac{y^2}{16} = 1$ is an equation of an hyperbola ("difference of the distances").

SKETCHING A HYPERBOLA

The hyperbola

$$\frac{x^2}{9} - \frac{y^2}{16} = 1$$

meets the x-axis at the points (3,0) and (-3,0). It does not meet the y-axis at all. For, when x = 0, the equation becomes

$$\frac{0^2}{9} - \frac{y^2}{16} = 1$$

or $$-\frac{y^2}{16} = 1$$

or $$y^2 = -16.$$

For (x,y) to be a point on the graph, both x and y must be real numbers.

But a real number squared is never negative.

Since this hyperbola meets the axes in only two points, it is not as easy to graph as an ellipse. Fortunately, the hyperbola has a certain property that is quite helpful in graphing it. The following discussion introduces this property.

First rewrite the equation in such a way that y is expressed in terms of x:

$$\frac{x^2}{9} - \frac{y^2}{16} = 1$$

$$\frac{y^2}{16} = \frac{x^2}{9} - 1$$

$$\frac{y^2}{16} = \frac{1}{9}(x^2 - 9)$$

$$\frac{y}{4} = \pm \frac{1}{3}\sqrt{x^2 - 9}$$

$$y = \pm \frac{4}{3}\sqrt{x^2 - 9} .$$

Now consider the behavior of $\sqrt{x^2 - 9}$ when x is large.

x	$\sqrt{x^2 - 9}$	
3	0	
5	4	
6	5.196	(approximately)
10	9.539	
100	99.95	
1000	999.99549	

Clearly, for a large positive x, the $\sqrt{x^2 - 9}$ is very close to x. Thus,

$$\pm \frac{4}{3}\sqrt{x^2 - 9} \text{ is close to } \pm \frac{4}{3}x \quad \text{(for a large x).}$$

This means that for a large value of $|x|$, the points (x,y) on the hyperbola

$$\frac{x^2}{9} - \frac{y^2}{16} = 1$$

are very near the lines

$$y = \frac{4}{3}x \quad \text{or} \quad y = -\frac{4}{3}x .$$

This observation provides a short cut to graphing the hyperbola of Example 8. These are the steps:

(a) Draw the lines $y = \frac{4}{3} x$ and $y = -\frac{4}{3} x$ (shown as dotted lines in the figure below).

(b) Plot the points $(3,0)$ and $(-3,0)$, through which the hyperbola passes.

(c) Sketch the hyperbola with the aid of (a) and (b).

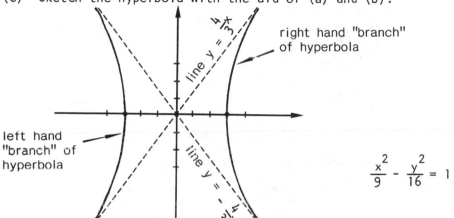

right hand "branch" of hyperbola

left hand "branch" of hyperbola

$$\frac{x^2}{9} - \frac{y^2}{16} = 1$$

ALTERNATE METHOD: A "trick" or even shorter method to graph $\frac{x^2}{9} - \frac{y^2}{16} = 1$ is to graph the four points: $(3,4)$, $(3,-4)$, $(-3,4)$, and $(-3,-4)$. They describe the vertices of a rectangle. The diagonals extended form the asymptotes of the hyperbola, $\frac{x^2}{9} - \frac{y^2}{16} = 1$. Let $y = 0$ and you have the vertices of the branches. With just the asymptotes and the vertices a good sketch can be drawn.

The two lines that the hyperbola approaches arbitrarily closely are called its <u>asymptotes</u>. <u>They are reference lines</u> (<u>dotted lines</u>), <u>not a part of the hyperbola</u>.

SHORT CUT FOR FINDING THE ASYMPTOTES

(a) Start with the equation

$$\frac{x^2}{9} - \frac{y^2}{16} = 1 .$$

(b) Replace "1" on the right side by "0",

$$\frac{x^2}{9} - \frac{y^2}{16} = 0 .$$

(c) Factor the left side of the resulting equation:

$$\left(\frac{x}{3} - \frac{y}{4}\right) \left(\frac{x}{3} + \frac{y}{4}\right) = 0 .$$

(d) The asymptotes are given by the equations, using the zero product rule:

$$\frac{x}{3} - \frac{y}{4} = 0 \qquad \text{and} \qquad \frac{x}{3} + \frac{y}{4} = 0,$$

that is $y = \frac{4}{3} x$ and $y = -\frac{4}{3} x$.

The following description of a hyperbola is justified in Projects 14-20.

THE HYPERBOLA

The graph of the equation

$$\frac{x^2}{a^2} - \frac{y^2}{b^2} = 1$$

is a hyperbola in standard position. It crosses the x-axis at $(a,0)$ and $(-a,0)$. The asymptotes are found by factoring the left side of the equation,

$$\frac{x^2}{a^2} - \frac{y^2}{b^2} = 0,$$

or

$$\left(\frac{x}{a} - \frac{y}{b}\right)\left(\frac{x}{a} + \frac{y}{b}\right) = 0,$$

which gives the lines

$$\frac{x}{a} - \frac{y}{b} = 0 \quad \text{and} \quad \frac{x}{a} + \frac{y}{b} = 0 .$$

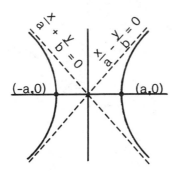

$$\frac{x^2}{a^2} - \frac{y^2}{b^2} = 1$$

The foci of the hyperbola are on the x-axis, symmetrically placed with respect to the origin.

Using the "trick" (shorter) method you plot (a,b), $(a, -b)$, $(-a,b)$ and $(-a,-b)$ to find the asymptotes. Then let $y = 0$ to find the inner-most points of the branches (vertices).

Show the asymptotes as dashed lines, which means they are not a part of the graph, but merely reference lines.

48. Sketch each of these hyperbolas. Show its asymptotes as dotted lines and label the points where it crosses the x-axis.

(a) $\frac{x^2}{16} - \frac{y^2}{9} = 1$ (b) $\frac{x^2}{25} - \frac{y^2}{4} = 1$ (c) $x^2 - \frac{y^2}{4} = 1$

(d) $\frac{x^2}{16} - \frac{y^2}{25} = 1$ (e) $x^2 - y^2 = 1$ (f) $4x^2 - 25y^2 = 100$

49. Sketch the graph of each of the following hyperbolas; show the asymptotes and where the branches cross the y-axis.

(a) $\dfrac{y^2}{16} - \dfrac{x^2}{9} = 1$ (b) $\dfrac{y^2}{36} - x^2= 1$

The branches open upward and downward, they do not cross the x-axis.

INTERSECTIONS OF CONICS

50. Draw a freehand sketch showing the maximum number of points a line may intersect

    (a) a line other than itself   (b) a parabola

    (c) a circle   (d) an ellipse

    (e) a hyperbola

51. Draw a freehand sketch showing the maximum number of points a circle may intersect

    (a) a circle other than itself   (b) a hyperbola

    (c) a parabola   (d) an ellipse

52. Draw a freehand sketch showing the maximum number of points a hyperbola may intersect

    (a) a hyperbola other than itself   (b) a parabola

    (c) an ellipse

53. Draw a freehand sketch showing the maximum number of points a parabola may intersect

    (a) a parabola other than itself   (b) an ellipse

54. Draw a freehand sketch showing the maximum number of points an ellipse may intersect an ellipse other than itself.

* * *

SOLVING QUADRATIC EQUATIONS (OPTIONAL)

The following examples show how to estimate the coordinates of the intersection points of two conics by sketching the graphs. Those points may be computed precisely by solving the two equations simultaneously.

Example 9.   (a) Draw a rough sketch of the graphs of the following equations and estimate the intersection points.

$$\begin{cases} x^2 + y^2 = 1 \\ 2x + y = 1 \end{cases}$$

(b) Compute the intersection points algebraically.

(continued)

Solution.   (a)   The sketch:

$$\begin{cases} x^2 + y^2 = 1 \\ 2x + y = 1 \end{cases}$$

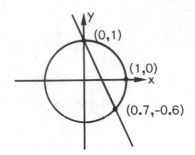

(0,1)

(1,0)
x

(0.7,-0.6)

Estimate: Intersections seem to be near (0,1) and (0.7,-0.6).

(b)   Computing algebraically:

$$\begin{cases} x^2 + y^2 = 1 \\ 2x + y = 1 \end{cases}$$

$$\begin{cases} x^2 + y^2 = 1 \\ y = -2x + 1 \end{cases}$$

(Substituting the y-value of the lower equation into the upper:)

$$x^2 + (-2x + 1)^2 = 1$$

$$x^2 + 4x^2 - 4x + 1 = 1$$

$$5x^2 - 4x = 0$$

$$x(5x - 4) = 0$$

$$x = 0 \quad \text{or} \quad 5x - 4 = 0$$

$$x = 0 \quad \text{or} \quad x = \frac{4}{5}$$

Substituting these values for x in the original equation of the line, the intersection points are

$$(0,1) \quad \text{and} \quad (\frac{4}{5}, -\frac{3}{5}).$$

Example 10.   (a)   Draw a rough sketch of the graph of the following equations and estimate the coordinates of the intersection points.

$$\begin{cases} x^2 + y^2 = 36 \\ \frac{x^2}{16} - \frac{y^2}{9} = 1 \end{cases}$$

(b)   Compute the points algebraically.

Solution.   (a)   The sketch:

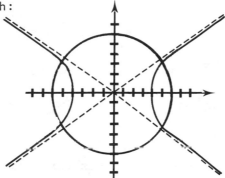

Estimate: Intersections seem to be near $(5,3)$, $(5,-3)$, $(-5,3)$, and $(-5,-3)$.

(b)   Solving algebraically:

$$\begin{cases} x^2 + y^2 = 36 \\ \dfrac{x^2}{16} - \dfrac{y^2}{9} = 1 \end{cases}$$

$$\begin{cases} x^2 + y^2 = 36 \\ 9x^2 - 16y^2 = 144 \end{cases}$$

$$\begin{cases} 9x^2 + 9y^2 = 324 \\ 9x^2 - 16y^2 = 144 \end{cases}$$

$$25y^2 = 180$$

$$y^2 = \frac{180}{25}$$

$$y = \pm \frac{6}{5}\sqrt{5}$$

Substituting $y = \frac{6}{5}\sqrt{5}$ into $x^2 + y^2 = 36$,

$$x^2 + \left(\frac{180}{25}\right) = 36$$

$$x^2 = 36 - \frac{180}{25}$$

$$x^2 = \frac{720}{25}$$

$$x = \pm \frac{12}{5}\sqrt{5} \, ;$$

(continued)

This gives the points,

$$(\frac{12}{5}\sqrt{5}, \frac{6}{5}\sqrt{5}) \text{ and } (\frac{12}{5}\sqrt{5}, -\frac{6}{5}\sqrt{5}).$$

Similarly, $y = -\frac{6}{5}\sqrt{5}$ gives the other two points,

$$(\frac{12}{5}\sqrt{5}, -\frac{6}{5}\sqrt{5}) \text{ and } (-\frac{12}{5}\sqrt{5}, -\frac{6}{5}\sqrt{5}),$$

which the rough sketch approximates.

55. Sketch each pair of equations, estimate the points of intersection, then compute the intersection points precisely using algebra.

(a) parabola and line at (0,1) and (-1,0)
(c) circle and ellipse at (4,0) and (-4,0)
(d) circle and line at
$(\frac{3 + \sqrt{41}}{2}, \frac{3 - \sqrt{41}}{2})$
and
$(\frac{3 - \sqrt{41}}{2}, \frac{3 + \sqrt{41}}{2})$

(a) $\begin{cases} y = x^2 + 2x + 1 \\ y = x + 1 \end{cases}$

(b) $\begin{cases} x^2 - y^2 = 16 \\ x - 2y = -1 \end{cases}$

(c) $\begin{cases} x^2 + y^2 = 16 \\ 9x^2 + 16y^2 = 144 \end{cases}$

(d) $\begin{cases} x^2 + y^2 = 25 \\ x + y = 3 \end{cases}$

* * * SUMMARY * * *

The plight of a lost prospector introduced analytic geometry, which uses algebra to deal with geometric problems. The following table summarizes much of the chapter:

(continued)

GEOMETRIC CONCEPT	ALGEBRAIC TRANSLATION
Distance between points on the x-axis (or on the y-axis)	$\|a - b\|$
Distance between the points (x_1, y_1) and (x_2, y_2) in the plane	$d = \sqrt{(x_1 - x_2)^2 + (y_1 - y_2)^2}$
Circle with center at (h, k) and radius r	$\begin{cases} (x - h)^2 + (y - k)^2 = r^2 \\ \qquad \text{or} \\ x^2 + y^2 + Ax + By + C = 0 \end{cases}$
Parabola (vertex may not be at the origin, but directrix is parallel to a coordinate axis)	$y = ax^2$ or $x = ay^2$ In more general terms: $y = ax^2 + bx + c$ or $x = ay^2 + by + c$
Line	$ax + by + c = 0$
Ellipse (symmetric with respect to both coordinate axes)	$\dfrac{x^2}{a^2} + \dfrac{y^2}{b^2} = 1$ ("standard position")
Hyperbola (symmetric with respect to both coordinate axes)	$\dfrac{x^2}{a^2} - \dfrac{y^2}{b^2} = 1$ ("standard position")

The distance formula was useful in finding the equation of a <u>circle</u> with radius r and the center at (h,k):

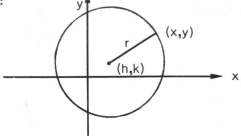

$$(x - h)^2 + (y - k)^2 = r^2$$

For the definitions of the other conic sections, see the text. Briefly:

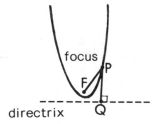

parabola

"distance to fixed point equals distance to fixed line"

$$PF = PQ$$

ellipse

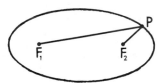

"constant sum of distances to two fixed points"

$$PF_1 + PF_2 = constant$$

hyperbola

"constant difference of distances to two fixed points"

$$|PF_1 - PF_2| = constant$$

In graphing the hyperbola, two lines, called "asymptotes", are of use.

* * * WORDS AND SYMBOLS * * *

absolute value $|x|$

Pythagorean Theorem

parabola

vertex of parabola

distance formula: focus (foci)
  (i) on a line, $|a - b|$ directrix
  (ii) in the plane, $\sqrt{(x_1 - x_2)^2 + (y_1 - y_2)^2}$ ellipse
circle   hyperbola
completing the square   asymptotes
line   conic section

* * *   CONCLUDING REMARKS   * * *

    The ancient Greeks studied the parabola, ellipse, and hyperbola
because they found them fascinating. They had no practical application
in mind. But, as happens so often with knowledge acquired only out of
curiosity, the conic sections have had many applications over the centuries.

    Consider, for instance, the parabola. It has an amazing reflecting
property. All rays of light coming in perpendicular to its directrix are
reflected off the parabola to the focus:

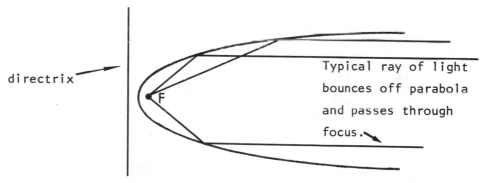

directrix

F

Typical ray of light
bounces off parabola
and passes through
focus.

There is evidence that some 2300 years ago Archimedes applied this property to burn the ships of an invading Roman fleet. Modern microwave relay
stations (there are about 100 at 30 mile intervals) transmit TV signals
across the nation; they have parabolic reflectors because of this property.
A flashlight or an automobile headlight reflector is parabolic for the
same reason. So is a receiver for signals from satellites.

    In the early part of the 17th century, Galileo discovered that the
path of a thrown ball is a parabola. Newton, later in the same century,
showed that the path of a comet that does not return periodically is
either a parabola or part of a hyperbola.

    A freely hanging rope or chain does <u>not</u> take a parabolic shape, as

Galileo thought. It forms a far more complicated curve, called a catenary, whose equation involves exponentials, such as

$$y = 2^x + 2^{-x} .$$

However, Huygens, in the middle of the 17th century, showed that a cable from which weights are suspended uniformly with respect to horizontal distance is indeed a parabola. For this reason, parabolas appear in the design of some bridges.

The ellipse (the path of a planet or a satellite) and hyperbola have also had many applications over the centuries. In general, it is impossible to predict which mathematical discoveries will have later theoretical or practical repercussions. After all, the future cannot be predicted. The future, instead, tantalizes. Is that not one of the charms of life, its unpredictability?

*   *   *   CONIC SECTIONS SELF QUIZ   *   *   *

1. Fill in the table:

Hint: See table in Summary section. Allow more space on your paper for this problem than is indicated here.

Conic	Geometric definition	General form of the equation
ellipse		
circle		
hyperbola		
parabola		

2. Without drawing the graph, name the type of conic section:

Exercises 2 and 3 are a good check of your understanding of this chapter.

(a) $\dfrac{x^2}{4} + \dfrac{y^2}{4} = 1$ (b) $4x^2 + y^2 = 1$

(c) $x^2 = 4y^2 + 4$ (d) $x + 4y = 1$

(e) $4x^2 + y^2 = 0$ (f) $x^2 + 4y = 1$

(g) $x^2 - 4y^2 = 0$ (h) $y^2 = x + 4$

(i) $\dfrac{x}{4} + \dfrac{y}{9} = 1$ (j) $y = 4x^2$

3. Match each member of the first column with a member of the second:

(1) $\dfrac{x^2}{9} + \dfrac{y^2}{16} = 1$ (a) ellipse, foci on y-axis

(2) $\dfrac{x^2}{25} - \dfrac{y^2}{4} = 1$ (b) parabola, directrix parallel to x-axis

(3) $y = x^2 + 2x + 1$ (c) circle, center at (-2,7)

(4) $\dfrac{x^2}{16} + \dfrac{y^2}{9} = 1$ (d) hyperbola, crossing x-axis

(5) $x^2 + y^2 = 15$ (e) parabola, directrix parallel to y-axis

(6) $9x^2 = 4y^2$ (f) the point (0,0)

(7) $y^2 = x + y + 1$ (g) ellipse, foci on x-axis

(8) $y = 2x - 3$ (h) hyperbola, crossing y-axis

(9) $(x + 2)^2 + (y - 7)^2 = 1$ (i) circle, center at origin

(10) $x^2 + y^2 = 0$ (j) two intersecting lines through the origin

(11) $\dfrac{y^2}{25} - \dfrac{x^2}{4} = 1$ (k) a line

4. Write a general form of an equation of a line.

ANSWERS TO SELF QUIZ

1. Ellipse: all points in a plane such that the sum of the distances to two fixed points is a constant; $\dfrac{x^2}{a^2} + \dfrac{y^2}{b^2} = 1$ in "standard position".

Circle: all points in a plane that are equidistant from a fixed point; $(x - h)^2 + (y - k)^2 = r^2$ [center at (h,k), radius r].

Hyperbola: all points in a plane such that the difference of the distances to two fixed points is a constant; $\dfrac{x^2}{a^2} - \dfrac{y^2}{b^2} = 1$ (or $\dfrac{y^2}{b^2} - \dfrac{x^2}{a^2} = 1$) in "standard position".

Parabola: all points equidistant from a line and a fixed point not on that line; $y = ax^2 + bx + c$ (or $x = ay^2 + by + c$).

2. (a) circle, (b) ellipse, (c) hyperbola, (d) line, (e) point, (0,0), (f) parabola, (g) two lines, (h) parabola, (i) line, (j) parabola.

3.   (1), (a); (2), (d); (3), (b); (4), (g); (5), (i); (6), (j); (7), (e);
     (8), (k); (9), (c); (10), (f); (11), (h).

4.   ax + by + c = 0.

The next chapter has more information concerning the equations of a line.

* * * CONIC SECTIONS REVIEW * * *

1.   Suppose Nevada Joe says, "I'm <u>twice</u> as far from the mine as I am from
     Dead Man's Oasis . . . "

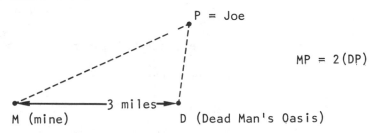

     MP = 2(DP)

     (a)   Assume that the mine and the oasis are 3 miles apart. Use a
           ruler and compass to find several points where Joe might be.
     (b)   Draw a smooth curve through the points in (a).
     (c)   Discuss the curve in (b) with other students.

2.   (Continuing Review Exercise 1.) Now introduce a coordinate system such
     that the mine is at (0,0) and the oasis is at (3,0). Recall that Joe
     is twice as far from the mine as from the oasis.

     (a)   Use the distance formula to find an equation that describes the
           points where Joe might be.
     (b)   Write the equation in (a) free of radicals, if you have not
           already done so.
     (c)   Simplify the equation in (b) to a form such that the class may be
           able to look at it and immediately recognize the graph.

3.   Suppose that there are tacks at fixed points A and B and that you have
     a <u>loop</u> of string whose length is greater than twice the distance be-
     tween A and B. Stretch the string taut with a pencil. What curve do
     you get when the pencil moves along the string?

loop of string   pencil

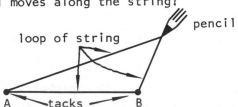

     A   tacks   B

4. What is the formula for the distance between the two points x_1 and x_2 on the x-axis?

5. What is the formula for the distance between the two points (x_1, y_1) and (x_2, y_2) in the coordinate plane?

6. (a) Is the point $(9, -8)$ inside, on, or outside the circle

   $$x^2 + y^2 = 144 ?$$

   (b) If it is not on the circle, how far is it from the circle? Express the distance as a three place decimal.

   Use your calculator.

7. What is the equation of the circle with the center at $(-5, 4)$ and a radius of 9?

8. Consider the following parabolas:

   (a) $y = -x^2$
   (b) $x - y^2$
   (c) $y = 3x^2$
   (d) $x = -5y^2$

   Without graphing, answer the following two questions for each of the above:

   (i) Is the directrix parallel to the x-axis or the y-axis?
   (ii) Does the parabola open upward, downward, to the right, or to the left?

9. Match each equation in the left column with the name of its graph in the right column:

   (a) $4x - 9y = 1$ (1) parabola (vertical directrix)
   (b) $4x^2 - 9y^2 = 0$ (2) point
   (c) $4x - 9y^2 = 1$ (3) circle
   (d) $4x^2 + 9y^2 = 0$ (4) no point (x,y) satisfies the equation
   (e) $4x^2 - 9y = 1$ (5) ellipse
   (f) $4x^2 + 4y^2 = 9$ (6) parabola (horizontal directrix)
   (g) $4x^2 - 9y^2 = 1$ (7) two intersecting lines
   (h) $4x^2 + 9y^2 = 1$ (8) hyperbola
   (i) $4x^2 + 9y^2 = -1$ (9) line

   Hint: (g) Write as
   $$\frac{x^2}{\frac{1}{4}} - \frac{y^2}{\frac{1}{9}} = 1$$

10. Sketch the graph of $\frac{x^2}{49} + \frac{y^2}{25} = 1$. What is this geometric figure?

11. (a) Sketch the graph of $\dfrac{x^2}{49} - \dfrac{y^2}{25} = 0$.

    (b) What is the geometric figure in (a)?

What is the relationship between Exercise 11 and 12?

12. (a) Sketch the graph of $\dfrac{x^2}{49} - \dfrac{y^2}{25} = 1$.

    (b) What is the geometric figure in (a)?

13. Graph the asymptotes of the hyperbolas:

    (a) $\dfrac{x^2}{1} - \dfrac{y^2}{4} = 1$ and   (b) $\dfrac{y^2}{4} - \dfrac{x^2}{1} = 1$.

14. Graph: $6x^2 + 6y^2 = 0$

15. Devise a specific equation of your own that has not appeared in this chapter whose graph is

    (a) a parabola   (b) a hyperbola

    (c) a point   (d) a circle

    (e) an ellipse   (f) a line

16. Explain why a circle may be considered a special case of the ellipse.

17. In a plane **without** coordinates, what is the geometric definition of each of the following?

    (a) a circle   (b) an ellipse

    (c) a hyperbola   (d) a parabola

18. Which of the following has no points in its graph?

    (a) $\dfrac{x^2}{9} + \dfrac{y^2}{4} = 1$ (b) $\dfrac{x^2}{9} + \dfrac{y^2}{4} = -1$

    (c) $\dfrac{x^2}{9} - \dfrac{y^2}{4} = 1$ (d) $\dfrac{x^2}{9} - \dfrac{y^2}{4} = -1$

19. (a) Sketch the graphs of both equations on the same axes and estimate the coordinates of the intersection points:

$$\begin{cases} x^2 - y^2 = 16 \\ x + y = 8 \end{cases}$$

    (b) Find the intersection points algebraically.

20. Do Review Exercise 19 for these equations:

$$\begin{cases} x^2 + y^2 = 25 \\ x^2 + 4y^2 = 36 \end{cases}$$

21. Find the equation of the parabola whose focus is (0,2) and whose directrix is the x-axis.

22. Each of the following is an equation of a circle. Without graphing it, give the location of the center and the radius.

    (a) $x^2 + y^2 = 3$

    (b) $x^2 + 2x + y^2 = 3$

    (c) $x^2 + 2x + y^2 + 4y = 3$

    (d) $x^2 + 6x + y^2 - 8y = 0$

ANSWERS TO CONIC SECTIONS REVIEW

1. Circle (find several points!)

2. (a) $\sqrt{(x-0)^2 + (y-0)^2} = 2\sqrt{(x-3)^2 + (y-0)^2}$

    (b) $x^2 + y^2 = 4[(x-3)^2 + y^2]$

    (c) $(x-4)^2 + y^2 = 4$ [a circle, center at (4,0), radius 2]

3. Ellipse

4. $|x_1 - x_2|$

5. $\sqrt{(x_1 - x_2)^2 + (y_1 - y_2)^2}$

6. (a) Outside; $\sqrt{145} > \sqrt{144}$

    (b) $\sqrt{145} - \sqrt{144} \doteq 12.042 - 12 = 0.042$

7. $(x+5)^2 + (x-4)^2 = 81$

8. (a) (i) parallel to x-axis,   (ii) downward;

    (b) (i) parallel to y-axis,   (ii) right;

    (c) (i) parallel to x-axis,   (ii) upward;

    (d) (i) parallel to y-axis,   (ii) left.

9. (a) 9, (b) 7, (c) 1, (d) 2, (e) 6, (f) 3, (g) 8, (h) 5, (i) 4

10. Ellipse that meets axes at (7,0), (-7,0), (0,5), and (0,-5).

11. Two lines: $\frac{x}{7} - \frac{y}{5} = 0$ and $\frac{x}{7} + \frac{y}{5} = 0$.

12. Hyperbola (asymptotes from Ex. 11), crosses x-axis at (7,0) and (-7,0).

13. y = 2x and y = -2x are the asymptotes for both curves.

14. Point: (0,0)

15. These student-devised equations should be a source of discussion.

16. When the two foci of the ellipse coincide, the ellipse becomes a circle.

17. (a) In a plane, all points that are the same distance from a fixed point (center).
    (b) In a plane, all points such that the sum of the distances from each point to two fixed points (foci) is a constant.
    (c) In a plane, all points such that the difference of the distances from each point to two fixed points (foci) is of constant absolute value.
    (d) In a plane, all points such that the distance from each point to a fixed line (directrix) is the same as the distance to a fixed point (focus).

18. (b)

19. (a) Estimate (hyperbola and line),   (b) (5,3)

20. (a) Estimate (circle and ellipse),   (b) $(\frac{8}{3}\sqrt{3}, \frac{1}{3}\sqrt{33})$,
    $(\frac{8}{3}\sqrt{3}, -\frac{1}{3}\sqrt{33})$, $(-\frac{8}{3}\sqrt{3}, \frac{1}{3}\sqrt{33})$, $(-\frac{8}{3}\sqrt{3}, -\frac{1}{3}\sqrt{33})$.

21. $y = \frac{1}{4}x^2 + 1$

22. (a) center (0,0), radius $\sqrt{3}$,   (b) center (-1,0), radius 2,
    (c) center (-1,-2), radius $\sqrt{8}$ or $2\sqrt{2}$, (d) center (-3,4), radius 5.

* * * CONIC SECTIONS REVIEW SELF-TEST * * *

1. Suppose Joe says, "I'm the same distance from the mine as I am from Dead Man's Oasis . . . "
   (a) If the mine is at M = (0,0) and the oasis is at D = (2,1), find the equation describing Joe's possible location in the coordinate plane.
   (b) What type of conic section describes where Joe might be in (a)?

2. Sketch the graph of the equation:
   (a) $y^2 - 4x^2 = 0$ (b) $y^2 = x^2$
   (c) $(x+2)^2 + (y-1)^2 = 4$ (d) $x^2 + 4y^2 = 4$
   (e) $y = x^2 - 1$ (f) $y^2 - x^2 = 1$

3. Without graphing, find the center and radius of each of the following circles:
   (a) $x^2 + 10x + y^2 - 2y = -2$ (b) $y^2 + 12y + x^2 - 8x = 10$

4. Without graphing, name the type of conic section given by the equation:

   (a) $9y^2 - x^2 = 0$ (b) $4x^2 + 25y^2 = 100$

   (c) $y = 2x^2 - 3$ (d) $y = 2x + 3$

   (e) $4x^2 - y^2 = 100$ (f) $x^2 + y^2 = 2x + 2y$

ANSWERS TO CONIC SECTIONS REVIEW SELF TEST

1. (a) $4x + 2y - 5 = 0$, (b) line.

2. (a) (b) two intersecting lines, (c) circle, center at $(-2,1)$,
   radius 2, (d) ellipse, (e) parabola, (f) hyperbola

3. (a) center at $(-5,1)$, radius is $\sqrt{24}$, or $2\sqrt{6}$, (b) center at
   $(4,-6)$, radius is $\sqrt{62}$.

4. (a) two intersecting lines, (b) ellipse, (c) parabola, (d) line,
   (e) hyperbola, (f) circle (center at $(1,1)$, radius is $\sqrt{2}$).

* * * CONIC SECTIONS PROJECTS * * *

1. (a) By completing the square, show that the equation

   $$x^2 + y^2 + Ax + By + C = 0$$

   is equivalent to

   $$(x + \tfrac{A}{2})^2 + (y + \tfrac{B}{2})^2 = \frac{A^2 + B^2 - 4C}{4}$$

   (b) Show that if $A^2 + B^2 < 4C$, then there are no points on the graph.

   (c) Show that if $A^2 + B^2 = 4C$, there is exactly one point on the
   graph.

   (d) Show that if $A^2 + B^2 > 4C$, then the graph is a circle.

THE ELLIPSE

2. In Example 7 of the central section of this chapter, the equation of a
   particular ellipse was obtained. This project obtains the equation of
   the ellipse whose foci are at $(c,0)$ and $(-c,0)$, and whose "constant
   sum" is 2a, where $a > 0$.

   (a) Show that $\sqrt{(x + c)^2 + y^2} + \sqrt{(x - c)^2 + y^2} = 2a$.

   (b) Following all the steps of Example 7 carefully, show that the equation in part (a) can be changed to

$$\frac{x^2}{a^2} + \frac{y^2}{a^2-c^2} = 1.$$

(c) Considering a point P that is simultaneously on the y-axis and on the ellipse, show that $a > c$. (Draw a picture.)

Since $a^2 - c^2 > 0$, it can be expressed as the square of a number,

$$a^2 - c^2 = b^2 \qquad \text{(for convenience, } b > 0\text{)}.$$

The equation of the ellipse then becomes

$$\frac{x^2}{a^2} + \frac{y^2}{b^2} = 1.$$

3. (See Project 2.) Show that $a > b$.

4. (See Projects 2 and 3.) Let a and b be positive, and $a > b$.
   (a) Sketch the ellipse

   $$\frac{x^2}{a^2} + \frac{y^2}{b^2} = 1.$$

   (b) What are the x-coordinates of the foci of the ellipse in (a)?

5. (See Projects 2, 3, and 4.) If $0 < a < b$; where are the foci of the ellipse

   $$\frac{x^2}{a^2} + \frac{y^2}{b^2} = 1 ?$$

6. Suppose an ellipse is inscribed in a rectangular garden:

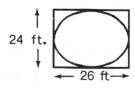

24 ft.

← 26 ft →

   (a) If you use string and two stakes in the ground to draw the above ellipse, how long should the string be?
   (b) Where are the foci?
   (c) Draw the ellipse to scale.

7. Using a board (or cardboard), string, and two tacks, construct an ellipse in the manner of Exercise 41 in the Central Section of this

chapter.

(a) Select a point P_1 on the ellipse and draw a line t tangent at point P_1 as well as you can. Let the foci be F_1 and F_2.

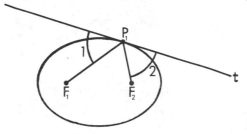

(b) Measure ∡1 and ∡2.

(c) Select other points (P_2, P_3, P_4, and P_5) on the ellipse, draw the tangent line at each point, then fill in the table:

	P_2	P_3	P_4	P_5
∡1				
∡2				

(d) What would you guess is true about ∡1 and ∡2 for any point P?

(e) Why would a billiards ball shot from F_1 reflect off the ellipti- *Assume no "english" is put on the ball.*
cal cushion at P_1 and hit a ball at F_2 in the figure above?

8. Some "whisper chambers" (a whisper can be heard across the room), such as the Mormon Tabernacle in Salt Lake City, are based on the ellipse. How does this explain their "whisper" characteristics? (See Project 7.)

9. Find a stick with a circular cross section, such as a broomstick or a dowel (which can be bought at a lumber yard).

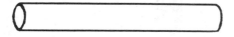

(a) Saw the stick at a right angle to its length. What figure is the flat cross-section that results?

(b) If you saw the stick, but <u>not</u> at a right angle, what kind of figure do you get?

(continued)

10. The light rays from a flashlight form a cone. How could you shine a
     flashlight against a wall so that the lighted part of the wall is
     (a) a circle   (b) an ellipse
     (c) a parabola   (d) part of a hyperbola?
     (Carry out the experiment and sketch the results.)

11. This project shows that the cross section formed in Project 9 (b) is
     indeed an ellipse, as defined in the chapter. In the following fig-
     ure, the cylinder is cut by plane X. Two spheres have the same
     diameter as the cylinder and fit exactly inside it. Each sphere is
     tangent to plane X at points F_1 and F_2, respectively. Point P is on
     the surface of the cylinder and lies in plane X. Points A and B are
     also on the surface of the cylinder such that points A, P, and B lie
     in a vertical straight line and on the surface of the spheres.

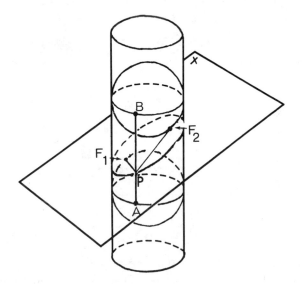

     Assume that all tangents to a sphere from a point outside it are equal in length. Show that

(a) $PA = PF_1$.   (b) $PB = PF_2$.

(c)   What happens to the length AB as point P moves around the inter-
      section of the plane and the cylinder?

(d)   Show that $PF_1 + PF_2$ is independent of P.

(e)   From (d), show that the cross section by plane X is indeed an
      ellipse.

THE PARABOLA

12. Example 5 in the Central Section of this chapter obtained the equation
     of a specific parabola. Following the same steps, show that the equa-
     tion of the parabola whose focus is at $(0,c)$ and whose directrix is
     the line $y = -c$ is

$$4cy = x^2$$

or

$$y = \frac{1}{4c} x^2.$$

It follows that for any nonzero constant $\underline{a}$, the graph of $y = ax^2$ is a
parabola. It is shown in analytic geometry that the graph of

$y = ax^2 + bx + c$ is a parabola for any constants a, b, c, with $a \neq 0$.

13. This project concerns the reflecting property of the parabola.

(a)   Make an accurate graph of the parabola $y = x^2$.

(b)   Select a point P on the parabola. Draw a line L_1 such that
      $\angle 2$ equals $\angle 1$, as in this diagram.

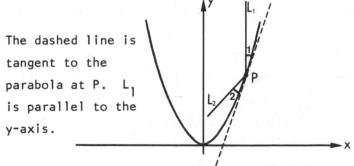

The dashed line is
tangent to the
parabola at P. L_1
is parallel to the
y-axis.

(c)   Do (b) for two more points on the parabola.

The three lines "L_2" drawn in (b) and (c) should intersect in a single

point (which is the focus of the parabola: $(0,\frac{1}{4})$.) (See the Projects

in the next chapter (SLOPE) for a proof that they do.)

THE HYPERBOLA

14. The hyperbola is used in electronic navigational equipment called "Loran" (short for "long range navigation").

    (a) Read about Loran in an encyclopedia.

    (b) Give a report on (a) to the class.

    The next project shows a method of drawing the hyperbola similar to that used to draw the ellipse.

15. The hyperbola can be drawn using a board, a ruler, a string, and two tacks:

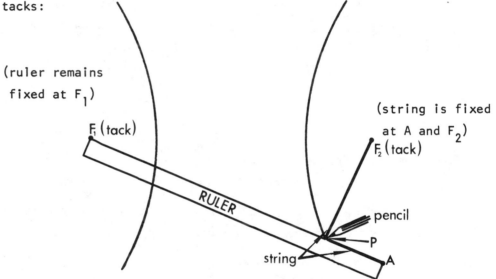

    (a) Place the tacks at the foci, F_1 and F_2.

    (b) The length of string is secured at F_2 and at one end of the ruler at point A. The string must have a length less than the distance $F_1 F_2$.

    (c) The pencil keeps the string taut as it moves along the ruler. (The endpoint of the ruler will stay fixed at F_1 while the end of the ruler at A will move.)

    (d) Draw the lower right piece of the hyperbola as you see in the figure. When points F_1, P, and F_2 are in line along the ruler, flip over the ruler and continue, making the upper right part of the hyperbola.

    (e) Reverse the roles of F_1 and F_2 and draw the left piece of the hyperbola.

16. Explain why the curve produced in Project 15 is a hyperbola.

17. Example 8 in the Central Section of this chapter develops the equation of a specific hyperbola. This project treats the general case.

   Consider the hyperbola with foci $F_1 = (c,0)$ and $F_2 = (-c,0)$. Assume that

$$\left|PF_1 - PF_2\right| = 2a, \text{ thus } a > 0.$$

(a) Show that the equation

$$\sqrt{(x+c)^2 + y^2} - \sqrt{(x-c)^2 + y^2} = \pm\, 2a$$

describes the hyperbola.

(b) Following the steps in Example 8 of the central section, show that the equation of the hyperbola can be transformed to

$$\frac{x^2}{a^2} - \frac{y^2}{c^2 - a^2} = 1.$$

(c) Show that $c > a$.

(d) From (c) it follows that $c^2 - a^2 = b^2$ for some $b > 0$. Thus the equation of the hyperbola can be written in the form

$$\frac{x^2}{a^2} - \frac{y^2}{b^2} = 1.$$

18. (See Project 17.) Find the coordinates of the foci of the hyperbola

(a) $\dfrac{x^2}{16} - \dfrac{y^2}{9} = 1$ (b) $\dfrac{y^2}{16} - \dfrac{x^2}{9} = 1.$

THE ASYMPTOTES OF A HYPERBOLA

   In the Central Section of this chapter, the discussion called Sketching a Hyperbola presented numerical evidence for the assertion that for large x, x and $\sqrt{x^2 - 9}$ differ very little. The next project discusses this. Incidentally, it provides a case where we "rationalize the numerator" rather than the denominator.

19. (a) Show that

$$x - \sqrt{x^2 - 9} = \frac{9}{x + \sqrt{x^2 - 9}}$$

Hint: Multiply $x - \sqrt{x^2 - 9}$ by $\dfrac{x + \sqrt{x^2 - 9}}{x + \sqrt{x^2 - 9}}$

(b) Use (a) to show that as x gets large, $x - \sqrt{x^2 - 9}$ approaches 0.

20. In analytic geometry it is shown that the graph of

$$Ax^2 + Bxy + Cy^2 + Dx + Ey + F = 0$$

is a conic section. Moreover, if

$$\underline{B^2 - 4AC \text{ is}} \begin{cases} \text{negative,} \\ \text{positive,} \\ \text{zero,} \end{cases} \underline{\text{then the graph is}} \begin{cases} \text{an ellipse.} \\ \text{a hyperbola.} \\ \text{a parabola.} \end{cases}$$

(Remember the code word: NEPH - negative, ellipse; positive, hyperbola.) There are certain dull exceptions, called the degenerate conics, illustrated in the cases below.
Show that the graph of

(a) $2x^2 + 3y^2 + 5 = 0$ is the empty set (no points at all).

(b) $2x^2 + 3y^2 = 0$ has exactly one point.

(c) $x^2 - y^2 = 0$ consists of two lines.

(d) $(y - 2x + 3)^2 = 0$ consists of one line.

21. Read about Descartes and coordinate geometry in at least one of the following:

(a) Bell, E. T. Men of Mathematics, Chapter 3, Simon and Schuster, Inc., 1961.

(b) Kline, Morris. Mathematics: A Cultural Approach, Chapter 13, Addison-Wesley Publishing Company, Inc., 1963.

(c) Kline, Morris. Mathematical Thought From Ancient to Modern Times, Oxford University Press, 1972.

(d) van der Waerden, B. L. Science Awakening, pp. 241-48. New York: Oxford University Press, 1961.

22. Read The Watershed by Arthur Koestler for a fascinating study of Kepler as a man and as a scientist. It was he who discovered that the orbit of each planet is an ellipse.

23. Work with another student to draw a parabola in the following manner. [You will need a board, a string, a tack, and a plastic right triangle (as used in mechanical drawing), or some suitable instrument for maintaining a right angle.]

(a) Let a tack be at F (the focus). The string is secured at F and at A and is kept taut by the pencil at point P. Point B, the

right angle vertex, is on the directrix. Let the length of the
string be the same as the length AB.

(b) Why is P a point on the parabola?

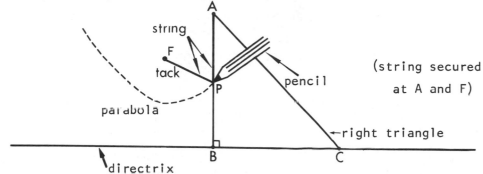

(c) Keeping points A, P, and B in line and moving the triangle so
that side BC is always on the directrix, work with another
student to make a careful drawing of a parabola. (It may help to
place a ruler or some other guide along the directrix.)

24. Suppose at fixed points A, B, and C there are tacks and that you have
a loop of string with a length greater than the sum of the lengths of
the sides of triangle ABC. Stretch the string taut. Explain how this
curve is related to an ellipse.

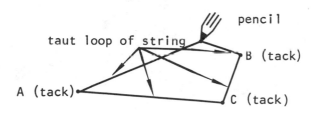

25. The graph of

$$x^2 + y^2 + Ax + By + C = 0$$

is

a circle, if $A^2 + B^2 > 4C$.

a point, if $A^2 + B^2 = 4C$.

the empty set , if $A^2 + B^2 < 4C$.

(Hint: use completing the square.) Prove this statement.

11

Algebra II

Slope

* * * CENTRAL SECTION * * *

The symbolism and the techniques of algebra are of major importance in calculus, a subject which this chapter will introduce. One of the main problems of calculus will be treated: <u>how to measure the steepness of a curve at any point on that curve.</u> Both "rationalizing a square root" and the binomial theorem will be of use in solving special cases of this problem. The Projects include some applications of the ideas developed in the chapter.

THE SLOPE OF A LINE DETERMINED BY TWO POINTS

Let $P = (x_1, y_1)$ and $Q = (x_2, y_2)$ be two points in the coordinate plane. They determine a line, as shown in this diagram.

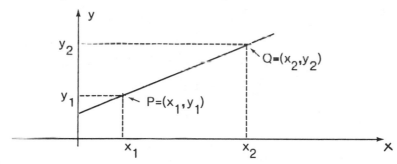

DEFINITION. The <u>slope of the line through the points P and Q is defined to be the quotient.</u>

$$\frac{y_2 - y_1}{x_2 - x_1} \qquad (\text{assume } x_2 \neq x_1)$$

The slope is also
$$\frac{y_1 - y_2}{x_1 - x_2}$$

In the following diagram both $y_2 - y_1$ and $x_2 - x_1$ are positive and are the lengths of the legs of the right triangle as shown. (They are labelled "rise" and "run.")

It can be shown that the slope of a line does not depend on which pair of points P and Q are chosen.

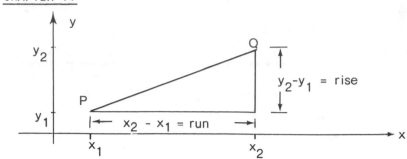

Example 1. Find the slope of the line through $P = (1,2)$ and $Q = (3,7)$.

Solution. The slope is

$$\frac{7 - 2}{3 - 1} = \frac{5}{2}$$

$$= 2.5$$

It may happen that $x_2 - x_1$ or $y_2 - y_1$ is negative, as in this diagram, which shows a case where $x_2 - x_1$ is positive and $y_2 - y_1$ is negative.

REMINDER!
If the line rises going left to right, the slope is positive. If it falls going left to right, the slope is negative.

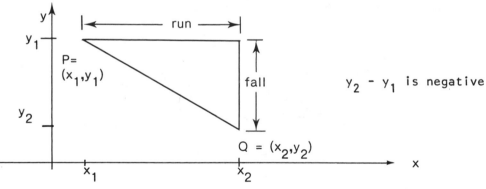

In this case the "run" is positive and there is a "fall" from P to Q rather than a "rise." The quotient

$$\frac{y_2 - y_1}{x_2 - x_1}$$

in this case is negative. As a pencil traces out this line from left to right it moves <u>downward</u>. That is the meaning of a negative slope.

EXERCISES

1. In each case plot the two points, draw the line that they determine, and compute the slope.

    (a) (1,2) and (3,5)   (b) (1,2) and (3,6)

    (c) (4,3) and (7,3)   (d) (2,3) and (5,0)

(a) 1.5
(c) 0

Observe, as in Exercise 1(c), that the slope of a line parallel to the x-axis is 0. The slope of a line parallel to the y-axis is not defined.

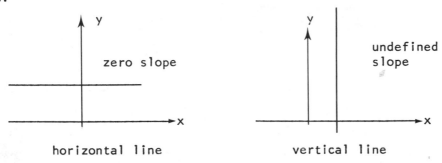

zero slope

horizontal line

undefined slope

vertical line

Why would trying to define the slope of a vertical line get us involved with division by 0?

The slope measures how steep a line is. The four lines in these diagrams are typical.

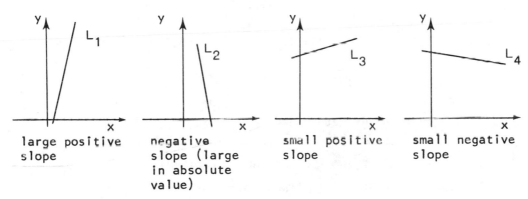

L_1

large positive slope

L_2

negative slope (large in absolute value)

L_3

small positive slope

L_4

small negative slope

2. Draw a line of slope

(a) 5   (b) $\frac{1}{5}$ (c) -5   (d) $-\frac{1}{5}$

TWO - POINT EQUATION OF A LINE

Consider a point (x,y) on the line through the points (2,3) and (5,8).

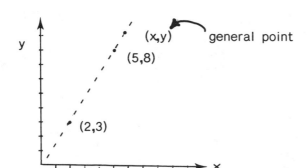

(x,y)   general point

(5,8)

(2,3)

The three points (2,3), (5,8) and (x,y) are on a line.

Any two points on a line determine the same slope.

The slope determined by the points (2,3) and (5,8) equals the slope determined by the points (2,3) and (x,y). Thus

$$\frac{8 - 3}{5 - 2} = \frac{y - 3}{x - 2}$$

hence

$$\frac{5}{3} = \frac{y - 3}{x - 2}$$

The last equation provides a way of telling quickly whether a third point (x,y) is on the line through the points (2,3) and 5,8).

3. Which of these points is on the line through (2,3) and (5,8)?

In each case plot the points (2,3) and 5,8) as well as the given point. Use the following equation to decide.

$$\frac{5}{3} = \frac{y - 3}{x - 2}$$

(a) yes
(c) no

(a) (8,13)   (b) (11,17)   (c) (0,0)   (d) (14,23)

* * *

Comparison of slopes establishes this general criterion for a third point to lie on the line determined by $P_1 = (x_1, y_1)$ and $P_2 = (x_2, y_2)$.

THE TWO - POINT EQUATION OF A LINE

The general point (x,y) lies on the line through the points (x_1, y_1) and (x_2, y_2) if

$$\frac{y - y_1}{x - x_1} = \frac{y_2 - y_1}{x_2 - x_1}$$

In this formula (x,y) is a third point on the line through (x_1, y_1) and (x_2, y_2). We assume that neither denominator is zero.

Example 2. Find the two-point equation of the line through (1,4) and (2,3).

Solution. The equation is

Note: For the equation the general point must be used.

$$\frac{y - 4}{x - 1} = \frac{3 - 4}{2 - 1}$$

$$\frac{y - 4}{x - 1} = \frac{-1}{1}$$

Clearing of fractions lets (x,y) assume the value (1,4); this avoids division by 0.

$$y - 4 = -x + 1$$

This can be written as

$$y = -x + 5$$

(As a check, see whether (1,4) and (2,3) satisfy the final equation.
Since $4 = -(1) + 5$

the point (1,4) does satisfy the equation. Similarly since
                                   $3 = -(2) + 5$

(2,3) does also.)

4. Write a two-point equation of the line through the points,
    (a) (2,1) and (3,7)   (b) (4,1) and (2,5)
    (c) (1,1) and (1.1, 1.21)   (d) (-1,5) and (4,-3)

5. Change each of the equations obtained in Exercise 4 to the form
    $y = mx + b$ for appropriate constants m and b.

(a) $\frac{y-1}{x-2} = \frac{6}{1}$

(c) $\frac{y-1}{x-1} = \frac{0.21}{0.1}$

(a) $y = 6x - 11$

THE POINT-SLOPE EQUATION OF A LINE

The general point (x,y) lies on the line through (x_1,y_1) with slope m if

$$\frac{y - y_1}{x - x_1} = m$$

Example 3. Find an equation of the line with slope $-\frac{2}{3}$ and containing the point (1,4).

Solution. The general point (x,y) must be considered. Then,

$$\frac{y-4}{x-1} = -\frac{2}{3}$$

or
$$y - 4 = -\frac{2}{3}(x - 1)$$

This can be written as
$$y = -\frac{2}{3}x + \frac{14}{3}$$

6. Write a point-slope form of an equation for the line with the given slope and point.
    (a) slope $\frac{1}{2}$; (2,1)   (b) slope $-\frac{3}{4}$; (3,1)
    (c) slope $-\frac{2}{3}$; (-1,2)   (d) slope $\frac{3}{5}$; (-3,-2)

(a) $\frac{y-1}{x-2} = \frac{1}{2}$
or $y - 1 = \frac{1}{2}(x-2)$
(c) $\frac{y-2}{x+1} = -\frac{2}{3}$
or $y - 2 = -\frac{2}{3}(x+1)$

7. Change the equations in Exercise 6 to the form $y = mx + b$ for appropriate constants m and b.

(a) $y = \frac{1}{2}x$
(c) $y = -\frac{2}{3}x + \frac{4}{3}$

THE SLOPE - INTERCEPT EQUATION OF A LINE

The graph of $y = mx + b$ is a line, for any choice of constants m and b.   The next example treats a specific illustration.

Example 3.   Graph the line

$$y = 2x + 1$$

Find its slope.

Solution.   Find two points on the line by picking a couple of values for x, for instance let x = 1 and x = 4.   Then find the corresponding y-values.   A table may be useful.

x	1	4
y = 2x + 1	3	9

Then plot the two points obtained, namely (1,3) and (4,9).   These determine the line.   However, to catch a possible error, plot a third point, choosing say x = 2, hence y = 5.   This gives the point (2,5).

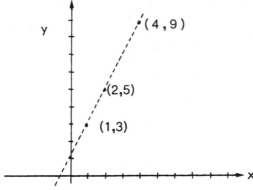

The line clearly has a positive slope.   Any two of the three known points can be used to find the slope.   The points (1,3) and (2,5) determine the slope as follows:

Check that the points (1,3) and (4,9) give the same slope.

$$\frac{5 - 3}{2 - 1} = \frac{2}{1}$$
$$= 2$$

The slope is 2.

(a) slope 2

(c) slope $\frac{1}{2}$

8.   Graph each line and find its slope.

(a) $y = 2x - 1$ (b) $y = -2x + 3$

(c) $y = \frac{1}{2} x$ (d) $y = x - 2$

As Exercise 6 illustrates, the slope of the line

$$y = mx + b$$

is m, the coefficient of x.

This is proved in Projects.

The number b in the equation y = mx + b has a simple interpretation. Note that when x = 0, y = m · 0 + b = b. Thus the point (0,b) is on the line y = mx + b. In other words, the line y = mx + b meets the y-axis at the point (0,b). The number b is called <u>the y-intercept</u>.

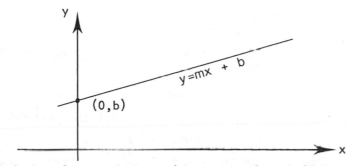

m = slope
b = y-intercept

9. Find the slope and the y-intercept of each line:

    (a) y = 3x + 2   (b) y = 2x + 3

    (c) y = 4x - 1   (d) y = -x + 5

10. Write an equation of the line that has

    (a) slope -2 and y-intercept 2   (b) slope 3 and y-intercept 0

    (c) slope 0.23 and y-intercept 5 (d) slope -1 and y-intercept $\frac{1}{2}$

    (e) slope $-\frac{2}{5}$ and y-intercept $\frac{5}{2}$ (f) slope 0 and y-intercept 4

    (a) $y = -2x + 2$
    (c) $y = 0.23x + 5$

    (e) $y = -\frac{2}{5}x + \frac{5}{2}$

*   *   *

These last few examples and exercises are summarized in the following formula

> THE SLOPE - INTERCEPT EQUATION OF A LINE
> The slope of the line whose equation is
> $$y = mx + b$$
> is m.   The y-intercept is b.

The problem of computing the slope of a line thus has two easy solutions. Either compute the slope after choosing two points on the line or write the equation in the form y = mx + b (the slope is m). Note that it does not matter which two points are picked on the line. The slope is the same everywhere on the line.

THE SLOPE OF A CURVE

Finding the slope of a curve is a different matter.   Consider as an example, the curve

$$y = \sqrt{x} \qquad (\text{for } x \geq 0)$$

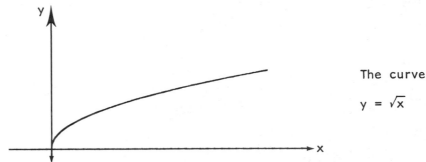

This is the top half of $y^2 = x$. (See Chapter 10)

The curve

$$y = \sqrt{x}$$

It is very steep near (0,0). As the pencil traces out the curve from left to right, the curve becomes less steep. In fact, far to the right it becomes almost horizontal, like the x-axis. In particular, the slope of the curve $y = \sqrt{x}$ changes from point to point. This is not the case with the line.

Put your pencil at (1,1), press the ruler up to it and rotate the ruler until its direction seems to match that of the curve near (1,1).

11.   (a)   Plot the curve $y = \sqrt{x}$ accurately, with the same scale on both axes.

(b)   Draw as well as you can the tangent line to the curve at (1,1).

(c)   Estimate the slope of the tangent line in (b) by measuring the rise and run of two points on the tangent line.

*11.   (c)
A centimeter ruler is best. Choose (1,1) as one point. Compare your answer with that of other students.*

12.   Carry out the steps of Exercise 11 for the point $(\frac{1}{4}, \frac{1}{2})$.

* * *

The accuracy of the approach in Exercises 9 and 10 is limited by at least two experimental "errors:"

(a)   guessing the angle of the tangent line

and

(b)   reading the ruler.

This method probably cannot give even two-decimal accuracy. Hence a better method is needed.

THE METHOD OF THE NEARBY POINT

Return to the problem of estimating how steep is the curve $y = \sqrt{x}$ at the point $P = (1,1)$. That is, estimate the slope of the tangent line to the curve at the point $(1,1)$. Rather than try to draw the tangent line, choose a second point Q, on the curve $y = \sqrt{x}$ near $(1,1)$. Consider the line through P and Q.

Note that both P and Q are on the curve. $P = (1,1)$ is fixed, Q will vary.

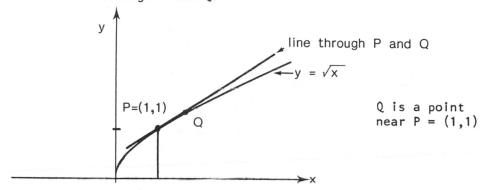

Q is a point
near P = (1,1)

When Q <u>is near</u> P the line determined by P and Q is a close approximation to the tangent line at P:

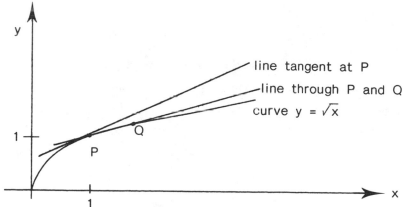

The slope of the line through P and Q is thus an <u>estimate</u> of the slope of the tangent line at P.

Example 4.   A calculator (or a table of square roots) gives

$$\sqrt{1.5} \doteq 1.2247$$

Use this information to estimate the slope of the tangent to the curve at $(1,1)$.

Solution.   The points $P = (1,1)$ and $Q = (1.5, \sqrt{1.5})$ determine a line. The slope of this line is

$$\frac{\sqrt{1.5} - 1}{1.5 - 1}$$

(continued)

Since

$$\sqrt{1.5} \doteq 1.2247,$$

the slope of the one is approximately

How does the estimate 0.4494 compare with the one you obtained in Exercise 11?

$$\frac{1.2247 - 1}{1.5 - 1} = \frac{0.2247}{0.5}$$

$$= 0.4494$$

This is an <u>estimate</u> of the slope of the tangent line at $(1,1)$.

13. The following data are given by a calculator or a square root table:

x	1.4	1.3	1.2	1.1	1.01
$\sqrt{x}$	1.1832	1.1402	1.0954	0.0488	1.00499

As in Example 4, estimate the slope of the tangent at $P = (1,1)$ using the line through P and

(a) $Q = (1.4, \sqrt{1.4})$ (b) $Q = (1.3, \sqrt{1.3})$

(a) 0.458
(c) 0.477
(e) 0.499

(c) $Q = (1.2, \sqrt{1.2})$ (d) $Q = (1.1, \sqrt{1.1})$

(e) $Q = (1.01, \sqrt{1.01})$

* * *

At first glance, the method of Exercise 13 may seem as imperfect as the method of drawing a tangent "by eye" and reading a ruler. At this point, all that either method provides is an <u>estimate</u> of the slope of the tangent line at $(1,1)$. But by introducing algebra, we can modify "the method of the nearby point" in such a way as to get a <u>precise</u> answer to the problem--not just an estimate.

ALGEBRA AND THE NEARBY-POINT METHOD

Rather than taking a <u>specific</u> point Q near P, consider a <u>typical</u> general point Q near P. This point is

$$Q = (1 + h, \sqrt{1 + h}).$$

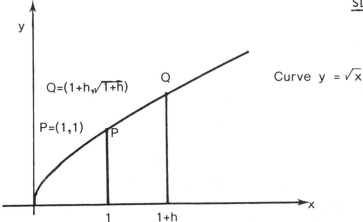

Curve $y = \sqrt{x}$

The letter "h" stands for a small "horizontal" change. Note that $h \neq 0$; for if $h = 0$, P and Q would be the same point. Two <u>distinct</u> points are needed to determine the slope of a line.

For instance in reference to P = (1,1) if Q = (1.4, $\sqrt{1.4}$) then h = 0.4.

The two-point formula for slope of a line yields this expression for the slope of the line through $P = (1,1)$ and $Q = (1 + h, \sqrt{1 + h}\,)$:

$$\text{slope} = \frac{\sqrt{1 + h} - 1}{1 + h - 1}$$

$$= \frac{\sqrt{1 + h} - 1}{h} \quad ?$$

As Q gets closer and closer to P, h gets near 0. But what happens to the quotient

$$\frac{\sqrt{1 + h} - 1}{h} \quad ?$$

Look back at your answers to Exercise 13, which evaluates the quotient for $h = 0.4$, 0.3, 0.2, 0.1 and 0.01. What do you think happens to the quotient $\frac{\sqrt{1 + h} - 1}{h}$ as h gets small?

In order to find out, <u>rationalize the numerator</u>:

$$\frac{\sqrt{1 + h} - 1}{h} = \frac{(\sqrt{1 + h} - 1)}{h} \cdot \frac{(\sqrt{1 + h} + 1)}{(\sqrt{1 + h} + 1)}$$

$$= \frac{1 + h - 1}{h(\sqrt{1 + h} + 1)}$$

$$= \frac{1}{\sqrt{1 + h} + 1}$$

Rationalizing the numerator may be a new technique for you.

Surprise! It is <u>easy</u> to tell what happens to

$$\frac{1}{\sqrt{1 + h} + 1}$$

as h approaches 0. First of all,

$$1 + h \text{ approaches } 1;$$

so

$$\sqrt{1 + h} \text{ approaches } \sqrt{1} = 1 .$$

Finally

$$\frac{1}{\sqrt{1 + h} + 1} \text{ approaches } \frac{1}{1 + 1} = \frac{1}{2}.$$

This implies that <u>the slope of the tangent line to the curve</u>,

$y = \sqrt{x}$ <u>at</u> (1,1) <u>is precisely</u> $\frac{1}{2} = 0.500$.

The nearby-point method, together with an algebraic trick, provides the <u>exact</u> answer. No diagram, square root table, or calculator is needed. The above type of argument which shows that the slope of the curve $y = \sqrt{x}$ at the point (1,1) is $\frac{1}{2}$ is fundamental to the calculus. This approach applies to all points on the curve $y = \sqrt{x}$ (as Exercises 14 and 15 show), as well as other curves. Usually the method is called "finding the derivative."

14. Use the "nearby-point" method to find the slope of the curve
    $y = \sqrt{x}$ at
    (a) (4,2)   (b) (9,3)   (c) $(\frac{1}{4},\frac{1}{2})$ (d) (25,5)

15. Use the nearby-point method to find the slope of the curve at
    $P = (x, \sqrt{x})$. The nearby point is $Q = (x + h, \sqrt{x + h})$.

(a) $\frac{1}{4}$

(c) 1

$\frac{1}{2\sqrt{x}}$

THE NEARBY-POINT METHOD APPLIED TO OTHER CURVES

There is nothing special about the curve $y = \sqrt{x}$. The nearby-point method applies to any curve given by an algebraic formula. The next example shows that the method applies equally well to the curve $y = x^2$.

Example 5. Find the slope of the tangent line to the curve

$$y = x^2$$

at the point $P = (2,4)$

Solution. Since the curve is $y = x^2$, the "nearby-point" Q, which has x-coordinate $2 + h$ has the y-coordinate equal to $(2 + h)^2$.

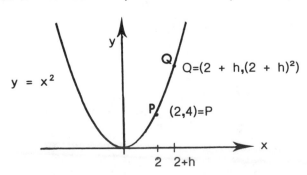

The slope of the line through P and Q is

$$\frac{(2 + h)^2 - 4}{2 + h - 2} = \frac{(2 + h)^2 - 4}{h}$$

$$= \frac{4 + 4h + h^2 - 4}{h}$$

$$= \frac{4h + h^2}{h}$$

$$= 4 + h$$

As Q gets nearer and nearer P, h gets nearer to 0. Thus $4 + h$ approaches 4.

From this is follows that the <u>slope of the tangent line to</u> $y = x^2$ <u>at</u> $(2,4)$ <u>is</u> 4.

16. Use the nearby-point method of Example 5 to find the slope of the tangent line to $y = x^2$ at

   (a) $(1,1)$ (b) $(3,9)$

   (c) $(4,16)$ (d) (x,x^2) for any x

Hint: Draw a sketch for each.

(a) 2
(c) 8

USING THE BINOMIAL THEOREM

The "slope of a curve" at the point P shall mean the slope of the tangent line to the curve at P. The slope of that tangent line is found by the nearby-point method.

The method is quite general, and can be used to find the slope of the curve $y = x^3$

at any point on it.

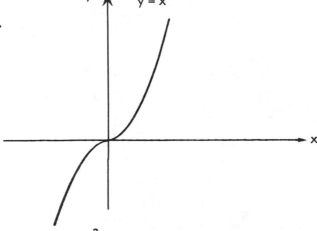

17. Inspect the graph of $y = x^3$. Do you think that

   (a) the slope is ever negative?

   (b) the slope is ever 0?

The next example finds the slope of $y = x^3$ at the point $(2,8)$.

Example 6. Find the slope of $y = x^3$ at $(2,8)$

Solution. The nearby point is

$$Q = (2 + h, (2 + h)^3).$$

The slope of the line through P and Q is

$$\frac{(2 + h)^3 - 8}{2 + h - 2} = \frac{(2 + h)^3 - 8}{h}$$

From the binomial theorem for exponent 3,

*Hint: See Ch. 7
(Binomial Theorem)*

$$(a + b)^3 = a^3 + 3a^2b + 3ab^2 + b^3,$$

it follows that

$$(2 + h)^3 = 2^3 + 3 \cdot 2^2 \cdot h + 3 \cdot 2 \cdot h^2 + h^3$$

$$= 8 + 12h + 6h^2 + h^3$$

Hence the slope of the line through P and Q equals

$$\frac{(8 + 12h + 6h^2 + h^3) - 8}{h}$$

$$= \frac{12h + 6h^2 + h^3}{h}$$

$$= 12 + 6h + h^2$$

Since h approaches 0.

As Q gets nearer $P = (2,8)$, the slope of the line through P and Q approaches 12.

The slope of the tangent line to $y = x^3$ at $(2,8)$, is therefore 12.

18. Use the nearby-point method, illustrated in Example 6, to find the slope of the tangent line to $y = x^3$ at

(a) 3

(c) $\frac{3}{4}$

(a) $(1,1)$ (b) $(0,0)$ (c) $(\frac{1}{2},\frac{1}{8})$ (d) (x,x^3) for any x

* * *

The nearby-point method can also be applied to the curves $y = x^4$, $y = x^5$ and so on. The next exercise concerns $y = x^4$.

19. (a) Graph $y = x^4$.

(b) For which values of x do you guess the slope is positive? zero? negative?

(c) $4x^3$

(c) Find the slope of the curve at $P = (x,x^4)$ by using the nearby point $Q = (x + h, (x + h)^4)$, then letting h approach 0. The binomial expansion
$(a + b)^4 = a^4 + 4a^3b + 6 a^2b^2 + 4 ab^3 + b^4$,   will be of use.

(d) Does the result in (c) agree with your guess in (b)?

20.   Carry out the analog of Exercise 17 for the curve $y = x^5$. $5x^4$

* * *

The following table summarizes the results obtained so far in the chapter.

CURVE	SLOPE FORMULA
$y = mx + b$ (line)	m
$y = \sqrt{x}$ $(= x^{\frac{1}{2}})$ (parabola)	$\frac{1}{2} \cdot \frac{1}{\sqrt{x}}$ $(= \frac{1}{2} x^{-\frac{1}{2}})$
$y = x^2$ (parabola)	$2x$
$y = x^3$	$3x^2$
$y = x^4$	$4x^3$
$y = x^5$	$5x^4$

Note the pattern in the last five rows. The slope formula for $y = x^a$ in each case is ax^{a-1}. That this holds for all numbers $\underline{a}$ is proved in calculus.

Check the assertion for $a = \frac{1}{2}$ and $a = 5$.

SLOPE FORMULA FOR A POLYNOMIAL

The next challenge is to find the slope formula for a polynomial. Example 7 illustrates a typical case.

Example 7. Find the slope of the curve

$$y = 3x^2 + 5x - 7$$

at the typical point $P = (x, 3x^2 + 5x - 7)$.

Solution. The "nearby point" is

$$Q = (x + h, 3(x + h)^2 + 5(x + h) - 7)$$

Thus, the slope of the line through P and Q is

$$\frac{[3(x + h)^2 + 5(x + h) - 7] - (3x^2 + 5x - 7)}{h}$$

$$= \frac{3(x + h)^2 - 3x^2 + 5(x + h) - 5x - 7 + 7}{h}$$

$$= \frac{3(x^2 + 2xh + h^2) - 3x^2 + 5x + 5h - 5x}{h}$$

$$= \frac{6xh + 3h^2 + 5h}{h}$$

$$= 6x + 3h + 5 \qquad \text{(continued)}$$

As h approaches 0, the quotient thus approaches 6x + 5.
Thus the slope formula for the curve (parabola)

$$y = 3x^2 + 5x - 7 \text{ is}$$
$$\text{slope} = 6x + 5.$$

*That is, the
slope at the
point (x,y) is
6x + 5.*

21. (a) Graph the curve $y = 3x^2 + 5x - 7$ that was discussed in
          Example 7.

*(b) Base your
answer on
inspection of the
graph in (a).
(c) positive if
$x > -\frac{5}{6}$, negative
if $x < -\frac{5}{6}$,
zero if $x = -\frac{5}{6}$*

     (b) For which x do you think the slope is positive? negative?
          zero?

     (c) According to Example 7, the slope formula for $y = 3x^2 + 5x - 7$
          is 6x + 5. Use this result to determine precisely where the
          slope is positive, negative or zero.

22. Use the nearby-point method to find the slope formula for
     $y = 5x^3 + 7x^2 + 2x + 4$.

*In the language
of calculus one
says "the
derivative of
$5x^3 - 7x^2 + 2x + 4$ is
$15x^2 - 14x + 2$."*

23. Use the nearby-point method to find the slope formula for
     $y = 3x^4 - 6x + 2$.

*22. $15x^2 - 14x + 2$
23. $12x^3 - 6$*

$$* \quad * \quad *$$

The results in Example 7 and Exercises 20 and 21 illustrate the
following general formula, proved in calculus:

*Study this table
carefully until
you know the
pattern.*

CURVE	SLOPE FORMULA (Slope at (x,y))
$y = ax^2 + bx + c$	$2ax + b$
$y = ax^3 + bx^2 + cx + d$	$3ax^2 + 2bx + c$
$y = ax^4 + bx^3 + cx^2 + dx + e$	$4ax^3 + 3bx^2 + 2cx + d$

The above table illustrates that the curve $y = ax^2 + bx + c$ at a
point (x,y) has slope 2ax + b. Exercise 24 continues this idea.

*See the above
table.*

24. Describe in words to a classmate how to find the slope formula
     of a curve at a point (x,y) if the curve is described by the equation

     (a) $y = ax^3 + bx^2 + cx + d$ (b) $y = ax^4 + bx^3 + cx^2 + dx + e$

The slope formula gives the slope of a polynomial curve for any point (x,y). Thus, for any specific point on the curve the slope is easy to compute.

Example 8.   Compute the slope of $y = -3x^2 + x - 5$ at the point $(1,-7)$.

Solution.   The slope formula is

$$-3(2x) + 1 \quad \text{or} \quad -6x + 1$$

Thus at $(1,-7)$ the slope is

$$-6(1) + 1 = -5 .$$

25.   Compute the slope formula for each of the following:

(a) $y = -5x + 3$ (b) $y = 4x^2 - 3x + 1$

(c) $y = x^3 - x^2 - x + 1$ (d) $y = 2x^4 - x^3 + x^2 - 4x + 9$

26.   What is the slope at the point (x,y) for each part of Exercise 25?

27.   Find the slope when $x = -1$ for each part of Exercise 25.

28.   Compute the slope at the point $(2,-5)$ on the graph of

$$y = -x^3 + x^2 - x + 1.$$

29.   Describe to a classmate how to compute the answer in Exercise 28.

(a) -5
(c) $3x^2 - 2x - 1$

(a) -5
(c) $3x^2 - 2x - 1$

(a) -5
(c) 4

-9

$* * *$ SUMMARY $* * *$

The chapter opened with a definition of the slope of the line through the points (x_1,y_1) and (x_2,y_2):

$$\text{slope} = \frac{y_2 - y_1}{x_2 - x_1}$$

The slope of a line parallel to the x-axis is 0. The slope of a line parallel to the y-axis is not defined. A line that slopes upward to the right has a positive slope; if it slopes downward to the right, the slope is negative.

Why?

positive slope   negative slope

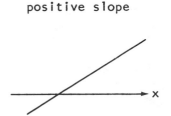

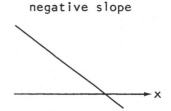

(x_1,y_1) and
(x_2,y_2) may be any
two points on a
line where (x,y)
is a general third
point.

An equation of the line through the points (x_1,y_1) and (x_2,y_2) is

$$\frac{y - y_1}{x - x_1} = \frac{y_2 - y_1}{x_2 - x_1}$$ ("two point" equation of a line)

In order to allow (x,y) to be (x_1,y_1) and avoid division by zero, we write the two point equation as

$$y - y_1 = (\frac{y_2 - y_1}{x_2 - x_1}) \ (x - x_1)$$

An equation of the line through the point (x_1,y_1) with slope m is

(x,y) is a general
point and (x_1,y_1)
is any fixed point
on the line.

$$\frac{y - y_1}{x - x_1} = m$$ ("point-slope" equation of a line)

To avoid division by zero, we write the point-slope form as

$$y - y_1 = m(x - x_1) \ .$$

Algebraic manipulation can change these equations into the form

$$y = mx + b$$ ("slope-intercept" equation of a line)

The slope of the line is m; the y-intercept is b.

The slope of a curve varies from point to point. It can be estimated by making an accurate graph and guessing the slope of the tangent line. But the nearby-point method gives the exact slope, with no need to graph the curve. This method depends on the fact that if points P and Q are on the given curve, and Q is very near to P, then the line through P and Q has a direction close to that of the tangent line at P:

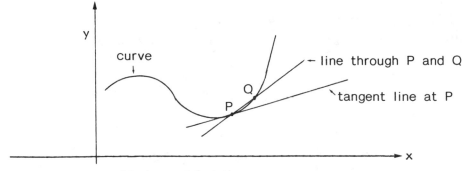

The method was applied to the curves

$$y = \sqrt{x} \qquad\qquad y = x^2 \qquad\qquad y = x^3 \qquad\qquad \text{and } y = 3x^2 + 5x - 7$$

This suggested a general approach for finding the <u>slope formula</u>:

CURVE (OR LINE)	SLOPE FORMULA
$y = mx + b$	m
$y = ax^2 + bx + c$	$2ax + b$
$y = ax^3 + bx^2 + cx + d$	$3ax^2 + 2bx + c$
and so on	

(These formulas are obtained in a calculus course.) Applications of the slope formula are included in the Projects.

* * * WORDS AND SYMBOLS * * *

slope of a line	y-intercept
two-point form	"nearby-point" method
slope-intercept form	slope formula
$y = mx + b$	point-slope form

* * * CONCLUDING REMARKS * * *

The method of the nearby point was first used by Fermat in a 1637 manuscript entitled <u>Methodus ad Disquirendam Maximam et Minimam</u>. This Latin title reads in English, <u>A</u> <u>Method</u> <u>for</u> <u>Finding</u> <u>Largest</u> <u>and</u> <u>Smallest</u> <u>Values</u>. Let us describe how he used the slope formula to find the "peaks" and "valleys" on a curve.

Let $P(x)$ be a polynomial. Consider the graph of $y = P(x)$:

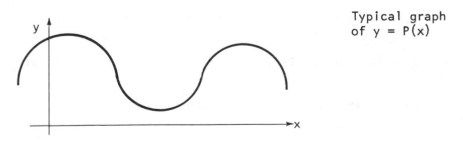

Typical graph
of $y = P(x)$

Fermat noticed that at the peaks and at the bottoms of the valleys in the graph the tangent line is horizontal.

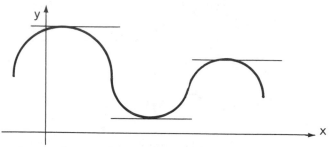

(The tangent line
is horizontal at
the peaks and
valleys.)

Now, a horizontal tangent line has slope 0. So Fermat observed that in
order to find the peaks and valleys, one merely searches among the points
of the curve where the slope is 0. For example, suppose you wish to find
a maximum (or minimum) value for

$$y = P(x),$$

where P(x) is a polynomial in x. Simply compute the slope formula for
P(x) and solve for x in the equation,

$$\text{slope formula} = 0$$

This procedure is one of the most widely applied techniques of the
calculus and has applications in technology. Further examples are to be
found in the Projects.

* * * SLOPE SELF QUIZ * * *

1. (a) Compute the slope of the line through the points (-4,2) and
         (3,7).
    (b) Write a two-point form of the equation of the line in part (a).
    (c) Write the slope-intercept (y = mx + b) form of the equation of
         the line in (a).
    (d) Check that the two given points in (a) satisfy the equation in
         (c).

2. Does the tangent line to the curve $y = \sqrt{x}$ at (4,2) pass through the
    point (11,5)?

3. What is the slope of each of the following lines?
    (a) y = -x + 4   (b) y = 0.14x - 9
    (c) y = 8x - 0.7   (d) y + x - 3 = 0
    (e) What is the y-intercept for each of the above lines?

4. What is the slope at the point (x,y) on the graph of each of the
    following?

(a) $y = \frac{2}{3}x + 1$ (b) $y = -x^2 + 5x - 2$

(c) $y = -5x^3 + x^2 - x + 9$ (d) $y = 2x^4 - x^3 + x^2 - 5x + 4$

5. What is the slope of the graph of $y = x^3 - 2x^2 + x - 7$ at the point
    $(2,-5)$?

6. (a) Write a point-slope form of an equation of the line with slope
        $-\frac{4}{5}$ and containing the point $(1,3)$.

    (b) Write the slope-intercept $(y = mx + b)$ form of the equation in
        part (a).

7. (a) Find the slope of the line containing the points $(1,-2)$ and
        $(-3,0)$.

    (b) Find an equation of the line in (a) in slope-intercept form.

    (c) Find an equation of the line in (b) in $ax + by + c = 0$ form.

ANSWERS TO SLOPE SELF QUIZ

1. (a) $\frac{5}{7}$ (b) $\frac{y-2}{x+4} = \frac{5}{7}$ or $\frac{y-7}{x-3} = \frac{5}{7}$ (c) $y = \frac{5}{7}x + \frac{34}{7}$ (d) Check

2. No   3. (a) -1 (b) 0.14   (c) 8   (d) -1 (e) 4, -9, -0.7, 3

4. (a) $\frac{2}{3}$ (b) $-2x + 5$ (c) $-15x^2 + 2x$ (d) $8x^3 - 3x^2 + 2x - 5$

5. Slope = 5   6. (a) $\frac{y-3}{x-1} = -\frac{4}{5}$ (b) $y = -\frac{4}{5}x + \frac{19}{5}$

7. (a) $-\frac{1}{2}$ (b) $y = -\frac{1}{2}x - \frac{3}{2}$ (c) $2y + x + 3 = 0$

* * * SLOPE REVIEW * * *

1. Compute the slope of the line through the points
    (a) $(2,1)$ and $(3,4)$ (b) $(4,-5)$ and $(6,-7)$
    (c) $(5,-3)$ and $(6,-3)$ (d) $(2,7)$ and $(2,-5)$

2. For what lines is the slope
    (a) zero?   (b) undefined?

3. If (x_1,y_1) and (x_2,y_2) are points on a line, why does

$$\frac{y_1 - y_2}{x_1 - x_2} = \frac{y_2 - y_1}{x_2 - x_1} \ ?$$

4. Find a two-point form of an equation of the line containing the points

    (a) (-1,5) and (3,-4)   (b) (-2,3) and (6,-4)

5. Find the slope-intercept ($y = mx + b$) form of an equation of the line containing the points

    (a) (2,1) and (3,5)   (b) (-1,7) and (-4,-5)

6. How could you check each equation in Exercise 5?

7. If two lines have the same slope, then they are _____?_____.

8. In the $y = ax + b$ form of an equation of a line the coefficient a is the _____?_____ and the number b is the _____?_____.

9. Use the nearby-point method to find the slope of the curve

    $y = 2x^3 - 6x$ at the point (2,4).

10. Compute the slope at the point (x,y) for each of the following by using the appropriate slope formula.

    (a) $y = 7x - 9$ (b) $y = 5x^2 - 4x + 27$

    (c) $y = 7x^3 - 5x^2 + 3x - 10$ (d) $y = x^4 + x^3 - x^2 + x - 13$

11. Compute the slope when $x = -1$ for each part of Exercise 10.

12. Compute the slope at the point (-2,-5) for the graph of

    $y = x^4 - x^3 - 3x^2 + 5x - 7$.

13. Find the slope of the lines containing the points

    (a) (3,4) and (-1,2)   (b) (0,2) and (4,-3)

    (c) (2,1) and (3,-2)

14. Find an equation in slope-intercept form for each line in Exercise 13.

ANSWERS TO SLOPE REVIEW

1. (a) 3   (b) -1   (c) 0   (d) undefined

2. (a) horizontal line   (b) vertical line

3. Multiply either expression by $1 = \frac{-1}{-1}$ to get the other.

4. (a) $\frac{y + 4}{x - 3} = -\frac{9}{4}$ or $\frac{y - 5}{x + 1} = -\frac{9}{4}$

    (b) $\frac{y - 3}{x + 2} = -\frac{7}{8}$ or $\frac{y + 4}{x - 6} = -\frac{7}{8}$

5. (a) $y = 4x - 7$ (b) $y = 4x + 11$

6. Substitute the (x,y) values of each of two given points.

7. parallel   8. slope; y-intercept   9. 18

10. (a) 7   (b) $10x - 4$ (c) $21x^2 - 10x + 3$ (d) $4x^3 + 3x^2 - 2x + 1$

11. (a) 7   (b) -14 (c) 34   (d) 2   12. -27

13. (a) $\frac{1}{2}$ (b) $-\frac{5}{4}$ (c) -3

14. (a) $y = \frac{1}{2}x + \frac{5}{2}$ (b) $y = -\frac{5}{4}x + 2$ (c) $y = -3x + 7$

* * * SLOPE REVIEW SELF TEST * * *

1. (a) What is the slope of the line through the points $(2,-7)$ and $(-5,1)$?

(b) Write a two-point form of an equation of the line in (a).

(c) Write the equation in (b) in the slope-intercept $(y=mx+b)$ form.

(d) Where does the line cross the y-axis?

(e) Check that your answer for the slope in (a) agrees with your equation in (c).

(f) Check that the two points in (a) satisfy your equation in (c).

2. If two lines have the same _____?_____ , then they are parallel.

3. Consider the graph of $y = 2x^3 - 4x^2 + 3x - 5$. Where is the slope greater, at the point $(1,-4)$ or the point $(-1,-14)$?

4. At which points on the curve $y = 2x^3 - 9x^2 + 12x$ is the tangent parallel to the x-axis?

5. Find the slope of the line through the points $(1,0)$ and $(-2,3)$.

6. Find an equation of the line in Exercise 5 in point-slope form.

ANSWERS TO SLOPE REVIEW SELF TEST

1. (a) $-\frac{8}{7}$ (b) $\frac{y+7}{x-2} = -\frac{8}{7}$ or $\frac{y-1}{x+5} = -\frac{8}{7}$ (c) $y = -\frac{8}{7}x - \frac{33}{7}$

(d) $y = -\frac{33}{7}$ (e) Check   (f) Check   2. slope

3. $(-1,-14)$ 4. When $x = 1$ or 2, hence at $(1,5)$ and $(2,4)$

5. -1 6. $y = -x + 1$

* * * SLOPE PROJECTS * * *

1. Let a and b be two nonzero numbers. Show that the line

$$\frac{x}{a} + \frac{y}{b} = 1$$

crosses the x-axis at (a,0) and the y-axis at (0,b). (The
y-intercept is b; the "x-intercept" is a.) The equation

$$\frac{x}{a} + \frac{y}{b} = 1$$

is called the <u>intercept form</u> of the line.

2. It was asserted in the chapter that the line y = ax + b has
slope a. Prove that this is the case, following these steps.

(a) Consider two distinct values x_1 and x_2. Let P_1 be the point
on the line with x-coordinate x_1; let P_2 be the point on
the line with x-coordinate x_2. Show that

$$P_1 = (x_1, ax_1 + b) \quad \text{and} \quad P_2 = (x_2, ax_2 + b)$$

(b) Show that the slope determined by P_1 and P_2 is simply a.

3. A calculator or a four decimal cube root table gives the following:

x	2	1.8	1.2	1.1	1.01
$\sqrt[3]{x}$	1.2599	1.2164	1.0627	1.0323	1.0033

(a) Use the data in the table to graph $y = \sqrt[3]{x}$ for $1 \le x \le 2$.

(b) Estimate the slope of the tangent line to the curve at (1,1)
by drawing the tangent line "by eye" and measuring rise and
run with a centimeter ruler.

(c) 0.314
(rounded off)

(c) Estimate the slope of the tangent line at (1,1) using the
nearby point (1.2, 1.0627).

(d) Estimate the slope of the tangent line at (1,1) using the
nearby point (1.01, 1.0033).

(e) What do you <u>think</u> is the precise value of the slope of the
tangent line at (1,1)?

4. A calculator or a four decimal table of reciprocals gives the following:

x	2.4	2.3	2.2	2.1
$\frac{1}{x}$	0.4167	0.4348	0.4545	0.4762

(a) Graph $y = \frac{1}{x}$ for $1 \leq x \leq 2.4$, using the above data.

Estimate the slope of the tangent line at $P = (2,\frac{1}{2})$ using P and as the nearby point:

(b) $Q = (2.4, 0.4167)$ (c) $Q = (2.3, 0.4348)$

(d) $Q = (2.2, 0.4545)$ (e) $Q = (2.1, 0.4762)$

(f) What do you <u>think</u> the slope of the tangent line at $(2, \frac{1}{2})$ is precisely?

(c) -0.2173

(e) -0.238

5. This Project concerns the slope formula for the curve $y = \frac{1}{x}$ for any $x \neq 0$.

A small amount of simple algebra will be needed.

(a) Let $P = (x, \frac{1}{x})$ be a fixed point on the curve. Let $Q = (x + h, \frac{1}{x + h})$ be the typical nearby point. Show that the slope of the line through P and Q is

$$-\frac{1}{x(x + h)}$$

(b) From (a) deduce that the slope formula for $y = \frac{1}{x}$ is $-\frac{1}{x^2}$.

(c) Use (b) to obtain the precise answer for the slope of the tangent line in Project 4.

6. This Project concerns the slope formula for the curve $y = x^{\frac{1}{3}}$.

Let $P = (x, x^{\frac{1}{3}})$ be a fixed point on the curve and $Q = (x + h, (x + h)^{\frac{1}{3}})$ be the typical "nearby" point.

(a) Show that the slope of the line through P and Q is

$$\frac{(x + h)^{\frac{1}{3}} - x^{\frac{1}{3}}}{h}$$

(b)   Multiply numerator and denominator by

$$(x + h)^{\frac{2}{3}} + (x + h)^{\frac{1}{3}} x^{\frac{1}{3}} + x^{\frac{2}{3}},$$

and show that the quotient in (a) equals

This procedure resembles rationalizing an expression that has a square root.

$$\frac{1}{(x + h)^{\frac{2}{3}} + (x + h)^{\frac{1}{3}} x^{\frac{1}{3}} + x^{\frac{2}{3}}}$$

(c)   Use (b) to show that as h approaches 0, the quotient in (a) approaches

The general rule is proved in a calculus course.

$$\frac{1}{3x^{\frac{2}{3}}}$$

By (c) the slope formula for $y = x^{\frac{1}{3}}$ is $\dfrac{1}{3x^{\frac{2}{3}}}$ or $\dfrac{1}{3} x^{-\frac{2}{3}}$

This is another instance of the general rule that $y = x^a$ has slope formula ax^{a-1}.

(d)   Use the formula in (c) to find the slope of the tangent line in Project 3.

The remaining Projects show how the slope formula is applied in practical problems.

7.   This Project concerns the graph of

$$y = 3x^2 + 2x + 5$$

(a)   Copy and fill in this table:

x	-2	-1	0	1	2
$3x^2 + 2x + 5$					

(b)   Plot the five points in (a) and use them to graph the parabola $y = 3x^2 + 2x + 5$.

The next Project continues this one.

(c)   By inspecting the graph you made in (b) estimate the x-coordinate of the lowest point on the graph.

8.   At the lowest point on the graph of Project 7 the tangent line is horizontal. Thus the slope must be zero at the lowest point.

(a)   What is the slope formula for the curve $y = 3x^2 + 2x + 5$?

(b) For what value of x does the slope formula give the value zero?

(c) Compare the answer to (b) with the answer to Project 7(c).

9. (a) Graph the curve $y = 5x^2 - 3x + 2$.

(b) Use the slope formula to find the x-coordinate of the lowest point on the graph in (a).

(b) $x = \frac{3}{10}$

10. (a) Graph the curve $y = -3x^2 + x + 1$.

(b) Use the slope formula to find the x-coordinate of the highest point on the graph in (a).

(b) $x = \frac{1}{6}$

11. This Project concerns the curve $y = x^3 - 3x$.

(a) Copy and fill in this table.

x	-2	-1	0	1	2
$x^3 - 3x$					

(b) Using the five points in (a), graph the curve $y = x^3 - 3x$.

(c) The graph in (b) has one summit and one valley. By inspection of the graph, estimate the x-coordinate of the summit and also the x-coordinate of the bottom of the valley.

(d) Using the slope formula, find the precise answers to (c).

The graph is shaped something like the letter N with corners rounded.

12. Carry out the analog of Project 11 for the curve $y = x^3 - x$.

13. Carry out the analog of Project 11 for the curve

$$y = 4x^3 - 18x^2 + 20x.$$

12. Summit when $x = \frac{-\sqrt{3}}{3}$, bottom when $x = \frac{\sqrt{3}}{3}$

13. Summit when $x = \frac{9 - \sqrt{21}}{6}$, bottom when $x = \frac{9 + \sqrt{21}}{6}$

14. A tray is to be made out of a rectangular sheet of metal of dimensions 4 feet by 5 feet by removing four equal squares, one from each corner and folding up the flaps.

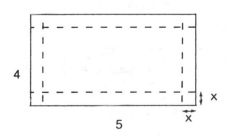
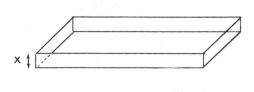

This problem suggests two questions:

What is the largest volume that such a tray can have?

What must be the length of the sides of the cut-out squares?

(a)   Show that if the side of the cut-out square is x, then for all trays x is between 0 and 2.

(b)   Show that if the sides of the cut-out squares have length x, then the the area of the base of the tray is $(5 - 2x)(4 - 2x)$ and that the height of the tray is x.

(c)   Show that the tray in (b) has volume

$$y = 4x^3 - 18x^2 + 20x.$$

(d)   Using Project 13, show that the cut that corresponds to the largest tray is $x = \dfrac{9 - \sqrt{21}}{6}$ (feet).

15.   Show that the slope formula for $y = x^5$ is $5x^4$. (Use the nearby-point method.)

16.   Show that the slope formula for $y = \dfrac{1}{x^2}$ is $-\dfrac{2}{x^3}$. (Use the nearby-point method.)

* * *

The following is a definition of variation.

> VARIATION
>
> Let k be a nonzero constant.
>
> 1.   We say y varies <u>directly</u> as x varies if
> $$y = kx.$$
>
> 2.   We say y varies <u>inversely</u> as x varies if
> $$y = \frac{k}{x} .$$
>
> 3.   We say y varies <u>jointly</u> as x and z vary if
> $$y = kxz.$$

17.   (a)   Suppose y varies directly as x varies and that x = -3 when y = 7. Find k.

(b)   Suppose y varies inversely as x varies and that x = 9 when $y = -\dfrac{2}{3}$. Find k.

(c)   Suppose y varies jointly as x and z vary and that y = -2 when $x = \dfrac{3}{4}$ and z = 6. Find k.

18. Refer to each comparable part of Exercise 17 for the value of k.

(a) Find y when x = $\frac{3}{7}$

(b) Find y when x = -5

(c) Find y when x = 2 and z = -1.

* * * CENTRAL SECTION * * *

A person on a pier is pulling in a fish line. A fish, swimming at the surface of the water, is caught at the end of the line. When the person reels in one foot of line, does the fish move one foot also? Less than one foot? More than one foot?

The authors give credit to Prof. Bill Leonard, CSU Fullerton, for this great problem!

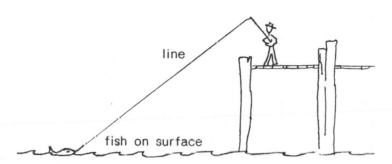

line

fish on surface

Think about this. At the end of this central section we will return to this problem.

DO NOT GIVE UP TOO SOON IN SOLVING THIS PROBLEM. Think about it over a period of time.

"LESS THAN" AND "GREATER THAN"

When x and y are numbers, and x is to the left of y on the number line, we say "x is less than y" and write $x < y$. We also say y is to the right of x, or "y is greater than x," and write $y > x$.

Many people prefer "<" rather than ">" because it may be referred to the number line, where the smaller number is on the left.

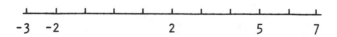

x   y

$x < y$ or $y > x$

Thus, $-3 < -2$, $2 < 5$, and $7 > -3$, as illustrated in the following figure.

-3 -2   2   5   7

The assertion "x is less than or equal to y" is written $x \leq y$. We can write $3 < 4$, $3 \leq 4$, and $4 \leq 4$. The assertion "y is greater than or equal to x" is denoted $y \geq x$. For instance, the assertion, "If x is a positive number, then so is its square, x^2," is abbreviated to "$x > 0$ implies $x^2 > 0$." Similarly, the claim, "The square of a number is positive or 0" is abbreviated to "For all x, $x^2 \geq 0$." Also, we may write "5 is <u>not</u> less than 4" as 5 $\not<$ 4, where the slash is short for "not."

To some extent, inequalities behave just like equalities. But, in several respects, they do not. First, we will review the rules that are similar to those for equalities.

RULES FOR INEQUALITIES

STUDY CAREFULLY THE RULES FOR INEQUALITIES. Any of the rules may be checked by starting with an obvious example, say -1 < 5.

1. If a $\leq$ b and c $\leq$ d, then a + c $\leq$ b + d. (It is safe to add two inequalities that are in the same direction: both $\leq$ or both $\geq$.)

    For example,

    $$\text{since } 2 < 4 \text{ and } -7 < 1,$$
    $$\text{then } 2 + (-7) < 4 + 1$$
    $$\text{or} \qquad -5 < 5.$$

2. If a $\leq$ b and p is a <u>positive</u> <u>number</u>, then pa $\leq$ pb and $\frac{a}{p} \leq \frac{b}{p}$. (It is safe to multiply or divide both sides of an inequality by the same <u>positive</u> number.)

    For example,

    $$\text{since } -3 < 5,$$
    $$\text{then } \frac{-3}{7} < \frac{5}{7} .$$
    $$\text{Also, } (-3)(2) < (5)(2),$$
    $$\text{or} \qquad -6 < 10.$$

3. If a $\leq$ b and c is any number, positive, negative, or 0, then a + c $\leq$ b + c and a - c $\leq$ b - c. (It is safe to add or subtract the same number from both sides of an inequality.)

    For example,

    $$\text{since} \qquad -3 \leq 4,$$
    $$\text{then } -3 + (-7) \leq 4 + (-7)$$
    $$\text{or} \qquad -10 \leq -3.$$
    $$\text{Also, } -3 - (5) \leq 4 - (5)$$
    $$\text{or} \qquad -8 \leq -1$$

But there are further important rules for inequalities that are quite different from those for equalities.

The first three rules are easy to accept. You must be very <u>aware</u> of the <u>fourth rule</u>.

4. If a $\leq$ b and n is a negative number, then na $\geq$ nb and $\frac{a}{n} \geq \frac{b}{n}$.

    (Multiplying or dividing both sides of an inequality by the same negative number <u>reverses</u> the <u>direction</u> of the inequality.)

For example, if we multiply both sides of the inequality

$$-2 < 4$$

by the negative number, -3, we <u>reverse</u> the direction of the inequality.

$$(-3)(-2) > (-3)(4)$$

$$\text{or} \quad 6 > -12$$

(Indeed, 6 is to the right of -12 on the number line, hence 6 is larger than -12.)

We now use these ideas to show how you may solve for a variable in a statement of inequality.

Example 1.   If $2x - 3 \le 5x - 7$, what possible values can x have?

Solution. $2x - 3 \le 5x - 7$ (given)

$\qquad\qquad\quad -3 \le 3x - 7$ (subtract 2x from both sides)

$\qquad\qquad\quad\ 4 \le 3x$ (add 7 to both sides)

$\qquad\qquad\quad \dfrac{4}{3} \le x$ (divide both sides by 3)

Thus, $\dfrac{4}{3} \le x$. (Moreover, the steps are reversible. So, if $\dfrac{4}{3} \le x$, then $2x - 3 \le 5x - 7$.) The set of solutions of the original inequality consists of all numbers such that $\dfrac{4}{3} \le x$, as sketched on the x-axis:

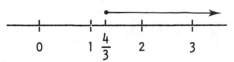

$$0 \qquad 1 \ \tfrac{4}{3} \quad 2 \qquad 3$$

(The solid dot at $\dfrac{4}{3}$ indicates that $\dfrac{4}{3}$ is included.)

Example 2.   Sketch all numbers x such that $-2x > x + 3$.

Solution. $-2x > x + 3$ (given)

$\qquad\qquad\quad -3x > 3$ (subtract x from both sides)

$\qquad\qquad\quad \dfrac{-3x}{-3} < \dfrac{3}{-3}$ (dividing both sides by the same

$\qquad\qquad\qquad\qquad\qquad\qquad$ negative number <u>reverses</u> the

$\qquad\qquad\quad\ x \ < -1$ inequality)

Thus, x can be any number to the left of -1 on the x-axis:

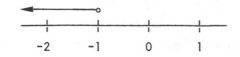

The hollow dot at -1 indicates that -1 is not a solution.

EXERCISES

1. In each case, solve the inequality. That is, determine all x that
    satisfy the inequality and sketch the set of solutions on the number
    line.

(a) $\frac{1}{3} \le x$

(c) $-3 \le x$

(e) $10 < x$

(a)   3x + 5 $\ge$ 6   (b)   5x - 2 < 4

(c)   2x - 7 $\le$ 6x + 5   (d)   3x + 8 > 5x - 2

(e)   x + 8 < 3x - 12   (f)   -5x $\le$ 2x - 21

$$* \quad * \quad *$$

Sometimes the possible values of x may be described by two inequalities simultaneously. For instance, the inequality,

$$2 < x < 5,$$

The expression
$2 < x < 5$ may also
be written
$2 < x$ and $x < 5$

asserts that 2 < x and x < 5 simultaneously. In words,

"2 is less than x and x is less than 5."

In short, x is between 2 and 5, but cannot equal 2 or 5:

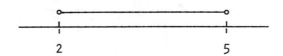

Example 3. Sketch all x such that -x < -12 and 3x < 60.
Solution. Multiplying the inequality -x < -12 by -1 gives x > 12, which
is 12 < x. Dividing the inequality 3x < 60 by 3 yields x < 20. Thus,
12 < x and x < 20. In short, we have 12 < x < 20:

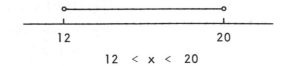

$$12 < x < 20$$

NOTE: Using < in inequalities (rather than >) may be easier, since the
       smaller number is on the left, just as it is on the number line.

2. Sketch the numbers x that satisfy each of the following inequalities, then describe the solutions algebraically.

(a) $3 + x < 4x < 8$ (This is short for the two inequalities
$$3 + x < 4x \quad \text{and} \quad 4x < 8.)$$

(b) $5x - 2 \leq 7x + 3 \leq 4x + 9$ (c) $4x - 3 < 6x + 2 < 3x + 8$ *(c)* $-\frac{5}{2} < x < 2$

(d) $-6 + x > -2x > -10$ (e) $x + 5 < 2x + 3 < 10$ *(d)* $2 < x < 5$

3. Sketch the numbers that satisfy each inequality, then describe the solution algebraically.

(a) $4 - x > 2x + 5 > -3x - 9$ (b) $3 - 2x < 2 \leq 10 - x$ *(a)* $-\frac{14}{5} < x < -\frac{1}{3}$

(c) $x - 5 \leq 3x \leq -2x + 3$ (d) $6x - 2 \leq 8x + 3 \leq 5x + 9$ *(c)* $-\frac{5}{2} \leq x \leq \frac{3}{5}$

(e) $1 - x < x - 2 < -2x + 10$ (f) $15 \geq 3x + 4 > x + 2$ *(e)* $\frac{3}{2} < x < 4$

*   *   *

ABSOLUTE VALUE

The absolute value of a number describes its size. If $x \geq 0$, the absolute value of x, denoted $|x|$, is simply x itself. But if $x < 0$, the absolute value of x is -x. Thus $|x| \geq 0$ for all numbers x. In short, $|x|$ tells the distance that x is from 0 on the number line. Observe that

$$|x| = x \text{ if } x \geq 0,$$

$$\text{but } |x| = -x \text{ if } x < 0.$$

For instance, $|5| = 5$ and $|-7| = 7$, since 5 is 5 units from zero and -7 is 7 units from zero on the number line.

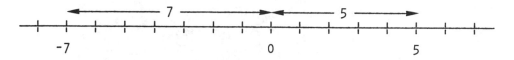

The inequality $|x| \leq 3$ tells that x is somewhere between -3 and 3, inclusive:

The number x is 3 or less units from zero.

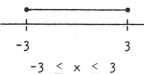

The inequality $|x| > 5$ describes all numbers x such that $x > 5$ or $x < -5$:

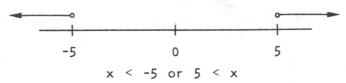

The number x is more than 5 units from zero.

$$x < -5 \text{ or } 5 < x$$

The set of possible solutions consists of two separate infinite rays.

An inequality involving absolute values can be phrased in terms of inequalities in which absolute values do not appear.

Example 4.   Sketch the set of numbers x such that $|x - 3| \leq 4$.

Solution. We have immediately,

$$-4 \leq x - 3 \leq 4.$$

To get x by itself, we add 3 throughout, obtaining

$$-4 + 3 \leq x \leq 4 + 3$$
$$\text{or} \quad -1 \leq x \leq 7.$$

So the set of solutions consists of all x from -1 through 7 inclusive:

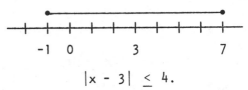

$$|x - 3| \leq 4.$$

Thus, we may also say that the set of solutions consists of "all x that are within a distance of 4 units from the number 3." Note that 3 is the midpoint of the interval from -1 to 7.

4.   Sketch the set of solutions of these inequalities.

(a) $-3 \leq x \leq 3$

(c) $-2 \leq x \leq 2$

(a) $|3x| \leq 9$ (b) $|-2x| \leq 8$

(c) $|x| \leq 2$ (d) $|3x + 5| \leq 8$

REMARK: In solving inequalities involving absolute values, it is useful to find those x that make the inequality an equality. These become useful guidepoints or markers for deciding where the inequality is true and where it is false. The numbers at which the equality holds are called <u>critical numbers</u>.

STUDY THIS SECTION CAREFULLY

Example 5. Sketch the set of solutions of $|2x - 3| \leq 7$.

Solution. First solve the equation $|2x - 3| = 7$, which holds either when

$$2x - 3 = 7 \qquad \text{or when} \qquad 2x - 3 = -7$$

We have then

$$2x - 3 = 7 \quad \text{or} \quad 2x - 3 = -7$$
$$2x = 10 \qquad\qquad 2x = -4$$
$$x = 5 \qquad\qquad x = -2$$

The critical numbers are 5 and -2, as shown in the diagram.

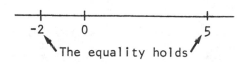

The equality holds

at these critical numbers.

To determine what happens between -2 and 5, pick any convenient number in that interval, say 0. Does the inequality $|2x - 3| \leq 7$ hold when $x = 0$? In other words, is $|2(0) - 3| \leq 7$? That is, is $|-3| \leq 7$? Yes, since $3 \leq 7$. The inequality $|2x - 3| \leq 7$ then holds for all x such that $-2 \leq x \leq 5$.

   What if $5 < x$ or $x < -2$? To find out, test convenient values of x. When $x = 10$, say, is $|2x - 3| \leq 7$? That is, is $|2(10) - 3| \leq 7$? In other words, is $|17| \leq 7$? No. The inequality is then false for all $x > 5$.   Similarly, it is false for $x < -2$. (Pick a convenient $x < -2$ to check this statement.) So the solutions form the interval from -2 to 5 inclusive:

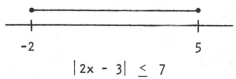

$$|2x - 3| \leq 7$$

Example 6. The inequality $|x - 1| > 4$ behaves quite differently. Sketch the set of x's for which it is true.

Solution. There are two cases: $x - 1 \geq 0$ or $x - 1 < 0$. In the first case, $|x - 1| = x - 1$. In the second case, $|x - 1| = -(x - 1)$.

The first case: $x - 1 > 4$,

then $x > 5$.

The second case: $-(x - 1) > 4$,

then $x - 1 < -4$ (multiply by -1)

or $x < -3$ (add 1 to both sides)

So, there are two ways for the inequality $|x - 1| > 4$ to hold. Either $x > 5$ or $x < -3$:

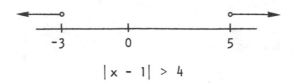

$$|x - 1| > 4$$

REMARK: There is another way of looking at Example 6.   The assertion "$|x - 1| > 4$" means "the distance from x to 1 is greater than 4." So x must be more than 4 units to the right of 1 or more than 4 units to the left of 1.

5.   In each case, sketch the set of solutions, if there are any.

(a) $|3x| \leq 9$ (b) $|x| \geq -1$

(c) $|2x + 5| \leq 11$ (d) $|x + 2| > 3$

(e) $|x| \leq -1$ (f) $|2x + 1| > 4$

(g) $|2x - 4| > 6$ (h) $|2x - 3| \leq 5$

(i) $|x| \geq 1$ (j) $|x| < 3$

(a) $-3 \leq x \leq 3$
(c) $-8 \leq x \leq 3$
(g) $x < -1$ or $x > 5$
(i) $x \leq -1$ or $1 \leq x$

* * *

Another type of inequality in which the same method of solution is useful involves quadratic polynomials.   For instance,

$$x^2 - x - 6 > 0$$
$$\text{or}\ (x - 3)(x + 2) > 0.$$

The next example is of this type.

Example 7.   Graph all x such that $(x - 3)(x + 2) > 0$.   Describe the solutions algebraically.

Solution.   Consider first the equation $(x - 3)(x + 2) = 0$.   Then

$$x - 3 = 0 \qquad \text{or} \qquad x + 2 = 0$$
$$\text{and} \qquad x = 3 \qquad \text{or} \qquad x = -2.$$

Hence, the critical numbers are 3 and -2.   Now, any x that satisfies $(x - 3)(x + 2) > 0$ cannot be one of these values.   So, indicate those two critical numbers by hollow dots:

Next, test a number to the left of -2, say, x = -10. Does the inequality
(x - 3)(x + 2) > 0 hold for x = -10? Let us see:

$$(x - 3)(x + 2) > 0$$

$$(-10 - 3)(-10 + 2) \overset{?}{>} 0$$

$$(-13)(-8) \overset{?}{>} 0 \qquad \text{Yes.}$$

Next, test a number between -2 and 3, say, choose x = 0. Again, substitute
in the inequality (x - 3)(x + 2) > 0:

$$(0 - 3)(0 + 2) \overset{?}{>} 0 \qquad \text{No.}$$

Finally, test a number to the right of 3, say x = 10. We ask,

$$(10 - 3)(10 + 2) \overset{?}{>} 0 \qquad \text{Yes.}$$

Then, we combine this information to graph the possible solutions of the
inequality (x - 3)(x + 2) > 0:

In short,

$$x < -2 \qquad \text{or} \qquad 3 < x.$$

The word "or" in this description tells us that x satisfies at least
one of the two inequalities. (Note that it <u>cannot</u> satisfy both at the same
time.)

Example 8. Graph the set of numbers x that satisfy the following inequality, then describe the solutions algebraically.

$$x^2 \le x + 6$$

Solution.

$$x^2 - x - 6 \le 0 \qquad \text{(subtract x + 6 from both sides)}$$

$$(x - 3)(x + 2) \le 0 \qquad \text{(factor)}$$

The critical numbers are found to be 3 and -2, as in the previous example.
Testing numbers in the same manner as before, we see that the solutions
form the interval $-2 \le x \le 3$:

$$-2 \le x \le 3$$

The solid dots show that the numbers -2 and 3 are included in the solutions.

6.   Sketch the solutions and describe them algebraically.

(a) $x^2 + 1 \leq 2x$

(b) $x^2 + 2x > 3$

(c) $x^2 + 2 > 3x$

(d) $2x^2 < 1 - x$

(e) $(x - 2)(x + 2) < 5$

(f) $(x - 3)(x + 3) > 7$

(g) $(2x + 1)(x - 2) > 0$

(h) $2x^2 + x > 3$

(a) $x = 1$
(c) $x < 1$ or $2 < x$
(e) $-3 < x < 3$

(g) $x < -\frac{1}{2}$ or $2 < x$

*   *   *

GRAPHING INEQUALITIES IN THE PLANE

Inequalities involving both x and y as variables lead us to graphs in the plane.

Example 9.   Graph (a) $x + y = 0$

(b) $x + y > 0$

(c) $x + y < 0$.

Solution.   (a) $x + y = 0$ is the straight line shown in this figure.

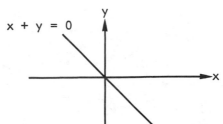

We use (a) to solve (b) and (c).   The line $x + y = 0$ cuts the plane into two regions, called "half planes."   In one of these, $x + y < 0$.   In the other, $x + y > 0$.   To decide which half plane belongs to which inequality, pick any point in the plane that is not on the line $x + y = 0$.   We can choose (1,1), for instance (or (-2,0), or (0,5), etc.).   Observe in which of the two half planes determined by the line $x + y = 0$ the point (1,1) lies.   It lies in the upper one.   Then check which inequality it satisfies, $x + y > 0$ or $x + y < 0$.   Substituting the point (1,1), we see that it satisfies the inequality $x + y > 0$, since $1 + 1 > 0$.   Thus, the <u>upper</u> half plane corresponds to $x + y > 0$.   The lower half plane corresponds to $x + y < 0$.

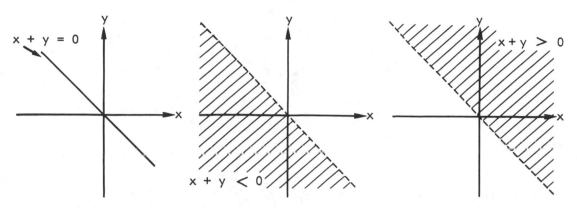

Note that in the two figures to the right, the reference line $x + y = 0$ is shown as a <u>dashed</u> line. In those last two cases, if the inequality had been $x + y \leq 0$ or $x + y \geq 0$, the reference line would have been drawn solid, to indicate its inclusion.

7. Graph the inequalities:

   (a) $y - 2x > 0;$ $y - 2x < 0$ (b) $2x + 3y > 0;$ $2x + 3y < 0$

   (c) $x + y > 1;$ $x + y < 1$ (d) $2x - 3y > 0;$ $2x - 3y < 0$

   (e) $2x - 3y > 6;$ $2x - 3y < 6$

8. Graph in the plane:

   (a) $y > 2x + 3$ (b) $y \geq -\frac{1}{2}x + 1$

   (c) $3x - 4y \leq 12$ (d) $x > 3y + 3$

   (e) $x > 0$ (f) $y \leq -2$

* * *

The next few exercises illustrate the kind of problems met in "linear programming," which deals with solutions of simultaneous inequalities.
Example 10. Sketch all points in the plane that meet the three following conditions simultaneously:

$$x \leq 0, \quad y \geq 0, \quad \text{and} \quad y < x + 4$$

Solution. The set of solutions consists of all points below the line $y = x + 4$, that are on or to the left of the line $x = 0$, and that are on or above the line $y = 0$. In other words, the solutions are all points inside the right triangle ABC shown in the diagram on the next page, together with all points on its border except points on its hypotenuse AB.

below this line   on or left of this line

A

B

C

on or above this line

9. In each case, sketch the set of points (x,y) in the plane that meet the three conditions simultaneously.

   (a) $x \geq 0$, $y \geq 0$, $2x + y \leq 2$

   (b) $x \geq 0$, $y \geq 1$, $x + y \leq 7$

   (c) $x \geq 0$, $y \geq x$, $x + 2y \leq 2$

   (d) $x \geq 1$, $y \geq 1$, $2x + 3y \leq 6$

10 In each case, sketch the set of points (x,y) in the plane that meet the conditions simultaneously.

   (a) $y \geq x$, $y \leq 2x$, $y \leq 3$

   (b) $x \geq 0$, $y \geq 0$, $y \leq 3 - x$, $y \geq x$

   (c) $|x| \leq 1$, $|y| \leq 2$

   (d) $y - x + 1 \geq 0$, $x \geq 1$, $y \leq -x + 2$

* * *

From geometry, recall the triangle inequality:

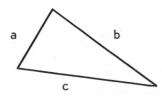

a

b

c

"The sum of two sides of a triangle is greater than the third side."

$c < a + b$

11. Let us return to the problem of the fish. The following figure shows the initial position of the fish and its position after one foot of line is reeled in. Now, when one foot of line is reeled in, does the fish move one foot also? Less than one foot? More than one foot?

Let L equal the original amount of line.

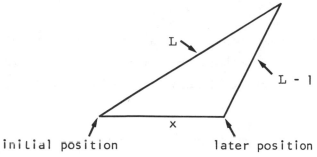

$*$ $*$ $*$ SUMMARY $*$ $*$ $*$

This chapter concerns the concepts "less than", denoted " < ", and "greater than", denoted " > ". For instance, x < y indicates that number x is left of number y on the number line. The relation x $\leq$ y permits x < y or x = y.

Inequalities resemble equalities to some extent. However, care must be exercised when multiplication or division is involved, especially if by a negative number.

These rules hold:

1. If a $\leq$ b and c $\leq$ d, then a + c $\leq$ b + d.

2. If a $\leq$ b and p > 0, then pa $\leq$ pb.

3. If a $\leq$ b, then a + c $\leq$ b + c and a - c $\leq$ b - c.

4. If a $\leq$ b and n < 0, then na $\geq$ nb and $\frac{a}{n} \geq \frac{b}{n}$. (Note the reverse direction of inequality after multiplying or dividing by a negative number.)

These rules enable us to solve inequalities. The set of solutions (if there are any) may occupy large parts of the number line. In finding these parts, it is helpful to use critical numbers, where the corresponding equation holds. The solid dot ($\bullet$) and hollow dot ($\circ$) help in sketching the set of solutions.

For a number x, whether positive, negative, or 0, $|x|$, the absolute value of x, is a number that describes its distance from the origin. (Formally: If x $\geq$ 0, $|x|$ = x; but if x < 0, $|x|$ = -x.) To solve inequalities involving absolute values, translate them into inequalities without absolute values.

Inequalities involving both x and y are graphed in the plane instead of on the number line. The methods are similar to those in the case of one variable. The key technique is considering the line where equality holds.

*   *   * WORDS AND SYMBOLS *   *   *

< ("less than")   inequality
> ("greater than")   absolute value
≤ ("less than or equal")   triangle inequality
≥ ("greater than or equal")   graphing inequalities
≮ ("not less than")
≯ ("not greater than")

*   *   * CONCLUDING REMARKS *   *   *

The most famous inequality was first studied in the time of the Greeks. It is called the "isoperimetric inequality" (iso = same, perimeter = perimeter). It says that the largest area A that can be surrounded by a loop of string of length L satisfies the inequality

$$A \leq \frac{L^2}{4\pi}$$

(For a circle of perimeter L, the inequality becomes an equality.) So, the circle is the largest area that can be surrounded by a loop of a given length. Or, to put it another way, of all regions of a given area, the circle has the smallest perimeter.

Similarly, it can be shown that of all solids with a given volume, the sphere has the least surface area. This is why raindrops are spherical and why houses should be as close to spherical as possible to save on heating and cooling.

Entire books are written on inequalities. Here is a simple one you might like to experiment with. Take any four positive numbers, a, b, c, d. How small can you make the expression $\frac{a}{b} + \frac{b}{c} + \frac{c}{d} + \frac{d}{a}$? When is the sum as small as possible?

* * * CENTRAL SELF QUIZ * * *

1.   State an inequality satisfied by x.

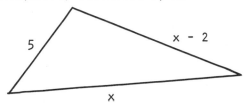

2. Sketch the graph on the number line and describe the solutions algebraically.

(a) $-3x + 7 > 9$ (b) $-x \leq -10$ and $2x < 50$

(c) $2 + x < 4x \leq 6$ (d) $-4 + x > -3x > -9$

(e) $|x - 2| \leq 3$ (f) $|5x| \geq 10$

(g) $|2x - 1| < 5$ (h) $x^2 + 4 > 5x$

3. Graph in the plane:

(a) $2x - 3y < 6$ (b) $2x - 3y > 6$ and $2x < y$

ANSWERS TO SELF QUIZ

1. $x > \frac{7}{2}$

2. (a) $x < -\frac{2}{3}$ (b) $10 \leq x < 25$ (c) $\frac{2}{3} < x \leq \frac{3}{2}$

(d) $1 < x < 3$ (e) $-1 \leq x \leq 5$ (f) $x \leq -2$ or $2 \leq x$

(g) $-2 < x < 3$ (h) $x < 1$ or $4 < x$

3. (a)   (b)

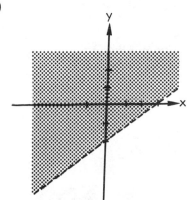

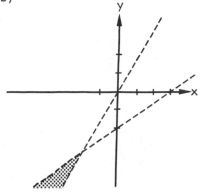

* * * INEQUALITIES REVIEW * * *

1. My farm consists of all land that is east of Highway 99, north of
    Road 707, and on the south side of Clearwater Canal.

    [Highway 99 runs N-S and
    Road 707 runs E-W. They
    intersect at Yoville.
    Clearwater Canal inter-
    sects Highway 99 four
    miles north of Yoville
    and Road 707 three miles
    east of Yoville.]

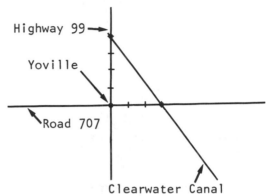

    Describe my farm using the (x,y)-coordinate system.

2. In each case below, use the triangle inequality to find an inequality
    satisfied by x. Solve for x.

    (a)   (b)   (c)

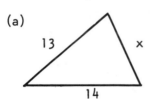
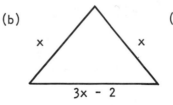
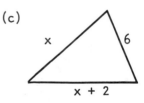

3. Sketch the solutions on the x-axis and describe them algebraically.

    (a) $-7x + 4 \leq 11$ (b) $|x| > 6$
    (c) $|x + 1| < 3$ (d) $-x \leq 8$ and $5x < 25$
    (e) $|x| > -2$ (f) $|x| \leq -5$
    (g) $(x - 3)(x + 1) < 0$ (h) $(2x - 3)(x + 2) \geq 0$
    (i) $-5 + x > -2x > -10$ (j) $|2x + 3| > 4$
    (k) $|x - 3| \leq 2$ (l) $x^2 > x + 2$

4. Graph in the (x,y)-plane.

    (a) $y < 3$ (b) $x \geq -2$

(c) $y < 3$ and $x \geq -2$ (d) $|x| < 1$

(e) $|y| > -3$ (f) $x + y > -2$ and $x + y < 2$

(g) $|x| < 2$ and $|y| \leq 3$

ANSWERS TO INEQUALITIES REVIEW

1. $3y + 4x < 12;$ $x > 0;$ $y > 0$

2. (a) $x > 1$ (b) $x < 2$ (c) $x > 2$

3. (a) $x \geq -1$ (b) $x < -6$ or $6 < x$ (c) $-4 < x < 2$

   (d) $-8 \leq x < 5$ (e) all real numbers x   (f) no x

   (g) $-1 < x < 3$ (h) $x \leq -2$ or $\frac{3}{2} < x$ (i) $\frac{5}{3} < x < 5$

   (j) $x < -\frac{7}{2}$ or $\frac{1}{2} < x$ (k) $1 < x < 5$ (l) $x < -1$ or $2 < x$

4. (a)
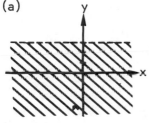

   (b)
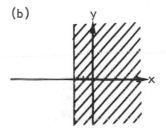

   (c)
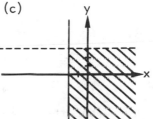

   (d)
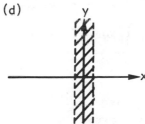

   (e) all points in
        the plane

   (f)
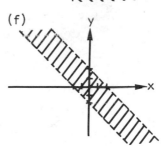

   (g)
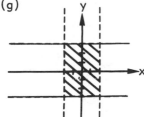

* * *   INEQUALITIES REVIEW SELF TEST   * * *

1. Suppose in the opening fish problem that the fish stays on the sur-
    face of the water, but <u>pulls</u> one foot of line from the reel. Does
    the fish move through the water less than one foot? Exactly one
    foot? More than one foot?

2. Solve for all possible x:

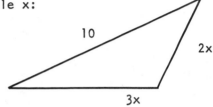

3. Sketch the graph and write the algebraic solution.

    (a) $|x - 7| > 5$ (b) $-6 + x < -3x < -12$

    (c) $2x^2 > x + 1$ (d) $(x - 5)(x + 6) < 0$

4. Graph the simultaneous inequalities, $x + y < 2$; $x - y < 2$;
    $x + y > -2$; $x - y > -2$.

ANSWERS TO INEQUALITIES REVIEW SELF TEST

1. More than one foot. Why?

2. $2 < x < 10$

3. (a) $x < 2$ or $12 < x$ (b) No solution. Why?

    (c) $x < -\frac{1}{2}$ or $1 < x$ (d) $-6 < x < 5$

4.

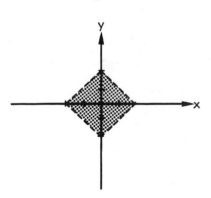

* * *   INEQUALITIES PROJECTS   * * *

1. (a) Why is $(2x - 3)^2 \geq 0$ for all x?

    (b) Deduce that $4x^2 \geq 12x - 9$ for all x.

2. (a) Is $x^2 + 9 \geq 5x$ for all x?

    (b) Is $x^2 + 9 \geq 7x$ for all x?

    (c) Is $x^2 + 9 \geq 6x$ for all x?

3. For all positive numbers x, then $x + \frac{1}{x} > 0$. How small can $x + \frac{1}{x}$ be?

    (a) Experiment with various choices for x.

    (b) Make a conjecture.

    (c) Prove it.

4. Let $x \geq 0$ and $y \geq 0$.

    (a) For various choices of x and y, compare the quantities $\frac{x + y}{2}$ and $\sqrt{xy}$.

    (b) Make a conjecture.

    (c) Prove it.

    (Incidentally, $\frac{x + y}{2}$ is called the <u>arithmetic mean</u> of x and y; $\sqrt{xy}$ is the <u>geometric mean</u> of x and y.)

    (d) See another proof on Page 442 in <u>Geometry</u>: <u>A Guided Inquiry</u> by Chakerian, Crabill, and Stein; Sunburst, 1986.

5. (a) Give an example of an inequality that has no solutions.

    (b) Give an example of an inequality that has exactly <u>one</u> solution.

6. A triangle has the vertices (1,1), (2,2), and (3,1). Describe the set of points in this triangle by means of three simultaneous inequalities involving the (x,y) coordinate system.

7. (a) Sketch the set S of solutions of these four simultaneous inequalities: $x \geq 0$, $y \geq 0$, $x + y \leq 2$, $x + 3y \leq 4$.

    (b) What is the largest possible value of $5x + 5y$ for points (x,y) in S? (Hint: Experiment with a few points in S.)

    NOTE: This is a "linear programming" problem. In such a problem, which in practice usually involves hundreds or thousands of variables, one wishes to maximize one expression, when the variables are subject to many inequalities.

8. Is $x^2 - 4xy + 5y^2$ ever negative?

(a) Experiment with various choices of x and y.

(b) Analyze the problem by using the fact that
$$(x - 2y)^2 = x^2 - 4xy + 4y^2$$

9. Graph on the number line and solve for x:

(a) $1 < x^2 < 9$ (b) $-2 \le x^2 - 3 < 13$

(c) $(x - 1)(x + 2)(x - 3) < 0$

10. Graph on the number line and solve for x:

(a) $\dfrac{x - 1}{x + 4} > 0$ (b) $\dfrac{x + 3}{x - 2} < 0$ (c) $\dfrac{1}{x - 1} \le 1$

11. Sketch the portion of the (x,y)-plane where

(a) $|x| < 1$ (b) $|y| \le 5$

(c) $|x + y| \le 1$ (d) $|2x + 3y| < 6$

(e) $|x| < 1$ and $|y| \le 5$ (f) $|x + y| \le 1$ and $|2x + 3y| < 6$

12. Sketch the portion of the (x,y)-plane where

(a) $x^2 + y^2 \le 4$ (b) $x^2 + y^2 \ge 1$

(c) $1 \le x^2 + y^2 \le 4$ (d) $(x - 1)^2 + (y - 2)^2 \le 16$

13. Sketch the points in the (x,y)-plane which satisfy the following simultaneous inequalities.

(a) $x^2 + y^2 \le 1$ and $y \ge 0$ (b) $y \ge x^2$ and $y < x$

(c) $y \ge x^2 - 1$ and $x^2 + y^2 \le 4$

14. Sketch the portion in the (x,y)-plane where $|x| + |y| \le 2$. [Compare your graph with Exercise 4 in the Review Self Test.]

15. Solve the inequality:

(a) $x^2 > 2x + 7$ (b) $|x - 5| < |2x + 1|$

16. Use your calculator to check that

(a) $0 < |\pi - \frac{22}{7}| < (2)(10^{-3})$ (b) $0 < |\pi - \frac{355}{115}| < (3)(10^{-7})$

17. <u>Without</u> a calculator, find which is larger: $\sqrt{2} + \sqrt{3}$ or $\sqrt{10}$. (Hint: Remember that when you square $\sqrt{a}$ you get <u>a</u>, assuming a $\geq$ 0.)

18. Graph:

(a) $y > x^2$ (b) $y < x + 2$ (c) $y > x^2$ and $y < x + 2$

* * * CUMULATIVE REVIEW CHAPTERS 10-12 * * *

1. Identify the graph of each of the following as a circle, parabola, hyperbola, ellipse, point, line, two intersecting lines, or no graph at all in the real number plane.

(a) $x^2 + y = 4$ (b) $25y^2 + 4x^2 = 100$

(c) $4x^2 + 4y^2 = 16$ (d) $(x - y)(x - y) = 0$

(c) $4x^2 - 25y^2 = 100$ (f) $x^2 + 2x + y^2 = 0$

(g) $y^2 - x = 1$ (h) $x^2 + y^2 = -1$

(i) $x^2 = y^2$ (j) $x^2 + y^2 = 0$

2. Sketch the graph:

(a) $y = x^2 - 4x + 4$ (b) $\frac{x^2}{36} + \frac{y^2}{25} = 1$

(c) $y^2 - \frac{x^2}{25} = 1$ (d) $\frac{x^2}{4} + \frac{y^2}{4} = 4$

(e) $x^2 - y^2 = 0$ (f) $x^2 + 2y^2 = 0$

(g) $\frac{x^2}{4} - \frac{y^2}{36} = 1$ (h) $x^2 + 2xy + y^2 = 0$

3. Each of the following is a circle. Without graphing it, give the location of the center and the radius.

(a) $x^2 + 4x + y^2 = 0$ (b) $x^2 - 2x + y^2 + 6y = 4$

(c) $x^2 + y^2 - 10y = 7$ (d) $x^2 + y^2 = 15$

4. (a)   Sketch the graphs and estimate the intersection points.

$$x^2 + y^2 = 25$$
$$y = \frac{3}{4} x$$

(b)   Compute the intersection points algebraically.

5. Find the slope of each line.

(a) $2x + 3y = 7$ (b) $y = \frac{7}{8} x - 4$

6. (a)   Compute the slope of the line through the points $(-1,7)$ and $(-3,-4)$.

(b)   Find an equation in $ax + by + c = 0$ form for the line containing the points in (a).

(c)   Find an equation in $y = mx + b$ form for the line in (b).

7. Compute the slope at the point (x,y) for each of the following by using the appropriate slope formula.

(a) $y = -5x + 4$ (b) $y = 3x^2 - x + 9$

(c) $y = x^3 - x^2 + 2x - 5$ (d) $y = 2x^4 - x^3 + x^2 - x + 2$

8. Compute the slope when $x = 1$ for each part of Exercise 7.

9. Compute the slope of the graph of $y = 3x^2 - x + 5$ at the point $(0,5)$.

10. Solve for all possible x:

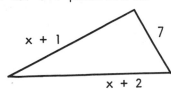

11. Solve the inequality and sketch the solutions on the number line.

(a) $-2x + 7 < x + 9$ (b) $-\frac{2}{3}x + 5 \geq 10$

12. Sketch the numbers x that satisfy each inequality and describe the solution algebraically.

(a) $2 + x < 3x < 7$ (b) $4x - 2 \leq 6x + 3 \leq 4x + 8$

(c) $|x| > 5$ (d) $|x| \leq 5$ (continued)

(e) $|2x - 1| \leq 3$ (f) $|3x + 1| \geq 2$

(g) $(x - \frac{1}{2})(x + \frac{2}{3}) < 0$ (h) $x^2 + 12 \geq 7x$

13. Graph in the (x,y)-plane.

(a) $2x - 3y > 12$ (b) $y < \frac{2}{3}x + 4$

(c) $|x| > 1$ (d) $|y| < \frac{1}{2}$

(e) $x > -1$ (f) $y < -\frac{1}{2}$

14. Sketch the set of points (x,y) in the plane that meet the given simultaneous conditions.

(a) $x > 0$; $y < 0$; $2x > y + 3$

(b) $|x| < 3$; $|y| \leq 2$

(c) $y + x \leq 2$; $y + x \geq -2$

ANSWERS TO CUMULATIVE REVIEW

1. (a) parabola (b) ellipse (c) circle (d) one line (e) hyperbola
    (f) circle (g) parabola (h) no graph (i) two intersecting lines
    (j) point

2. (a)   (b)   (c)

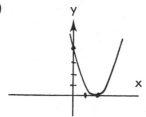   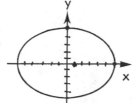   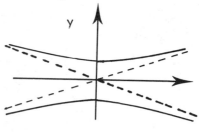

(d)   (e)   (f)

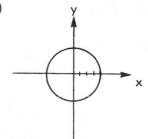   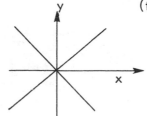   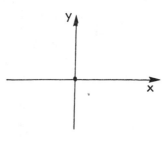

(g) 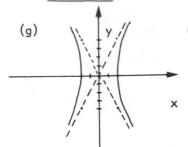   (h)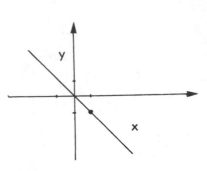

3. (a) center = (-2,0), radius = 2   (b) center = (1,-3), radius = $\sqrt{14}$
    (c) center = (0,5), radius = $\sqrt{32}$ (d) center = (0,0), radius = $\sqrt{15}$

4. (a) 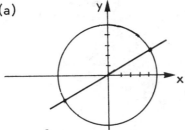   (b) (4,3) and (-4,-3)

5. (a) $-\dfrac{2}{3}$ (b) $\dfrac{7}{8}$

6. (a) $\dfrac{11}{2}$ (b) $-11x + 2y - 25 = 0$ (c) $y = \dfrac{11}{2}x + \dfrac{25}{2}$

7. (a) -5   (b) $6x - 1$ (c) $3x^2 - 2x + 2$ (d) $8x^3 - 3x^2 + 2x - 1$

8. (a) -5   (b) 5   (c) 3   (d) 6

9. -1

10. $x > 2$

11. (a) $x > -\dfrac{2}{3}$ (b) $x \le -\dfrac{15}{2}$

12. (a) $1 < x < \dfrac{7}{3}$ (b) $-\dfrac{5}{2} \le x \le \dfrac{5}{2}$ (c) $x < -5$ or $5 < x$

    (d) $-5 \le x \le 5$ (e) $-1 \le x \le 2$ (f) $x \le -1$ or $\dfrac{1}{3} \le x$

    (g) $-\dfrac{2}{3} < x < \dfrac{1}{2}$ (h) $x \le 3$ or $4 \le x$

13. (a)   (b)   (c)

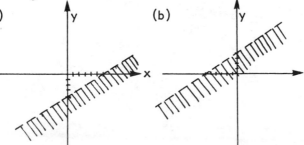 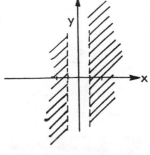

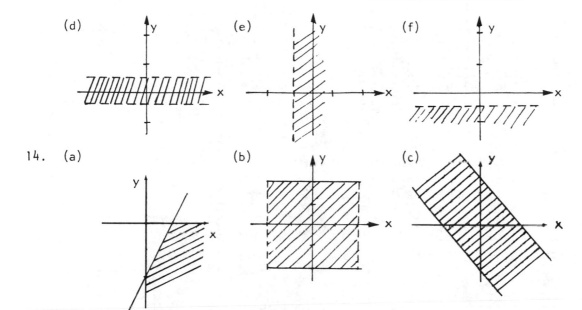

(d)

(e)

(f)

14.   (a)

(b)

(c)

FOUR-PLACE TABLE OF COMMON LOGARITHMS

N	0	1	2	3	4	5	6	7	8	9
10	0000	0043	0086	0128	0170	0212	0253	0294	0334	0374
11	0414	0453	0492	0531	0569	0607	0645	0682	0719	0755
12	0792	0828	0864	0899	0934	0969	1004	1038	1072	1106
13	1139	1173	1206	1239	1271	1303	1335	1367	1399	1430
14	1461	1492	1523	1553	1584	1614	1644	1673	1703	1732
15	1761	1790	1818	1847	1875	1903	1931	1959	1987	2014
16	2041	2068	2095	2122	2148	2175	2201	2227	2253	2279
17	2304	2330	2355	2380	2405	2430	2455	2480	2504	2529
18	2553	2577	2601	2625	2648	2672	2695	2718	2742	2765
19	2788	2810	2833	2856	2878	2900	2923	2945	2967	2989
20	3010	3032	3054	3075	3096	3118	3139	3160	3181	3201
21	3222	3243	3263	3284	3304	3324	3345	3365	3385	3404
22	3424	3444	3464	3483	3502	3522	3541	3560	3579	3598
23	3617	3636	3655	3674	3692	3711	3729	3747	3766	3784
24	3802	3820	3838	3856	3874	3892	3909	3927	3945	3962
25	3979	3997	4014	4031	4048	4065	4082	4099	4116	4133
26	4150	4166	4183	4200	4216	4232	4249	4265	4281	4298
27	4314	4330	4346	4362	4378	4393	4409	4425	4440	4456
28	4472	4487	4502	4518	4533	4548	4564	4579	4594	4609
29	4624	4639	4654	4669	4683	4698	4713	4728	4742	4757
30	4771	4786	4800	4814	4829	4843	4857	4871	4886	4900
31	4914	4928	4942	4955	4969	4983	4997	5011	5024	5038
32	5051	5065	5079	5092	5105	5119	5132	5145	5159	5172
33	5185	5198	5211	5224	5237	5250	5263	5276	5289	5302
34	5315	5328	5340	5353	5366	5378	5391	5403	5416	5428
35	5441	5453	5465	5478	5490	5502	5514	5527	5539	5551
36	5563	5575	5587	5599	5611	5623	5635	5647	5658	5670
37	5682	5694	5705	5717	5729	5740	5752	5763	5775	5786
38	5798	5809	5821	5832	5843	5855	5866	5877	5888	5899
39	5911	5922	5933	5944	5955	5966	5977	5988	5999	6010
40	6021	6031	6042	6053	6064	6075	6085	6096	6107	6117
41	6128	6138	6149	6160	6170	6180	6191	6201	6212	6222
42	6232	6243	6253	6263	6274	6284	6294	6304	6314	6325
43	6335	6345	6355	6365	6375	6385	6395	6405	6415	6425
44	6435	6444	6454	6464	6474	6484	6493	6503	6513	6522
45	6532	6542	6551	6561	6571	6580	6590	6599	6609	6618
46	6628	6637	6646	6656	6665	6675	6684	6693	6702	6712
47	6721	6730	6739	6749	6758	6767	6776	6785	6794	6803
48	6812	6821	6830	6839	6848	6857	6866	6875	6884	6893
49	6902	6911	6920	6928	6937	6946	6955	6964	6972	6981
50	6990	6998	7007	7016	7024	7033	7042	7050	7059	7067
51	7076	7084	7093	7101	7110	7118	7126	7135	7143	7152
52	7160	7168	7177	7185	7193	7202	7210	7218	7226	7235
53	7243	7251	7259	7267	7275	7284	7292	7300	7308	7316
54	7324	7332	7340	7348	7356	7364	7372	7380	7388	7396

**THESE TABLES ARE INCLUDED FOR STUDENTS
WHO DO NOT HAVE ACCESS TO A CALCULATOR.**

N	0	1	2	3	4	5	6	7	8	9
55	7404	7412	7419	7427	7435	7443	7451	7459	7466	7474
56	7482	7490	7497	7505	7513	7520	7528	7536	7543	7551
57	7559	7566	7574	7582	7589	7597	7604	7612	7619	7627
58	7634	7642	7649	7657	7664	7672	7679	7686	7694	7701
59	7709	7716	7723	7731	7738	7745	7752	7760	7767	7774
60	7782	7789	7796	7803	7810	7818	7825	7832	7839	7846
61	7853	7860	7868	7875	7882	7889	7896	7903	7910	7917
62	7924	7931	7938	7945	7952	7959	7966	7973	7980	7987
63	7993	8000	8007	8014	8021	8028	8035	8041	8048	8055
64	8062	8069	8075	8082	8089	8096	8102	8109	8116	8122
65	8129	8136	8142	8149	8156	8162	8169	8176	8182	8189
66	8195	8202	8209	8215	8222	8228	8235	8241	8248	8254
67	8261	8267	8274	8280	8287	8293	8299	8306	8312	8319
68	8325	8331	8338	8344	8351	8357	8363	8370	8376	8382
69	8388	8395	8401	8407	8414	8420	8426	8432	8439	8445
70	8451	8457	8463	8470	8476	8482	8488	8494	8500	8506
71	8513	8519	8525	8531	8537	8543	8549	8555	8561	8567
72	8573	8579	8585	8591	8597	8603	8609	8615	8621	8627
73	8633	8639	8645	8651	8657	8663	8669	8675	8681	8686
74	8692	8698	8704	8710	8716	8722	8727	8733	8739	8745
75	8751	8756	8762	8768	8774	8779	8785	8791	8797	8802
76	8808	8814	8820	8825	8831	8837	8842	8848	8854	8859
77	8865	8871	8876	8882	8887	8893	8899	8904	8910	8915
78	8921	8927	8932	8938	8943	8949	8954	8960	8965	8971
79	8976	8982	8987	8993	8998	9004	9009	9015	9020	9025
80	9031	9036	9042	9047	9053	9058	9063	9069	9074	9079
81	9085	9090	9096	9101	9106	9112	9117	9122	9128	9133
82	9138	9143	9149	9154	9159	9165	9170	9175	9180	9186
83	9191	9196	9201	9206	9212	9217	9222	9227	9232	9238
84	9243	9248	9253	9258	9263	9269	9274	9279	9284	9289
85	9294	9299	9304	9309	9315	9320	9325	9330	9335	9340
86	9345	9350	9355	9360	9365	9370	9375	9380	9385	9390
87	9395	9400	9405	9410	9415	9420	9425	9430	9435	9440
88	9445	9450	9455	9460	9465	9469	9474	9479	9484	9489
89	9494	9499	9504	9509	9513	9518	9523	9528	9533	9538
90	9542	9547	9552	9557	9562	9566	9571	9576	9581	9586
91	9590	9595	9600	9605	9609	9614	9619	9624	9628	9633
92	9638	9643	9647	9652	9657	9661	9666	9671	9675	9680
93	9685	9689	9694	9699	9703	9708	9713	9717	9722	9727
94	9731	9736	9741	9745	9750	9754	9759	9763	9768	9773
95	9777	9782	9786	9791	9795	9800	9805	9809	9814	9818
96	9823	9827	9832	9836	9841	9845	9850	9854	9859	9863
97	9868	9872	9877	9881	9886	9890	9894	9899	9903	9908
98	9912	9917	9921	9926	9930	9934	9939	9943	9948	9952
99	9956	9961	9965	9969	9974	9978	9983	9987	9991	9996

THESE TABLES ARE INCLUDED FOR STUDENTS WHO DO NOT HAVE ACCESS TO A CALCULATOR.

N	N²	√N	N	N²	√N
1	1	1.000	51	2601	7.141
2	4	1.414	52	2704	7.211
3	9	1.732	53	2809	7.280
4	16	2.000	54	2916	7.348
5	25	2.236	55	3025	7.416
6	36	2.449	56	3136	7.483
7	49	2.646	57	3249	7.550
8	64	2.828	58	3364	7.616
9	81	3.000	59	3481	7.681
10	100	3.162	60	3600	7.746
11	121	3.317	61	3721	7.810
12	144	3.464	62	3844	7.874
13	169	3.606	63	3969	7.937
14	196	3.742	64	4096	8.000
15	225	3.873	65	4225	8.062
16	256	4.000	66	4356	8.124
17	289	4.123	67	4489	8.185
18	324	4.243	68	4624	8.246
19	361	4.359	69	4761	8.307
20	400	4.472	70	4900	8.367
21	441	4.583	71	5041	8.426
22	484	4.690	72	5184	8.485
23	529	4.796	73	5329	8.544
24	576	4.899	74	5476	8.602
25	625	5.000	75	5625	8.660
26	676	5.099	76	5776	8.718
27	729	5.196	77	5929	8.775
28	784	5.292	78	6084	8.832
29	841	5.385	79	6241	8.888
30	900	5.477	80	6400	8.944
31	961	5.568	81	6561	9.000
32	1024	5.657	82	6724	9.055
33	1089	5.745	83	6889	9.110
34	1156	5.831	84	7056	9.165
35	1225	5.916	85	7225	9.220
36	1296	6.000	86	7396	9.274
37	1369	6.083	87	7569	9.327
38	1444	6.164	88	7744	9.381
39	1521	6.245	89	7921	9.434
40	1600	6.325	90	8100	9.487
41	1681	6.403	91	8281	9.539
42	1764	6.481	92	8464	9.592
43	1849	6.557	93	8649	9.644
44	1936	6.633	94	8836	9.695
45	2025	6.708	95	9025	9.747
46	2116	6.782	96	9216	9.798
47	2209	6.856	97	9409	9.849
48	2304	6.928	98	9604	9.899
49	2401	7.000	99	9801	9.950
50	2500	7.071	100	10000	10.000

Index for Algebra II

TRIGONOMETRY

Contents for Trigonometry

| Trigonometry | # Prelude |

It is with pleasure that we offer TRIGONOMETRY, an extensively tested text, to the mathematical public at the secondary and college level. We believe that it represents a practical and effective option, more in tune with present-day student and teacher attitudes than the many texts rooted in reforms of the previous generation.

A glance at its pages will show that technical vocabulary is kept minimal without sacrificing mathematical content. The arrangement of exposition and exercises encourages active student participation in the learning process. Moreover, this text is compatible with the lecture, individual, or small-group learning method. Because of the chapter origanization, students of a wide range of abilities in the same classroom will feel comfortable with the approach.

Selected answers are given in the margins of the central (main) section of each chapter. All answers for the self quizzes, tests, and reviews are given at the end of the section.

Hints and learning suggestions are in the margins. For a short course, time may be saved by omitting some or all of the review exercises; however, most classes will be helped by them.

The Solutions Key contains teaching suggestions for establishing and maintaining a small group learning structure. Of course, the more traditional lecture or individualized approaches may be used.

We wish to acknowledge the contributions of the many students and teachers who have used earlier versions of this book. In particular we wish to thank the following: Dr. Robert Van Dyne, Maria Wong, Archie White, Joanne Moldenhauer, Eugene Tashima, Valerie McIntyre, James O'Keefe, Ken Gelatt, Joan Graf, Ronnie James, and Craig Armstrong, Davis, Ca.; Mary Gorse, Palatine, Il.; Dr. Donald McKinley, Sacramento, Ca.; George Koonce, Dixon, Ca.; Ann Carroll, Santa Monica, Ca.; Gerry Benoit, Newport, R.I.; Henri Picciotto, San Francisco, Ca.; Sue Roosenvaad, Northfield, Mn.; Paul Zeitz, Colorado Springs, Co.; Jerry Weil, Gonzales, Ca.; Donna Jean Wilson, Superior Az. We especially appreciate the assistance of Hal Faulkner in the final preparation of the manuscript.

Introduction

The chapters that follow are devoted to "trigonometry". Though the word, meaning "triangle measurement", was coined early in the seventeenth century, trigonometry as a subject had been studied for more than 2000 years before that. The astronomers of Babylon, Greece, India and the Arab lands were the main users of trigonometry. They needed it for the calculation of the position of the heavenly bodies, information critical to their astrology, to their reckoning of time, and to their navigation.

For all those years, trigonometry was concerned with certain lines in the circle. Extensive tables were made showing how the lengths of OA and AP in the following diagram varied as the size of the angle AOP varied:

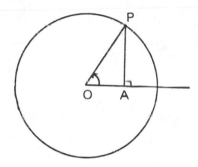

In Renaissance Europe trigonometry separated from astronomy, becoming a discipline in its own right. One of its major developing applications was in surveying, which was concerned with the lengths and angles in a triangle, rather than with lengths in a circle. Surveying was a common occupation in the settlement and expansion of the United States. For instance, George Washington was a trained surveyor; Thoreau earned his living from his surveying, not from his writing. Even Lincoln at one time earned his living as a surveyor and was known for his care and accuracy as he was called on to settle boundary disputes.

Both these approaches to trigonometry-- that of the circle and that of the triangle-- will be developed in the pages that follow, for both are used today. The trigonometry of the triangle, still important in surveying, is also a tool of the shop worker who calculates how to shape complicated pipes, pipe elbows, and hoppers out of sheet metal or steel. The trigonometry of the circle is important in the study of calculus, traditionally part of the physical science curriculum but now also commonly introduced to the future psychologist, doctor, economist, or biologist.

1

Trigonometry

The Unit Circle

This chapter introduces the trigonometry of the circle. The following materials will be needed: graph paper, protractor, compass, and centimeter ruler.

In this chapter it is assumed that these materials are available.

EXERCISES

1. In the following diagram, $\angle$ QOB = 70° and $\angle$ POA = 35°. (Thus $\angle$ QOB is twice $\angle$ POA.)

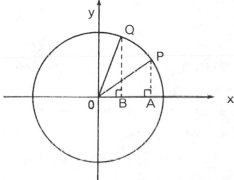

(a) Copy the figure with the aid of a protractor and compass.

(b) How do the lengths of the line segments PA and QB compare? Is QB equal to twice PA? Is QB greater than twice PA? Is QB less than twice PA?

At the end of the chapter an exercise will return to this question.

THE UNIT CIRCLE

Throughout this chapter the basic unit of length will always be ten centimeters long:

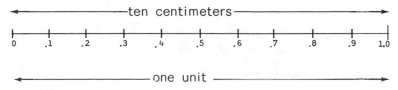

Each centimeter on the centimeter ruler is divided into ten smaller parts called millimeters. Thus it is easy to measure lengths to two decimal places in terms of the basic unit. For instance, the length of this diagram is 0.72 basic units:

← —————— 0.72 basic units ——————→

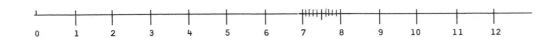

0   1   2   3   4   5   6   7   8   9   10   11   12

2. Using a centimeter ruler, find the lengths in basic units of each of these line segments.

(a) ——————————————

(a) 0.35 basic units

(b) ——

(b) 1.43 basic units

All lengths will be expressed in terms of a basic unit length of ten centimeters. For simplicity the words "basic units" will be omitted. Thus, the length of the segment shown before Exercise 2 is just 0.72.

The circle of radius 1 and with center at the origin of the (x,y)-coordinate system is called the <u>unit circle</u>. (The radius is ten centimeters.

Sharpen the pencil of your compass before beginning Exercise 3.

3. (a) Draw x- and y-axes on a sheet of graph paper. (Place the origin in the center of the page.)

   (b) Draw a circle of radius 1 (that is, ten centimeters) with center at the origin.

(a) (1,0) and (-1,0)

Keep the unit circle you drew in Exercise 3. It will be used in the following exercises.

4. Find the coordinates of the points where the circle in Exercise 3 meets

   (a) the x-axis.   (b) the y-axis

* * *

The counter-clockwise <u>direction</u> is used for positive angles. Later in the chapter negative angles will be defined; their second arm is obtained by a <u>clockwise</u> motion.

This chapter concerns angles drawn in the unit circle. These angles are always placed according to these directions:

1. The vertex of the angle is at the center of the unit circle.

2. One arm (or side) of the angle coincides with the positive part of the x-axis.

3. The second arm of the angle is obtained by rotating the first arm <u>counterclockwise</u>.

This diagram shows how a 60° angle is drawn.   The circle is, of course, to be drawn to a larger scale; that is, <u>the radius is 10 centimeters and equals one unit</u>.

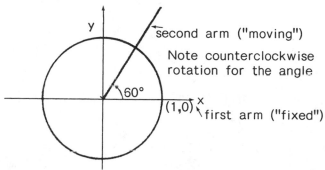

5.   On the unit circle you constructed in Exercise 3, draw each of the following angles in the manner illustrated in the previous example. (a) 40 degrees, (b) 135 degrees, (c) 190 degrees, (d) 305 degrees.

6.   (a)   In Exercise 5, let P be the point where the second arm (the arm which rotates) crosses the circle, as seen in the diagram below.

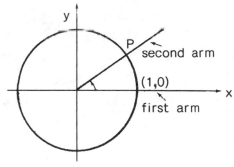

(b)   For each angle of Exercise 5, use a centimeter ruler to estimate the x and y coordinates of point P to two decimal places.   (The second decimal place may vary slightly, due to experimental error.)

Since ten centimeters equal one unit, two decimal place accuracy is possible.

(a) (0.77, 0.64)
(c) (-0.98, -0.17)

THE SINE AND COSINE

The sine and cosine of an angle are the basic functions of trigonometry.   They are defined with reference to the unit circle.

DEFINITION. An angle, θ ("theta"), is drawn with vertex at the origin and with the fixed arm along the x-axis; the second arm of the angle is rotated counterclockwise and crosses the unit circle at a single point, which we call P.

The Greek letter theta provides us with the "th" in words such as arithmetic, mathematics and theology.

The

x-coordinate of P is called the cosine of θ.

The

y-coordinate of P is called the sine of θ.

It is most important to observe that the sine and cosine are defined in terms of a point on the unit circle, a circle with radius 1.

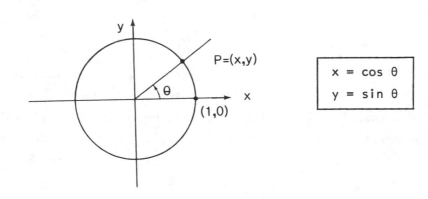

$$x = \cos \theta$$
$$y = \sin \theta$$

NOTATION. The sine of θ is also written as sin θ and the cosine of θ as cos θ. If the coordinates of P are (x,y), we write

$$P = (x,y)$$

and x is the cosine of θ and y is the sine of θ.

Example 1. Use the unit circle and your protractor to find cos 150° and sin 150°.

Solution. Use your unit circle to verify:

P ≐ (-0.87, 0.50)

Hence: cos 150° ≐ -0.87
          sin 150° ≐ 0.50

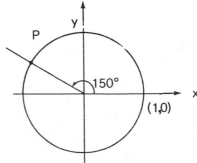

The symbol ≐ is to be read as "is approximately." It turns out that sin 150° is exactly $\frac{1}{2}$ = 0.50 but cos 150° is not -0.87, but $\frac{-\sqrt{3}}{2}$ ≐ -0.866.

QUADRANTS

The two axes cut the coordinate plane into four parts, called "quadrants:"

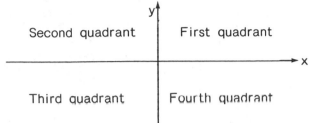

Second quadrant   First quadrant

Third quadrant   Fourth quadrant

If the second arm (the moving side) of an angle is not along one of the axes, the point P, where it meets the unit circle, lies in one of the four quadrants. The <u>quadrant</u> <u>in</u> <u>which</u> <u>the</u> <u>angle</u> <u>lies</u> is defined as the quadrant in which P is situated. As the four following diagrams show,

an angle of 30° is in the first quadrant;

an angle of 135° is in the second quadrant;

an angle of 200° is in the third quadrant;

an angle of 327° is in the fourth quadrant.

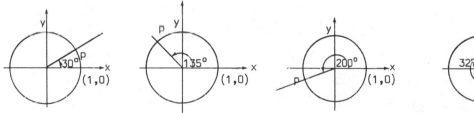

7. In what quadrant is each of these angles?

(a) 80°   (b) 100° (c) 179°   (d) 181°   (e) 260°   (f) 300°

The angles 0°, 90°, 180°, 270°, and 360° are not assigned a quadrant.

*(a) first
(c) second
(e) third*

* * *

The sine and cosine, being coordinates of a point, can be positive, negative, or zero. The next exercise examines this important fact.

8. (a) Referring to the unit circle shown, fill in the table with the words "positive" or "negative."

When answering this question make a quick sketch of the unit circle and the angle in each case.

Quadrant	cos θ	sin θ
first		
second		
third		
fourth		

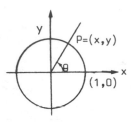

(b)   For what θ is cos θ = 0?

(c)   For what θ is sin θ = 0?

(a) 0.77
(c) 0.71
(e) -0.98
(g) -0.82

9.   Referring to your work in Exercise 5, use a centimeter ruler to estimate each of the following:

(a)   cos 40°   (b)   sin 40°   (c)   sin 135°   (d)   cos 135°

(e)   cos 190°   (f)   sin 190°   (g)   sin 305°   (h)   cos 305°

10. (a) 1
*   (c) -1*
*   (e) 0*
*   (g) 0*
*   (i) 0*

10.   Using the unit circle, find:

(a)   cos 0°   (b)   sin 90°   (c)   cos 180°

(d)   sin 270°   (e)   cos 90°   (f)   sin 0°

(g)   sin 180°   (h)   cos 270°   (i)   sin 360°

11.   In which quadrant

a) first

(a)   are sin θ and cos θ both positive?

(b)   are sin θ and cos θ both negative?

(c) fourth

(c)   is cos θ positive and sin θ negative?

(d)   is cos θ negative and sin θ positive?

(a) 0.73
(c) -0.73

12.   Using the unit circle, find:

(a)   cos 43°   (b)   sin 157°   (c)   cos 223°   (d)   sin 286°

13.   Using the unit circle, find all angles between 0° and 360° such that

$$\cos \theta = 0.5.$$

First decide in which quadrants θ can be. (Note that cos θ is positive). Compare your answers with those of other students.

14.   Find all angles θ between 0 and 360 degrees, <u>if there are any</u>, such that

(a)   sin θ = -0.64   (b)   cos θ = 1.50   (c)   sin θ = -3.01

(d)   cos θ = 0.29   (e)   sin θ = 0.5   (f)   sin θ = -0.5

14. (a) 220° and 320°
*   (c) no angle*
*   (e) 30° and 150°*

15.   Consider point P in this diagram

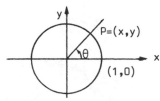

$$x = \cos \theta$$
$$y = \sin \theta$$

How many values of θ, where $0° \leq θ < 360°$, are there such that

(a) cos θ = 0.6   (b) sin θ = -0.29

(c) cos θ = -1   (d) sin θ = 5.27

(e) cos θ = -3.00   (f) sin θ = 0.

Hint: There may be none!

(a) two
(c) one
(e) none

16. If $0° \leq θ < 360°$, find all θ such that

(a) cos θ = sin θ   (b) cos θ = 1   (c) sin θ = -1   (d) cos θ = -sin θ

(e) cos θ = -1   (f) sin θ = 1   (g) sin θ = 0   (h) cos θ = 0

(a) 45°, 225°
(c) 270°
(e) 180°
(g) 0°, 180°

17. When θ increases from 0 to 90 degrees,

(a) does cos θ increase or decrease?   (b) does sin θ increase or decrease?

18. Explain why it is always true that

(a) $-1 \leq \cos θ \leq 1$ (b) $-1 \leq \sin θ \leq 1$

ANGLES GREATER THAN 360°

So far the second arm of the angle, the "moving arm," has been permitted to rotate from 0° to 360°. If this second arm is allowed to rotate more than once around the unit circle it will describe angles larger than 360°. For instance, this diagram shows an angle of 495°:

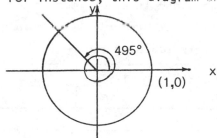

(The angle is 360° + 135°.) By the time the second arm rotates twice around the unit circle it has swept out an angle of 720° (= 360° + 360°). Three rotations produce an angle of 1080°, and so on.

19. Draw an angle of

(a) 400°   (b) 450°   (c) 550°   (d) 695°

(e) Compare your drawings with those of other students.

(f) Using a picture, explain why

    (i) cos 400° = cos 40°   (ii) sin 400° = sin 40°

    (iii) cos (360° + θ) = cos θ   (iv) sin (360° + θ) = sin θ.

NEGATIVE ANGLES

A <u>clockwise</u> rotation of the second arm of the angle is associated with a <u>negative angle</u>. For instance, this diagram shows an angle of -30°.

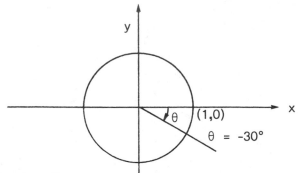

Remember: a
clockwise angle
rotation is
negative; a
counterclockwise
angle rotation
is positive.

20. Draw an angle of

(a) -40°   (b) -175°   (c) -315°   (d) -475°

(e) Compare your drawings with those of other students.

(f) Explain why

(i) cos (-40°) = cos 320°   (ii) sin (-40°) = sin 320°

(iii) cos (360° - θ) = cos (-θ)   (iv) sin (360° - θ) = sin (-θ)

THE RELATION BETWEEN cos θ AND sin θ

There is a relation between cos θ and sin θ. In fact, it turns out
that if cos θ is known, then sin θ is almost completely determined. The
next exercise outlines this fundamental relation between cos θ and sin θ.

21. (a) Copy and fill in this table to two decimal places.

angle	30°	129°	205°	317°
sin θ				
cos θ				
$\sin^2\theta$				
$\cos^2\theta$				
$\sin^2\theta + \cos^2\theta$				

NOTE: The
notation $\sin^2\theta$ is
short for
$(\sin\theta)^2$. We read
it aloud as "sine
squared of θ;" but
it means $(\sin\theta)^2$.

(b) Is your work in (a) in approximate agreement with the

assertion that for all θ,

$$\sin^2\theta + \cos^2\theta = 1?$$

(c) Use the unit circle and the Pythagorean theorem to show that for

all θ

$$\boxed{\sin^2\theta + \cos^2\theta = 1}$$

This equation in Exercise 21 (c) is fundamental. If you know sin θ
and you know the quadrant, you can find cos θ, and vice versa. It is an

example of a <u>trigonometric identity</u>: an equation concerning trigonometry functions that is true for all angles θ. An <u>algebraic identity</u>, such as

$$x^2 - 1 = (x + 1)(x - 1),$$

is an algebraic equation true for all numbers. Exercise 19 includes the trigonometric identities

$$\cos (360° + θ) = \cos θ \qquad \text{and} \qquad \sin (360° + θ) = \sin θ.$$

Exercise 20 includes the trigonometric identities

$$\cos (360° - θ) = \cos (-θ) \qquad \text{and} \qquad \sin (360° - θ) = \sin (-θ).$$

22. Use the identity in Exercise 21 and a rough sketch of the unit circle to find cos θ if

(a) sin θ = 0.6 and 0°<θ<90   (b) sin θ = $-\dfrac{7}{25}$ and 180°<θ<270°

(c) sin θ = 0.8 and 90°<θ<180°   (d) sin θ = $-\dfrac{5}{13}$ and 270°<θ<360°

(e) sin θ = $-\dfrac{3}{5}$ and 180°<θ<270° (f) sin θ = $\dfrac{\sqrt{3}}{2}$ and 90°<θ<180°

23. Do Exercise 22, but this time assume that in each case 0 ≤θ<360°. There may be more than one answer.

USING THE UNIT CIRCLE FOR IDENTITIES

In the next few exercises freehand sketches of the unit circle will be helpful. Do not memorize the identities obtained. Just remember that they exist and be able to obtain them quickly with a rough sketch of the unit circle.

24. Use a rough sketch of the unit circle to compare

(a) sin (-25°) with sin 25°

(b) sin (-140°) with sin 140°

(c) sin (-θ) with sin θ for a specific θ of your choice

(d) Explain why the following is true for all angles θ.

$$\boxed{\sin (-θ) = -\sin θ}$$

25. Make a rough sketch of the unit circle and compare

(a) cos (-40°) with cos 40°

(b) cos (-150°) with cos 150°

(c) cos (-θ) with cos θ for a specific θ of your choice.

(d) Explain why the following is true for all angles θ.

$$\boxed{\cos (-θ) = \cos θ}$$

A rough sketch of the unit circle is needed to determine the quadrant and whether cos θ is positive or negative.

22. (a) 0.8
(c) -0.6
(e) $-\dfrac{4}{5}$

23. (a) 0.8 if first quadrant; -0.8 if second quadrant
(c) 0.6 if first quadrant; -0.6 if second quadrant;
(e) $-\dfrac{4}{5}$ if third quadrant; $\dfrac{4}{5}$ if fourth quadrant.

Do not memorize the identities cos (-θ) = cos θ and sin (-θ) = -sin θ. It is too easy to get them confused. Draw a unit circle each time one of these identities is needed.

LET θ BE A SMALL ACUTE ANGLE.

* * *

It may be tempting to think that any trigonometric equation that "looks nice" is an identity. The next exercise should serve as warning that this is not always the case.

26. (a) For which of the following θ does sin 2θ = 2 sin θ?

θ = 180°,   θ = 360°,   θ = 90°.

(b) Show that the equation

$$\sin 2\theta = 2 \sin \theta$$

is <u>not</u> an identity. [Hint: One exception proves it false.]

For an equation to be an <u>identity</u> it must be true for <u>all</u> θ for which both sides of the equation are defined.

27. Make a rough sketch of the unit circle to compare

(a) sin 25° with sin 155° (note: 180° - 25° = 155°)

(b) sin 60° with sin 120°

(c) sin θ with sin (180° - θ) for a specific θ of your choice

(d) Explain why the following is true for all angles θ.

$$\boxed{\sin (180° - \theta) = \sin \theta}$$

Omission of parentheses around 180 - θ can lead to confusion. Hence, they must be used.

28. Make a rough sketch of the unit circle to show that

(a) cos 30° = -cos 150°

(b) Explain why the following is true for all angles θ.

$$\boxed{\cos (180° - \theta) = - \cos \theta}$$

29. Make a rough sketch of the unit circle to compare

(a) sin 205° with sin 25° (note: 205° = 180° + 25°)

(b) sin 250° with sin 70°

(c) sin (180° + θ) with sin θ for a specific θ of your choice.

(d) Explain why the following is true for all angles θ.

$$\boxed{\sin (180° + \theta) = - \sin \theta}$$

HELPFUL SUGGESTION: Let θ be a <u>small angle</u> in the first quadrant in order to recall the identities in Exercises 25-30. Draw a rough sketch of the unit circle.

30. Use a rough sketch of the unit circle to compare

(a) cos (180° + θ) with cos θ for θ = 30°

(b) cos (180° + θ) with cos θ for a specific θ of your choice

(c) Explain why

$$\boxed{\cos (180° + \theta) = - \cos \theta}$$ for all θ.

31. Use the unit circle below to compare

    (a) cos 30° with sin 60°   (b) sin 10° with cos 80°

    (c) cos 70° with sin 20°   (d) sin 40° with cos 50°

    (e) In the figure that follows, explain why triangles OBQ and PAO
        are congruent.

    (f) Use part (e) to explain the following statement

$$\cos (90° - \theta) = \sin \theta$$

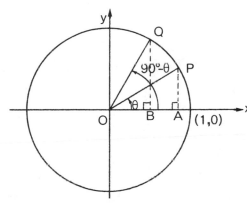

If one acute angle in a right triangle is θ, what is the other acute angle?

32. Use the figure in Exercise 31 to explain why

$$\sin (90° - \theta) = \cos \theta$$

MEMORIZE the identities in Exercises 31 and 32. They will be used often.

    The next exercise gives practice in using the identities obtained so far. Make a quick sketch of the unit circle whenever it is of use.

33. If $\sin 30° = \frac{1}{2}$ find

    (a) cos 60°   (b) sin 60°   (c) cos 30°   (d) sin (-30°)

    (e) cos (-30°)   (f) sin 150°   (g) cos 150°   (h) sin 210°

    (i) cos 210°   (j) cos (120°)

34. Use the unit circle to compare

    (a) sin 115° and cos 25° (note: 115° = 90° + 25°)

    (b) cos 115° with sin 25°.

(a) $\frac{1}{2}$

(c) $\frac{\sqrt{3}}{2}$

(e) $\frac{\sqrt{3}}{2}$

(g) $-\frac{\sqrt{3}}{2}$

(i) $-\frac{\sqrt{3}}{2}$

(c) Use the accompanying diagram to show that for all θ

$$\boxed{\sin (90 + \theta) = \cos \theta}$$

and that

$$\boxed{\cos (90 + \theta) = -\sin \theta}$$

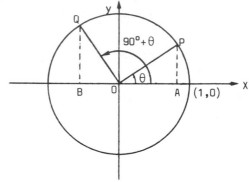

The next exercise applies the various identities obtained so far.

35. If sin 37° = 0.602 and cos 24° = 0.914 find

(a) cos 53°   (b) sin 66°   (c) cos 156°   (d) sin 143°

(e) sin 217°   (f) cos 204°   (g) sin 323°   (h) cos 336°

(i) sin (-37°) (j) cos (-24°) (k) sin 397°   (l) cos 384°

(a) 0.602
(c) -0.914
(e) -0.602
(g) -0.602
(i) -0.602
(k) 0.602

COMMENT. This is the time to pause and reflect about what has been accomplished in the preceding exercises. Various trigonometric identities have been obtained. Pause to make a list of all the "boxed" identities. Memorize the following identities:

$$\cos^2\theta + \sin^2\theta = 1 \qquad\qquad \cos (90° - \theta) = \sin \theta$$
$$\cos (90° - \theta) = \sin \theta \qquad\qquad \sin (90° - \theta) = \cos \theta$$
$$\sin (90° - \theta) = \cos \theta \qquad\qquad \cos (90° + \theta) = -\sin \theta$$
$$\sin (90° + \theta) = \cos \theta$$

Use the unit circle to reconstruct the others quickly. They all depend on the symmetry of the circle, together with some attention to the quadrant of θ.

When deriving an identity, sketch θ as a small first-quadrant angle (a small acute angle) This makes the work easier. However, the identities hold for all θ, including negative θ.

These identities provide a means of finding cos θ or sin θ for any angle θ if the cosine (or sine) is known for angles between 0° and 45°. The next exercise illustrates this important application.

36. (OPTIONAL) If cos θ and sin θ are known for 0° ≤ θ ≤ 45°

explain how to find cos θ and sin θ

(a) for all θ, 0° ≤ θ ≤ 90°. [Hint: see Exercise 31 and 32.]

(b) for all θ, 90° ≤ θ ≤ 180°. [Hint: see (a) and Exercises 27, 28, and 34.]

(c) for all θ, 180° ≤ θ ≤ 270°. Hint: [see Exercises 29 and 30.]

(d) for all θ, 270° ≤ θ ≤ 360°. Hint: [see Exercises 24 and 25.]

THE TANGENT OF θ

The chapter has so far been devoted to two trigonometric functions, cos θ and sin θ. The next most important trigonometric function is called "the tangent of θ", written tan θ. It also is defined with reference to the unit circle.

DEFINITION. An angle θ is drawn with vertex at the origin and with fixed arm along the x-axis. Draw the fixed vertical line that is tangent to the unit circle at the point (1,0). The line determined by the second arm of the angle meets this tangent line at a point P whose x-coordinate is 1. The y-coordinate of P is called the tangent of 0, or briefly, tan θ.

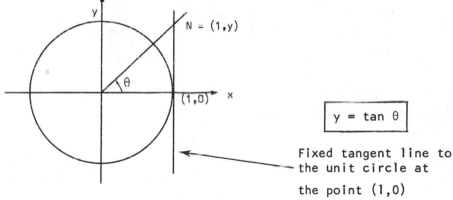

$$y = \tan \theta$$

Fixed tangent line to the unit circle at the point (1,0)

(If the second arm of the angle coincides with the y-axis, then tan θ is not defined.)

The next few exercises concern tan θ in the first and fourth quadrants only.

37. In the first and fourth quadrants, when is tan θ

(a) positive?   (b) negative?

(a) first quadrant

38. (a) Using the unit circle (radius = 10 centimeters) and the definition of tan θ, estimate tan θ to two decimal places and fill in this table.

You may have to tape several sheets of paper together for some of the angles.

angle θ	0 degrees	20 degrees	40 degrees	60 degrees	80 degrees
tan θ					

(a) tan 20° ≐ 0.36
tan 60° ≐ 1.73

(b) As θ increases from 0° to 90° what happens to tan θ?

39. Use the unit circle of Exercise 38 to estimate θ if $0° \leq \theta < 90°$ and

(a) $\tan \theta = 2$ (b) $\tan \theta = 1$ (c) $\tan \theta = \frac{1}{2}$

40. Estimate $\tan \theta$ to two decimal places:

Continue to use the unit circle of Exercise 38.

angle θ	0°	-20°	-40°	-60°	-80°
$\tan \theta$					

For second and third quadrant angles extend the moving arm back-wards to meet the line that is tangent to the unit circle at (1,0).

$\tan \theta = y$

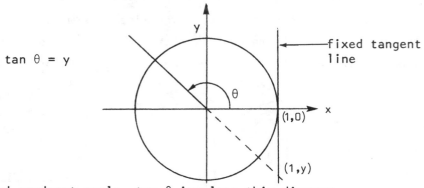

For a third quadrant angle, $\tan \theta$ involves this diagram.

$\tan \theta = y$

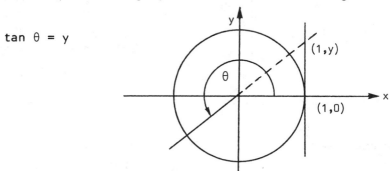

41. Fill in this table to two decimal place accuracy.

angle θ	135°	225°	-45°	315°	10°	190°	180°
$\tan \theta$							

42. Make a freehand sketch of the unit circle and the fixed tangent line to show that

(a) $\tan 210° = \tan 30°$ (b) $\tan 135° = \tan 315°$

(c) $\tan 160° = - \tan 20°$ (d) $\tan 40° = - \tan 140°$

43. If $\tan 45° = 1$ and $\tan 30° \doteq 0.577$, use a sketch of the unit circle
     to compute

    (a) $\tan (-45°)$ (b) $\tan 135°$ (c) $\tan 225°$ (d) $\tan 315°$

    (e) $\tan 150°$ (f) $\tan (-30°)$ (g) $\tan 210°$ (h) $\tan 330°$

    (a) -1
    (c) 1
    (e) -0.577
    (g) 0.577

44. (a) Show that if θ is near 90° but <u>less</u> than 90°, $\tan \theta$ is a large
          positive number.

    (b) Show that if θ is near 90° but <u>more</u> than 90°, $\tan \theta$ is a large
          negative number.

45. (a) There are two values of θ when $0° \le \theta < 360°$ for which $\tan \theta$
          is not defined. What are they?

    (b) There are two values of θ when $0° \le \theta < 360°$ for which
          $\tan \theta = 0$. What are they?

46. What is the relation between

    (a) $\tan 50°$ and $\tan (-50°)$?   (b) $\tan 120°$ and $\tan (-120°)$?

    (c) Explain to another student, with the aid of a diagram, why

$$\boxed{\tan (-\theta) = - \tan \theta}$$

47. Explain to another student why

$$\boxed{\tan (180° - \theta) = - \tan \theta}$$

48. Explain to another student why

$$\boxed{\tan (180° + \theta) = \tan \theta}$$

Do not memorize these identities. Remember that they exist, but each time you need one, draw a unit circle to help you recall it. Otherwise you may easily get the sign (- or +) wrong.

THE RELATION BETWEEN $\tan \theta$, $\sin \theta$, AND $\cos \theta$

There is a close relation between $\sin \theta$, $\cos \theta$, and $\tan \theta$. For
example, if you know $\sin \theta$ and $\cos \theta$, you can easily compute $\tan \theta$. The
next three exercises explore this relation.

49. (a) Using a unit circle (the radius is ten centimeters) and a
          centimeter ruler, fill in this table to two decimals.

angle θ	60°	50°	20°
sin θ			
cos θ			
$\dfrac{\sin\ \theta}{\cos\ \theta}$			
tan θ			

Compute the quotient $\dfrac{\sin\ \theta}{\cos\ \theta}$

(b)   What relation does there seem to be between

$$\frac{\sin\ \theta}{\cos\ \theta}\ \text{ and } \tan\theta?$$

50.   For $0° \le \theta < 90°$ use similar right triangles OAQ and OBP to explain why

$$\boxed{\tan\ \theta = \frac{\sin\ \theta}{\cos\ \theta}}\qquad \text{MEMORIZE!}$$

Hint: Remember it is the unit circle.

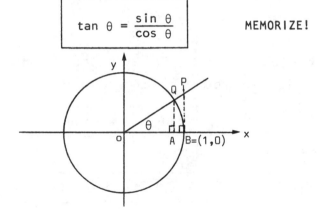

The equation of Exercise 50 defines tan θ in terms of sin θ and cos θ whenever cos θ is not 0. It is an important link between the three trigonometric functions.

In the next chapter it will be shown that

$\cos\ 30° = \dfrac{\sqrt{3}}{2}$ *and*

$\sin\ 30° = \dfrac{1}{2}$.

If you want to study this now, see Project 2.

(a) $\dfrac{1}{\sqrt{3}}$

(c) $\dfrac{1}{\sqrt{3}}$

(e) $-\dfrac{1}{\sqrt{3}}$

51.   If $\cos\ 30° = \dfrac{\sqrt{3}}{2}$ and $\sin\ 30° = \dfrac{1}{2}$ use the identity in Exercise 50 to find

(a)   tan 30°   (b)   tan 150°   (c)   tan 210°

(d)   tan 330°   (e)   tan (-30°)   (f)   tan 510°

THE SECANT, COSECANT, AND COTANGENT FUNCTIONS

All problems in trigonometry can be solved using the trigonometric functions introduced so far. (In fact, the sine and cosine are sufficient.) However, to make the notation in some problems easier, and to avoid complicated denominators, it has been customary historically to introduce three other trigonometric functions. They are:

(1)   secant of θ, written sec θ,

(2)   cosecant of θ, written csc θ, and

(3)   cotangent of θ, written cot θ.

They are defined as follows:

$$\sec \theta = \frac{1}{\cos \theta}$$ $$\csc \theta = \frac{1}{\sin \theta}$$ $$\cot \theta = \frac{1}{\tan \theta}$$ *MEMORIZE!*

52.   Show that whenever $\sin \theta \neq 0$,

$$\cot \theta = \frac{\cos \theta}{\sin \theta} .$$ MEMORIZE!

53.   Which of the following quantities tend to become arbitrarily large when θ is near 90° but less than 90°?

(a) tan θ   (b) cot θ   (c) sec θ   (d) csc θ

(a) yes
(c) yes

54.   Which of these quantities tend to become arbitrarily large when θ is near 0°, but positive?

(a) tan θ   (b) cot θ   (c) sec θ   (d) csc θ

(a) no
(c) no

55.   For which values of θ, where 0° ≤ θ < 360°, are these functions undefined?

(a) tan θ   (b) cot θ   (c) sec θ   (d) csc θ

(a) 90° and 270°
(c) 90° and 270°

* * *

The next five exercises show that information about sin θ or cos θ can be used to calculate the values of the other five trigonometric functions.

56.   If 0° < θ < 90° and $\cos \theta = \frac{4}{5}$, use identities to find

(a) sin θ   (b) tan θ   (c) cot θ   (d) sec θ   (e) csc θ

57.   Do Exercise 56 if 270° < θ < 360°.

58.   If 0° < θ < 90° and $\sin \theta = \frac{5}{13}$, use identities to find

(a) cos θ   (b) tan θ   (c) cot θ   (d) sec θ   (e) csc θ

59.   Do Exercise 58 if 90° < θ < 180°.

60.   If 180° < θ < 270° and $\sin \theta = -\frac{24}{25}$, use identities to find

(a) cos θ   (b) tan θ   (c) cot θ   (d) sec θ   (e) csc θ

56. (a) $\frac{3}{5}$ (c) $\frac{4}{3}$
(e) $\frac{5}{3}$

57. (a) $-\frac{3}{5}$ (c) $-\frac{4}{3}$
(e) $-\frac{5}{3}$

58. (a) $\frac{12}{13}$ (c) $\frac{12}{5}$
(e) $\frac{13}{5}$

59. (a) $-\frac{12}{13}$ (c)$-\frac{12}{5}$
(e) $\frac{13}{5}$

60. (a)$-\frac{7}{25}$ (c) $\frac{7}{24}$
(e)$-\frac{25}{24}$

THE PYTHAGOREAN IDENTITIES

The Pythagorean Theorem implies the identity involving sin θ and cos θ:

$$\sin^2\theta + \cos^2\theta = 1$$

(See Exercise 21.) There are two other trigonometric identities that are consequences of this identity. The next exercise derives them

61. (a) Divide each side of the $\sin^2\theta + \cos^2\theta = 1$ by $\cos^2\theta$ to show that

$$\tan^2\theta + 1 = \sec^2\theta$$

(b) Divide each side of the identity $\sin^2\theta + \cos^2\theta = 1$ by $\sin^2\theta$ to show that

$$\cot^2\theta + 1 = \csc^2\theta$$

MEMORIZE the three Pythagorean identities.

62. (See Exercise 61.) For first quadrant θ which is larger,

(a) sec θ

(a) tan θ or sec θ?   (b) cot θ or csc θ?

63. Find sec θ if tan θ $= \frac{3}{4}$ and $0° < \theta < 90°$.

64. Find csc θ if cot θ $= -\frac{5}{12}$ and $270° < \theta < 360°$.

Relate this question to cos θ

65. In what quadrants is sec θ

65. (a) first and fourth

(a) positive?   (b) negative?

66. In what quadrants is csc θ

Relate this question to sin θ

(a) positive?   (b) negative?

67. Find tan θ if sec θ $= \frac{5}{4}$ and $270° < \theta < 360°$.

68. Can you find cot θ if csc θ $= \frac{\sqrt{2}}{2}$ and $180° < \theta < 270°$? Explain.

68. Hint: it is impossible. Why?

69. Show that sec θ is never between -1 and 1. Equivalently, show that $\sec^2\theta \geq 1$ whenever sec θ is defined.

70. In Chapter 3 it will be shown that sin 2θ = 2 sin θ cos θ. What does this imply about Exercise 1, which concerns sin 2θ and sin θ?

* * * SUMMARY * * *

This chapter has examined primarily three quantities that are associated with an angle θ in a unit circle:

cos θ,   sin θ,   tan θ.

They are defined by these two diagrams:

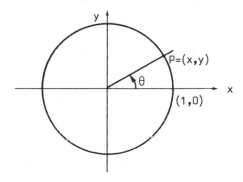

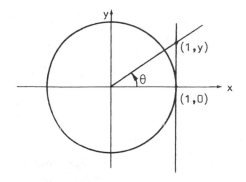

$$x = \cos \theta$$
$$y = \sin \theta$$

$$y = \tan \theta$$

Similar triangles showed that

$$\tan \theta = \frac{\sin \theta}{\cos \theta}.$$

Three other functions were defined in terms of $\sin \theta$ and $\cos \theta$:

$$\sec \theta = \frac{1}{\cos \theta}, \qquad \csc \theta = \frac{1}{\sin \theta}, \qquad \cot \theta = \frac{\cos \theta}{\sin \theta} \left(= \frac{1}{\tan \theta}\right)$$

Memorize these definitions

   If the value of one of the trigonometric functions of θ and the quadrant of θ are known, then the remaining five functions can be computed.

* * * WORDS AND SYMBOLS * * *

θ	$\csc \theta$
$\sin \theta$	$\sec \theta$
$\cos \theta$	$\cot \theta$
$\tan \theta$	quadrant

* * * IDENTITIES * * *

$$\sin(-\theta) = - \sin \theta$$
$$\cos(-\theta) = \cos \theta$$
$$\tan(-\theta) = - \tan \theta$$
$$\sin(90° - \theta) = \cos \theta$$
$$\cos(90° - \theta) = \sin \theta$$
$$\sin(90° + \theta) = \cos \theta$$
$$\cos(90° + \theta) = - \sin \theta$$

} MEMORIZE

$$\sin(180° - \theta) = \sin \theta$$
$$\cos(180° - \theta) = - \cos \theta$$
$$\tan(180° - \theta) = - \tan \theta$$
$$\sin(180° + \theta) = - \sin \theta$$
$$\cos(180° + \theta) = - \cos \theta$$
$$\tan(180° + \theta) = \tan \theta$$

Do not memorize the other identities. In each case, a quick sketch of the unit circle is an adequate reminder.

Memorize and be able to derive the Pythagorean identities.

Pythagorean identities:

$$\sin^2\theta + \cos^2\theta = 1 \qquad \tan^2\theta + 1 = \sec^2\theta \qquad \cot^2\theta + 1 = \csc^2\theta$$

* * * CONCLUDING REMARKS * * *

The sine as now defined in terms of the unit circle goes back to the Hindu mathematician Varahamihira around the year 500. It was introduced to the Arabs in the 9th century. About this time the Arab astronomers introduced the tangent and cotangent of an angle. By the end of the 10th century they had introduced the secant and cosecant. It was the Arabs who, preserving the astronomy and mathematics of the Greeks and combining it with their own discoveries and those of Hindu scientists, brought these sciences to Europe after the Dark Ages.

In the 19th century it was discovered that almost any function could be expressed in terms of sines and cosines. This will be described in the Concluding Remarks of a later chapter.

* * * UNIT CIRCLE SELF QUIZ * * *

1. Copy and fill in the table with the word "positive" or "negative."

	$\cos\theta$	$\sin\theta$	$\tan\theta$	$\sec\theta$	$\csc\theta$	$\cot\theta$
1st Quadrant						
2nd Quadrant						
3rd Quadrant						
4th Quadrant						

2. If $\cos\theta = -\dfrac{8}{17}$ and $90° < \theta < 180°$, find

   (a) $\sin\theta$ (b) $\tan\theta$ (c) $\sec\theta$ (d) $\csc\theta$ (e) $\cot\theta$

3. If $\sin\theta = -\dfrac{1}{\sqrt{3}}$ and $270° < \theta < 360°$, find

   (a) $\csc\theta$ (b) $\cot\theta$

4. If $\cos\theta = \dfrac{-2}{\sqrt{5}}$ and $180° < \theta < 270°$, find

   (a) $\sec\theta$ (b) $\tan\theta$

5. If $0° \le \theta < 360°$, state for which θ (if any) each of the following are undefined:

   (a) $\cos\theta$ (b) $\sin\theta$ (c) $\tan\theta$

   (d) $\cot\theta$ (e) $\sec\theta$ (f) $\csc\theta$

6. Which of the six trigonometric functions can never be

   (a) larger than 1 ?   (b) less than -1 ?

ANSWERS TO UNIT CIRCLE SELF QUIZ

		cos θ	sin θ	tan θ	sec θ	csc θ	cot θ
1.	1st Q.	+	+	+	+	+	+
	2nd Q.	−	+	−	−	+	−
	3rd Q.	−	−	+	−	−	+
	4th Q.	+	−	−	+	−	−

2 (a) $\dfrac{15}{17}$ (b) $-\dfrac{15}{8}$ (c) $-\dfrac{17}{8}$ (d) $\dfrac{17}{15}$ (e) $-\dfrac{8}{15}$

3. (a) $-\sqrt{3}$ (b) $-\sqrt{2}$

4. (a) $-\dfrac{\sqrt{5}}{2}$ (b) $\dfrac{1}{2}$

5. (a),(b) everywhere defined   (c) 90°, 270°   (d) 0°, 180°

   (e) 90°, 270°   (f) 0°, 180°

6. (a),(b) cosine and sine.

* * * UNIT CIRCLE REVIEW * * *

1. Explain, using diagrams, how sin θ and cos θ are defined.

2. Explain, using diagrams, how tan θ is defined.

3. (a) In what quadrants is cos θ positive? negative?

   (b) In what quadrants is sin θ positive? negative?

   (c) In what quadrants is tan θ positive? negative?

4. State the (Pythagorean) identity relating

   (a) sin θ and cos θ   (b) tan θ and sec θ   (c) cot θ and csc θ

5. What identity relates sin θ, cos θ, and tan θ?

6. Express each of the following in terms of sin θ and/or cos θ.

   (a) cot θ   (b) sec θ   (c) csc θ

7. Find all θ where 0° ≤ θ < 360° such that   *Hint: Use the*
                                                      unit circle.
   (a) tan θ = cot θ   (b) sin θ = cos θ

   (c) sin θ = csc θ   (d) cos θ = sec θ

8. If 0° ≤ θ < 360°, for what θ will sin θ be largest? smallest? equal
   to zero?

9. If $0° \le \theta < 360°$, for what θ will $\cos \theta$ be largest? smallest? equal to zero?

10. (a) If $0° \le \theta < 360°$, for what θ is $\tan\theta$ undefined? equal to zero?

    (b) Work part (a) in two ways:

    (1) using the unit circle   (2) using $\tan \theta = \dfrac{\sin \theta}{\cos \theta}$.

11. If $0° \le \theta < 360°$, for what θ is $\cot \theta$ undefined? equal to zero?

12. If $0° \le \theta < 360°$, for what θ is

    (a) $\sec \theta$ undefined   (b) $\csc \theta$ undefined   (c) $\sec \theta = 1$

    (d) $\csc \theta = 1$ (e) $\sec \theta = -1$ (f) $\csc \theta = -1$

    (g) $\sec \theta = 0$ (h) $\csc \theta = 0$ (i) $-1 < \sec \theta < 1$

    (j) $-1 < \csc \theta < 1$?

13. If $0° \le \theta < 360°$, for what θ is

    (a) $\sec \theta = 2$?   (b) $\csc \theta = 2$?

14. (a) Can $\sec \theta$ ever equal zero? Explain, with reference to $\cos \theta$.

    (b) Can $\csc \theta$ ever equal zero? Explain, with reference to $\sin \theta$.

15. Find the remaining trigonometric functions if

    (a) $\sin \theta = \dfrac{8}{17}$ and $90° < \theta < 180°$ (b) $\cos \theta = -.5$ and $180° < \theta < 270°$

    (c) $\sec \theta = 1.25$ and $270° < \theta < 360°$ (d) $\csc \theta = 2$ and $0° < \theta < 90°$

16. If $\sin 35° = .574$ and $\cos 35° = .819$ find

    (a) $\sin 55°$ (b) $\cos 55°$ (c) $\sin 125°$

    (d) $\cos 125°$ (e) $\sin 145°$ (f) $\cos 145°$

    (g) $\sin 215°$ (h) $\cos 215°$ (i) $\sin 325°$

    (j) $\cos 325°$ (k) $\sin (-35°)$ (l) $\cos (-35°)$

17. Find $\tan \theta$ if $270° < \theta < 360°$ and $\sec \theta = \sqrt{5}$.

18. Find $\csc \theta$ if $\cot \theta = -\sqrt{7}$ and $90° < \theta < 180°$.

19. For each trigonometric function in the first column, find <u>all</u> the statements in the second column which apply, if $0° \le \theta \le 90°$.

    (1) $\sin \theta$ (a)   goes through all positive numerical values

    (2) $\cos \theta$ (b)   never less than 0

    (3) $\tan \theta$ (c)   can never be 2

                               (d)   decreases as θ increases

                               (e)   increases as θ increases

                               (f)   is undefined for some θ

                               (g)   goes through all negative numerical values

ANSWERS TO THE UNIT CIRCLE REVIEW

1. See text.   2. See text.

3. (a) I, IV positive; II, III negative   4. (a) $\sin^2\theta + \cos^2\theta = 1$

   (b) I, II positive; III, IV negative   (b) $\tan^2\theta + 1 = \sec^2\theta$

   (c) I, III positive; II, IV negative   (c) $\cot^2\theta + 1 = \csc^2\theta$

5. $\dfrac{\sin\theta}{\cos\theta} = \tan\theta$

6. (a) $\cot\theta = \dfrac{\cos\theta}{\sin\theta}$ 7. (a) $\theta = 45°, 135°, 225°, 315°$

   (b) $\sec\theta = \dfrac{1}{\cos\theta}$ (b) $\theta = 45°, 225°$

                                                          (c) $\theta = 90°, 270°$

   (c) $\csc\theta = \dfrac{1}{\sin\theta}$ (d) $\theta = 0°, 180°$

8. largest: $\theta = 90°$ smallest: $270°$ equal 0: $0°, 180°$

9. largest: $\theta = 0°$ smallest: $180°$ equal 0: $90°, 270°$

10. (a) $\tan\theta$ undefined: $\theta = 90°, 270°$; equal 0: $\theta = 0°, 180°$

    (b) consult text

11. $\cot\theta$ undefined: $\theta = 0°, 180°$; equal 0: $90°, 270°$

12. (a) $\theta = 90°, 270°$ (b) $\theta = 0°, 180°$ (c) $\theta = 0°$ (d) $\theta = 90°$

    (e) $\theta = 180°$ (f) $\theta = 270°$ (g) none   (h) none

    (i) none   (j) none

13. (a) $60°, 300°$ 14. (a) no - see text definition

    (b) $30°, 150°$ (b) no - see text definition

15. (a) $\sin\theta = \dfrac{8}{17}$, $\cos\theta = -\dfrac{15}{17}$; $\tan\theta = -\dfrac{8}{15}$, $\cot\theta = -\dfrac{15}{8}$, $\sec\theta = -\dfrac{17}{15}$,

    $\csc\theta = \dfrac{17}{8}$ (b) $\sin\theta = -\dfrac{\sqrt{3}}{2}$, $\tan\theta = \sqrt{3}$, $\cot\theta = \dfrac{\sqrt{3}}{3}$, $\sec\theta = -2$,

    $\csc\theta = -\dfrac{2\sqrt{3}}{3}$ (c) $\sin\theta = -\dfrac{3}{5}$, $\cos\theta = \dfrac{4}{5}$, $\tan\theta = -\dfrac{3}{4}$, $\cot\theta = -\dfrac{4}{3}$,

    $\sec\theta = \dfrac{5}{4}$, $\csc\theta = -\dfrac{5}{3}$ (d) $\sin\theta = \dfrac{1}{2}$, $\cos\theta = \dfrac{\sqrt{3}}{3}$, $\tan\theta = \dfrac{\sqrt{3}}{3}$,

    $\cot\theta = \sqrt{3}$, $\sec\theta = \dfrac{2\sqrt{3}}{3}$, $\csc\theta = 2$

16. (a) .819   (b) .574   (c) .819   (d) -.574   (e) .574   (f) -.819

    (g) -.574   (h) -.819   (i) -.574   (j) 0.819   (k) -.574   (1) .819

17. $\tan\theta = -2$ 18. $\csc\theta = 2\sqrt{2}$

19. 1. (b), (c), (e)   2. (b), (c), (d)   3. (a), (b), (e), (f)

* * * UNIT CIRCLE REVIEW SELF TEST * * *

1. Match the two columns.

   trig function   second side of the angle is

   (i) $\cos \theta > 0$ (a) above the x axis

   (ii) $\csc \theta < 0$ (b) below the x axis

   (iii) $\sin \theta > 0$ (c) to the right of the y axis

   (iv) $\sec \theta < 0$ (d) to the left of the y axis

2. Find all the other trigonometric functions of θ if $\sin \theta = -\frac{\sqrt{2}}{2}$ and $270° < \theta < 360°$.

3. Use a sketch of the unit circle to determine which is the larger value, $\sin \theta$ or $\tan \theta$, if

   (a) $0° < \theta < 90°$ (b) $90° < \theta < 180°$

   (c) $180° < \theta < 270°$ (d) $270° < \theta < 360°$

4. As θ goes from $0°$ to $90°$, does each of the following increase or decrease in value?

   (a) $\cos \theta$ (b) $\sin \theta$ (c) $\tan \theta$

5. Write the trigonometric identity which relates

   (a) $\csc \theta$ and $\cot \theta$ (b) $\tan \theta$ and $\sec \theta$

6. If θ_1 and θ_2 are angles between 0 and 90 degrees, determine which angle is larger and explain your answers if:

   Hint: (b) and (c) Use the Pythagorean identities

   (a) $\cos \theta_1 = 0.7$ and $\cos \theta_2 = 0.5$

   (b) $\cos \theta_1 = 0.6$ and $\sin \theta_2 = 0.6$

   (c) $\cos \theta_1 = 0.4$ and $\sin \theta_2 = 0.8$

ANSWERS TO UNIT CIRCLE SELF TEST

1. (i) -(c), (ii) -(b) (iii) -(a)   (iv) -(d)

2. $\cos \theta = \frac{\sqrt{2}}{2}$, $\tan \theta = -1$, $\cot \theta = -1$, $\sec \theta = \sqrt{2}$, $\csc \theta = -\sqrt{2}$

3 (a) $\tan \theta$ (b) $\sin \theta$ (c) $\tan \theta$ (d) $\sin \theta$

4. (a) decrease (b) increase (c) increase

5. (a) $\cot^2 \theta + 1 = \csc^2 \theta$ (b) $\tan^2 \theta + 1 = \sec^2 \theta$

6. Use the fact that $\cos \theta$ decreases from $0°$ to $90°$, then

   (a) $\theta_1 < \theta_2$ (b) $\cos \theta_2 = 0.8$, thus $\theta_1 > \theta_2$.

   (c) $\cos \theta_2 = 0.6$, thus $\theta_1 > \theta_2$.

* * * THE UNIT CIRCLE PROJECTS * * *

With the aid of geometry it is possible to compute precisely the sine and cosine of certain frequently used angles. This is outlined in Projects 1.

1.   The diagram below shows an equilateral triangle AOB in the unit circle.

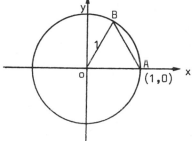

(a)   Show that angle AOB is 60°.

(b)   Show that coordinates of B are $x = \frac{1}{2}$ and $y = \frac{\sqrt{3}}{2}$.

(c)   Deduce that $\cos 60° = \frac{1}{2}$ and $\sin 60° = \frac{\sqrt{3}}{2}$.

2.   Using the results of Project 1 find $\cos 30°$ and $\sin 30°$.

3.   This diagram shows an isosceles right triangle AOB in the unit circle:

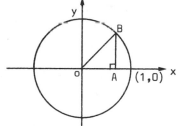

(a)   Show that angle AOB is 45°.

(b)   Show that the coordinates of B are $x = \frac{\sqrt{2}}{2}$ and $y = \frac{\sqrt{2}}{2}$.

(c)   Deduce that
$$\cos 45° = \frac{\sqrt{2}}{2} \quad \text{and} \quad \sin 45° = \frac{\sqrt{2}}{2}.$$

4.   Consider the following diagram for a positive θ near 0:

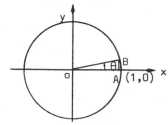

(a)   What happens to $\sin \theta$ (= AB) as θ gets closer and closer to 0?

(b)   What happens to 1 - cos θ (= AC) as θ gets closer and closer to 0, where point C = (1,0).

(c)   What do you think happens to

$$\frac{1 - \cos \theta}{\sin \theta} = \frac{AC}{AB}$$

as θ gets closer and closer to 0?

(d)   What do you think happens to

$$\frac{1 - \cos \theta}{\sin^2 \theta} = \frac{AC}{(AB)^2}$$

as θ gets closer and closer to 0?

5.   (a)   Show that $(1 - \cos \theta)(1 + \cos \theta) = \sin^2 \theta$.

     (b)   Use (a) to answer Project 4(d).

6.   A practical, quick estimate of sin θ for 0° ≤ θ ≤ 45° is given by this formula:

$$\sin \theta \doteq 0.017\theta - \frac{(0.017\theta)^3}{6}.$$

     (a)   What estimate does it give for θ = 10°, 30°, and 40°?

     (b)   Compare the estimates in (a) to those obtained with the aid of a ruler (or a calculator, or the table of sin θ at the back of the text.)

7.   The Pythagorean theorem, which played a critical role in this chapter, deserves to be proved again. Here is an outline of a very short proof, which depends on this diagram in which there are three right triangles.

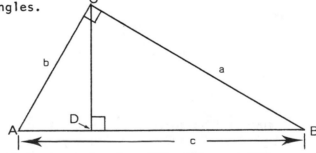

     (a)   Prove that the triangles ADC, CDB, and ACB are similar.

     (b)   Deduce that there is a constant k ≠ 0 such that

$$\text{area } ACD = kb^2$$

$$\text{area } CDB = ka^2$$

and $$\text{area } ACB = kc^2$$

(c)   Deduce that
$$c^2 = a^2 + b^2 .$$

8. For $0° < \theta < 45°$ how do tan 2θ and 2 tan θ compare? Are they ever equal? Is one always larger than the other? (Experiment with various angles, using a unit circle of radius ten centimeters, protractor, and centimeter ruler.)

9. For the early history of trigonometry see Science Awakening by B.L. Van der Waerden, Oxford, New York, 1961.

10. For an extensive discussion of the history of trigonometry see Mathematical Thought from Ancient to Modern Times, by Morris Kline, Oxford, New York, 1972.

11. Define the function f by

$$f(\theta) = \frac{\cot^2\theta \ \sec^2\theta - 1}{\csc\theta \ \cot^2\theta \ \sin\theta} .$$

Show that f is constant.

<table>
<tr><td>**2**
Trigonometry</td><td>**Triangles**</td></tr>
</table>

Any triangle has three angles and three sides. In the figure below, the angles are labelled A, B, and C, while the sides have length a, b, and c:

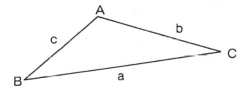

NOTATION: Side c and angle C are opposite each other, as are side b and angle B, and side a and angle A.

Not all three angles and three sides are required to describe completely the size and shape of the triangle. For instance, two sides and the angle included between them (say c, A, and b) provide enough information.

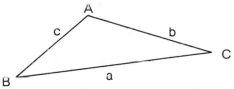

Similarly, when given A, c, and B the triangle is determined. (Draw it).
Also, the sides a, b, and c specify the triangle completely.

These are the SAS, ASA and SSS cases from geometry.

The above geometric facts raise a practical problem:

When given the minimum number of sides and angles
to determine a triangle (SAS, ASA, or SSS), how
do you calculate the unknown angles and sides?

This is the problem that is solved in this chapter. The formulas developed have applications in the everyday world, such as surveying and shop work. They are also required in elementary physics courses, analytic geometry, and calculus.

EXERCISES

1. In the following figure points A and B are on a flat valley and P is

the top of a mountain whose elevation is 10° at point A and 25° at point B.

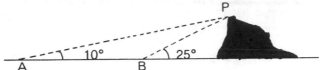

Points A, B, and P are in the same vertical plane. Points A and B are one mile apart.

(a) Make an accurate scale diagram and estimate the height of the mountain.

(b) Compare your answer with those of other students.

Later in this chapter Exercise 1 will be solved precisely with the aid of trigonometry.

THREE USEFUL GEOMETRIC FACTS

The following three theorems from geometry will be used frequently in this chapter:

(i) The area of a triangle is one-half the base times the altitude,

Area=½bh

h

b

(ii) (The Pythagorean Theorem) For all right triangles,

$$c^2 = a^2 + b^2$$

c

a

b

(iii) Corresponding sides of similar triangles are proportional:

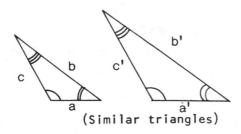

$$\frac{a}{a'} = \frac{b}{b'} = \frac{c}{c'}$$

(Similar triangles)

THE PLANE AND THE UNIT CIRCLE

In the previous chapter the sine, cosine, and tangent functions were defined in terms of the unit circle. Now they will be related to any point in the coordinate plane (other than the origin).

2. Let θ be an acute angle and let P be the point where the second arm of the angle meets the unit circle. Let Q be any other point on this second arm (other than the origin).

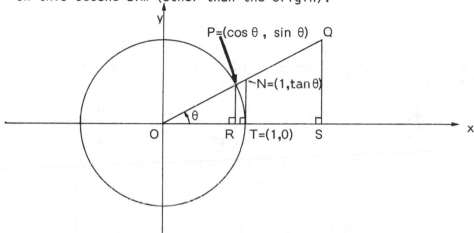

Use similar triangles to show that

(a) $\cos \theta = \dfrac{OS}{OQ}$

(b) $\sin \theta = \dfrac{SQ}{OQ}$

(c) $\tan \theta = \dfrac{SQ}{OS}$

Observe that Exercise 2 shows the relation of cos θ, sin θ and tan θ to a point <u>not</u> on the unit circle in the first quadrant. (The other quadrants will be considered shortly.)

3. Use the results of Exercise 2 to compute cos θ, sin θ, and tan θ if the second arm of the angle θ passes through the point Q = (2,1).

4. As in Exercise 3, compute cos θ, sin θ, and tan θ if the second arm of the angle θ passes through

(a)   Q = (3,4)   (b)   Q = (5,12)   (c)   Q = (1,5)

* * *

The next exercise is a generalization of Exercises 3 and 4. Though the formulas are stated only for first quadrant angles, it is not difficult to check that they hold in other quadrants also.

Remember that the circle has radius 1.

$cos\ \theta = \dfrac{2}{\sqrt{5}}$

$sin\ \theta = \dfrac{1}{\sqrt{5}}$

$tan\ \theta = \dfrac{1}{2}$

(a) $cos\ \theta = \dfrac{3}{5}$

$sin\ \theta = \dfrac{4}{5}$

$tan\ \theta = \dfrac{4}{3}$

(c) $cos\ \theta = \dfrac{1}{\sqrt{26}}$

$sin\ \theta = \dfrac{5}{\sqrt{26}}$

$tan\ \theta = 5$

7. (a) $cos\ \theta = \dfrac{-2}{\sqrt{5}}$

$sin\ \theta = \dfrac{1}{\sqrt{5}}$

$tan\ \theta = -\dfrac{1}{2}$

(c) $cos\ \theta = \dfrac{2}{\sqrt{13}}$

$sin\ \theta = \dfrac{-3}{\sqrt{13}}$

$tan\ \theta = \dfrac{-3}{2}$

8. (a) $cos\ \theta = \dfrac{3}{\sqrt{34}}$

$sin\ \theta = \dfrac{5}{\sqrt{34}}$

$tan\ \theta = \dfrac{5}{3}$

(c) $cos\ \theta = -\dfrac{7}{25}$

$sin\ \theta = -\dfrac{24}{25}$

$tan\ \theta = \dfrac{24}{7}$

(e) $cos\ \theta = 1$
$sin\ \theta = 0$
$tan\ \theta = 0$

9. (a) $cos\ \theta = \dfrac{1}{\sqrt{2}}$

$sin\ \theta = \dfrac{1}{\sqrt{2}}$

$tan\ \theta = 1$

(c) $cos\ \theta = -\dfrac{\sqrt{3}}{2}$

$sin\ \theta = -\dfrac{1}{2}$

$tan\ \theta = \dfrac{1}{\sqrt{3}}$

(e) $cos\ \theta = \dfrac{1}{2}$

$sin\ \theta = \dfrac{-\sqrt{3}}{2}$

$tan\ \theta = -\sqrt{3}$

5. If Q is any point (a,b), show that:

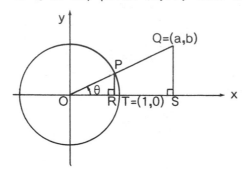
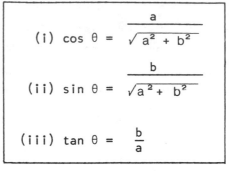

(i) $cos\ \theta = \dfrac{a}{\sqrt{a^2 + b^2}}$

(ii) $sin\ \theta = \dfrac{b}{\sqrt{a^2 + b^2}}$

(iii) $tan\ \theta = \dfrac{b}{a}$

6. Find cos θ, sin θ, and tan θ if the second arm of θ passes through the point

(a) (15,8)   (b) $(\frac{3}{7}, \frac{4}{7})$ (c) $(\frac{1}{4}, \frac{3}{4})$

Keep in mind when working the next three exercises that the formulas in Exercise 5 are valid in all four quadrants.

7. Find sin θ, cos θ, and tan θ if the second arm of θ goes through the point:

(a) (-2,1)   (b) (-3,-5)   (c) (4,-6)   (d) (-1,1)

8. Find cos θ, sin θ, and tan θ if the second arm of θ passes through the point:

(a) (3,5)   (b) (-4,3)   (c) (-7,-24)

(d) (m,n)   (e) (3,0)   (f) (0,2)

9. Find sin θ, cos θ, and tan θ if the second arm of θ passes through point:

(a) (1,1)   (b) (-1,√3)   (c) (-√3, -1)

(d) (3,-3)   (e) (1,-√3)   (f) (-2,-2)

TWO IMPORTANT REFERENCE TRIANGLES

The angles 30°, 45°, and 60°occur so often in trigonometry that it is valuable to know the trigonometric functions of these angles. From these values, related trigonometric values of other angles, such as

120°, 135°, 150°, 210°, 215°, . . .

may be found easily. The trigonometric values for angles of 30°, 45°, and 60° are obtainable from an inspection of the following two triangles.

The 30°-60°-90° triangle:

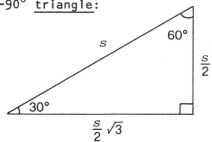

$$\frac{S}{2}$$ short leg = $(\frac{1}{2})$(hypotenuse)

long leg = $(\sqrt{3})$(short leg)

The 45°-45°-90° triangle:

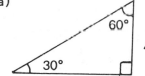

hypotenuse = $(\sqrt{2})$(leg)

Memorize the relation among the sides for each of these triangles.

The next exercise offers practice in dealing with these two triangles.

10. Solve for the remaining sides in each triangle.

(a)

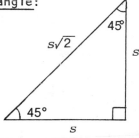

(b)

(c)

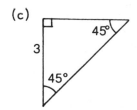

(d)

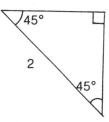

*(a) hyp. = 8
long leg = 4 $\sqrt{3}$*

*(c) hyp. = 3$\sqrt{2}$
     leg = 3*

The next two examples illustrate how the 30°-60°-90° and 45°-45°-90° triangles are frequently used to find the angle θ if the second arm passes through a "convenient" point Q.

Example 1. Find θ if the second arm of the angle θ passes through the point Q = $(-\sqrt{3}, 1)$

Solution.   Sketch a diagram:

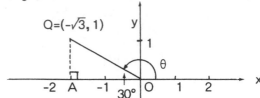

The point $(-\sqrt{3}, 1)$ gives a reference right triangle with sides of length 1, $\sqrt{3}$ and 2.

Note the reference triangle OAQ is a 30°-60°-90° triangle and angle QOA is 30°, thus,

$$30° + \theta = 180°$$

or $$\theta = 180° - 30°$$

hence $$\theta = 150°$$

Example 2.   Find θ if the second arm of the angle θ passes through the point Q = (5,-5).

Solution.   A sketch shows a 45°-45°-90° reference triangle (an isosceles right triangle).

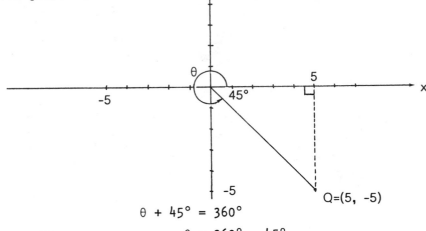

The point (5, -5) gives a 45° - 45° - 90° reference triangle with two equal legs of length 5.

Thus $$\theta + 45° = 360°$$

or $$\theta = 360° - 45°$$

hence $$\theta = 315°.$$

11.   Plot each point in Exercise 9 and using a suitable reference triangle, find θ in each case.

Conversely if $\tan \theta$ and the quadrant of the angle are known, it is possible to compute other points on the second arm, since $\tan \theta = \frac{b}{a}$ $= \frac{rb}{ra}$ $(r \neq 0)$.

When the point (a,b) lies on the second arm of the angle θ,

$$\tan \theta = \frac{b}{a}$$

$$= \frac{rb}{ra}. \qquad (r \neq 0)$$

For any positive real number r, the point (ra,rb) also lies on the same arm that (a,b) does.   Example 3 illustrates this.

Example 3.   Let $\tan \theta = \frac{3}{7}$ and $180° < \theta < 270°$.   Find two points on the second arm of the angle.

Solution.   A sketch shows a third quadrant angle whose second arm passes through $(-7,-3)$,

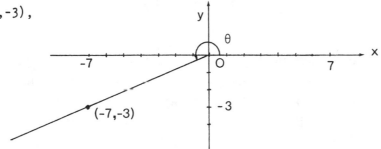

Hence $(-14,-6)$ is also on the second arm.   In fact, for any positive real number r the point $(-7r,-3r)$ is also on the second arm.

12.   If $0° < \theta < 90°$ and $\tan \theta = \frac{2}{3}$, which of the following points is on the second arm of θ?

(a) $(3,2)$ (b) $(1,\frac{2}{3})$ (c) $(\frac{1}{6},\frac{1}{3})$ (d) $(\frac{2}{5},\frac{3}{5})$ (e) $(\sqrt{54},\sqrt{24})$

13.   Find the coordinates of at least one point on the second arm of θ, other than the origin if $\tan \theta = \frac{1}{2}$ and

(a) $0° < \theta < 90°$ (b) $180° < \theta < 270°$

14.   Find coordinates of at least two points on the second arm of θ if

(a) $\tan \theta = \frac{1}{3}$ and $180° < \theta < 270°$ (b) $\tan \theta = \sqrt{3}$ and $0° < \theta < 90°$

(c) $\tan \theta = -\frac{1}{4}$ and $270° < \theta < 360°$ (d) $\tan \theta = -\frac{4}{5}$ and $90° < \theta < 180°$

15.   For each of the angles in Exercise 14, use the coordinates of a point you found and the results in Exercise 5 to find all the other trigonometric functions.

TRIANGLES WITHOUT THE COORDINATE SYSTEM

The sine, cosine, and tangent were defined in the previous chapter in terms of the coordinates of certain points associated with the unit circle.   This chapter has shown how these three functions can also be calculated in terms of a point (other than the origin) anywhere in the plane and in terms of certain right triangles whose legs are parallel to the axes.   In other words, for the time being, the unit circle is less important.   Next you will learn how to express the trigonometric functions

*14. (a)   (-3,-1) or
(-6,-2)
(c)   (4,-1) or
(8,-2)
(There are many
others)*

15.(a) $\cos \theta = -\frac{3}{\sqrt{10}}$

$\sin \theta = -\frac{1}{\sqrt{10}}$

$\cot \theta = 3$

$\sec \theta = -\frac{\sqrt{10}}{3}$

$\csc \theta = -\sqrt{10}$

(c) $\cos \theta = \frac{4}{\sqrt{17}}$

$\sin \theta = \frac{-1}{\sqrt{17}}$

$\cot \theta = -4$

$\sec \theta = \frac{\sqrt{17}}{4}$

$\csc \theta = -\sqrt{17}$

of a right triangle independently of a coordinate system. Only a right triangle will be needed. The chapter now begins to present the surveyor's view of trigonometry. The angle θ to be considered will be restricted to an acute angle,

$$0° < \theta < 90°$$

Let the second arm of the angle θ be in the first quadrant and pass through the point P = (a,b), other than the origin. Consider the right triangle OAP that P determines in this diagram:

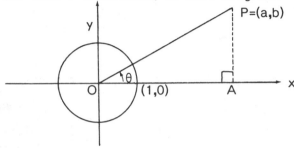

Now disregard the coordinate axes and the unit circle. All that is left is this right triangle:

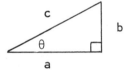

$$(c^2 = a^2 + b^2)$$

$$c = \sqrt{a^2 + b^2}$$

In short, the six trigonometric functions may be expressed in terms of the sides of the triangle:

Study these six definitions carefully. Memorize them. Say them aloud to another student.

Observe that the trig functions here refer to a right triangle.

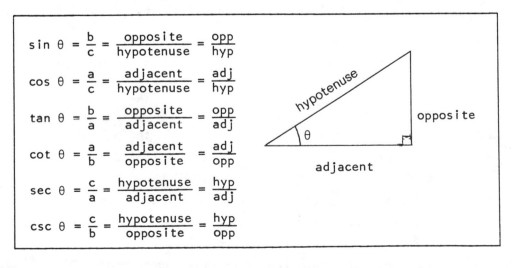

$$\sin \theta = \frac{b}{c} = \frac{\text{opposite}}{\text{hypotenuse}} = \frac{\text{opp}}{\text{hyp}}$$

$$\cos \theta = \frac{a}{c} = \frac{\text{adjacent}}{\text{hypotenuse}} = \frac{\text{adj}}{\text{hyp}}$$

$$\tan \theta = \frac{b}{a} = \frac{\text{opposite}}{\text{adjacent}} = \frac{\text{opp}}{\text{adj}}$$

$$\cot \theta = \frac{a}{b} = \frac{\text{adjacent}}{\text{opposite}} = \frac{\text{adj}}{\text{opp}}$$

$$\sec \theta = \frac{c}{a} = \frac{\text{hypotenuse}}{\text{adjacent}} = \frac{\text{hyp}}{\text{adj}}$$

$$\csc \theta = \frac{c}{b} = \frac{\text{hypotenuse}}{\text{opposite}} = \frac{\text{hyp}}{\text{opp}}$$

Example 4.   Find the trigonometric functions of θ in the triangle,

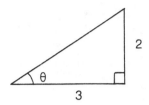

Solution.   By the Pythagorean Theorem the hypotenuse is

$\sqrt{2^2 + 3^2} = \sqrt{13}$. Therefore,

$$\cos \theta = \frac{adj}{hyp} \qquad \sin \theta = \frac{opp}{hyp} \qquad \tan \theta = \frac{opp}{adj}$$

$$= \frac{3}{\sqrt{13}} \qquad\qquad = \frac{2}{\sqrt{13}} \qquad\qquad = \frac{2}{3}$$

$$\sec \theta = \frac{1}{\cos \theta} \qquad \csc \theta = \frac{1}{\sin \theta} \qquad \cot \theta = \frac{1}{\tan \theta}$$

$$= \frac{\sqrt{13}}{3} \qquad\qquad = \frac{\sqrt{13}}{2} \qquad\qquad = \frac{3}{2}$$

Either of the two acute angles of a right triangle can be named θ. "Adjacent" still means "adjacent to the angle." The next example illustrates this important idea.

Example 5.   Calculate the six trigonometric functions of θ in this triangle.

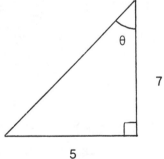

Solution.   First of all, the hypotenuse is equal to $\sqrt{5^2 + 7^2} = \sqrt{74}$. Therefore,

$$\cos \theta = \frac{adj}{hyp} \qquad \sin \theta = \frac{opp}{hyp} \qquad \tan \theta = \frac{opp}{adj}$$

$$= \frac{7}{\sqrt{74}} \qquad\qquad = \frac{5}{\sqrt{74}} \qquad\qquad = \frac{5}{7}$$

(continued)

$$\sec \theta = \frac{1}{\cos \theta} \qquad\qquad \csc \theta = \frac{1}{\sin \theta} \qquad\qquad \cot \theta = \frac{1}{\tan \theta}$$

$$= \frac{\sqrt{74}}{7} \qquad\qquad\qquad = \frac{\sqrt{74}}{5} \qquad\qquad\qquad = \frac{7}{5}$$

16. Using a ruler and graph paper, draw a right triangle with an acute angle θ such that

(a) $\sin \theta = \frac{1}{3}$ (b) $\cos \theta = \frac{2}{5}$ (c) $\tan \theta = 4$

(d) $\csc \theta = \frac{4}{3}$ (e) $\cot \theta = \frac{2}{5}$ (f) $\sec \theta = 4$

Always copy the figure. Show all work.

(a) $\cos \theta = \dfrac{5}{\sqrt{34}}$

      $\sin \theta = \dfrac{3}{\sqrt{34}}$

      $\tan \theta = \dfrac{3}{5}$

      $\cot \theta = \dfrac{5}{3}$

      $\sec \theta = \dfrac{\sqrt{34}}{5}$

      $\csc \theta = \dfrac{\sqrt{34}}{3}$

(c) $\cos \theta = \dfrac{2}{\sqrt{5}}$

      $\sin \theta = \dfrac{1}{\sqrt{5}}$

      $\tan \theta = \dfrac{1}{2}$

      $\cot \theta = 2$

      $\sec \theta = \dfrac{\sqrt{5}}{2}$

      $\csc \theta = \sqrt{5}$

(e) $\cos \theta = \dfrac{\sqrt{3}}{2}$

      $\sin \theta = \dfrac{1}{2}$

      $\tan \theta = \dfrac{1}{\sqrt{3}}$

      $\cot \theta = \sqrt{3}$

      $\sec \theta = \dfrac{2}{\sqrt{3}}$

      $\csc \theta = 2$

17. Find the trigonometric functions of θ in each of the following right triangles in terms of the lengths of the sides.

(a)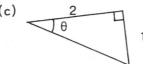

(b)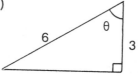

(c)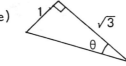

(d)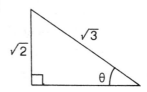

(e)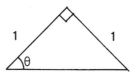

(f)

CONCERNING SQUARE ROOTS IN THE DENOMINATOR

One of the answers for Exercise 17   (e) is expressed in the form

$$\tan \theta = \frac{1}{\sqrt{3}} \; .$$

There is no need to rationalize the denominator. Indeed, the answer may be clearer if the denominator is left <u>not</u> rationalized. However, for some reason we might want to rationalize the denominator. Then

$$\frac{1}{\sqrt{3}} = \frac{1}{\sqrt{3}} \cdot \frac{\sqrt{3}}{\sqrt{3}}$$

$$= \frac{\sqrt{3}}{3}$$

$$\doteq \frac{1.732}{3}$$

or $\doteq 0.577$

18. Suppose that θ is an acute angle in a right triangle. Make a rough sketch of the triangle and compute the six trigonometric functions of θ if

(a) $\theta = 30°$ (b) $\theta = 45°$ (c) $\theta = 60°$

THE CALCULATOR AND TRIGONOMETRIC FUNCTIONS

Most angles, unfortunately, are not 30°, 45°, and 60°, common though they may be. However, the trigonometric functions of other angles have been calculated to many decimal places. (Even a thousand years ago Arab astronomers tabulated them to six decimal accuracy.) A glance at the back of the book will reveal a table of sine, cosine, and tangent for angles from 0° to 90° at 1° intervals. Such tables are built into scientific calculators. We use the calculator to solve two kinds of trigonometric problems:

(1) given the angle θ, then find $\cos \theta$, $\sin \theta$, or $\tan \theta$,

(2) given one of $\cos \theta$, $\sin \theta$, or $\tan \theta$, then find θ.

For instance if $\theta = 35°$, the calculator shows

$$\cos \theta = 0.819152 \ldots$$
$$\sin \theta = 0.573576 \ldots$$
$$\tan \theta = 0.700207 \ldots$$

On the other hand, if we know that

$$\cos \theta = 0.438371 \ldots$$

we can use the INV and Cos calculator keys (in that order) to find that

$$\theta = 64°$$

The next example offers more than one approach to solving for θ involving a right triangle:

Example 6. Estimate the angle θ to the nearest tenth of a degree in this right triangle:

It may be helpful to look back at the discussion just before Exercise 10.

(a) $\cos 30° = \frac{\sqrt{3}}{2}$

$\sin 30° = \frac{1}{2}$

$\tan 30° = \frac{\sqrt{3}}{3}$

$\cot 30° = \sqrt{3}$

$\sec 30° = \frac{2}{\sqrt{3}}$

$\csc 30° = 2$

(c) $\cos 60° = \frac{1}{2}$

$\sin 60° = \frac{\sqrt{3}}{2}$

$\tan 60° = \sqrt{3}$

$\cot 60° = \frac{1}{\sqrt{3}}$

$\sec 60° = 2$

$\csc 60° = \frac{2}{\sqrt{3}}$

For now, disregard the "radians" column in the table. Later it will be important. This is another way of measuring angles which you will learn about in Chapter 4.

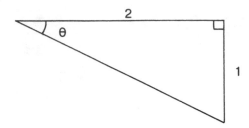

Solution.   Because the Pythagorean Theorem gives the hypotenuse, any one of $\sin\theta$, $\cos\theta$, or $\tan\theta$ could be used.   The first two involve the hypotenuse, which is $\sqrt{5}$:

$$\sin\theta = \frac{1}{\sqrt{5}} \doteq 0.447213 \ldots$$

and $$\cos\theta = \frac{2}{\sqrt{5}} \doteq 0.894427 \ldots$$

In both cases enter the data $\sin\theta$ (or $\cos\theta$) and then use the INV sin (or INV cos) keys to find θ.

However, it is more direct to use

$$\tan\theta = \frac{1}{2} = 0.5$$

then use the INV tan keys to get

$$\theta \doteq 26.56505°$$

We may round off this number to 26.57° or some value consistent with the accuracy of our original data.

Example 7.   Estimate θ in the diagram to the nearest tenth of a degree.

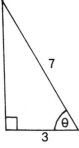

Solution.   First of all,

$$\cos\theta = \frac{3}{7} \doteq 0.428571$$

Using the INV Cos keys, we find

$$\theta = 64.6°$$

19. Using a calculator, find θ to the nearest tenth of a degree in each of the following:

(a)

(b)

(c)

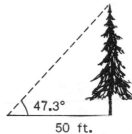

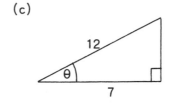

(a) 11.5°
(c) 54.4°

20. If the elevation of the sun is 47.3° and the shadow of a tree is 50 feet, how tall is the tree?

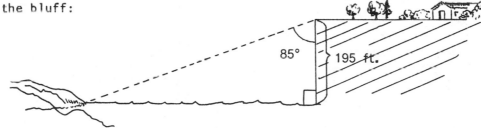

54.2 ft.

21. From a point on top of a 195 foot vertical bluff, a point on the opposite side of a river determines an angle of 85° with the side of the bluff:

85°   195 ft.

Find x, the width of the river.

21. 2228.85 ft.

If there were no error in the 85° and the 195 feet, the answer would be exact. In reality the data are only approximate. From a practical point, the answer should be rounded off to three significant figures. In scientific notation, this would be $(2.23)10^3$ feet. Although the calculator may give greater digit accuracy, the answer must be rounded off to agree with digit accuracy of the original data.

SOLVING RIGHT TRIANGLES

Some exercises have already involved finding some unknown side or angle of a right triangle, given some information about other angles or sides. To <u>solve</u> a triangle means "to find all of the unknown sides and angles." Thus, all three sides and three angles are then known.

NOTATION. In a right triangle, letters a, b, c denote the lengths of the three sides. The hypotenuse has length c. The three angles are denoted A, B, and C, with angle A opposite side a, and angle B opposite side b, and angle C opposite the hypotenuse c. (Thus angle C = 90°.)

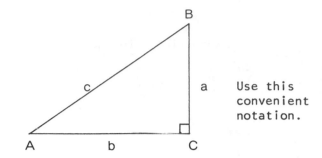

Use this
convenient
notation.

The next exercise illustrates the complete solution of a right
triangle.

Example 8.   Solve for the remaining sides and angles of the right
triangle in this diagram:

*When data are
given to only
three digit
accuracy, it is
best to round off
answers to three
digits. This is
especially true
when a calculator
is used.*

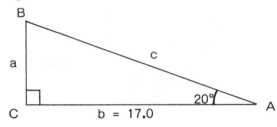

Solution.   Since $\angle$ C = 90°, $\angle$ B = 90° - 20° = 70°.

Next solve for the two unknown lengths:

$$\frac{a}{17.0} = \tan 20°$$

$$a = (17.0)\tan 20°$$

$$= (17.0)(0.364)$$

$$= 6.188$$

$$\doteq 6.19$$

$$\cos 20° = \frac{17.0}{c}$$

$$c = \frac{17.0}{\cos 20°}$$

$$c = \frac{17.0}{0.940}$$

$$\doteq 18.1$$

The triangle is solved, for all three sides and three angles are
known.

22. Use tables to solve for the remaining angles and sides:

(a)   (b)

*DRAW THE FIGURE
ON YOUR PAPER!*

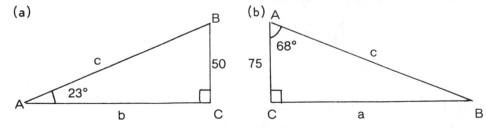

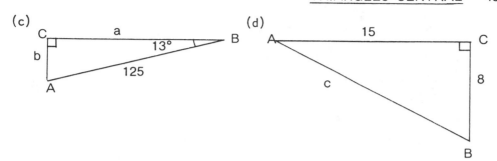

(c)

(d) *Find the angles to one decimal place*

(a) ∢B = 67°;
b = 117.8≈118;
c = 127.9≈128
(c) a = 121.8≈122
b = 28.125≈28.1
∢A = 77°

23. Find h in the following diagram,

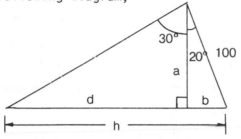

(a) by drawing an accurate scale diagram,

(b) by trigonometry.

    (i) find b   (ii) find a   (iii) find d

(c) Finally, find h.

HINT: First find a and b, the other two sides of the right-hand triangle. Use a to find d, the side opposite the 30° angle in the left-hand triangle. Now, find h.

* * *

With the trigonometry already available in this chapter it is now possible to solve accurately the problem posed in Exercise 1. The next exercise outlines this solution.

24. Recall that the problem in Exercise 1 concerns the height of a distant mountain. Points A and B are one mile apart and point P is the top of the mountain; A, B, and P are points in a vertical plane. When viewed from A the point P determines the angle of 10°; when viewed from B it determines an angle of 25°. Find the height of the mountain. This height is denoted h in this following diagram:

Try to solve it without looking ahead. Hint: Use tan 10° and tan 25°.

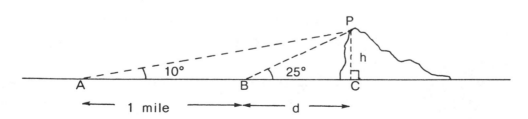

The solution involves trigonometry and a little algebra.

(a)   Express tan 25° in terms of d and h.

(b)   Express tan 10° in terms of d and h.

*Hint: (c) Solve
for h in (a)
and (b).*

(c)   Solve for h in (a) and (b) and deduce that

$$d(\tan 25°) = (1 + d)\tan 10°$$

(a) tan 25° = $\frac{h}{d}$

*(e) h is about
1497 feet.*

(d)   Solve for d in (c) and express it as a decimal.

(e)   Having found d, find h, the height of the mountain.

SOLVING TRIANGLES THAT ARE NOT RIGHT TRIANGLES

Right triangles are special in that one of their angles is 90°. The chapter now shows the solution of any triangle. For this purpose two important relations are developed, these are known as "the law of sines" and the "law of cosines."

THE LAW OF SINES

The next two exercises introduce the law of sines.

25.   Consider the triangle with

    ✱ A = 50°   ✱ B = 60°   ✱ C = 70°

    a = 88.4   b = 100   c = 108.5

*At most, four
decimal place
accuracy is
possible with the
given data.*

Compute to four decimal places:

(a) $\dfrac{\sin A}{a}$ (b) $\dfrac{\sin B}{b}$ (c) $\dfrac{\sin C}{c}$

(d)   What would you guess is the relation among $\dfrac{\sin A}{a}$, $\dfrac{\sin B}{b}$, and $\dfrac{\sin C}{c}$?

                    *   *   *

In the next exercise a centimeter ruler, a well sharpened pencil, and a protractor will be needed.

26.   Draw a triangle such that all of the sides are at least 10 centimeters long. Make sure it is <u>not</u> a right triangle; draw the three angles of different sizes.

(a)   Label the angles A, B, and C. Measure them with a protractor.

(b)   Find sin A, sin B, and sin C with a calculator.

(c)   Label the corresponding sides a, b, and c. (Be sure that it is labeled properly; that is, side a is opposite angle A, etc.) Measure the sides with a centimeter ruler.

(continued)

(d)   Check your triangle for the following relationships:

$$\frac{\sin A}{a} = \frac{\sin B}{b} = \frac{\sin C}{c}$$

NOTE: *In practice the Law of Sines is used as*

$$\frac{\sin A}{a} = \frac{\sin B}{b}$$

or

$$\frac{\sin B}{b} = \frac{\sin C}{c}$$

or

$$\frac{\sin A}{a} = \frac{\sin C}{c}$$

rather than the form in part (d)

* * *

The next exercise contains a proof of the Law of Sines, which asserts that the relations in Exercise 26. (d) hold for all triangles.

27.   Consider triangle A, B, C with sides a, b, c. Let h be the length of the altitude from vertex C to side c.

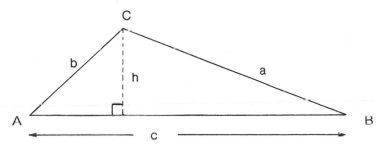

(a)   Show that

h = b sin A and that h = a sin B.

(b)   Eliminate h from the equations in (a) and show that

$$\frac{\sin A}{a} = \frac{\sin B}{b}$$

Hint: (b) Show that
b sin A = a sin B

(c)   Next draw an altitude from A to side a (which in this case has to be extended):

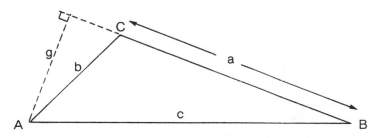

Let g be the length of this altitude.   Show that

$$g = b \sin (180° - C)$$
$$= b \sin C.$$

Recall from the unit circle that
sin(180° − θ) = sin θ

(d)   Show that g = c sin B.

(e)   From (c) and (d) deduce that

$$\frac{\sin B}{b} = \frac{\sin C}{c}.$$

Hint: (e) show that
c sin B = b sin C.

(continued)

Taking (b) and (e) together yields the important

LAW OF SINES
$\dfrac{\sin A}{a} = \dfrac{\sin B}{b} = \dfrac{\sin C}{c}$

The next example shows how the Law of Sines is used when solving a triangle.

Example 9.   Solve for the other parts of the following triangle if

   A = 115°,   C = 23°, and b = 12.3.

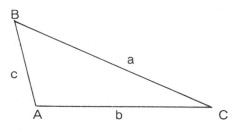

∡ is a common abbreviation for "angle." We often use ∡ A to both name and to show its measure. The context should make the meaning clear.

Solution.   ∡ B = 180° − (115° + 23°) = 42°

$$\frac{\sin 42°}{12.3} = \frac{\sin 23°}{c} \qquad \text{and} \qquad \frac{\sin 42°}{12.3} = \frac{\sin 115°}{a}$$

$$c = \frac{(\sin 23°)(12.3)}{\sin 42°} \qquad\qquad a = \frac{(\sin 65°)(12.3)}{\sin 42°}$$

$$= \frac{(0.391)(12.3)}{0.669} \qquad\qquad\quad = \frac{(0.906)(12.3)}{0.669}$$

$$\doteq 7.19 \qquad\qquad\qquad\qquad\quad \doteq 16.7$$

Again, recall that sin(180°−θ) = sinθ, thus sin 115° = sin 65°

The next exercise, like Example 9, illustrates the use of the Law of Sines.

28.   Solve for the other angles and sides if:

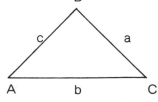

(a)   ∡ A = 33°,   ∡ B = 40°, and c = 15

(b)   ∡ C = 55°,   ∡ A = 110°, and b = 8.3

(c)   ∡ B = 74°,   ∡ C = 29°, and a = 2.1

(a) ∡ C = 107°
* a = 8.54*
* b = 10.1*
(c) ∡ A = 77°
* b = 2.07*
* c = 1.04*

WHEN TO USE THE LAW OF SINES

The Law of Sines helps solve a triangle if the given information consists of a side and the two adjacent angles (the ASA condition in geometry) It also takes care of the case where any two angles and a side are known (but AAS is actually ASA).

However, the Law of Sines will <u>not</u> help solve a triangle if the three sides are known (SSS), nor will it help in the case when an angle and the two adjacent sides are known (the SAS case in geometry).

Make a sketch of a triangle to convince yourself that the Law of Sines will not solve these cases.

THE LAW OF COSINES

The next exercise introduces the other basic tool for solving triangles, the Law of Cosines. It will be useful in solving the SAS and SSS cases.

29. Consider a triangle with a = 5, b = 7, and c = 6.

   (a) Is the Law of Sines of use in finding ∠ C?

   (b) It turns out that ∠ C ≐ 57.1° . Using this value for ∠ C, compute and compare the values of the following numbers:

$$c^2 \quad \text{and} \quad a^2 + b^2 - 2ab \cos C$$

Note that ∠ C is approximately 57.1°

   (c) This is the Law of Cosines which Exercises 30 and 31 will show valid for any triangle:

> ### THE LAW OF COSINES
> $$c^2 = a^2 + b^2 - 2ab \cos C$$

MEMORIZE!

Note. If the triangle is a right triangle and C is the 90° angle, then $\cos 90° = 0$, and the Law of Cosines is reduced to the Pythagorean Theorem. The next two exercises outline a proof of the Law of Cosines. The proof, which depends heavily on the Pythagorean Theorem, must be broken into two cases: when angle C is less than 90° (acute), and the case when angle C is more than 90° (obtuse). That is why the proof takes Exercises 30 and 31 to complete.

* * *

30. This exercise proves the Law of Cosines when angle C is acute. (The proof uses the Pythagorean Theorem twice.) Consider triangle ABC with 0° < C < 90°. Note the lengths h and k; they are important in the proof.

Do not rush this important exercise.

EXERCISE 30, THE ACUTE TRIANGLE CASE, CALLS FOR A TOTAL CLASS DISCUSSION.

State the reason that justifies each of these six steps:

Step 1. $c^2 = h^2 + (a - k)^2$

Make sure you understand why these steps are true.

Step 2. $h^2 = b^2 - k^2$

Step 3. Thus $c^2 = b^2 - k^2 + (a - k)^2$

Step 4. Hence $c^2 = a^2 + b^2 - 2ak.$

Step 5. But $k = b \cos C.$

Step 6. Use steps 4 and 5 to get the Law of Cosines:

$$c^2 = a^2 + b^2 - 2ab \cos C.$$

*   *   *

Exercise 30 provides a means for solving an acute triangle for a, b, c, or $\angle$ C if the other three values are known. The next exercise proves the case for an obtuse triangle, where angle C is larger than 90°.

31. Consider triangle ABC with $90° < C$. Note the lengths h and k.
Observe side BC must be extended for the altitude AD to be drawn.

AGAIN, A TOTAL CLASS DISCUSSION IS NEEDED IN THE OBTUSE TRIANGLE CASE.

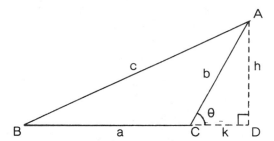

State the reason that justifies each of these seven steps:

Step 1. $c^2 = h^2 + (a + k)^2$

Step 2. $h^2 = b^2 - k^2$

Step 3. Thus $c^2 = b^2 - k^2 + (a + k)^2$

Step 4. Hence $c^2 = a^2 + b^2 + 2ak$

Step 5. But $k = b \cos \theta$

Hint: Step 6. Cos (180° - C) = - cos C from the unit circle in Chapter 1.

Step 6. $\theta = 180° - C.$ Thus $\cos \theta = - \cos C.$

Step 7. Show by steps 4, 5, and 6, that the Law of Cosines follows:

$$c^2 = a^2 + b^2 - 2ab \cos C.$$

Note. Though the arguments in Exercise 30 and 31 differ slightly, both lead to the same formula. In the Law of Cosines there are four quantities: three sides and one angle. Hence if three of these values are known, the Law of Cosines can be used to find the fourth. It is specially useful in the SSS and SAS cases. For emphasis the Law of Sines and the Law of Cosines are repeated here. They permit the solution of any triangle for which sufficient data are given.

Use Law of Sines when given ASA

LAW OF SINES	LAW OF COSINES
$\dfrac{\sin A}{a} = \dfrac{\sin B}{b} = \dfrac{\sin C}{c}$	$c^2 = a^2 + b^2 - 2ab \cos C$

MEMORIZE these two laws.

32. (a) Rewrite the Law of Cosines, so that if you knew b, c, and angle A, you could find side a. Draw a figure.

(b) Rewrite the Law of Cosines, so that if you knew a, c, and angle B, you could find side b. Draw a figure.

* * *

The next two examples illustrate the application of the Law of Cosines, which sometimes is used together with the Law of Sines when solving a triangle.

Example 10. Solve for the remaining angles and side if ∡ A = 130°, b = 21.6, and c = 17.

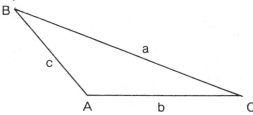

From the unit circle:
cos (180° - θ) = - cos θ. Thus cos 130°
= cos(180°-50°)
* = -cos 50°*

Recall the order of operations in simplifying:
(1) exponents
(2) multiplication (and division)
(3) addition (and subtraction)
Do not skip steps
sin (180° - θ) = sin θ Thus sin 130°
* = sin (180° - 50°)*
* = sin 50°*
* = 0.766*

Solution. $a^2 = b^2 + c^2 - 2bc \cos A$

$\qquad = (21.6)^2 + (17)^2 - 2(21.6)(17) \cos 130°$

$\qquad$ But cos 130° = -cos 50°

so

$a^2 = (21.6)^2 + (17)^2 + 2(21.6)(17)(0.643)$

$\qquad = 466.56 + 289 + 2(21.6)(17)(0.643)$

$\qquad = 466.56 + 289 + 472.2192$

$\qquad = 1227.7792$

$a = \sqrt{1227.7792} \qquad$ or a ≐ 35.0

The Law of Sines can now be used:

$$\frac{\sin C}{c} = \frac{\sin A}{a}$$

$$\sin C = \frac{c \sin A}{a}$$

$$= \frac{17(0.766)}{35.0}$$

$$\doteq 0.372$$

Using the INV Sin keys we get

$$\measuredangle C \doteq 21.8°$$

Also $\measuredangle B = 180° - (130° + 21.8°) \doteq 28.2°$

$$*\quad*\quad*$$

Example 11.   Find $\measuredangle B$ when a = 7, b = 8, and c = 9:

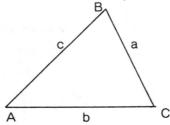

Solution:

$$b^2 = a^2 + c^2 - 2ac \cos B$$

$$2ac \cos B = a^2 + c^2 - b^2$$

$$\cos B = \frac{a^2 + c^2 - b^2}{2ac}$$

$$= \frac{(7)^2 + (9)^2 - (8)^2}{2(7)(9)}$$

$$\doteq 0.524$$

$$\measuredangle B \doteq 58.4°$$

$$*\quad*\quad*$$

Always draw the triangle on your paper with the sides and angles properly labelled. Do not merely show computation.

Show all work in the following exercises, as illustrated in Examples 10 and 11.

33.   Compute $\measuredangle C$ in Example 11.

33. 73.4°

*Use the Law of
Sines when given
ASA and the Law
of Cosines when
given SAS or SSS.*

34. Label the triangle appropriately; then compute side a in the diagram.

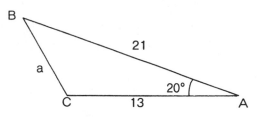

34. $\sqrt{96.76}$

$\doteq 9.84$

35. Solve for the remaining angles and sides.

(a)   C   (b)

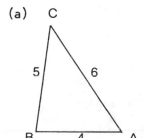

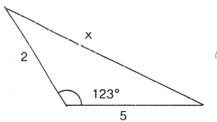

(a) $\angle A = 55.8°$

$\angle B = 82.8°$

$\angle C = 41.4°$

36. Solve each of these triangles for x:

(a)   (b)

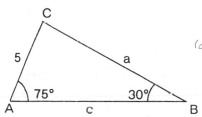

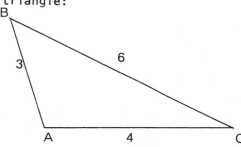

(a) 2.42

37. Solve for angle A in this triangle:

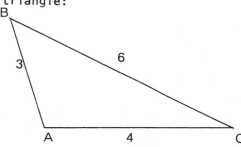

117.3°

THE AMBIGUOUS CASE AND THE LAW OF SINES

In solving triangles sometimes the data given lead to impossible or
ambiguous situations. It may be that we know two sides of a triangle and
an angle that is <u>not</u> included between them. <u>This</u> <u>may</u> <u>mean</u> <u>that</u> <u>there</u> <u>is</u>
<u>no</u> <u>triangle</u> <u>whatsoever</u> <u>for</u> <u>which</u> <u>the</u> <u>data</u> <u>are</u> <u>true.</u> On the other hand,

there <u>may</u> <u>be</u> <u>two</u> <u>different</u> <u>triangles</u>, <u>or</u>, <u>perhaps</u>, <u>exactly</u> <u>one</u> <u>triangle</u>. It is the well known SSA case from geometry.

Assume that sides a and c and angle A are known. It may be that the length a is simply too short to be part of a triangle with given side c and ⟩ A:

CASE I (no triangle)

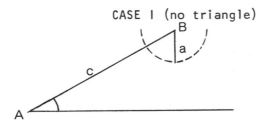

(a is too small to reach to the third side)

a is too small (a < c sin A)

Or a may be long enough, but could determine two triangles:

CASE II (two triangles)

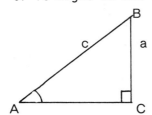
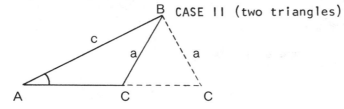

(⟩C can be acute or obtuse)

c sin A < a < c

Or it might be one triangle:

CASE III (exactly one triangle)

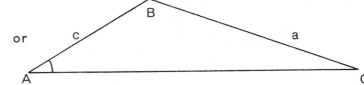

or

a = c sin A, or ⟩ C = 90°

Awareness of SSA ambiguity and a careful sketch usually help when working with ambiguous (SSA) problems. The two-triangle possibility is illustrated in the next example.   The other cases cause little trouble.

Example 12.   Solve triangle ABC, given that ⟩ A = 20°, c = 8, and a = 5.

Solution.   Begin with careful drawings, which suggest the possibility of two triangles.

If ⊀ C is acute: If ⊀ C is obtuse

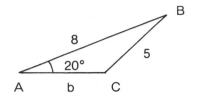

$$\frac{\sin C}{8} = \frac{\sin 20°}{5}$$

Let θ = 33.1° (acute ⊀ C)

$$\sin C \doteq \frac{8(0.342)}{5}$$

Since sin (180° - θ) = sin θ

then obtuse ⊀ C = 180° - 33.1°

$$= 0.547$$

$$= 146.9°$$

⊀ C ≐ 33.1°

⊀ B ≐ 180° - (33.1° + 20°)   ⊀ B = 180° - (146.9° + 20°)

$$\doteq 126.9°$$ $$\doteq 13.1°$$

$$\frac{\sin B}{b} = \frac{\sin C}{c}$$ $$\frac{\sin B}{b} = \frac{\sin C}{c}$$

$$b = \frac{c \sin B}{\sin C}$$ $$b = \frac{c \sin B}{\sin C}$$

$$= \frac{(8)(0.800)}{0.547}$$ $$= \frac{8(0.227)}{0.547}$$

$$\doteq 11.7$$ $$\doteq 3.32$$

38. Solve triangle ABC if ⊀ C = 50°, b = 5, and c = 4.

39. In which of the following is there exactly one answer for x? Explain.

(a)
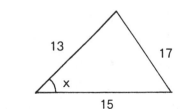

(b)
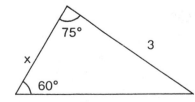

(c)
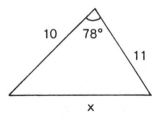

(d)
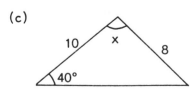

Hint: 38. Recall sin (180° -θ) = sin θ

Hint: Watch for ambiguity in ⊀ A (⊀ B acute) ⊀ B = 73.2° ⊀ A = 56.8° a = 4.37 (⊀ B obtuse) ⊀ B = 106.8° ⊀ A = 23.2° a = 2.06

40. Is there a triangle satisfying the following conditions? If so, how many? Explain by using a careful sketch.

(a)   a = 20, c = 8, ⊀ C = 30°   (b)   a = 20, c = 12, ⊀ C = 30°

(c)   a = 20, c = 25, ⊀ C = 30°   (d)   a = 20, c = 10, ⊀ C = 30°

41. Is there a triangle with sides

(a)   2, 4, and 7?   (b)   4, 5, and 7 ?

(c)   4, $\sqrt{33}$, and 7?   (d)   4, 6, and 7?

(e)   Explain you answers to (a), (b), (c), and (d).

(f)   In parts (b), (c), and (d) is the angle opposite the side of length 7 acute? right? obtuse? Explain using the Law of Cosines.

(a) no
(c) yes

Hint: (f) : consider cos θ in the first and second quadrants.

*   *   *

The final topic in this chapter is not directly linked to solving triangles. Rather, it concerns the solution of trigonometric equations.

SOLVING TRIGONOMETRIC EQUATIONS

In some applications of trigonometry, for instance in physics and calculus, it is necessary to find the angle θ, given information in the form of an equation such as

$$\cos^2\theta = \frac{1}{2} \qquad \text{or} \qquad \tan^2\theta = -\tan\theta$$

or

$$\cos^2\theta + 3\cos\theta + 2 = 0$$

Some examples illustrate how such equations are solved. As you solve them keep in mind the following:

(1)   the quadratic formula

(2)   the factoring method for solving a quadratic

(3)   the 30°-60°-90° and the 45°-45°-90° triangles

For convenience, only angles θ <u>in the range</u>, $0° \leq \theta < 360°$ <u>will be considered</u>. After all, if θ is a solution of the equation, so is θ + 360°, or more generally, θ + n·360° for any integer n.

The next example illustrates these ideas.

Example 13. If $0° \leq \theta < 360°$, solve for θ:

(a) $\cos^2\theta = \frac{1}{2}$ (b) $\tan^2\theta = -\tan\theta$

Solution.

(a) $\cos^2\theta = \frac{1}{2}$

$\cos\theta = \pm\frac{1}{\sqrt{2}}$

then

$\cos\theta = \frac{1}{\sqrt{2}}$ or $\cos\theta = -\frac{1}{\sqrt{2}}$

and   and

$\theta = 45°$ or $\theta = 135°$ or

315°   225°

Hence $\theta = 45°$, 135°, 225°, or 315°

(b) $\tan^2\theta = -\tan\theta$

$\tan^2\theta + \tan\theta = 0$

$\tan\theta(\tan\theta + 1) = 0$

then

$\tan\theta = 0$ or $\tan\theta + 1 = 0$

and

$\theta = 0°$ or $\tan\theta = -1$

180°   and

= 135° or

315°

Hence $\theta = 0°$, 180°, 135°, or 315°

Exercise 42 has problems similar to Example 13.

42. Solve for all θ such that $0 \leq \theta < 360°$:

(a) $\tan\theta = 1$

(b) $\cos\theta = -\frac{1}{2}$

(c) $4\sin^2\theta - 3 = 0$

(d) $\cos^2\theta - \cos\theta = 0$

(e) $2\cos^2\theta = 1$

(f) $\csc^2\theta = 2$

(g) $\sin\theta = \cos\theta$

(h) $\tan\theta = -0.577$

(a) *45°,225°*
(c) *60°,120°*
   240°,300°
(e) *45°,135°*
   225°,315°
(g) *45°,225°*

*   *   *

An equation such as

$$\cos^2\theta + 3\cos\theta + 2 = 0$$

is of the form,

$$x^2 + 3x + 2 = 0$$

where cos θ replaces x. Hence, if cos θ is known, then θ can be computed. The next example shows how to solve this type of equation.

Example 14. Solve for θ, where $0 \leq \theta < 360°$:

$$\cos^2\theta + 3\cos\theta + 2 = 0$$

Solution.

$$\cos^2\theta + 3\cos\theta + 2 = 0$$

$$(\cos\theta + 1)(\cos\theta + 2) = 0$$

(continued)

then

$$\cos\theta + 1 = 0 \qquad\qquad or \qquad\qquad \cos\theta + 2 = 0$$
$$\cos\theta = -1 \qquad\qquad\qquad\qquad \cos\theta = -2$$
$$\theta = 180° \qquad\qquad\qquad\qquad (impossible)$$

* * *

43. Solve for   where $0° \le \theta < 360°$:

(a) $\tan^2\theta - 3 = 0$ $\qquad\qquad$ (b) $\sin^2\theta + 2\sin\theta + 1 = 0$

(c) $2\cos^2\theta + \cos\theta - 1 = 0$ $\qquad$ (d) $\tan^2\theta - \tan\theta = 0$

44. Solve for θ where $0° \le \theta < 360°$:

(a) $2\cos\theta\sin\theta = \cos\theta$ $\qquad$ (b) $4\sin^2\theta = 1$

(c) $\sec^2\theta = 1$ $\qquad\qquad\qquad$ (d) $\sin\theta = \csc\theta$

(e) $\sin\theta + 1 = 2\cos^2\theta$ $\qquad$ (f) $2\sin^2\theta + 3\cos\theta = 0$

(g) $\sin\theta = -\cos\theta$ $\qquad\qquad$ (h) $\cos\theta\sin\theta = \cos\theta$

* * *

COMMENT: In a more complicated quadratic equation it may be necessary to use the quadratic formula to solve for the trigonometric function. Then a calculator may be needed to find θ. (See Project 19.)

* * *

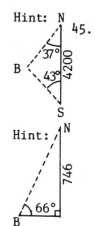

Hint:

Hint:

45. The towers of the Golden Gate Bridge are 4200 feet apart. An observer at the north tower sees a small boat that forms an angle of 37° with the south tower. An observer at the south tower sees the same boat at an angle of 43° with the north tower.

(a) How far is the boat from each tower?

(b) On another small boat, an observer sees an angle of 66° between the top of the north tower (746 feet above the water) and its base at the water line. How far is that observer from the base of the north tower?

* * * SUMMARY * * *

The previous chapter focussed on the trigonometry of a circle; this chapter concentrated on the trigonometry of triangles. Both perspectives have practical and theoretical importance.

The chapter began by relating $\cos\theta$, $\sin\theta$, and $\tan\theta$ to any point (a,b) other than the origin, in the coordinate plane:

$$\cos \theta = \frac{a}{\sqrt{a^2 + b^2}}$$

$$\sin \theta = \frac{b}{\sqrt{a^2 + b^2}}$$

$$\tan \theta = \frac{b}{a}$$

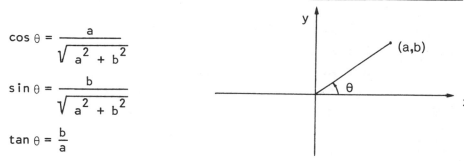

When $0° < \theta < 90°$, it was a small step to relate the (six) trigonometric functions to a right triangle:

$$\cos \theta = \frac{adj}{hyp} \qquad \sec \theta = \frac{hyp}{adj}$$

$$\sin \theta = \frac{opp}{hyp} \qquad \csc \theta = \frac{hyp}{opp}$$

$$\tan \theta = \frac{opp}{adj} \qquad \cot \theta = \frac{adj}{opp}$$

After illustrating the solution of right triangles, the chapter showed how to solve any triangle. Two results were needed:

THE LAW OF SINES: and   THE LAW OF COSINES:

$$\frac{\sin A}{a} = \frac{\sin B}{b} = \frac{\sin C}{c} \qquad\qquad c^2 = a^2 + b^2 - 2ab \cos C$$

In one type of triangle (SSA), the solution may be ambiguous; there may be two, one, or no solutions in such cases.

The chapter concluded by solving trigonometric equations.

* * * WORDS AND SYMBOLS * * *

hypotenuse	ASA
adjacent	SSS
opposite	SAS
ambiguous case	Law of Sines
A, B, C (angles)	Law of Cosines
a, b, c (sides)	solving a triangle

* * * CONCLUDING REMARKS * * *

Why do we measure angles in such a way that the number 360 is assigned to the angle that corresponds to "once around the circle?" A king's thumb is the origin of the inch. One-ten millionth of the distance

from the equator of the Earth to the North Pole provides the meter.
But where did "360" come from?

This number was used by Babylonian astronomers, in part because it is
close to the number of days in a year, in part, probably, because it has
so many divisors. But there is no need to measure angles this way. For
years a proposal has circulated to replace 360 by 400. Then a right angle
would have the measure 100.

Notice that the number 360 is as arbitrary as the king's "inch."
We could choose any number we wish to measure the angle that we now call
"360 degrees." In fact, in calculus it turns out to be best to measure
angles in "radians." (An angle of 360° is called an angle of $2\pi \doteq 6.28$
"radians.") This type of angle-measurement will be discussed in a later
chapter. You might devise your own way to measure angles, just to convince yourself that there is nothing inevitable or divine about the
number 360.

*   *   *   TRIANGLES SELF QUIZ *   *   *

1.   Find all the trigonometric functions of θ:

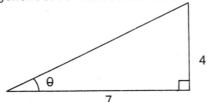

2. If the moving arm of θ goes through the point $(-1,5)$, find all the
    trigonometric functions of θ.

3. Find all the parts of the triangle not given:
    (a)   (b)

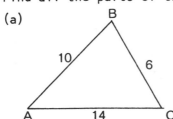
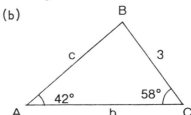

4. Solve for θ, where $0° \leq \theta < 360°$
    (a) $\tan \theta = \cot \theta$ (b) $2 \sin^2\theta + 3 \sin\theta + 1 = 0$

ANSWERS TO SELF QUIZ

1. $\sin \theta = \dfrac{4}{\sqrt{65}}$, $\cos \theta = \dfrac{7}{\sqrt{65}}$, $\tan \theta = \dfrac{4}{7}$, $\cot \theta = \dfrac{7}{4}$, $\sec \theta = \dfrac{\sqrt{65}}{7}$, $\csc \theta = \dfrac{\sqrt{65}}{4}$

2. $\cos \theta = -\dfrac{1}{\sqrt{26}}$, $\sin \theta = \dfrac{5}{\sqrt{26}}$, $\tan \theta = -5$, $\cot \theta = -\dfrac{1}{5}$, $\sec \theta = -\sqrt{26}$, $\csc \theta = \dfrac{\sqrt{26}}{5}$

3. (a) $\angle A = 21.8°$, $\angle B = 120°$, $\angle C = 38.2°$ (b) $\angle B = 80°$, $c = 3.8$, $b = 4.42$

4. (a) $45°$, $135°$, $225°$, $315°$,   (b) $210°$, $270°$, $330°$

* * * TRIANGLES REVIEW * * *

1. In terms of the sides of a right triangle (opposite, adjacent, hypotenuse) define each of the following:

    (a) sine   (b) cosine   (c) tangent

    (d) cotangent   (e) secant   (f) cosecant

2. Use graph paper and ruler to make an accurate drawing of a right triangle for each acute angle θ (label θ):

    (a) $\cos \theta = \dfrac{2}{3}$ (b) $\tan \theta = \dfrac{3}{2}$ (c) $\sin \theta = \dfrac{5}{7}$

    (d) $\cot \theta = 3$ (e) $\csc \theta = 2$ (f) $\sec \theta = \dfrac{4}{3}$

3   If the second arm of θ goes through the point P, find the six trigonometric functions of θ if point P has the coordinates

    (a) $(-5,7)$ (b) $(-8,-15)$ (c) (m,n)

4. (a) Fill in the table with <u>precise</u> values (such as $\sqrt{3}$, $\dfrac{1}{\sqrt{2}}$, etc.):

θ	$\cos \theta$	$\sin \theta$	$\tan \theta$
0°			
30°			
45°			
60°			
90°			
120°			
135°			
180°			
210°			
225°			

Draw the 30°-60°-90° and 45°-45°-90° triangles for a reference.

(continued)

θ	cos θ	sin θ	tan θ
270°			
300°			
315°			

5. For triangle ABC, the Law of Cosines

$$c^2 = a^2 + b^2 - 2ab \cos C$$

may be written in two other forms:

$$b^2 = ?$$

$$a^2 = ?$$

6. What happens in the Law of Cosines when ⦜ C = 90° ?

$$c^2 = a^2 + b^2 - 2ab \cos C$$

7. Use Exercise 6 to compare c^2 with $a^2 + b^2$ if angle C is
    (a) obtuse,   (b) acute,   (c) a right angle.

8. Find the remaining angles and sides:

    (a)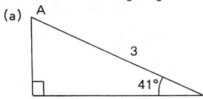

    (b)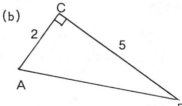

9. Find x:

    (a)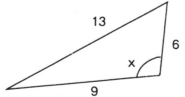

    (b)

    (c)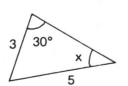

    (d)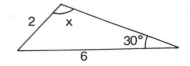

10. Explain why each of the following is an impossible case.

(a)

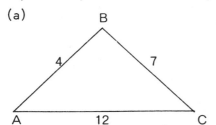

(b)

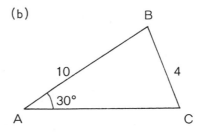

Hint: (b) Consider a certain right triangle.

11. (a) Find x, y, and z to the nearest tenth of a degree:

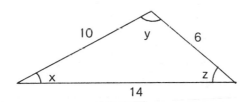

(b) Which is the largest angle of the triangle?

(c) What do you observe about the relation between the measure of
angles, as compared to the relation between the lengths of the
sides?

12. Would you use the Law of Sines, the Law of Cosines, or some other
method to find the 3 sides and the 3 angles of a triangle if you were
given:

(a) S S S   (b) S A S

(c) A S A   (d) A A S

13. If $\sin \theta = 0.800$ and $0° < \theta < 90°$, find

(a) $\cos \theta$ (b) $\cos(90° - \theta)$ (c) $\cos(90° + \theta)$

(d) $\sin(180° - \theta)$ (e) $\sin(180° + \theta)$ (f) $\tan \theta$

14. If $0° \le \theta < 360°$, for what θ is each of the following true?

(a) $\cos \theta = \sin \theta$ (b) $\sin \theta = \cos(90° - \theta)$

(c) $\tan \theta = \cot \theta$ (d) $\sin \theta = \tan \theta$

(e) $\cos \theta = \cot \theta$ (f) $\sec \theta = \cos \theta$

(g) $\csc \theta = \sin \theta$

15. Solve for θ, if $0° \le \theta < 360°$:

(a) $\csc^2 \theta - 2 = 0$ (b) $2\sin^2 \theta - \sin \theta - 1 = 0$

(c) $\tan^2 \theta = \tan \theta$ (d) $2\cos^2 \theta = \cos \theta + 1$

(e) $3\tan^2 \theta - 1 = 0$ (f) $\sin^2 \theta - \cos^2 \theta = 0$

16. (a) Consider:

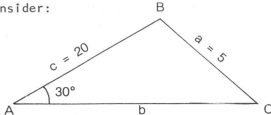

Can you solve for the remaining angles and side? Explain.

(b) Consider:

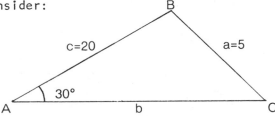

Can you solve for the remaining angles and side? Explain.

17. The Washington Monument is 555.4 feet high. If you are standing on level ground, how far are you from the center of the base of the monument if the angle of elevation of the top of the monument is 63°?

Hint: Find ∢ A or ∢ C then find h

18. A real estate developer wants to sell the following triangular lot. What is the area?

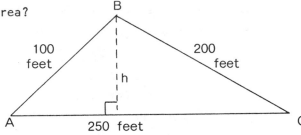

19. In Yosemite Valley, El Capitan (a sheer cliff) is about 3550 feet above the valley floor. A hiker, standing on the valley floor, must elevate the angle of his vision 78° to see the top of the cliff.

(a) How far is the hiker from the base of the cliff?

(b) After computing the answer to (a) on his note pad, the hiker again elevates his eyes to an angle of 61° and sees a climber on the side of El Capitan. How high is the climber above the valley floor?

20. If A is the area of a triangle with sides a, b, and c, and if

$$s = \frac{a + b + c}{2} \quad \text{then}$$

$$A = \sqrt{s(s - a)(s - b)(s - c)} \ .$$

For a proof of this formula, see p. 310 of Geometry: A Guided Inquiry Chakerian, Crabill, and Stein (Sunburst, 1986).

Use the above formula to compute the areas of the following triangles.

(a)   (b)

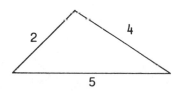   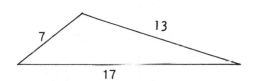

ANSWERS TO REVIEW

1. (a) $\text{sine} = \frac{\text{opp}}{\text{hyp}}$ (b) $\text{cosine} = \frac{\text{adj}}{\text{hyp}}$ (c) $\text{tangent} = \frac{\text{opp}}{\text{adj}}$

(d) $\text{cotangent} = \frac{\text{adj}}{\text{opp}}$ (e) $\text{secant} = \frac{\text{hyp}}{\text{adj}}$ (f) $\text{csc} = \frac{\text{hyp}}{\text{opp}}$

3. (a) $(-5,7)$ $\sin \theta = \frac{7}{\sqrt{74}}$, $\cos \theta = -\frac{5}{\sqrt{74}}$, $\tan \theta = -\frac{7}{5}$, $\cot \theta = -\frac{5}{7}$,

$\sec \theta = -\frac{\sqrt{74}}{5}$, $\csc \theta = \frac{\sqrt{74}}{7}$ (b) $(-8,-15)$, $\sin \theta = -\frac{15}{17}$,

$\cos \theta = -\frac{8}{17}$, $\tan \theta = \frac{15}{8}$, $\cot \theta = \frac{8}{15}$, $\sec \theta = -\frac{17}{8}$, $\csc \theta = -\frac{17}{15}$

(c) (m,n), $\sin \theta = \frac{n}{\sqrt{m^2 + n^2}}$, $\cos \theta = \frac{m}{\sqrt{m^2 + n^2}}$, $\tan \theta = \frac{n}{m}$,

$\cot \theta = \frac{m}{n}$, $\sec \theta = \frac{\sqrt{m^2 + n^2}}{m}$, $\csc \theta = \frac{\sqrt{m^2 + n^2}}{n}$

4. (a)

θ	$\cos \theta$	$\sin \theta$	$\tan \theta$
0°	1	0	0
30°	$\frac{\sqrt{3}}{2}$	$\frac{1}{2}$	$\frac{1}{\sqrt{3}}$
45°	$\frac{\sqrt{2}}{2}$	$\frac{\sqrt{2}}{2}$	1
60°	$\frac{1}{2}$	$\frac{\sqrt{3}}{2}$	$\sqrt{3}$
90°	0	1	undefined

(continued)

θ	$\cos\theta$	$\sin\theta$	$\tan\theta$
120°	$-\dfrac{1}{2}$	$\dfrac{\sqrt{3}}{2}$	$-\sqrt{3}$
135°	$-\dfrac{\sqrt{2}}{2}$	$\dfrac{\sqrt{2}}{2}$	-1
180°	-1	0	0
210°	$-\dfrac{\sqrt{3}}{2}$	$-\dfrac{1}{2}$	$\dfrac{1}{\sqrt{3}}$
225°	$-\dfrac{\sqrt{2}}{2}$	$-\dfrac{\sqrt{2}}{2}$	1
270°	0	-1	undefined
300°	$\dfrac{1}{2}$	$-\dfrac{\sqrt{3}}{2}$	$-\sqrt{3}$
315°	$\dfrac{\sqrt{2}}{2}$	$-\dfrac{\sqrt{2}}{2}$	-1

5. $b^2 = c^2 + a^2 - 2ac \cos B$ $a^2 = c^2 + b^2 - 2bc \cos A$

6. when $\angle\, C = 90°,\ c^2 = a^2 + b^2$

7. (a) $c^2 > a^2 + b^2$ (b) $c^2 < a^2 + b^2$ (c) $c^2 = a^2 + b^2$

8. (a) $b = 1.968$ (b) $c = \sqrt{29}$
         $a = 2.265$ $\angle\, A = 68.2°$
         $\angle\, A = 49°$ $\angle\, B = 21.8°$

9. (a) $\cos x = \dfrac{-13}{27}$ (b) $x = \sqrt{74 - 70\cos 32°}$ (c) $\dfrac{3}{10} = \sin x$
         $x = 118.8°$ $x = 3.83$
                                                $x = 17.5°$
    (d) no solution

10. (a) Triangle inequality   (b) not a triangle

11. (a) $y = 120°$ $z = 38.2°$ $x = 21.8°$ (b) angle y
    (c) angles opposite longest sides are largest

12. (a) Law of Cosines   (b) Law of Cosines
    (c) Law of Sines   (d) Law of Sines

13. (a) $\cos\theta = 0.600$ (b) 0.800   (c) -0.800
    (d) 0.800   (e) -0.800   (f) $\tan\theta = \dfrac{4}{3} = 1.\overline{33}$

14. (a) $\theta = 45°, 225°$ (b) all θ (c) $\theta = 45°, 135°, 225°, 315°$
    (d) $\theta = 0°, 180°$ (e) $90°, 270°$ (f) $0°, 180°$
    (g) $90°, 270°$

15. (a) θ = 45°, 135°, 225°, 315°   (b) 90°, 210°, 330°

(c) 0°, 45°, 180°, 225°   (d) 0°, 120°, 240°

(e) 30°, 150°, 210°, 330°   (f) 45°, 135°, 225°, 315°

16. (a) No violation of Law of Sines

(b) Yes, but this is ambiguous case--two sets of values possible

17. 283.0 feet

18. ∡ A = 49.5°, h = 76.0 feet, area = 9500 sq. ft.

19. (a) 755 feet   (b) 1362 feet

20. (a) 3.80   (b) 41.9

* * * TRIANGLES REVIEW SELF TEST * * *

1. Find the precise values of the trigonometric functions of θ for each
of the following triangles:

(a)   (b)   (c)

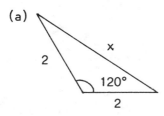   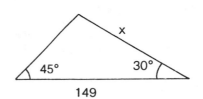   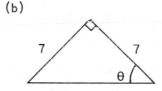

(d) Are (a), (b) and (c) "special triangles?" Explain.

2. Find all the trigonometric functions of θ if the moving arm of θ
goes through $(1, \sqrt{2})$.

3. Find the angles

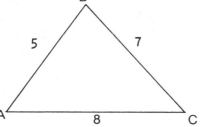

4. Find x:

(a)   (b)

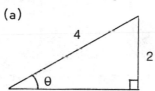   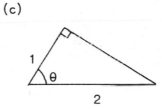

5. Find θ where 0° ≤ θ < 360°

(a) tan θ = sin θ   (b) $\sin^2 \theta = \cos^2 \theta - \cos \theta$

ANSWERS TO SELF TEST

1. (a) $\sin \theta = 0.5$, $\cos \theta = \dfrac{\sqrt{3}}{2}$, $\tan \theta = \dfrac{1}{\sqrt{3}}$, $\csc \theta = 2$, $\sec \theta = \dfrac{2}{\sqrt{3}}$,

$\cot \theta = \sqrt{3}$ (b) $\sin \theta = \dfrac{1}{\sqrt{2}}$, $\cos \theta = \dfrac{1}{\sqrt{2}}$, $\tan \theta = 1$, $\csc \theta = \sqrt{2}$

$\sec \theta = \sqrt{2}$, $\cot \theta = 1$ (c) $\sin \theta = \dfrac{\sqrt{3}}{2}$, $\cos \theta = 0.5$, $\tan \theta = \sqrt{3}$

$\cot \theta = \dfrac{1}{\sqrt{3}}$, $\sec \theta = 2$, $\csc \theta = \dfrac{2}{\sqrt{3}}$ 2. $\cos \theta = \dfrac{1}{\sqrt{3}}$

$\sin \theta = \sqrt{\dfrac{2}{3}}$, $\tan \theta = \sqrt{2}$, $\cot \theta = \dfrac{1}{\sqrt{2}}$, $\sec \theta = \sqrt{3}$, $\csc \theta = \sqrt{\dfrac{3}{2}}$

3. $B = 81.8°$, $C = 38.2°$, $A = 60.0°$,   4. (a) 3.46   (b) $109.1 \doteq 109$

5.   (a) 0°, 180°   (b) 0°, 120°, 240°

* * *   TRIANGLES PROJECTS   * * *

1.   The term "cosine" was derived from "sine of the complement."
Similarly, "cotangent" was derived from "tangent of the complement,"
and "cosecant" the "secant of the complement:"

$$\cos \theta = \sin(90° - \theta)$$
$$\cot \theta = \tan(90° - \theta)$$
$$\csc \theta = \sec(90° - \theta)$$

Use these ideas to help you solve the following problem.

In terms of the distance between two points in the figure,

Hint: Remember the unit circle and how the trig functions are defined in terms of sine and cosine.

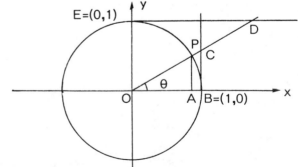

find each of the following.

(a) $\sin \theta =$ (b) $\cos \theta =$ (c) $\tan \theta =$

(d) $\cot \theta =$ (e) $\sec \theta =$ (f) $\csc \theta =$

2. In geometry it was proved that two sides of a triangle are equal if and only if their opposite angles are equal. Prove this by trigonometry.

3. Estimate θ, if $0° \leq \theta < 360°$ and

$$\sin^2\theta + 2\sin\theta - 1 = 0.$$

4. This project outlines a proof that $\cos 72° = \dfrac{\sqrt{5} - 1}{4}$.

    (a) Using geometry, find all angles in this figure, where AC = BC.

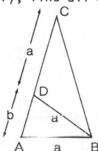

(It turns out that angle C is 36°)

    (b) Call the length AD = b. Show that $(a + b)^2 = 2a^2(1 + \cos 72°)$ and $a^2 = a^2 + b^2 - 2ab \cos 72°$.

    (c) From (b) show that

$$4 \cos^2 72° + 2 \cos 72° - 1 = 0.$$

    (d) Deduce that $\cos 72° = \dfrac{\sqrt{5} - 1}{4}$

5. Show that the area of a triangle with sides a and b and **included** angle C is $(\frac{1}{2})ab \sin C$.

6. A point (x,y) such that x and y are integers is called a "lattice point." Using elementary geometry, show that the area of any integer whose three vertices are lattice points is either an integer or half of an integer

7. (See Project 6 for "lattice point.") It is impossible for three lattice points to determine a triangle one of whose angles is 60°. This is an outline of one proof.

    Let the sides of the alleged triangle be a, b, and c, and C = 60°.

    (a) Show that $c^2 = a^2 + b^2 - ab$.

    (b) Show that a^2, b^2, and c^2 are integers .

    (c) Deduce that ab is an integer .

    (d) Combining (c) and Projects 5 and 6, deduce that $\sqrt{3}$ is a rational number.

8. (a) Study Project 7 carefully. Then show that if three lattice points determine the angle C, then tan C is rational.

   (b) If tan C is rational, show that angle C can be determined by three lattice points.

9. Find all triangles for which

$$\frac{\tan A}{a} = \frac{\tan B}{b} = \frac{\tan C}{c}$$

10. How would you go about measuring the diameter of the Earth? The Greek scientist Eratosthenes did it about 150 B.C. (with an error of less than 1%). To find out about his method, read pp. 109-110 in Volume I of History of Mathematics by D.E. Smith, (Dover Publications 1951). The general problem of earth measurements is also described on pp. 368-376 of Volume II of D.E. Smith's History of Mathematics (Dover, 1953).

11. Given the triangle with angles A, B, and C, and a, b, c and the opposite sides, respectively, show that

$$\frac{a - b}{a + b} = \frac{\sin A - \sin B}{\sin A + \sin B}.$$

    Hint: Show
    $a = \frac{b \sin A}{\sin B}$

    (This result will be used in the Projects section of the next unit to obtain the Law of Tangents.)

12. There are six elements associated with a triangle, the three angles and three sides. In many cases, if we know three of the six elements, we know the triangle completely. Is it possible for two triangles to have five of the six elements identical and yet not be congruent? Try some examples. (A brief description of this problem can be found on page 133 of Scientific American, June 1970.)

13. Read about the development of trigonometry and some of its uses in an encyclopedia, or in one of the following:

    Introduction to Mathematics by Cooley and Wahlert, Houghton-Mifflin, Boston, 1968.

    History of Mathematics by D.E. Smith, Vol. II, Dover Publications, New York, 1953.

    Mathematical Thought from Ancient to Modern Times, Morris Kline, Oxford University Press, New York, 1972.

    Science Awakening by B.L. van der Waerden, Oxford University Press, 1961.

Mathematics in Western Culture by Morris Kline, Oxford University Press, 1953.

14. The "sine" of an angle and the "sinus" of a sinus headache both come from the same word. Look up the origin of "sine," and explain its relation to "sinus."

15. Find x:

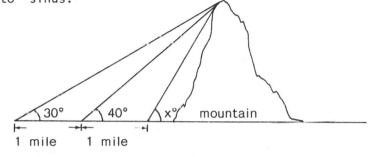

30°   40°   x°   mountain

1 mile   1 mile

16. A person climbs to the top of a 150 foot bluff overlooking the ocean.

(a) How far away is the horizon?

(b) An island that has a 150 foot hill rises out of the ocean 100 miles away. Could he see it? (Take the radius of the Earth as 4000 miles.)

17. A 1000 foot long straight pipe expands 1 foot on a hot day (it remains straight as it expands). There is a hinge at its midpoint M.

M

A   C   B

1000

on a hot day;
AM + MB = 1001

(a) Without computing, guess the height MC.

(b) Compute MC.

18. A real estate developer has two lots as in the diagrams below. What is the area of each?

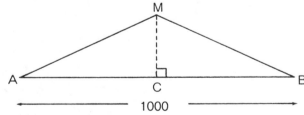

(a) B   600 ft.   C

400 ft.   500 ft.

A   300 ft.   D

(b) A   120°   B   100 ft.   125 ft.   C

175 ft.   150 ft.   D

Hint: (a) Draw BD (b) Draw AC. Then see Review Exercises 18 and 20.

19. Solve for θ: cos θ = tan θ

Trigonometric Identities

The last two chapters presented several identities relating the trigonometric functions, such as

$$\tan \theta = \frac{\sin \theta}{\cos \theta} \quad \text{and} \quad \cos^2\theta + \sin^2\theta = 1.$$

This chapter will develop some fundamental identities of a different character. Exercise 1 introduces a problem in physics which can be solved easily with the aid of a certain identity discussed later in the chapter. That is just one of the many applications of identities to physics, analytic geometry, and calculus, as well as everyday work, such as sheet-metal layout.

Suggestion: It will be helpful to review the identities in Chapter 1.

EXERCISES

1. Suppose a ball is thrown at an angle A with a velocity of 32 feet per second. It can be shown (see Project 11) that the ball travels a distance d = 64 sin A cos A feet:

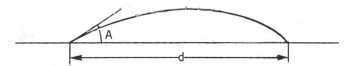

   (a) At what angle A do you think that the ball should be thrown to achieve the maximum distance?

   (b) Use a calculator to experiment with various choices of angle A to give a maximum distance.

* * *

Later in the chapter we will return to the problem raised in Exercise 1. First, however, the necessary theory must be developed.

THE DISTANCE FORMULA

The basis of this chapter is the **Pythagorean Theorem** which will now be expressed in terms of the coordinate plane. Consider two points (x_1, y_1) and (x_2, y_2), as in this diagram:

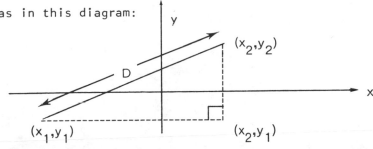

The diagram shows that the **Pythagorean Theorem** applied to the right triangle with dashed legs takes the form,

$$D^2 = (x_2 - x_1)^2 + (y_2 - y_1)^2$$

This yields

The distance formula is a very powerful tool; coordinate geometry is based upon it.

+---+
| THE DISTANCE FORMULA |
| $D = \sqrt{(x_2 - x_1)^2 + (y_2 - y_1)^2}$ |
+---+

MEMORIZE!

Example 1. Find the distance between (-3,-5) and (6,7).

Solution.

It makes no difference which of the two points is chosen to be (x_1, y_1) and which is (x_2, y_2).

$$distance = \sqrt{(6 - (-3))^2 + (7 - (-5))^2}$$

$$= \sqrt{(9)^2 + (12)^2}$$

$$= \sqrt{81 + 144}$$

$$= \sqrt{225}$$

$$= 15$$

2. Find the distance between each pair of points (simplify the expression):

(a) (2,1) and (3,4)   (b) (-5,7) and (3,-6)

(c) (-2,-3) and (-5, -8)   (d) (cos θ ,sin θ) and (1,0)

(a) $\sqrt{10}$

(c) $\sqrt{34}$

(d) $\sqrt{2 - 2\cos\theta}$

THE IDENTITY FOR cos (α + β)

The next exercise develops the basic identity of trigonometry. From it all the other identities will follow. It concerns two angles and their sum. The two angles will be labelled α ("alpha") and β ("beta"), the first two letters of the Greek alphabet. The identity will enable

Hence the origin of the word "alphabet."

$$\cos(\alpha + \beta)$$

to be computed if

$$\cos \alpha, \sin \alpha, \cos \beta, \text{ and } \sin \beta$$

are known. (See the end of Exercise 3 for the statement of this identity.)

3. Consider the following circle with the center at the origin:

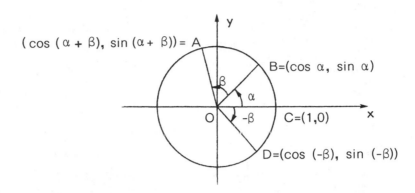

NOTE: This is the key exercise; the remainder of the chapter depends on it.

(a) Using the properties of a circle (or congruent triangles if you prefer), explain why

$$AC = BD.$$

AC = distance from A to C and BD = distance from B to D.

(b) Thus it follows that

$$(AC)^2 = (BD)^2 .$$

(Note. The rest of the argument consists of expressing $(AC)^2$ and $(BD)^2$ in terms of the coordinates of A, B, C, and D with the aid of the distance formula.)

(c) Use the distance formula and algebra to show that

$$(AC)^2 = 2 - 2\cos(\alpha + \beta)$$

(d) Use the distance formula and algebra to show that

$$(BD)^2 = 2 - 2\cos\alpha\cos\beta + 2\sin\alpha\sin\beta$$

It is necessary to simplify the formulas you get with the distance formula in (b) and (c). Recall that $\cos(-\theta) = \cos\theta$ and $\sin(-\theta) = -\sin\theta$.

(e) Combine (b), (c) and (d) to establish the important identity,

$$\boxed{\cos(\alpha + \beta) = \cos\alpha\cos\beta - \sin\alpha\sin\beta}$$ MEMORIZE!

Memorize this identity. Add it to your list of identities.

The next example is a check of this identity for some specific values of α and β.

Example 2. If $\alpha = 90°$ and $\beta = 45°$ compute the precise values of $\cos(\alpha + \beta)$ in terms of cosine and sine of α and β.

Solution.

$$\cos(\alpha + \beta) = \cos\alpha\cos\beta - \sin\alpha\sin\beta$$
$$\cos(90° + 45°) = \cos 90° \cos 45° - \sin 90° \sin 45°$$
$$\cos 135° = (0)\left(\frac{1}{\sqrt{2}}\right) - (1)\left(\frac{1}{\sqrt{2}}\right)$$

$$= -\frac{1}{\sqrt{2}}$$

4. Without using the trigonometric tables or a calculator, check
$$\cos(\alpha + \beta) = \cos \alpha \cos\beta - \sin\alpha \sin \beta$$
for each of the following:

(a) $\alpha = 90°, \beta = 0°$ (b) $\alpha = 45°, \beta = 45°$

(c) $\alpha = 30°, \beta = 60°$

5. This exercise shows how information about the sine and cosine of the angles 30°, 45°, and 60° can be used to find the cosine of other angles.

Note that 75° = 45° + 30°

(a) $\frac{\sqrt{6} - \sqrt{2}}{4}$

(a) Without using trigonometric tables or a calculator, compute the <u>precise</u> value of cos 75° and cos 105° using $\cos(\alpha + \beta)$.

(b) Use a calculator to check your answers in (a).

THE IDENTITY FOR $\cos(\alpha - \beta)$

The identity for $\cos(\alpha + \beta)$ is useful in finding the identity for $\cos(\alpha - \beta)$.

6. Replace β by -β throughout the $\cos(\alpha + \beta)$ identity to obtain the following important identity:

Hint: $\cos(\alpha - \beta) = \cos(\alpha + (-\beta))$, Memorize the $\cos(\alpha - \beta)$ identity.

$$\boxed{\cos(\alpha - \beta) = \cos \alpha \cos \beta + \sin \alpha \sin \beta}$$ MEMORIZE!

7. Without using trigonometric tables or a calculator, check
$$\cos(\alpha - \beta) = \cos \alpha \cos \beta + \sin \alpha \sin \beta$$
for each of the following:

(a) $\alpha = 90°, \beta = 0°$ (b) $\alpha = 45°, \beta = 45°$ (c) $\alpha = 90°, \beta = 45°$

(d) $\alpha = 120°, \beta = -30°$ (e) $\alpha = 60°, \beta = 30°$ (f) $\alpha = 45°, \beta = 90°$

8. Without using a trig table or calculator, find cos 15° precisely.

$\frac{\sqrt{3} + 1}{2\sqrt{2}}$ or $\frac{\sqrt{6} + \sqrt{2}}{4}$

$$* \quad * \quad *$$

The next identity concerns $\sin(\alpha + \beta)$. It is the second most important identity in the chapter. Together with the $\cos(\alpha + \beta)$ identity, it will lead to the rest of the identities.

THE IDENTITY FOR $\sin(\alpha + \beta)$

The identity for $\sin(\alpha + \beta)$ follows from the identity
$$\sin \theta = \cos(90° - \theta)$$
from the Unit Circle Chapter and the identity for $\cos(\alpha - \beta)$.

Exercise 9 outlines the proof.

9.   (a)   Show that $\sin(\alpha + \beta) = \cos[90° - (\alpha + \beta)]$.

Hint: (a)
$\sin \theta = \cos(90° - \theta)$.

(b)   Show that $\cos[90° - (\alpha + \beta)] = \cos[(90° - \alpha) - \beta]$.

(c)   From (a) and (b) deduce that

$$\boxed{\sin(\alpha + \beta) = \sin \alpha \; \cos \beta + \cos \alpha \; \sin \beta}$$ MEMORIZE!

Example 3.   If $\alpha = 180°$ and $\beta = 60°$, compute the precise values for $\sin(\alpha + \beta)$ in terms of cosine and sine of α and β.

Solution.

$$\sin(\alpha + \beta) = \sin \alpha \; \cos \beta \; + \cos \alpha \; \sin \beta$$
$$\sin(180° + 60°) = \sin 180° \; \cos 60° + \cos 180° \; \sin 60°$$
$$\sin 240° = (0)(\tfrac{1}{2}) + (-1)\,(\tfrac{\sqrt{3}}{2})$$
$$= -\frac{\sqrt{3}}{2}$$

10.   Without using trigonometric tables or a calculator, check the identity
$$\sin(\alpha + \beta) = \sin \alpha \; \cos \beta \; + \cos \alpha \; \sin \beta,$$
for each of the following:

(a) $\alpha = 90°, \beta = 0°$ (b) $\alpha = 45°, \beta = 45°$ (c) $\alpha = 30°, \beta = 60°$

11.   Compute the precise value of:

(a) $\sin 75°$ (b) $\sin 105°$

(c)   Use a calculator to check your answers in (a) and (b).

12.   Replace β with $-\beta$ in Exercise 9 to express $\sin(\alpha - \beta)$ in terms of sine and cosine of the angles α and β:

Write $\sin(\alpha - \beta) =$
$\sin(\alpha + (-\beta))$
Memorize the
identity.

$$\boxed{\sin(\alpha - \beta) = \sin \alpha \; \cos \beta - \cos \alpha \; \sin \beta}$$

13.   Without using the trigonometric tables or a calculator, check the identity
$$\sin(\alpha - \beta) = \sin \alpha \; \cos \beta - \cos \alpha \; \sin \beta$$
for each of the following:

(a) $\alpha = 60°, \beta = 30°$ (b) $\alpha = 0°, \beta = 30°$ (c) $\alpha = 120°, \beta = -30°$

14.   Compute $\sin 15°$ precisely. Show all work.

15.   If $\cos \alpha = \dfrac{4}{5}$, $\sin \alpha = \dfrac{3}{5}$, $\cos \beta = \dfrac{3}{5}$, and $\sin \beta = \dfrac{4}{5}$, compute as a rational number: (a) $\cos(\alpha + \beta)$ (b) $\sin(\alpha + \beta)$

(c) $\sin(\alpha - \beta)$ (d) $\cos(\alpha - \beta)$

(a) 0
(c) $-\frac{7}{25}$

THE IDENTITY FOR $\tan(\alpha + \beta)$

The next exercise derives the identity for $\tan(\alpha + \beta)$

16. (a) Explain each of the following steps in deriving the identity for $\tan(\alpha + \beta)$, which states $\tan(\alpha + \beta)$ in terms of $\tan \alpha$ and $\tan \beta$.

     Step 1. $\tan(\alpha + \beta) = \dfrac{\sin(\alpha + \beta)}{\cos(\alpha + \beta)}$

     Step 2. $= \dfrac{\sin \alpha \cos \beta + \cos \alpha \sin \beta}{\cos \alpha \cos \beta - \sin \alpha \sin \beta}$

TRICK

     Step 3. $= \dfrac{\dfrac{\sin \alpha \cos \beta + \cos \alpha \sin \beta}{\cos \alpha \cos \beta}}{\dfrac{\cos \alpha \cos \beta - \sin \alpha \sin \beta}{\cos \alpha \cos \beta}}$

     Step 4. $= \dfrac{\dfrac{\sin \alpha \cos \beta}{\cos \alpha \cos \beta} + \dfrac{\cos \alpha \sin \beta}{\cos \alpha \cos \beta}}{\dfrac{\cos \alpha \cos \beta}{\cos \alpha \cos \beta} - \dfrac{\sin \alpha \sin \beta}{\cos \alpha \cos \beta}}$

                $= \dfrac{\tan \alpha + \tan \beta}{1 - \tan \alpha \tan \beta}$

(b) The five steps in (a) lead to the identity,

$$\boxed{\tan(\alpha + \beta) = \dfrac{\tan \alpha + \tan \beta}{1 - \tan \alpha \tan \beta}}$$ MEMORIZE!

Example 4.   If $\tan \alpha = \dfrac{1}{2}$ and $\tan \beta = -\dfrac{2}{3}$, find $\tan(\alpha + \beta)$.

Solution. $\tan(\alpha + \beta) = \dfrac{\tan \alpha + \tan \beta}{1 - \tan \alpha \tan \beta}$

$$= \dfrac{\dfrac{1}{2} + \left(-\dfrac{2}{3}\right)}{1 - \left(\dfrac{1}{2}\right)\left(-\dfrac{2}{3}\right)}$$

$$= \dfrac{-\dfrac{1}{6}}{\dfrac{4}{3}} \quad \text{or} \quad -\dfrac{1}{8}$$

NOTE.   Observe that $\tan(\alpha + \beta)$ involves only $\tan \alpha$ and $\tan \beta$. However, both $\sin(\alpha + \beta)$ and $\cos(\alpha + \beta)$ involve $\cos \alpha$, $\sin \alpha$, and $\cos \beta$, $\sin \beta$.

17. Using the $\tan(\alpha + \beta)$ identity, compute each of the following without the use of tables or calculator:

     (a) $\tan 75°$ (b) $\tan 105°$

(a) $\dfrac{\sqrt{3} + 1}{\sqrt{3} - 1}$

18. Replacing β with -β in Exercise 16(b), show that tan(α - β) may be written in terms of tan α and tan β as

$$\tan(\alpha - \beta) = \frac{\tan \alpha - \tan \beta}{1 + \tan \alpha \tan \beta}$$ MEMORIZE!

19. Using the tan(α - β) identity, compute tan 15° without the use of tables or calculator.

$\frac{\sqrt{3} - 1}{\sqrt{3} + 1}$

20. If tan α = $\sqrt{3}$ and tan β = -$\sqrt{2}$, compute each of the following without the use of tables or calculator:

   (a) tan(α + β)   (b) tan(α - β)

$(a) \ \frac{\sqrt{3} - \sqrt{2}}{1 + \sqrt{6}}$

21. If tan α = -$\frac{3}{4}$ and tan β = $\frac{2}{3}$, compute:

   (a) tan(α + β)   (b) tan(α - β)

$(a) \ -\frac{1}{18}$

A LOOK BACKWARDS

At this point the chapter has developed six identities. It is now appropriate to stop and show how they all fit together. The following diagram shows their relationships. Note the critical roles that the unit circle and the distance formula played.

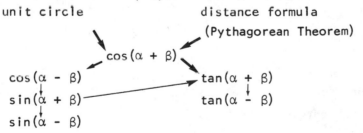

At this point the above identities should all be memorized.

Memorize all of these identities.

The "road map" shows how identities are linked together.

DOUBLE ANGLE IDENTITIES

When α = β, then sin(α + β) becomes sin(α + α) or sin 2α. Thus the identity for sin(α + β) becomes an identity concerning sin 2α, a so-called "double angle" identity. Even though they are just special cases of identities concerning α + β, the double angle identities are so frequently used that they are worth considering separately.

22. (a) Use the identity for sin(α + β), letting α = β, to show that

$$\sin 2\alpha = 2\sin \alpha \cos \alpha$$

(b) Use the identity for $\cos(\alpha + \beta)$, to show that

$$\cos 2\alpha = \cos^2\alpha - \sin^2\alpha$$

(c) Use the identity for $\tan(\alpha + \beta)$, to show that

$$\tan 2\alpha = \frac{2 \tan\alpha}{1 - \tan^2\alpha}$$

23. Without using tables or a calculator, check the identities for $\sin 2\alpha$, $\cos 2\alpha$, and $\tan 2\alpha$ when

(a) $\alpha = 30°$ (b) $\alpha = 60°$ (c) $\alpha = 150°$

24. (a) $\cos 2\alpha = -\frac{7}{25}$

$\sin 2\alpha = \frac{24}{25}$

$\tan 2\alpha = \frac{-24}{7}$

(c) $\cos 2\alpha = -\frac{3}{5}$

$\sin 2\alpha = -\frac{4}{5}$

$\tan 2\alpha = \frac{4}{3}$

24. Find $\cos 2\alpha$, $\sin 2\alpha$, and $\tan 2\alpha$ if

(a) $\cos \alpha = \frac{3}{5}$, $\sin \alpha = \frac{4}{5}$ (b) $\cos \alpha = -\frac{1}{2}$, $\sin \alpha = \frac{\sqrt{3}}{2}$

(c) $\cos \alpha = \frac{1}{\sqrt{5}}$, $\sin \alpha = -\frac{2}{\sqrt{5}}$

25. Hint: (c) $\cos 3° = \cos(2° + 1°)$

25. Let $c = \cos 1°$ and $s = \sin 1°$. Express each of the following in terms of the letters c and s:

(a) $\cos 2°$ (b) $\sin 2°$ (c) $\cos 3°$ (d) $\sin 3°$

(e) Explain how, knowing $\cos 1°$ and $\sin 1°$, you could find $\cos 16°$.

(a) $c^2 - s^2$
(b) $2cs$
(c) $c^3 - 3cs^2$
(d) $3c^2s - s^3$

OTHER FORMS OF THE IDENTITY FOR $\cos 2\alpha$

The identity for $\cos 2\alpha$ has three forms, all related to each other by the Pythagorean identity,

$$\cos^2\alpha + \sin^2\alpha = 1.$$

26. (a) In Exercise 22, the following form of $\cos 2\alpha$ was given:

$$\cos 2\alpha = \cos^2\alpha - \sin^2\alpha$$

(b) Show that the identity for $\cos 2\alpha$ in part (a) can also be written in terms of $\cos\alpha$ only, as follows:

$$\cos 2\alpha = 2 \cos^2\alpha - 1$$

(c) Show that the identity for $\cos 2\alpha$ in part (a) can also be written in terms of $\sin\alpha$ only:

$$\cos 2\alpha = 1 - 2 \sin^2\alpha$$

27. Without using tables, check that the three identities in Exercise 26 give the correct value for cos 2α when

    (a) α = 30°   (b) α = 45°   (c) α = 60°

28. If $\cos \alpha = \frac{5}{13}$ and $\sin \alpha = -\frac{12}{13}$, compute cos 2α in terms of

    (a) cosα and sinα   (b) sinα only   (c) cosα only

29. If $\cos\alpha = -\frac{\sqrt{5}}{3}$ and $\sin\alpha = -\frac{2}{3}$, compute cos 2α in terms of

    (a) cosα and sinα   (b) cosα only   (c) sinα only

All answers should be the same, namely, $\frac{-119}{169}$.

All answers should be the same.

Pause here to make sure you have memorized all the double angle (3α) identities.

HALF ANGLE IDENTITIES

The identity

$$\cos 2\alpha = 2 \cos^2 \alpha - 1$$

can be interpreted in two ways. So far we have calculated cos 2α when cosα is known. However suppose cos 2α is known, and cosα is to be found. A simple example of this would be the case α = 15°. For 2α is then 30° and cos 30° is known to be $\frac{\sqrt{3}}{2}$. The identity for cos 2α then reads

$$\cos 30° = 2 \cos^2 15° - 1$$

or

$$\frac{\sqrt{3}}{2} = 2 \cos^2 15° - 1$$

This equation could be solved for cos 15°.

   More generally, the double angle identities can be rewritten to produce what is called "half angle identities." For example, the identity

$$\cos 2\alpha = 2 \cos^2 \alpha - 1$$

may be rewritten as

$$\cos\alpha = 2 \cos^2 \left(\frac{\alpha}{2}\right) - 1$$

simply by replacing 2α with α and α with $\frac{\alpha}{2}$. Now it is possible to solve for $\cos \frac{\alpha}{2}$ in terms of cosα, as follows:

$$\cos\alpha = 2 \cos^2\left(\frac{\alpha}{2}\right) - 1$$

or $2 \cos^2\left(\frac{\alpha}{2}\right) - 1 = \cos\alpha$

$$2 \cos^2 \left(\frac{\alpha}{2}\right) = 1 + \cos\alpha$$

$$\cos^2 \left(\frac{\alpha}{2}\right) = \frac{1 + \cos \alpha}{2}$$

Memorize

Hence

$$\cos\left(\frac{\alpha}{2}\right) = \pm\sqrt{\frac{1 + \cos\alpha}{2}}$$

Note. The + or - in the $\cos\frac{\alpha}{2}$ identity has to be determined in each application; it depends in which quadrant the angle $\frac{\alpha}{2}$ lies.

Example 5. Given that $\cos 300° = \frac{1}{2}$, find $\cos\frac{300°}{2} = \cos 150°$.

Solution. By the $\cos\left(\frac{\alpha}{2}\right)$ identity, with $\alpha = 300°$,

$$\cos\left(\frac{\alpha}{2}\right) = \pm\sqrt{\frac{1 + \cos\alpha}{2}}$$

$$\cos\frac{300°}{2} = \pm\sqrt{\frac{1 + \cos 300°}{2}}$$

$$\cos 150° = -\sqrt{\frac{1 + \cos 300°}{2}}$$

The sign is negative because 150° is in the second quadrant, hence cos 150° is negative.

$$= -\sqrt{\frac{1 + \frac{1}{2}}{2}}$$

$$= -\frac{\sqrt{3}}{2}$$

The next exercise develops the identity for $\sin\left(\frac{\alpha}{2}\right)$.

30. (a) When 2α is replaced with α and α with $\left(\frac{\alpha}{2}\right)$, the identity

$$\cos 2\alpha = 1 - 2\sin^2\alpha$$

becomes

$$\cos\alpha = 1 - 2\sin^2\left(\frac{\alpha}{2}\right).$$

(b) Use the latter identity to show that

Memorize

$$\sin\left(\frac{\alpha}{2}\right) = \pm\sqrt{\frac{1 - \cos\alpha}{2}}$$

Again, the quadrant of $\frac{\alpha}{2}$ will determine whether $\sin\frac{\alpha}{2}$ is positive or negative.

31. Use $\cos\frac{\alpha}{2}$ and $\sin\frac{\alpha}{2}$ to show that $\tan\frac{\alpha}{2}$ in terms of $\cos\alpha$ is

The quadrant of the angle $\frac{\alpha}{2}$ determines the sign.

$$\tan\left(\frac{\alpha}{2}\right) = \pm\sqrt{\frac{1 - \cos\alpha}{1 + \cos\alpha}}$$

32. Use the identities for $\cos\left(\frac{\alpha}{2}\right)$, $\sin\left(\frac{\alpha}{2}\right)$ and $\tan\left(\frac{\alpha}{2}\right)$ to find the precise values for each of the following:

(a) $\cos 15°$ (b) $\sin 15°$ (c) $\tan 15°$

(d) $\cos 165°$ (e) $\sin 165°$ (f) $\tan 165°$

33. Compute without using tables or calculator:

(a) $\cos(22.5°)$ (b) $\sin(22.5°)$ (c) $\tan(22.5°)$

34. If $\cos 25° = 0.906$, without using tables or calculator, find:

(a) $\sin(12.5°)$ (b) $\cos(12.5°)$ (c) $\tan(12.5°)$

In Exercise 35, write the identity involved and then substitute the given values. Do not skip steps. Use the exercise as a review of the chapter thus far.

35. Suppose

$$\sin 70° = a \qquad \cos 70° = b \qquad \tan 70° = c$$

and

$$\sin 5° = d \qquad \cos 5° = e \qquad \tan 5° = f$$

Express in terms of a, b, c, d, e, or f (do not use tables or calculator):

(a) $\sin 75°$ (b) $\cos 75°$ (c) $\tan 75°$ (d) $\sin 65°$

(e) $\cos 65°$ (f) $\tan 65°$ (g) $\sin 10°$ (h) $\cos 10°$

(i) $\tan 10°$ (j) $\sin 35°$ (k) $\cos 35°$ (l) $\tan 35°$

PROVING IDENTITIES

The next two examples illustrate different methods for proving more complicated trigonometric identities. As you will observe, often there is a "trick" involved, as well as the problem of deciding which side of the assertion to use as a starting point in the proof. Study each example carefully, making sure you understand it before going on.

Method 1. Suppose you wish to prove A = B, where A and B are statements involving any of the six trigonometric functions. You can use the

32.
(a) $\sqrt{\dfrac{2+\sqrt{3}}{2}}$

(c) $\sqrt{\dfrac{2-\sqrt{3}}{2+\sqrt{3}}}$

(e) $\sqrt{\dfrac{2-\sqrt{3}}{2}}$

33.
(a) $\sqrt{\dfrac{2+\sqrt{2}}{2}}$

(c) $\sqrt{\dfrac{2-\sqrt{2}}{2+\sqrt{2}}}$

34. (a) $\sqrt{0.047}$

(c) $\sqrt{\dfrac{0.047}{0.953}} = \sqrt{0.0493}$

(a) ae + bd

(c) $\dfrac{c+f}{1-cf}$

(e) be + ad

(g) 2de

(i) $\dfrac{2f}{1-f^2}$

(k) $\sqrt{\dfrac{1+b}{2}}$

Study this section carefully.

familiar identities and the rules of algebra to modify one side so that it is exactly like the other side of the equation. That is, to prove

The vertical line is a good reminder that you do not accept the original statement as true until it has been proven.

(A is changed step

by step by using

algebra and trig

identities so that

it finally becomes B.)

$$A \overset{?}{=} B$$

B

Example 6 demonstrates this method.

Example 6. Show that $\dfrac{\sin \theta}{1 - \cos \theta} = \dfrac{1 + \cos \theta}{\sin \theta}$

Solution. We choose to work with the left side and show that it equals the right side:

We could have just as well have chosen to work only with the right side, changing it to agree with the left side.

$$\frac{\sin \theta}{1 - \cos \theta} \overset{?}{=} \frac{1 + \cos \theta}{\sin \theta}$$

$$\frac{\sin \theta}{1 - \cos \theta} \cdot \frac{1 + \cos \theta}{1 + \cos \theta}$$

$$\frac{\sin \theta (1 + \cos \theta)}{1 - \cos^2 \theta}$$

$$\frac{\sin \theta (1 + \cos \theta)}{\sin^2 \theta}$$

$$\frac{1 + \cos \theta}{\sin \theta}$$

NOTE. In Example 6 if $1 - \cos \theta = 0$ or $\sin \theta = 0$, then denominators equal to zero appear. The equation is meaningless in such a case. In general, in an identity, <u>values of</u> θ <u>that</u> <u>lead</u> <u>to a</u> <u>denominator</u> <u>equal</u> <u>to zero</u> <u>will</u> <u>be</u> <u>assumed</u> <u>to be</u> <u>excluded.</u>

Method 2. This method involves working on each side of the equation independently until arriving at an identity in which both sides are the same expression.

(Both A and B are modi-

fied independently using

algebra and trig identi-

ties until they both

equal D.)

$$A \overset{?}{=} B$$

D   D

Example 7 illustrates this second method.

Example 7. Prove that $\dfrac{\tan \theta - 1}{\tan \theta + 1} = \dfrac{1 - \cot \theta}{1 - \cot \theta}$.

Solution. Both sides are modified independently:

$$\dfrac{\tan \theta - 1}{\tan \theta + 1} \overset{?}{} \dfrac{1 - \cot \theta}{1 + \cot \theta}$$

$$\dfrac{\dfrac{\sin \theta}{\cos \theta} - 1}{\dfrac{\sin \theta}{\cos \theta} + 1} \qquad \dfrac{1 - \dfrac{\cos \theta}{\sin \theta}}{1 + \dfrac{\cos \theta}{\sin \theta}}$$

$$\dfrac{\cos \theta\left(\dfrac{\sin \theta}{\cos \theta} - 1\right)}{\cos \theta\left(\dfrac{\sin \theta}{\cos \theta} + 1\right)} \qquad \dfrac{\sin \theta\left(1 - \dfrac{\cos \theta}{\sin \theta}\right)}{\sin \theta\left(1 + \dfrac{\cos \theta}{\sin \theta}\right)}$$

$$\dfrac{\sin \theta - \cos \theta}{\sin \theta + \cos \theta} = \dfrac{\sin \theta - \cos \theta}{\sin \theta + \cos \theta}$$

NOTE. Example 7 also illustrates another useful technique: changing the expression to one involving only sines and cosines. If you cannot make any headway in a proof, you should express everything in terms of sines and cosines, for they are the fundamental trigonometric functions upon which the others depend.

36. Solve Example 6 by working the right side only.

37. Solve Example 7 by working the right side only.

COMMENT. Generally speaking, when working on one side only in proving identities it is easier to work with the more complicated side. The reason for this is that most students have had more experience simplifying algebraic expressions than making them more complicated.

A good understanding of algebra is essential in proving identities. This is as important as memorizing and using the standard trig identities correctly.

   The following exercises provide an opportunity to practice proving identities. There are more such exercises in the Review and in the Projects.

38. Prove the following identities:

(a) $2 \csc^2 \theta = \dfrac{1}{1 + \cos \theta} + \dfrac{1}{1 - \cos \theta}$ (b) $\sin^4 \theta - \cos^4 \theta = 2 \sin^2 \theta - 1$

(c) $\tan^4 \theta - \sec^4 \theta = 1 - 2 \sec^2 \theta$ (d) $\dfrac{\tan \theta - \cos \theta \cot \theta}{\csc \theta} = \dfrac{\sin \theta}{\cot \theta} - \dfrac{\cos \theta}{\sec \theta}$

(e) $\cos 2\theta = \dfrac{\csc^2 \theta - 2}{\csc^2 \theta}$ (f) $\dfrac{\sin 2\theta}{2 \sin^2 \theta} = \cot \theta$

(g) $\dfrac{1 - \sin 2\theta}{\cos 2\theta} = \dfrac{1 - \tan \theta}{1 + \tan \theta}$ (h) $1 + \cos \theta = \dfrac{\sin \theta}{\csc \theta - \cot \theta}$

39.   Prove the following identities:

(a) $\dfrac{\cot^2 \theta + 1}{\tan^2 \theta + 1} = \cot^2 \theta$ (b) $\csc^2 \left(\dfrac{\theta}{2}\right) = \dfrac{2 \sec \theta}{\sec \theta - 1}$

(c) $\dfrac{\sec \theta}{1 - \cos \theta} = \dfrac{\sec \theta + 1}{\sin^2 \theta}$ (d) $\dfrac{\sec \theta - \cos \theta}{\csc \theta} = \sin^2 \theta \tan \theta$

40.   Show that

Hint: sin 3θ = sin(2θ + θ)

$$\sin 3\theta = 3 \cos^2 \theta \sin \theta - \sin^3 \theta$$

41.   Show that

(a) $\cos 3\theta = \cos^3 \theta - 3 \sin^2 \theta \cos \theta$

(b) $\cos 3\theta = 4 \cos^3 \theta - 3 \cos \theta$

42.   (a)   Read Exercise 1 again.

(b)   Show that $d = 32 \sin 2A$

How large can the sine of an angle be?
(c) A = 45°

(c)   Use (b) to solve Exercise 1. Use properties of the sine function, not experiments based on the trigonometric tables.

* * * SUMMARY * * *

The unit circle and the distance formula were used to derive the key identity,

$$\cos(\alpha + \beta) = \cos\alpha \ \cos\beta - \sin\alpha \ \sin\beta.$$

Many other identities were derived from this one, as the chart shows.

Again the "road map" shows how the identities in this chapter are related.

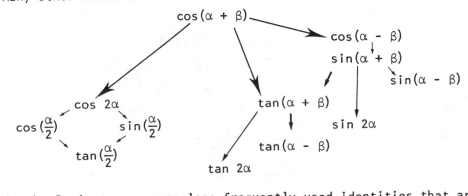

In the Projects are some less frequently used identities that are derived in the same manner from the basic identities. Some of them are important in the study of calculus.

SUM AND DIFFERENCE IDENTITIES

$$\cos(\alpha + \beta) = \cos \alpha \cos \beta - \sin \alpha \sin \beta$$

$$\cos(\alpha - \beta) = \cos \alpha \cos \beta + \sin \alpha \sin \beta$$

$$\sin(\alpha + \beta) = \sin \alpha \cos \beta + \cos \alpha \sin \beta$$

$$\sin(\alpha - \beta) = \sin \alpha \cos \beta - \cos \alpha \sin \beta$$

$$\tan(\alpha + \beta) = \frac{\tan \alpha + \tan \beta}{1 - \tan \alpha \tan \beta}$$

$$\tan(\alpha - \beta) = \frac{\tan \alpha - \tan \beta}{1 + \tan \alpha \tan \beta}$$

Memorize all the identities summarized here.

DOUBLE ANGLE IDENTITIES

$$\cos 2\alpha = \cos^2 \alpha - \sin^2 \alpha$$

$$= 2 \cos^2 \alpha - 1$$

$$= 1 - 2 \sin^2 \alpha$$

$$\sin 2\alpha = 2 \sin\alpha \cos\alpha$$

$$\tan 2\alpha = \frac{2 \tan \alpha}{1 - \tan^2 \alpha}$$

HALF ANGLE IDENTITIES

$$\cos\left(\frac{\alpha}{2}\right) = \pm \sqrt{\frac{1 + \cos \alpha}{2}}$$

$$\sin\left(\frac{\alpha}{2}\right) = \pm \sqrt{\frac{1 - \cos \alpha}{2}}$$

$$\tan\left(\frac{\alpha}{2}\right) = \pm \sqrt{\frac{1 - \cos \alpha}{1 + \cos \alpha}}$$

The above identities were used along with definitions, earlier identities, and the rules of algebra to prove identities as illustrated in Examples 6 and 7.

* * * WORDS AND SYMBOLS * * *

α, β
Proving trigonometric identities
double angle identity
half angle identity

* * * CONCLUDING REMARKS * * *

For hundreds of years the trigonometric identities of this chapter were the tools for computing the trigonometric tables. Let us follow the thinking of Claudius Ptolemy, a Greek mathematician who lived in Egypt in the second century A.D., as he constructed the first trigonometric table.

First of all, by geometric means he was able to find sin 72° and cos 72°. For the sake of brevity, we will use trigonometric identities to get cos 72°, as follows.

$$\cos 144° = -\cos 36° \qquad \text{(Why?)}$$

$$2 \cos^2 72° - 1 = -\sqrt{\frac{1 + \cos 72°}{2}}$$

Let cos 72° = x, then the expression becomes

$$2 x^2 - 1 = -\sqrt{\frac{1 + x}{2}}$$

Squaring, $$4x^4 - 4x^2 + 1 = \frac{1 + x}{2}$$

which simplifies to

$$8x^4 - 8x^2 - x + 1 = 0$$

The left side factors nicely, the equation is equivalent to

$$(4x^2 + 2x - 1)(2x^2 - x - 1) = 0$$

or finally

$$(4x^2 + 2x - 1)(2x + 1)(x - 1) = 0$$

The solutions $-\frac{1}{2}$ or 1 are irrelevant. The solutions of

$$4x^2 + 2x - 1 = 0$$

will include the solution, cos 72°. By the quadratic formula,

$$x = \frac{-2 \pm \sqrt{20}}{8}$$

or

$$x = \frac{-1 \pm \sqrt{5}}{4}$$

Since cos 72° is positive, the plus sign must be used, giving

$$\cos 72° = x = \frac{-1 + \sqrt{5}}{4}$$

It is then easy to get sin 72° by the relation, $\cos^2 72° + \sin^2 72° = 1$.

Having found cos 72° and sin 72°, and already knowing that

$$\cos 60° = \frac{1}{2} \quad \text{and} \quad \sin 60° = \frac{\sqrt{3}}{2}$$

Ptolemy found cos 12° and sin 12°; for instance

$$\cos 12° = \cos(72° - 60°)$$
$$= \cos 72° \cos 60° + \sin 72° \sin 60°$$

By the half angle formula for $\cos\left(\frac{\alpha}{2}\right)$, he then found

$$\cos 6° = \cos\left(\frac{12}{2}\right)°$$
$$= \sqrt{\frac{1 + \cos 12°}{2}}$$

Using the half angle formula repeatedly, he found cos 3°, $\cos\left(\frac{3}{2}\right)°$, and $\cos\left(\frac{3}{4}\right)°$. It is then easy to get the sine of these angles by the Pythagorean relationship, $\cos^2\theta + \sin^2\theta = 1$. By repeatedly adding the angle $\left(\frac{3}{4}\right)°$ he could build up a table. But he preferred to use intervals of $\left(\frac{1}{2}\right)°$. By other means which were quite ingenious he estimated that

$$\sin\left(\frac{1}{2}\right)° \doteq \frac{31}{60^2} + \frac{25}{60^3}$$

which is quite accurate. On this basis he made the first trigonometric table. Perhaps this laborious approach may help you appreciate your best friend in this course, your calculator!

* * * IDENTITIES SELF QUIZ * * *

1. If $\cos\alpha = \frac{5}{13}$, $\sin\alpha = \frac{12}{13}$, $\cos\beta = -\frac{3}{5}$, and $\sin\beta = -\frac{4}{5}$, compute:

    (a) $\sin(\alpha + \beta)$ (b) $\cos(\alpha - \beta)$ (c) $\sin(\beta - \alpha)$ (d) $\cos(\alpha + \beta)$

2. If $\tan\alpha = -\sqrt{5}$ and $\tan\beta = \frac{1}{\sqrt{2}}$ compute:

    (a) $\tan(\alpha + \beta)$ (b) $\tan(\beta - \alpha)$

3. If $90° < \alpha < 180°$ and $\cos\alpha = -0.8$ compute:

    (a) $\sin\left(\frac{\alpha}{2}\right)$ (b) $\tan\left(\frac{\alpha}{2}\right)$ (c) $\cos\left(\frac{\alpha}{2}\right)$

4. If $\cos \alpha = -\frac{7}{25}$ and $\sin \alpha = -\frac{24}{25}$ where $180° < \alpha < 270°$, compute:

    (a) $\sin 2\alpha$ (b) $\cos 2\alpha$ (c) $\tan 2\alpha$

5. Prove the identities:

    (a) $\csc^2 x = \frac{2}{1 - \cos 2x}$ (b) $\cot x = \frac{1 + \cos 2x}{\sin 2x}$

    (c) $(\sin x + \cos x)^2 = 1 + \sin 2x$

ANSWERS TO TRIGONOMETRIC IDENTITIES SELF QUIZ

1. (a) $-\frac{56}{65}$ (b) $-\frac{63}{65}$ (c) $\frac{16}{65}$ (d) $\frac{33}{65}$

2. (a) $\tan(\alpha + \beta) = \frac{1 - \sqrt{10}}{\sqrt{5} + \sqrt{2}}$ (b) $\frac{1 + \sqrt{10}}{\sqrt{2} - \sqrt{5}}$

3. (a) $\frac{3}{\sqrt{10}} = \frac{3\sqrt{10}}{10}$ (b) 3 (c) $\frac{1}{\sqrt{10}}$

4. (a) $\frac{336}{625}$ (b) $-\frac{527}{625}$ (c) $-\frac{336}{527}$ 5. Proofs

* * * TRIGONOMETRIC IDENTITIES REVIEW * * *

1. Consider the diagram in a unit circle where $0° < \theta < 90°$.

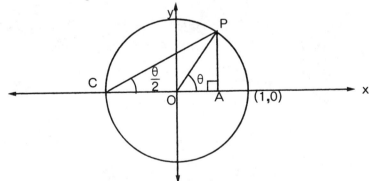

    (a) In terms of sine and cosine of θ, explain why

        $OA = \cos \theta$ and $AP = \sin \theta$

    (b) Using the right triangle CAP, explain in terms of $\sin \theta$ and $\cos \theta$ why

        $$\tan\left(\frac{\theta}{2}\right) = \frac{\sin \theta}{1 + \cos \theta}$$

2. Consider triangle PAT in the following diagram in a unit circle where $0° < \theta < 90°$.

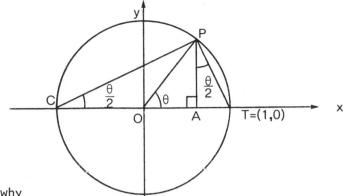

(a) Explain why

$$\tan\left(\frac{\theta}{2}\right) = \frac{1 - \cos\theta}{\sin\theta}$$

(b) Using part (a) and Exercise 1(b) prove that $\dfrac{1 - \cos\theta}{\sin\theta} = \dfrac{\sin\theta}{1 + \cos\theta}$

(c) Use (b) to conclude (again) that $\sin^2\theta + \cos^2\theta = 1$.

(d) Using previous identities, prove that

$$\tan\left(\frac{\theta}{2}\right) = \frac{1 - \cos\theta}{\sin\theta}$$

Hint:

$$\sqrt{\frac{1-\cos\theta}{1+\cos\theta} \cdot \frac{1-\cos\theta}{1-\cos\theta}} = ?$$

(e) Using previous identities, prove that

$$\tan\left(\frac{\theta}{2}\right) = \frac{\sin\theta}{1 + \cos\theta}$$

Hint:

$$\sqrt{\frac{1-\cos\theta}{1+\cos\theta} \cdot \frac{1+\cos\theta}{1+\cos\theta}} = ?$$

3. Using identities for $\sin(\alpha \pm \beta)$, express each of the following in terms of $\sin\theta$ and/or $\cos\theta$.

(a) $\sin(\theta + 90°)$ (b) $\sin(90° - \theta)$ (c) $\sin(\theta - 90°)$

(d) $\sin(\theta + 180°)$ (e) $\sin(180° - \theta)$ (f) $\sin(270° - \theta)$

4. Using identities for $\cos(\alpha \pm \beta)$, express each of the following in terms of $\sin\theta$ and/or $\cos\theta$.

(a) $\cos(\theta + 90°)$ (b) $\cos(90° - \theta)$ (c) $\cos(\theta - 90°)$

(d) $\cos(\theta + 180°)$ (e) $\cos(\theta + 270°)$ (f) $\cos(\theta - 360°)$

5. If $\sin 10° = 0.175$, $\cos 10° = 0.985$ and $\tan 10° = 0.176$, without using tables or calculator find:

(a) $\sin 20°$ (b) $\cos 20°$ (c) $\tan 20°$

6. (a) Write $\sin 3\theta = \sin(2\theta + \theta)$ in terms of $\sin\theta$ and $\cos\theta$.

(b) Show that $\sin 3\theta$ can be written in terms of $\sin\theta$ alone.

7. (a) Write $\cos 3\theta$ in terms of $\sin\theta$ and $\cos\theta$.

(b) Show that $\cos 3\theta$ can be written in terms of $\cos\theta$ alone.

8.   Prove the following identities:

(a) $\sin^2\theta + \cos^2\theta = \sec^2\theta - \tan^2\theta$ (b) $\dfrac{\sin\theta - 1}{\cos\theta} = \tan\theta - \sec\theta$

(c) $\tan\theta = \dfrac{\sin\theta + \tan\theta}{1 + \cos\theta}$ (d) $\dfrac{\sin^2\theta}{\cos^2\theta} = \dfrac{1}{\cos^2\theta} - 1$

(e) $2\sin^2\theta - 1 = \sin^4\theta - \cos^4\theta$ (f) $\tan\theta + \sec\theta = \dfrac{\sin\theta + 1}{\cos\theta}$

9.   Prove the identities:

(a) $\sin(45° + x) - \sin(45° - x) = \sqrt{2}\ \sin x$

(b) $\cos(30° + x)\cos(30° - x) - \sin(30° + x)\sin(30° - x) = \dfrac{1}{2}$

10.   Prove the identities:

(a) $\dfrac{\cot\theta - 1}{\cot\theta + 1} = \dfrac{1 - \sin 2\theta}{\cos 2\theta}$ (b) $\sec 2\theta = \dfrac{1}{1 - 2\sin^2\theta}$

(c) $\cot 2\theta = \dfrac{\csc\theta - 2\sin\theta}{2\cos\theta}$

11.   Solve for θ ($0° \le \theta < 360°$)

(a) $\tan\theta = 1$ (b) $2\sin\theta - 1 = 0$

(c) $2\sin\theta + 3 = 0$ (d) $3\tan\theta + 3 = 0$

(e) $2\cos^2\theta = \dfrac{1}{2}$ (f) $4\sin^2\theta - 3 = 0$

(g) $\cos\theta = -2$ (h) $2\cos^2\theta - \cos\theta = 1$

(i) $\cos\theta(2\cos\theta - 1) = 0$ (j) $\sin^2\theta - 2\sin\theta + 1 = 0$

12.   Prove:

(a) $\sin(\theta + 45°) = \cos(45° - \theta)$ (b) $\sin(\theta + 45°) = \cos(\theta - 45°)$

(c) $\dfrac{\sin 2\theta}{2\sin^2\theta} = \cot\theta$ (d) $\cos 2\theta = \dfrac{\csc^2\theta - 2}{\csc^2\theta}$

13.   Prove that

$\sin(\alpha + \beta)$ does <u>not</u> equal $\sin\alpha + \sin\beta$

Hint: Find a specific counter-example.

14.   Prove the identities

Hint: While Exercise 14 may look difficult it is quite easy.

(a) $\dfrac{\tan x - \tan(z - y)}{1 + \tan x \tan(z - y)} = \dfrac{\tan y - \tan(z - x)}{1 + \tan y \tan(z - x)}$

(b) $\dfrac{\tan x - \tan(y - z)}{1 + \tan x \tan(y - z)} = \dfrac{-\tan y + \tan(x + z)}{1 + \tan y \tan(x + z)}$

(c) $\dfrac{\tan x + \tan(y + z)}{1 - \tan x \tan(y + z)} = \dfrac{\tan y + \tan(x + z)}{1 - \tan y \tan(x + z)}$

15. Prove:

(a) $\sec \theta \csc \theta = \tan \theta + \cot \theta$

(b) $(1 + \tan \theta)^2 + (1 + \cot \theta)^2 = (\sec \theta + \csc \theta)^2$

(c) $\sin^2\theta - \cos^2\theta = \dfrac{\tan \theta - \cot \theta}{\tan \theta + \cot \theta}$

(d) $\sec \theta + \tan \theta = \dfrac{1}{\sec \theta - \tan \theta}$

(e) $\tan \alpha + \dfrac{\cos \alpha}{1 + \sin \alpha} = \sec \alpha$

16. Prove:

(a) $\dfrac{\sin^2 \alpha}{1 + \cos \alpha} = 1 - \dfrac{1}{\sec \alpha}$

(b) $\cos^4 x - \sin^4 x = 2 \cos^2 x - 1$

(c) $\sec^4 x + \tan^4 x - 2 \sec^2 x \tan^2 x - 1 = 0$

(d) $(\cos \alpha + \cos \beta)^2 + (\sin \alpha - \sin \beta)^2 = 2 + 2 \cos(\alpha + \beta)$

(e) $\sin^2\alpha \sec^2\alpha + \tan^2\alpha \cos^2\alpha = \sin^2\alpha + \tan^2\alpha$

(f) $\dfrac{1}{1 - \sin \alpha} + \dfrac{1}{1 + \sin \alpha} = 2 \sec^2\alpha$

17. Use identities to prove that:

(a) $\cos(90° + \theta) = - \sin \theta$ (b) $\cos(\alpha + 180°) = - \sin(90° - \alpha)$

(c) $\sin \alpha \cot \alpha + \cos(\alpha - 180°) = 0$

(d) $\sec \alpha \cos(90° - \alpha) = \tan \alpha$

18. Prove:

(a) $\dfrac{\tan \alpha - \tan \beta}{1 + \tan \alpha \tan \beta} = \dfrac{\sin \alpha \cos \beta - \cos \alpha \sin \beta}{\cos \alpha \cos \beta + \sin \alpha \sin \beta}$

(b) $\dfrac{\cot^2 x - 1}{2 \cot x} = \dfrac{\cos^2 x - \sin^2 x}{2 \sin x \cos x}$ (c) $\tan\left(\dfrac{\alpha}{2}\right) = \csc \alpha - \cot \alpha$

(d) $(1 - \cos^2 x)\csc^2 x = 1$ (e) $(\csc^2 x - 1)\tan^2 x = 1$

(f) $(\cot \theta + \csc \theta)^2 = \dfrac{1 + \cos \theta}{1 - \cos \theta}$

19. Prove:

(a) $\tan(\theta + 45°) = \dfrac{1 + \tan \theta}{1 - \tan \theta}$ (b) $\tan(45° - \theta) = \dfrac{1 - \tan \theta}{1 + \tan \theta}$

ANSWERS TO TRIGONOMETRIC IDENTITIES REVIEW

1. (a) Point P is on the unit circle.   (b) See triangle CAP.

2. (a) See triangle PAT   (b) See (a) and Ex. 1(b)   (c) Simplify

    (d) Proof, see hint.   (e) Proof, see hint.

3.   (a) $\cos \theta$ (b) $\cos \theta$ (c) $- \cos \theta$

      (d) $- \sin \theta$ (e) $\sin \theta$ (f) $- \cos \theta$

4.   (a) $- \sin \theta$ (b) $\sin \theta$ (c) $\sin \theta$

      (d) $- \cos \theta$ (e) $\sin \theta$ (f) $\cos \theta$

5.   (a) 0.345   (b) 0.940   (c) 0.363

6.   (a) $3 \sin \theta \cos^2 \theta - \sin^3 \theta$ (b) $\sin 3\theta = -4\sin^3 \theta + 3\sin \theta$

7.   (a) $\cos 3\theta = \cos^3 \theta - 3 \sin^2 \theta \cos \theta$ (b) $\cos 3\theta = 4\cos^3 \theta - 3 \cos \theta$

8. - 10. Prove   11. (a) 45°, 225° (b) 30°, 150° (c) impossible

      (d) 135°, 315° (e) 60°, 120°, 240°, 300° (f) 60°, 120°, 240°, 300°

      (g) impossible (h) 0°, 120°, 240° (i) 60°, 90°, 270°, 300° (j) 90°

12.   Prove 13. Let $\alpha = 30°$ and $\beta = 60°$ 14. - 19. Prove

* * * TRIGONOMETRIC IDENTITIES REVIEW SELF TEST * * *

1.   Find __all__ entries in the second column that match each entry in the
      first column. (There may be more than one item in the second
      column matching an item in the first column.)

(1) $\sin 2\alpha$ (a) $\sin\alpha \cos\beta - \cos\alpha \sin\beta$

(2) $\tan(\alpha + \beta)$ (b) $\pm \sqrt{\dfrac{1 - \cos \alpha}{1 + \cos \alpha}}$

(3) $\cos\left(\dfrac{\alpha}{2}\right)$ (c) $\cos\alpha \cos\beta - \sin\alpha \sin\beta$

(4) $\sin(\alpha - \beta)$ (d) $1 - 2 \sin^2\alpha$

(5) $\cos 2\alpha$ (e) $\pm \sqrt{\dfrac{1 - \cos\alpha}{2}}$

(6) $\cos(\alpha - \beta)$ (f) $2 \cos\alpha \sin\alpha$

(7) $\tan\left(\dfrac{\alpha}{2}\right)$ (g) $\pm \sqrt{\dfrac{\cos \alpha - 1}{2}}$

(8) $\cos(\alpha + \beta)$ (h) $\dfrac{\tan \alpha - \tan \beta}{1 + \tan \alpha \tan \beta}$

(9) $\tan 2\alpha$ (i) $\cos^2\alpha - \sin^2\alpha$

(10) $\sin(\alpha + \beta)$ (j) $\dfrac{\tan \beta + \tan\alpha}{1 - \tan \beta \tan\alpha}$

(11) $\tan(\alpha - \beta)$ (k) $\pm \sqrt{\dfrac{1 + \cos\alpha}{2}}$

(12) $\sin\left(\dfrac{\alpha}{2}\right)$ (l) $\sin \alpha \cos\beta + \cos\alpha \sin\beta$

(continue on next page)

(m) $\dfrac{2 \tan \alpha}{1 - \tan^2 \alpha}$

(n) $\cos \alpha \cos \beta + \sin \alpha \sin \beta$

(o) $2 \cos^2 \alpha - 1$

2. Prove the following identities:

(a) $\dfrac{\cos \theta}{1 - \sin\theta} = \sec \theta + \tan \theta$

(b) $\dfrac{\sec \theta}{1 - \cos \theta} = \dfrac{\sec \theta + 1}{\sin^2 \theta}$

(c) $\dfrac{1 + \cos \theta}{\cos \theta - 1} + \dfrac{1 + \sec \theta}{\sec \theta - 1} = 0$

(d) $\dfrac{\sin \theta}{\sin \theta + \cos \theta} = \dfrac{\sec \theta}{\sec \theta + \csc \theta}$

3. Prove the identities:

(a) $\sin 4\theta = 8 \sin\theta \, \cos^3\theta - 4 \sin\theta \, \cos\theta$

(b) $\cos 4\theta = 8 \cos^4\theta - 8 \cos^2\theta + 1$

4. Compute $\sin 75°$ in two ways:

(a) Using $\sin(30° + 45°)$.   (b) Using the half-angle formula.

(c) Show the answers to parts (a) and (b) are the same.

ANSWERS TO THE TRIGONOMETRIC IDENTITIES SELF TEST

1. (1)-(f),   (2)-(j),   (3)-(k),   (4)-(a),   (5)-(i)(d)(o),   (6)-(n),
(7)-(b),   (8)-(c),   (9)-(m),   (10)-(l),   (11)-(h),   (12)-(e)

2. - 3. Prove   4. (a) $\dfrac{\sqrt{2} + \sqrt{6}}{4}$ (b) $\dfrac{1}{2} \sqrt{2 + \sqrt{3}}$

(c) Show (a) and (b) are the same number. (Use no table or calculator!)

* * * TRIGONOMETRIC IDENTITIES PROJECTS * * *

The first seven projects concern a different type of identity, which is important in the theory of sound and in calculus.

1. Using the identities for $\sin(x + y)$ and $\sin(x - y)$, show that
$$\sin(x + y) + \sin(x - y) = 2 \sin x \cos y$$

2. In Project 1, let

$$x + y = A$$

and

$$x - y = B$$

Solve for x and y in terms of A and B. The results will be used in the next project.

3.   From Projects 1 and 2 conclude that

$$\sin A + \sin B = 2 \sin\left(\frac{A + B}{2}\right) \cos\left(\frac{A - B}{2}\right)$$

4.   Replacing B by -B in Project 3, show that

$$\sin A - \sin B = 2 \sin\left(\frac{A - B}{2}\right) \cos\left(\frac{A + B}{2}\right)$$

This particular identity is due to Vieta, a 16th century lawyer, considered by many to be the best mathematician of his time.

5.   Using the identities for $\cos(x + y)$ and $\cos(x - y)$, show that

$$\cos(x + y) + \cos(x - y) = 2 \cos x \cos y$$

and that

$$\cos(x + y) - \cos(x - y) = -2 \sin x \sin y$$

6.   Combine Projects 2 and 5 to show that

$$\cos A + \cos B = 2 \cos\left(\frac{A + B}{2}\right) \cos\left(\frac{A - B}{2}\right)$$

7.   Combine Projects 2 and 5 to show that

$$\cos A - \cos B = -2 \sin\left(\frac{A + B}{2}\right) \sin\left(\frac{A - B}{2}\right)$$

Projects 8 - 10 may be solved easily with the preceding identities.

8.   Write as a sum or difference:

(a) $2 \cos 50° \cos 23°$ (b) $2 \sin 134° \cos 15°$

(c) $-2 \sin 82° \sin 63°$

9.   Write as a product:

(a) $\cos 105° - \cos 25°$ (b) $\sin 50° + \sin 40°$ (c) $\cos 35° + \cos 55°$

10.   Prove:

(a) $\dfrac{\sin 4x - \sin 2x}{\cos 2x + \cos 4x} = \tan x$ (b) $\dfrac{\cos 3x + \cos x}{\cos x - 2 \sin^2 x \cos x} = 2$

The next Project gives the physical background of Exercise 1 in the main section of the chapter.

11. A ball is thrown at an angle A with velocity v_o feet per second. It rises and falls, sweeping out the path shown in this diagram.

(a) Using the triangle below, show that the initial horizontal speed is $v_o \cos A$ and the initial vertical speed is $v_o \sin A$.

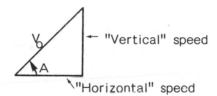

(b) Show that if there were no gravitational pull, the distance of the ball above the ground after t seconds would be

$$(v_o \sin A)t \text{ feet.}$$

(c) Gravity causes an object to fall $16t^2$ feet in t seconds (if it starts from rest.) Combine this fact with (b) to show that the height of the ball after t seconds is

$$(v_o \sin A)t - 16t^2 \text{ feet.}$$

(d) When the ball's height is zero, the ball strikes the ground. Show that this occurs when

$$t = \frac{v_o \sin A}{16} \text{ seconds.}$$

(e) Combining (a) and (d), show that the ball travels a horizontal distance of

$$(v_o \cos A)\left(\frac{v_o \sin A}{16}\right) \text{ feet}$$

This was the formula given, without proof, in Exercise 1 at the beginning of the chapter.

(f) In Exercise 1, the value of v_o was given as 32. Deduce from (e) that the ball travels

$$64 \sin A \cos A \text{ feet}$$

12. In Project 11 of the last chapter it was proved that in any triangle

$$\frac{a - b}{a + b} = \frac{\sin A - \sin B}{\sin A + \sin B}$$

Use this and Projects 3 and 4 of this chapter to prove

$$\frac{a - b}{a + b} = \frac{\tan \frac{1}{2} (A - B)}{\tan \frac{1}{2} (A + B)}$$

a result, known as the Law of Tangents, obtained by Vieta in 1579.

13.   The following identity has applications in engineering and advanced calculus: For any positive integer n

$$\cos x + \cos 2x + \ldots + \cos nx = \frac{\sin(n + \frac{1}{2})x}{2 \sin(\frac{1}{2} x)} - \frac{1}{2}$$

Prove this identity, as follows.

(a) First multiply by $2 \sin(\frac{1}{2} x)$ to clear denominators.

(b) Use Project 1 to express each product

$2 \sin(\frac{1}{2} x) \cos kx$, where k = 1, 2, . . ., n

as a difference of two sines.

14.   Prove:

(a) $\tan 2x = \dfrac{\sin 4x}{1 + \cos 4x}$ (b) $\tan 3x = \csc 6x - \cot 6x$

(c) $\sin 3\alpha = \sin \alpha [2 \cos(2\alpha) + 1]$

(d) $\cos 4x = 2(2 \cos^2 x - 1)^2 - 1$

(e) $\sin 4x = 4(\cos^2 x - \sin^2 x)\sin x \cos x$

(f) $2 - \sin^2(2x) = 2(\sin^4 x + \cos^4 x)$

15.   Prove:

(a) $\cos 3x = \cos x[2\cos(2x) - 1]$

(b) $\sin 4x = 4 \sin x \cos x \cos (2x)$

(c) $\sin 3x = (3 - 4 \sin^2 x)\sin x$

(d) $\sin 6x = 2[4 \cos^3 x - 3 \cos x] \sin 3x$

16.   Let (x_1, y_1) and (x_2, y_2) be two points in the plane. Let θ be the angle shown in this diagram.

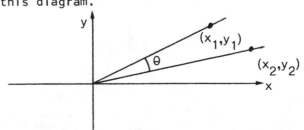

(a) Using the Law of Cosines, prove that

$$\cos \theta = \frac{x_1 y_1 + x_2 y_2}{\sqrt{x_1^2 + y_1^2} \; \sqrt{x_2^2 + y_2^2}}$$

(b) From (a) deduce that for all real numbers x_1, x_2, y_1, y_2

$$x_1 y_1 + x_2 y_2 \leq \sqrt{x_1^2 + y_1^2} \; \sqrt{x_2^2 + y_2^2}$$

(This is an important inequality in advanced applications of mathematics. It is a special case of Schwartz's inequality.)

17. The function $\cos 2x$ can be expressed as a polynomial in $\cos x$, namely

$$\cos 2x = 2 \cos^2 x - 1$$

(The polynomial is $2y^2 - 1$).

Show that it is impossible to express $\sin 2x$ as a polynomial in $\sin x$.

18. Find an equation of degree three (a cubic equation) of which $\cos 20°$ is a solution. The coefficients should be integers, not all zero.

19. Vieta, in 1591, was able to use the identity

(1) $$\cos 3\alpha = 4 \cos^3 \alpha - 3 \cos \alpha$$

to solve a cubic equation. Consider the equation
$$ax^3 + bx^2 + cx + d = 0.$$

Division by $\underline{a}$ yields an equivalent equation with leading coefficient 1,

(2) $$x^3 + \frac{b}{a} x^2 + \frac{c}{a} x + \frac{d}{a} = 0$$

Next let

$$y = x + \frac{b}{3a},$$

that is

$$x = y - \frac{b}{3a}$$

Replacing x in (2) by $y - \frac{b}{3a}$ yields an equation in which y^2 has coefficient 0. (Check this.) So Vieta begins with the equation

(3) $$y^3 + py + q = 0$$

without loss of generality, where p and q are "new" coefficients.
Now (1) is equivalent to

(4) $$\cos^3\alpha - \frac{3}{4}\cos\alpha - \frac{1}{4}\cos 3\alpha = 0$$

which resembles (3).

In (3) he replaces y by nz, where n will be determined later, obtaining

$$n^3z^3 + pnz + q = 0$$

or

(5) $$z^3 + \frac{p}{n^2}z + \frac{q}{n^3} = 0$$

Recall that p and q are given. Select n such that

$$\frac{p}{n^2} = -\frac{3}{4},$$

that is

$$n = \sqrt{-\frac{4p}{3}} \qquad \text{(Assume } p < 0\text{)}$$

Equation (5) becomes

$$z^3 - \frac{3}{4}z + \frac{q}{n^3} = 0$$

or

(6) $$z^3 - \frac{3}{4}z + \frac{q}{\left(\frac{-4p}{3}\right)^{\frac{3}{2}}} = 0$$

(Note that (5) is close to (4). By (4) a solution of

$$t^3 - \frac{3}{4}t - \frac{1}{4}\cos 3\alpha = 0$$

is given by t = cos α. Thus to solve (6) find α with a calculator such that

$$-\frac{1}{4}\cos 3\alpha = \frac{q}{\left(\frac{-4p}{3}\right)^{\frac{3}{2}}}$$

Then (6) will have the solution cos α (as well as cos(α + 120°) and cos(α + 240°).)

* * * CENTRAL SECTION * * *

This chapter is concerned with the graphs of the trigonometric functions. These graphs, which look quite different from those of any functions encountered in algebra, are important in physics and calculus. The definitions of the trigonometric functions in terms of the unit circle will play a key role throughout the chapter.

The chapter concludes by introducing a second way of measuring the size of angles, radian measurement. Students who go on to analytic geometry or calculus will find it is used there more often than the traditional degree measurement.

THE GRAPHS OF $y = \sin x$ AND $y = \cos x$

Whenever a function is graphed it is customary to use the x-axis for the "input" and the y-axis for the "output," or value of the function. Consider, for instance, the squaring functions, $y = x^2$. When the input is 3, the output is 3^2 or 9. Thus the point (3,9) lies on the graph of $y = x^2$, as shown in this diagram:

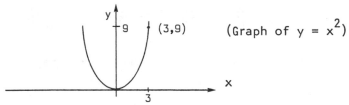

(Graph of $y = x^2$)

The graph of the sine function is obtained in the same way. It shows how the sine varies as the angle is changed. The input is the angle; the output is the sine of that angle. For this reason, the angle is given the symbol, x, and the sine of the angle is denoted y,

$$y = \sin x.$$

For instance, when $x = 45°$, then $y = \sin x$ is $\frac{\sqrt{2}}{2}$, which is about 0.707. Thus, the point $(45° , 0.707)$ is on the graph of $y = \sin x$.

EXERCISES

1.   This exercise develops the graph of y = sin x.

(a)   Using a calculator, copy and fill in the following table:

x	-360°	-315°	-270°	-225°	-180°	-135°	-90°	-45°	0°	45°	90°	135°	180°	225°	270°	315°	360°
sin x																	

(b)   Draw a coordinate system similar to the one shown below:

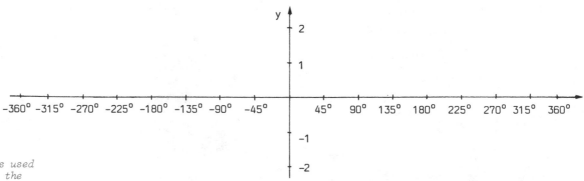

The x-axis is used to represent the size of angles (in degrees); the y-axis represents the sine of the angle (a real number). Note that the scales (units) are not the same on both axes.

(c)   Using the coordinate system drawn in (b), plot carefully the 17 points (x, sin x) tabulated in (a).

(d)   Draw a smooth curve through the points in (c). The graph you get in (c) should be a smooth curve and look something like this, ⌐⌐⌐   .

(e)   Compare your graph with the graphs of other students.

(f)   Describe how sin x varies as x goes from -360° to 360°. (When is it increasing? When is it decreasing? How large does it get? How small does it get?)

(g)   Label the graph: y = sin x.

The graph y = sin x is called the "sine curve."

2.   This exercise concerns the graph of the cosine function.

(a)   Using a trigonometric table or a calculator, copy and fill in the following table:

x	-360°	-315°	-270°	-225°	-180°	-135°	-90°	-45°	0°	45°	90°	135°	180°	225°	270°	315°	360°
cos x																	

(b)   Draw a coordinate system as in Exercise 1(b).

(c)   Plot carefully all points (x, cos x) from the table in (a).

(d)   Draw a smooth curve through the points in (c).

(e)   Describe in your own words how cos x varies as x goes from -360°
to 360°.

(f)   Label the graph: y = cos x.

3.   The graph in Exercise 1 concerns y = sin x for x between -360° and
360° only.   Using the identity

$$sin(x + 360°) = sin x,$$

sketch the graph of y = sin x for -360° ≤ x ≤ 1080°.

4.   Using the identity,

$$cos(x + 360°) = cos x,$$

sketch the graph of y = cos x for -360° ≤ x ≤ 1080°.

5.   Assume that you had made a very careful sketch of the graph of
y = cos x.   How could you use that graph to get a graph of y = sin x?

THE GRAPH OF y = tan x

The y values of the graphs of y = sin x and y = cos x never go above
1 nor below -1.   The graph of y = tan x is quite different since tan x
becomes arbitrarily large when x is near 90°.

6.   This exercise concerns the graph of y = tan x.

(a)   Copy the following table and fill it in with the aid of a calculator.

x	-90°	-80°	-70°	-60°	-45°	-30°	0°	30°	45°	60°	70°	80°	90°
tan x	undefined												undefined

(b)   Why is tan x undefined at x = 90° and x = -90°?

(c)   Copy the following diagram and scales accurately.   Allow a good
deal of space on your paper to extend the y-axis both above and
below the origin.   The graph will be used again.

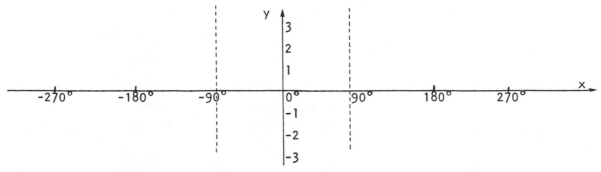

Changing an angle by 360° leaves the values of the sine and cosine unchanged. Thus, once you have graphed y = sin x or y = cos x for x in the range from 0° to 360°, just slide this part of the graph to the right or left 360°

Hint: (b) Recall the unit circle, and definition of the tangent.

Note: The graphs
of the sine and
cosine do not have
asymptotes. The
graphs of the
other four
trigonometric
functions do have
asymptotes.

NOTE. As you make your sketch you will find that the dashed lines at x = -90° and x = 90° are very helpful. The graph gets very close to these lines, without touching them. They are called <u>asymptotes</u> [from the Greek a(without) + syn(together) + piptein(to fall),].

(d) Plot all the points (x,tanx) from the table in (a) on the coordinate system you drew for (c).

(e) Draw a smooth curve through the points in (d), using the asymptotes in (c) as an aid when x is near 90° or -90°.

(f) Write the description of how tan x changes as x varies between -90° and 90°.

(g) Label the graph: y = tan x.

7. This exercise concerns the graph of y = tan x for 90° < x < 270°.

*The graph will
look something
like this.*

*Refer to your work
in Exercise 6 (a)
to save work.*

(a) Using the data in Exercise 6 and the unit circle definition of the tangent function, copy and fill in the following table:

x	90°	100°	110°	120°	135°	150°	180°	210°	225°	240°	250°	260°	270°
Tan x	Unde-fined												unde-fined

(b) Why is tan x undefined at x = 90° and x = 270°?

(c) Plot the points in (a) on the coordinate system you made for Exercise 6 (c).

(d) Draw a smooth curve through the points plotted in (c). Label the graph.

(e) Write a description of how tan x changes as x varies between 90° and 270°.

The next exercise concerns the graph of y = tan x for -270° < x < -90°.

8. (a) Use the unit circle together with the data from Exercise 6(a) to copy and fill in the table:

x	-270°	-260°	-250°	-240°	-225°	-210°	-180°	-150°	-135°	-120°	-110°	-100°	-90°
Tan x	unde-fined												unde-fined

(b) Why is tan x undefined when x = -270° and x = -90°?

(c) Plot the points in (a) on your coordinate system from Exercise 6(c).

(d) Draw a smooth curve through the points in part (c).

(e) Write a description of how tan x varies for x between -270° and -90°.

9. (a) At what intervals does the graph of the tangent function repeat itself?

    (b) At what values of x is the tangent function not defined?

    (c) What values of x correspond to the asymptotes?

The asymptotes are useful reference lines when graphing y = tan x. They are not a part of the graph. That is why they are indicated by dashed lines.

PERIOD OF A CIRCULAR FUNCTION

Observe that the graph of the tangent function repeats every 180°. We say, "the tangent function has a period of 180°." The graphs of the sine and cosine functions repeat every 360°. Thus, the sine and cosine functions have a period of 360°. The period of any trigonometric function is defined as the smallest positive number which when added to the inputs does not change the outputs.

The trigonometric functions are sometimes called the circular functions because their definition is related to the unit circle. The functions are periodic basically because the change of an angle by 360° leads to the same position of the second arm of the angle. While sine and cosine have a period of 360°, the tangent function has a smaller period, 180°.

10. What is the period of each of the following?

    (a) sec x   (b) csc x   (c) cot x

Hint: Begin by writing the definitions of these functions in terms of sine, cosine, or tangent.

MEASURING THE SIZE OF GEOMETRIC OBJECTS

How long is the line segment shown in this figure?

———————————————

A carpenter would say it is 2 and $\frac{1}{8}$ inches long. A physicist would say it is approximately 5.4 centimeters long. Both would be correct. Many people feel more comfortable with measurements in inches than with the centimeters of a metric system because of the units they used while growing up. The United States Congress has voted against the introduction of the metric system for over a century, the most recent vote being in 1974. For a long time, Congress wouldn't touch the metric system with a 12-foot pole, or should we say 3.66-meter pole?

There is just as much freedom in choosing how to measure angles as in measuring line segments. In daily life and in most scientific work the use of 360° to describe the complete circular angle is quite satisfactory. In calculus and physics, radian measure is much preferable and more convenient because of the close relation between radian measure and the real number system. At first glance you might find it awkward; however, this will change after you practice using the definition.

RADIAN MEASURE

Radian measure is defined as follows. Consider an angle, such as the one in this diagram:

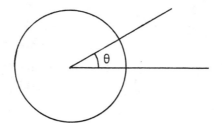

Draw a circle with the center at the vertex of the angle θ:

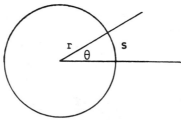

(circle with center
at vertex of angle)

Let the circle have radius r. Let s be the length of the arc cut off by the angle θ:

r = radius
s = arc length

The quotient

$$\frac{s}{r}$$

is the measure of the angle θ in <u>radians</u>. Briefly, this is the definition:

DEFINITION OF RADIAN MEASURE

$$\theta = \frac{s}{r}$$

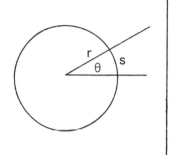

$\frac{s}{r}$ *is the measure*
of θ *in radians.*
Memorize this
definition.

EXAMPLE 1.   Find the radian measure of a right angle.

Solution.   The arc length (circumference) of a circle of radius r is 2πr. The arc cut off by a right angle is a quarter of the circumference,

$$s = \frac{2\pi r}{4}.$$

By the definition of radian measure, the right angle has measure

$$0 = \frac{s}{r}$$

$$= \frac{\frac{2\pi r}{4}}{r}$$

$$= \frac{2\pi r}{4} \cdot \frac{1}{r}$$

$$= \frac{\pi}{2} \qquad \text{(radians)}$$

The angle of 360° has radian measure

$$\frac{s}{r} = \frac{2\pi r}{r}$$

$$= 2\pi \text{ radians}$$

$\frac{\pi}{2}$ *is a perfectly*
fine number.
Since $\pi \doteq 3.14$,
then $\frac{\pi}{2} \doteq 1.57$.
Instead of "90
degrees," a right
angle now has the
measure
" $\frac{\pi}{2}$ *radians."*

This information is the basis for the following conversion formula that relates degree and radian measures. It is based on the proportion: "The measure of the angle in degrees is to the number of degrees in a full circle as the measure of the angle in radians is to the number of radians in a full circle."

DEGREE-RADIAN CONVERSION FORMULA

$$\frac{\text{angle in degrees}}{360°} = \frac{\text{angle in radians}}{2\pi}$$

Memorize this
formula.

Example 2.   (a)   Express an angle of $\frac{\pi}{6}$ radians in degree measure.

(b)   Express an angle of 60° in radian measure.

Solution.

(a) Let x = angle measure in degrees   (b) Let x = angle measure in radians

$$\frac{x}{360°} = \frac{\frac{\pi}{6}}{2\pi}$$

$$\frac{60°}{360°} = \frac{x}{2\pi}$$

$$x = \frac{\frac{\pi}{6}}{2\pi} \; (360°)$$

$$(2\pi) \frac{60°}{360°} = x$$

$$= \frac{\pi}{6} \cdot \frac{1}{2\pi} \; (360°)$$

$$\frac{\pi}{3} = x$$

$$= 30°$$

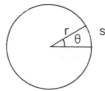

11. Referring to the figure at the right, compute each of the following:

(a) Find θ if r = 2 and s = 3   (b) Find r if s = 7 and θ = $\frac{\pi}{8}$

(a) $\frac{3}{2}$ (c) $\frac{3\pi}{2}$

(e) 15

(c) Find s if θ = $\frac{\pi}{2}$ and r = 3   (d) Find θ if s = 5π and r = 5

(e) Find r if s = 5π and θ = $\frac{\pi}{3}$ (f) Find s if θ = 2π and r = 3

12. Copy and fill in the table (use the degree-radian conversion formula):

π radians = 180°

45° = $\frac{\pi}{4}$ radians

23° = $\frac{23\pi}{180}$ radians

degrees	90°		30°		45°		300°		225°	315°		23°
radians	$\frac{\pi}{2}$	π		$\frac{\pi}{3}$		$\frac{3\pi}{2}$		$\frac{5\pi}{6}$			2π	

13. (a) Change 1 radian to degree measure.

(a) $\frac{180°}{\pi}$

(b) Use interpolation to check your answer with the table

(c) $\frac{\pi}{180}$

(b) Compute the answer in (a) to the nearest tenth of a degree and check it in the trigonometric table in the column labelled "radians."

(c) Change 1° to radian measure.

14. This figure shows an angle that cuts off an arc length exactly equal to the radius. Find:

(a) θ in radians

(b) θ in degrees

(c) Use (a) to write in your own words the definition of an angle of 1 radian.

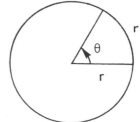

To give practice in working with radian measurement, let us graph the trigonometric functions when angles are measured in radians instead of degrees. Recall that when graphing earlier in this chapter, the scale (degrees) on the x-axis was compressed in order that the large numbers, such as 90, 360, and so on, would not require the use of long rolls of newsprint. If the scales on the x- and y-axes were the <u>same</u>, with angles measured in degrees, this diagram suggests how the graph of y = sin x would appear:

A magnifying glass would be needed to see that the graph is not simply a straight line.

The first gain from using radian measure is that the graph of y = sin x is easy to draw because the <u>same</u> <u>scale</u> is used on both axes. <u>Using</u> <u>radian</u> <u>measure</u>, <u>the</u> <u>graph</u> <u>is</u> <u>not</u> <u>distorted</u>. The next exercise illustrates this advantage.

RADIANS AND GRAPHING

15. (a) Copy and fill in the following table:

x (radians)	0	$\frac{\pi}{6}$	$\frac{\pi}{4}$	$\frac{\pi}{3}$	$\frac{\pi}{2}$	$\frac{2\pi}{3}$	$\frac{3\pi}{4}$	$\frac{5\pi}{6}$	π	2π
sin x										

(b) Draw a coordinate system similar to the following diagram. (Keep both scales as equal as possible; remember that $\pi \doteq 3.1$, which is a little larger than 3. Thus, $\frac{\pi}{3}$ is a little larger than 1.)

Note: For graphing purposes here, assume $\pi \simeq 3$ and use the scale: 1 cm = 1 unit.

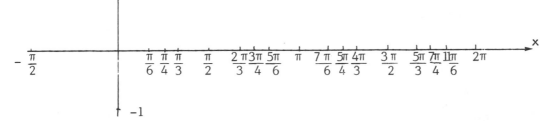

(c) Plot the points (x, sin x) from the table in (a) on your coordinate system in (b).

It is shown in calculus that the graph of y = sin x passes through the origin at an angle of 45°.

(d) Use the symmetry of the sine curve and the points in (a) to sketch other points where π < x < 2π.

(e) Draw a smooth curve through all points where 0 ≤ x ≤ 2π.

(f) Label the graph y = sin x for reference.

16. Do all parts of Exercise 15 for y = cos x.

NOTE. <u>For the remainder of this chapter and throughout the next chapter, all angles will be measured in radians, unless degree measurement is specified.</u>

17. (a) Copy and fill in the following table:

x	$\dfrac{-\pi}{2}$	$\dfrac{-\pi}{3}$	$\dfrac{-\pi}{4}$	$\dfrac{-\pi}{6}$	0	$\dfrac{\pi}{6}$	$\dfrac{\pi}{4}$	$\dfrac{\pi}{3}$	$\dfrac{\pi}{2}$
tan x									

(b) Draw a coordinate system similar to the following diagram. (Again, keep both scales equal and accurate; recall that π ≈ 3.)

Allow more space vertically on your paper than is shown here.

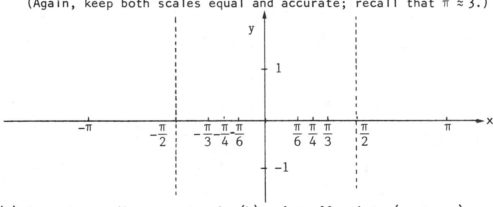

(c) On your coordinate system in (b), plot all points (x, tan x) from the table in (a).

(d) Using the asymptotes to help you draw the graph when x is near $\dfrac{\pi}{2}$ and $\dfrac{-\pi}{2}$, draw a smooth curve through the points in (c).

Draw the asymptotes for x = $\dfrac{\pi}{2}$ and x = - $\dfrac{\pi}{2}$.

(e) Label the graph y = tan x for reference.

(f) Sketch the graph of y = tan x for $\dfrac{-\pi}{2}$ < x < $\dfrac{3\pi}{2}$.

THE GRAPH OF y = COT x

Now, consider the definition

$$\cot x = \frac{1}{\tan x}$$

The definition shows that tan x and cot x are <u>both positive</u> or <u>both negative</u>, and that <u>as one increases the other decreases.</u> The next exercise concerns the graph of y = cot x.

18. (a) Copy and fill in the following table (Let $\sqrt{3} \doteq 1.7$):

x	0	$\frac{\pi}{6}$	$\frac{\pi}{4}$	$\frac{\pi}{3}$	$\frac{\pi}{2}$	$\frac{2\pi}{3}$	$\frac{3\pi}{4}$	$\frac{5\pi}{6}$	π
cot x									

Note that the angles are "nice." Use $\cot x = \frac{1}{\tan x}$ and a calculator.

(b) At which values for x are there asymptotes? Explain.

(c) Draw an (x,y)-coordinate system with equal scales for the axes (with π approximately 3, as in the graphs for Exercises 15 - 17).

(d) On the coordinate system in (c), plot the points (x,cot x) from your table in (a).

(e) Using asymptotes at x = 0 and π to help you graph, for x near 0 and π, draw a smooth curve through the points in (d).

(f) Label the graph y = cot x for reference.

What happens to cot x as tan x gets larger (smaller) and vice versa?

(g) The period of the tangent is π. Thus, the period of the cotangent is also π because of the definition $\cot x = \frac{1}{\tan x}$.

In view of this, where would you expect asymptotes in the graph of y = cot x? Explain.

(h) Use the periodic aspect of y = cot x to make a freehand sketch of the graph for -π < x < 2π.

19. (a) On another coordinate system make a freehand sketch of y = tan x for $\frac{-\pi}{2} < x < \frac{3\pi}{2}$.

(b) On the same coordinates as (a), use a pencil or pen of a different color to sketch y = cot x for 0 < x < 2π.

(c) Where do the two graphs cross?

(d) Near those values of x where tan x = 0, what happens to cot x?

(e) What happens to the graph of y = cot x when tan x is a very large positive number? A very large negative number?

(f) Label both graphs.

20. (a) Since $\csc x = \frac{1}{\sin x}$, the graph of y = csc x is the same as that for $y = \frac{1}{\sin x}$. Answer the following questions concerning sin x and csc x:

(i)   If sin x is positive (or negative) what is csc x?

(ii)   As csc x gets larger, what happens to sin x?

(iii) If sin x is positive and near 0, what happens to csc x?

Hint: If
$-1 < y < 1$, what
happens to
$\frac{1}{y}$? Experiment
with some numbers.

(iv)   If sin x is negative and near 0, what happens to csc x?

(v)   Can csc x have values such that -1 < csc x < 1? Explain.

(vi)   When sin x = 1, find csc x.

(vii)   When sin x = -1, find csc x.

(b)   Copy and fill in the table:

x	-2π	$\frac{-3\pi}{2}$	$-\pi$	$\frac{-\pi}{2}$	0	$\frac{\pi}{2}$	π	$\frac{3\pi}{2}$	2π
sin x									
csc x									

(c)   Draw a freehand sketch of the graph of y = sin x.

Hint: (d) The
graph of $y = csc\ x$
consists of "U's,"
alternately right-
side-up and up-
side-down.

(d)   On the same coordinate system as (c), use the table in (b) and asymptotes for suitable values of x to draw a freehand sketch of y = csc x.

(e)   What is the period of y = csc x?

(f)   Label each graph.

21.   Do all steps of Exercise 20 for the trigonometric functions

sec x = $\frac{1}{\cos x}$. That is, graph y = cos x and y = sec x relative to the same axes. Label both graphs.

MUSICAL TONES AND SINE CURVES

When a tuning fork is struck, a pure tone is heard. This sensation is caused by the air pressure rapidly rising and falling many times per second. The following graph is a record of the typical way that the pressure at a given point (say the ear) changes with time.

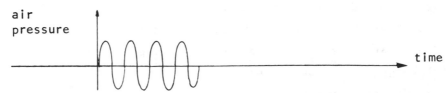

The pitch of a musical note is determined by the number of cycles per second. For instance, the pitch of A above middle C corresponds to 440 cycles per second. One cycle consists of the pressure rising, returning to "normal," then dipping below, and finally returning to "normal," as shown in this graph:

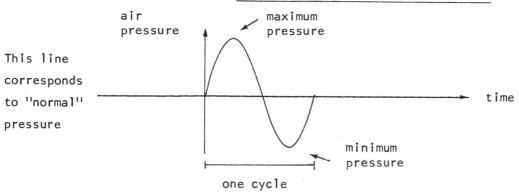

This line corresponds to "normal" pressure

one cycle

The graph corresponding to the musical note A above middle C would have 440 cycles between time t = 0 and time t = 1 second.

This is "concert A," the tuning note for a symphony orchestra.

A above middle C: 440 cycles in one second

Whether a note is played loudly or softly, the number of cycles per second (the pitch) remains the same. In a loud note the pressure changes more. Thus, the "pressure versus time" graph of an A above middle C that is louder than the one described by the last graph may look like this:

Note. In practice, the quality of a musical group often may be determined by how they control the pitch (tone) as the volume changes.

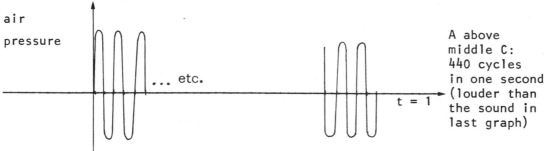

A above middle C: 440 cycles in one second (louder than the sound in last graph)

The sine curve or some magnification or compression of it, appears in the study of sound, light, a cork bobbing on the water, or a bridge swaying in the wind.

The graphs are different for each pitch. (The lowest pitch on any musical instrument is about 20 cycles per second. The highest sound that the human ear can hear has about 20 thousand cycles per second.) In all cases the corresponding graphs are related to the sine curve by a stretching or a squeezing along the x-axis and/or the y-axis. When the pitch changes, the cycle of the curve is stretched (lower pitch) or squeezed (higher pitch) along the x-axis. When the volume changes, the curve is stretched (higher volume) or squeezed (lower volume) along the y-axis.

When a pure tone is sounded by a tuning fork or an electronic instrument, the smooth, or pure sine curve is formed. Our ears can distinguish different instruments playing the same note at the same volume because the sine curve in each case will have different "blips" on it. For example a trumpet and a violin playing a 440 cycle note ("concert A") are distinguishable to the ear because of the different blips on the sine curve produced by the sound of each instrument.

Examples of sine curves are to be found everywhere in the natural world, from the path of a satellite with reference to the equator of the earth to the study of biorhythms in human beings.

The following exercises examine the equations for such graphs.

STRETCHING OR SQUEEZING THE SINE CURVE "VERTICALLY"

The next exercise concerns the graph of

$$y = A \sin x$$

Where A is some constant. When A = 1 the graph is simply the sine curve y = sin x. When working the exercise, note the effect of choosing various values for A.

22. Make a rough sketch of a graph for each of the following functions. Since there are six graphs, distinguish them in some way, perhaps by using different colors. Sketch them all on the same coordinate system and label each graph.

As you do this exercise, note the effect of the constant A on y = A sin x. In particular, how high does the graph go?

(a) $y = \sin x$ (b) $y = -\sin x$ (c) $y = 3 \sin x$

(d) $y = -3 \sin x$ (e) $y = \frac{1}{2} \sin x$ (f) $y = \frac{-1}{2} \sin x$

(a) It is stretched vertically; goes higher.
(c) It may be stretched or squeezed vertically, it is not turned upside down.

23. How does the graph of $y = A \sin x$ compare with the graph of $y = \sin x$? Consider the cases:

(a) $|A| > 1$ (b) $|A| < 1$ (c) $A > 0$ (d) $A < 0$

* * *

Study carefully the following definition of amplitude. Refer back to Exercises 22 and 23 to make sure of your understanding of the stretching (squeezing) and/or flipping upside down of the sine curve.

AMPLITUDE

DEFINITION. The <u>amplitude</u> of the function

$$y = A \sin x$$

is defined to be $|A|$.

Geometrically, the amplitude describes how high the graph of y = A sin x goes. For instance, the amplitude of y = -3 sin x is $|-3| = 3$. The largest y-coordinate of any point on the graph of y = -3 sin x is 3. The lowest is -3.

The number A may be positive or negative. Note that the <u>amplitude</u>, $|A|$, is always positive, for it is the absolute value of A.

(1) If A is negative, the graph of y = A sin x is obtained from the graph of y = sin x by a reflection on the x-axis (it is "upside-down"), followed by a stretching or squeezing in the vertical direction.

(2) If A is positive, the graph of y = A sin x is obtained from the graph of y = sin x by either a stretching or squeezing in the vertical direction.

24. Describe in words how the graphs of each of the following functions can be obtained from the graph of y = sin x. Do not draw them.

(a) y = 4 sin x

(b) y = $\frac{3}{2}$ sin x

(c) y = -4 sin x

(d) y = -$\frac{3}{2}$ sin x

25. Without sketching any graphs, describe how the graph of each of the following functions compares with the graph of y = cos x.

(a) y = 5 cos x

(b) y = -4 cos x

(c) y = $\frac{1}{10}$ cos x

(d) y = -$\frac{2}{3}$ sin x

The amplitude of the function y = A cos x is defined to be $|A|$, as in the case of the sine function.

We consider next the stretching or shrinking of the sine curve horizontally.

STRETCHING OR SHRINKING THE SINE CURVE HORIZONTALLY

The length of the period of the sine curve y = sin x is 2π

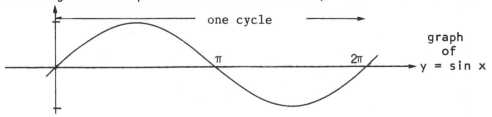

The function y = Bx, where B is some positive constant, may not have the same period as y = sin x. The next example illustrates this.

The amplitude in the case of a sound wave corresponds to loudness. In the case of a swaying bridge it describes how far the bridge sways from its normal position. AM radio refers to "amplitude modulation."

You should at this point be able to make quickly a reasonably accurate graph of y = sin x(the sine curve) freehand.

24. (a) right-side-up amplitude 4 (graph is 4 times as high as graph of sin x. (c) up-side-down amplitude 4

25. (a) right-side-up amplitude 5 (c) right-side-up amplitude $\frac{1}{10}$

Nothing is lost by insisting that B is positive. (In all applications it is positive.) If B is negative, say -3, note that sin Bx = sin (-3x) = -sin 3x. So the case of negative B can be obtained from the case of positive B. The length of the cycle is not changed.

Example 3.   Graph y = sin 2x and discuss its period.

Solution.   Note that as x goes from 0 to π, then 2x goes from 0 to 2π.

Thus, graph y = sin 2x as x goes from 0 to π.   To do so, make a

table of a few selected values.

x	0	$\frac{\pi}{4}$	$\frac{\pi}{2}$	$\frac{3\pi}{4}$	π
2x	0	$\frac{\pi}{2}$	π	$\frac{3\pi}{2}$	2π
sin 2x	0	1	0	-1	0

With the aid of the five points (x,sin 2x) just computed, a portion

of the graph of y = sin 2x may be sketched:

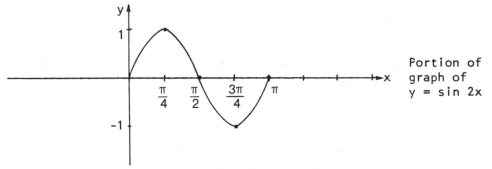

Portion of
graph of
y = sin 2x

Already one complete period has been graphed.   This cycle is com-

pleted when x changes by π.   In other words, for the graph of

y = sin 2x the period is π:

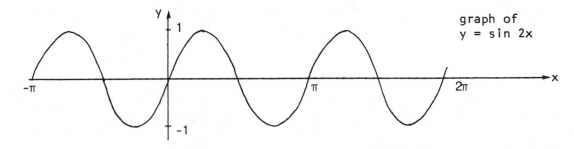

graph of
y = sin 2x

The graph of y = sin 2x has the same amplitude as the graph of

y = sin x.   However, the sine curve has been compressed (or squeezed)

horizontally to obtain the graph of y = sin 2x.

26. This exercise concerns the graph of y = sin 4x.

(a) Copy and fill in the table:

x				
4x	0	$\frac{\pi}{2}$	$\frac{3\pi}{2}$	2π
sin 4x				

(b) Plot the five points (x, sin 4x) in (a) on a copy of this coordinate system:

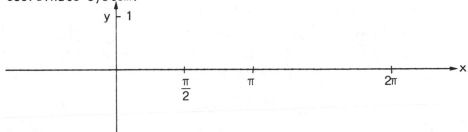

(c) What is the period of y = sin 4x? The amplitude?

(d) Graph y = sin 4x for $0 \le x \le \frac{\pi}{2}$.

(e) Graph y = sin 4x and label the graph.

(f) How many cycles does the graph of y = sin 4x have for $0 \le x \le 2\pi$?

27. This exercise concerns the graph of $y = \sin(\frac{1}{2}x)$.

(a) Copy and fill in the table:

x					
$\frac{1}{2}x$	0	$\frac{\pi}{2}$	π	$\frac{3\pi}{2}$	2π
$\sin (\frac{1}{2}x)$					

(b) Plot the five points $(x, \sin \frac{1}{2}x)$ in (a) on a copy of this co-ordinate system:

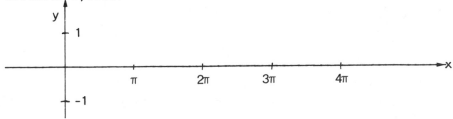

(c)   What is the period of $y = \sin \frac{1}{2} x$? The amplitude?

(d)   Graph $y = \sin(\frac{1}{2}x)$ for $0 \leq x \leq 4\pi$.

(e)   Graph $y = \sin(\frac{1}{2}x)$ and label the graph.

(f)   How many cycles does the graph of $y = \sin \frac{1}{2} x$ have for $0 \leq x \leq 2\pi$?

28.   Graph:

(a) $y = \sin 3x$ (b) $y = \sin(\frac{1}{3}x)$

*   *   *

As Example 3 and Exercises 26, 27, and 28 suggest, the number $|B|$ in $y = \sin Bx$ determines the <u>period</u> of $y = \sin Bx$. The <u>larger</u> $|B|$ is, the <u>smaller</u> is the period of $y = \sin Bx$. To find the period, solve the equation:

$$|B|x = 2\pi.$$

The solution is:

$$x = \frac{2\pi}{|B|}$$

y = sin 4x goes through the cycle <u>four</u> times as fast as y = sin x. Hence, the period is $\frac{2\pi}{4} = \frac{\pi}{2}$

In short,

the <u>period</u> of $y = \sin Bx$ is $\dfrac{2\pi}{|B|}$

(a) 12 and $\frac{\pi}{6}$

(c) 2π and 1

29.   Find B and the period in each case:

(a) $y = \sin 12x$ (b) $y = \sin \frac{1}{12} x$

(c) $y = \sin 2\pi x$ (d) $y = \sin 4\pi x$

*   *   *

NOTE.   Since cos x has the same period as sin x, namely 2π, then cos Bx has period

$$\frac{2\pi}{|B|}$$

(a) 100 and $\frac{\pi}{50}$

(c) 4π and $\frac{1}{2}$

30.   Find B and the period in each case:

(a) $y = \cos 100x$ (b) $y = \cos\frac{1}{10} x$

(c) $y = \cos 4\pi x$ (d) $y = \cos 10\pi x$

(a) $\frac{1}{440}$
This graph corresponds to the note A above middle C, 440 cycles per second.

31.   Consider the graph of $y = \sin 880\pi x$.

(a)   What is the period?

(b)   How many cycles does the graph complete for $0 \leq x \leq 1$?

THE GRAPH OF y = A sin Bx

The graph of y = A sin Bx is easily obtained by first graphing
y = sin Bx, then stretching (or squeezing) vertically by the factor |A|.

32. (a) Graph y = sin 4x (b) Graph y = 3 sin 4x $\quad$ *(c)* $\frac{\pi}{2}$

(c) What is the period of y = 3 sin 4x?

(d) What is the maximum y-coordinate on the graph of y = 3 sin 4x?

33. Graph:

(a) y = 3 cos 2x (b) y = -2 cos 3x

Summarizing:

In the functions y = A sin Bx or y = A cos Bx, where B is positive,

|A| is called the amplitude.

The period is

$$\frac{2\pi}{|B|}$$

34. What are the period and amplitude of:

(a) y = 5 sin 100x (b) y = -5 cos 100x $\quad$ *(a)* $\frac{\pi}{50}$ *and* 5

(c) y = 5 sin 100πx (d) y = 5 cos 40πx $\quad$ *(c)* $\frac{1}{50}$ *and* 5

THE GRAPH OF y = sin(x - C)

Let C be a constant, either positive or negative. How does the graph
of y = sin(x - C) compare with that of the sine curve? The next example
illustrates the comparison in the specific case C = $\frac{\pi}{3}$.

Example 4. Graph y = sin(x - $\frac{\pi}{3}$).

Solution. When x = $\frac{\pi}{3}$,

then x - $\frac{\pi}{3}$ = 0

and

sin(x - $\frac{\pi}{3}$) = 0

Thus the point ($\frac{\pi}{3}$, 0) lies on the graph.

Moreover, as x changes by 2π, so does x - $\frac{\pi}{3}$. Thus,

$$y = \sin(x - \frac{\pi}{3})$$

has period 2π, just like the sine curve y = sin x.

The graph of y = sin(x - $\frac{\pi}{3}$) is obtained by <u>shifting</u> the sine curve

<u>to the right</u> a distance of $\frac{\pi}{3}$.

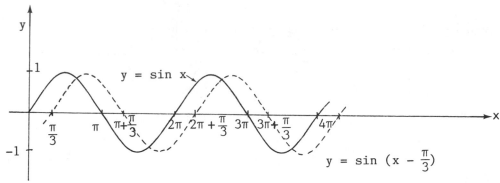

In the expression

$$y = \sin(x - C)$$

Phase shift is the basis of radar.

C is called the <u>phase</u> shift.   The graph of $y = \sin(x - C)$ is obtained by moving the graph of $y = \sin x$ horizontally right or left.   If C is <u>positive</u>, the sine curve is moved to the <u>right</u> a distance C.   If C is <u>negative</u>, the sine curve is moved to the <u>left</u> a distance $|C|$.   Similar remarks hold for $y = \cos(x - C)$.

35.   Graph:

(a) $y = \sin(x - \frac{\pi}{4})$ (b) $y = \sin(x + \frac{\pi}{4})$

(c) $y = \sin(x + \frac{\pi}{2})$ (d) $y = \cos(x + \pi)$

(e) $y = \cos(x - \frac{\pi}{3})$ (f) $y = \cos(x - \frac{\pi}{2})$,

THE GRAPH OF $y = \sin B(x - C)$

The graph of $y = \sin B(x - C)$, where B is positive, can be obtained from that of $y = \sin x$ in two steps.   The next example illustrates the process.

Example 5.   Graph $y = \sin 2(x - \frac{\pi}{6})$.

Solution.   The coefficient "2" will affect the period.   So, first graph

$$y = \sin(x - \frac{\pi}{6}).$$

The phase shift is $\frac{\pi}{6}$; hence the graph is obtained by shifting the

sine curve $\frac{\pi}{6}$ units to the <u>right</u>.

$y = \sin\left(x - \dfrac{\pi}{6}\right)$

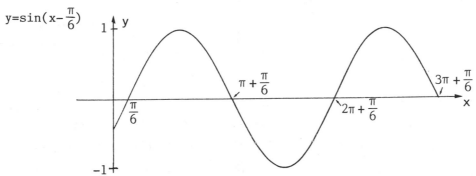

The factor "2" compresses the cycle by a factor of 2. (The period is

$\dfrac{2\pi}{2} = \pi$.) Thus the graph of $y = \sin 2\left(x - \dfrac{\pi}{6}\right)$ appears in this diagram:

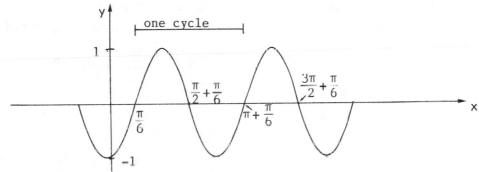

Observe that the function graphed in Example 5 can also be written

as

$$y = \sin\left(2x - \dfrac{\pi}{3}\right)$$

If it is necessary to graph $y = \sin\left(2x - \dfrac{\pi}{3}\right)$, first write it in the form

$$y = \sin 2\left(x - \dfrac{\pi}{6}\right).$$

As Example 5 illustrates, the graph of

$$y = \sin B(x - C)$$

can be obtained as follows:

    (a)   graph $y = \sin(x - C)$ and label it;

    (b)   modify the graph in (a) so that the period is $\dfrac{2\pi}{|B|}$. The result-

           ing graph passes through the point $(C,0)$. Sketch and label it.

36. Make a rough sketch of the graph of the following functions; label

    the graph in each case.

    (a) $y = \sin 2\left(x - \dfrac{\pi}{4}\right)$ (b) $y = \sin 2\left(x + \dfrac{\pi}{4}\right)$

    (c) $y = \sin\left(2x - \dfrac{\pi}{3}\right)$ (d) $y = \sin\left(2x + \dfrac{\pi}{3}\right)$

Study this section carefully. The equation must be in the form $y = \sin B(x - c)$ before you can graph it.

36. (a) period π, phase shift to the right $\dfrac{\pi}{4}$

(c) period π phase shift to the right $\dfrac{\pi}{6}$

37. Make a rough sketch of the graph of

    (a) $y = 3 \sin 2(x - \frac{\pi}{6})$ (b) $y = 3 \sin(\frac{1}{2}x + \pi)$

38. In the graph of

$$y = A \sin B(x - C)$$

    what is the

    (a) amplitude?   (b) period?   (c) phase shift?

39. With reference to the graph of $y = \sin x$, describe the graph of

    (a) $y = 5 + \sin x$ (b) $y = -2 + \sin x$

40. Describe the effect of D on the graph of $y = \sin x$ if $y = D + \sin x$

    and

    (a) D is positive,   (b) D is negative.

41. Describe the following graph in terms of the graph of $y = \sin x$:

$$y = 4 - 2 \sin(3x - \frac{\pi}{2})$$

* * *   SUMMARY   * * *

The chapter opened with a discussion of the graphs of the six
trigonometric ("circular") functions, where the angle is measured in
degrees.

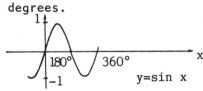

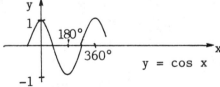

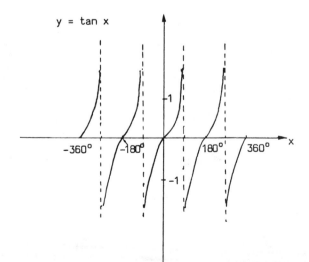

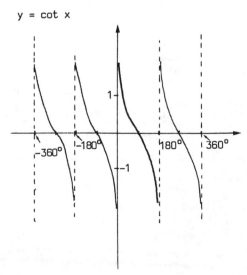

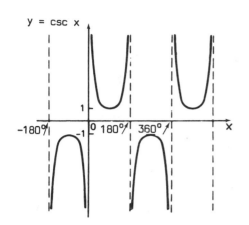

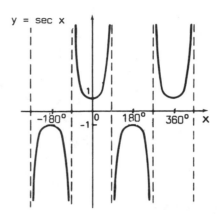

The graphs of four of the functions have asymptotes.

Radian measure was introduced in terms of a circle:

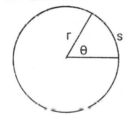 $\theta = \dfrac{s}{r}$

Translation between degrees and radians may be accomplished by using the equation

$$\frac{(\text{degrees})}{360°} = \frac{(\text{radians})}{2\pi}$$

In particular, an angle of one radian is about 57.3°.

Each trigonometric function is "periodic," repeating itself at a fixed interval. This table lists the periods of six functions.

function	sin x	cos x	tan x	cot x	sec x	csc x
period (in degrees)	360°	360°	180°	180°	360°	360°
period (in radians)	2π	2π	π	π	2π	2π

Modifications of the sine curve (or y = cos x) in the form

$$y = A \sin B(x - C)$$

were studied. The factor B determines the period, which is

$$\frac{2\pi}{|B|}.$$

The phase shift is indicated by C.

The phase is to the <u>right</u> C units if C is <u>positive</u>, and to the <u>left</u> |C| units, if C is <u>negative</u>.

The number D in

$$y = D + \sin x$$

indicates a raising or lowering (no distortion) of the graph of $y = \sin x$, depending whether D is positive or negative.

* * * WORDS AND SYMBOLS * * *

circular function   amplitude

sine curve   phase (shift)

period   asymptote

radian

* * * CONCLUDING REMARKS * * *

A modern computer calculates sin x (angle in radians) with the aid of a remarkable formula developed in the 17th century:

$$\sin x = x - \frac{x^3}{3!} + \frac{x^5}{5!} - \frac{x^7}{7!} + \ldots \qquad \text{(angle in radians)}$$

(n! is the product of all integers from 1 through n:

$n! = 1 \cdot 2 \cdot 3 \cdot \ldots \cdot (n - 1)(n)$.) For instance, the formula for sin 1 (the sine of an angle of one radian, about 57.3°.) is

$$\sin 1 = 1 - \frac{1}{3!} + \frac{1}{5!} - \frac{1}{7!} + \ldots$$

The first three summands give a very good estimate:

$$\sin 1 \doteq 1 - \frac{1}{3!} + \frac{1}{5!}$$

$$= 1 - \frac{1}{6} + \frac{1}{120}$$

$$\doteq -0.167 + 0.008$$

$$= 0.841$$

(Compare this with the value given by a calculator or in the table at the back of the book.)

If x is between 0 and 1 then

$$\sin x \doteq x - \frac{x^3}{3!} \qquad \text{(angle in radians)}$$

This formula provides an estimate also for angle in degrees. An angle of x degrees is the same as an angle of

$$\frac{\pi}{180} \cdot x \text{ radians.}$$

Thus,

$$\sin x \doteq \frac{\pi}{180} x - \frac{(\frac{\pi}{180} x)^3}{3!} \qquad \text{(angle in degrees)}$$

Taking

$$\frac{\pi}{180} \doteq 0.017$$

gives the estimate

(1) $$\qquad \sin x \doteq 0.017x - \frac{(0.017x)^3}{6} \qquad \text{(angle in degrees)}$$

For instance, this formula gives the following estimate of sin 20°:

$$\sin 20° \doteq (0.017)20 - \frac{[(0.017)(20)]^3}{6}$$

$$\doteq 0.34 - \frac{(0.34)^3}{6}$$

$$\doteq 0.34 - 0.007$$

$$\doteq 0.34$$

Hence

$$\sin 20° \doteq 0.34$$

The table lists

$$\sin 20° \doteq 0.342$$

Formula (1) is accurate to two decimal places for $0 \le x \le 30°$.

Such formulas are built into hand calculators as well. Little did the mathematicians of the 17th and 18th centuries who discovered such "polynomials of infinite degree" for a variety of functions suspect that in less than ten generations future scientists would find them so useful.

One further illustration: let e = 2.71828 . . . . , a number defined in calculus. Then for all x,

$$e^x = 1 + x + \frac{x^2}{2!} + \frac{x^3}{3!} + \frac{x^4}{4!} + \dots$$

* * * GRAPHING AND RADIANS SELF QUIZ * * *

1. On the same coordinate system make a <u>rough sketch</u> and label the graphs of y = sin x and y = cos x for $-2\pi \le x \le 2\pi$.

2. On the same coordinate system make a <u>rough sketch</u> and label the graphs of y = tan x and y = cot x for -π < x < π.

3. On the same coordinate system make a <u>rough sketch</u> and label the graphs of y = sin x and y = csc x for 0 ≤ x ≤ 2π.

4. Do Exercise 3 for y = cos x and y = sec x (instead of y = sin x and y = csc x).

5. Fill in the table:

degrees	23°		130°		240°		1°	
radians		$\frac{\pi}{6}$		2		$\frac{\pi}{12}$		1

6. (a) The graph of y = sin (x + $\frac{\pi}{2}$) is the same as the graph of y = cos___.

    (b) Give an argument for your answer in (a) using
         (i) an identity   (ii) phase

7. Describe the following graph in terms of y = cos x:

$$y = 3 - 4 \cos\left(2x + \frac{\pi}{3}\right)$$

8. Compute θ in radians

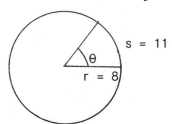

ANSWERS TO SELF QUIZ

1-4. graphs--see text

5.

degrees	23°	30°	130°	$\frac{360°}{\pi}$	240°	15°	1°	$\frac{180°}{\pi}$
radians	$\frac{23}{180}\pi$	$\frac{\pi}{6}$	$\frac{13}{18}\pi$	2	$\frac{4}{3}\pi$	$\frac{\pi}{12}$	$\frac{\pi}{180}$	1

6. (a) y = cos x
    (b)(i) cos θ = sin(θ + 90°)   (ii) phase shift $\frac{\pi}{2}$ to the left

7. (a) Amplitude = 4 (upside-down); phase shift = $\frac{\pi}{6}$ to the left; period = π; raised 3 units.

8. θ = $\frac{11}{8}$ radians

* * *   GRAPHING AND RADIANS REVIEW * * *

1. (a) Fill in the table:

degrees	90°	60°	180°	1°	27°	135°
radians						

(b) Fill in the table:

degrees						
radians	$\frac{\pi}{4}$	$\frac{\pi}{6}$	$\frac{7\pi}{6}$	4	1	$\frac{5\pi}{3}$

2. (a) What is the value of r if θ = 60° and the arc s has length 7?

(b) What is the length of the arc in a circle of radius r if the angle has x radians?

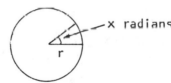

3. Describe each of the following graphs with reference to the graph of y = sin x:

(a) y = - sin x   (b) y = 3 sin x

(c) y = -3 sin x   (d) y = $\frac{1}{2}$ sin x

4. Describe the graph of y = -5 cos(3x - π) with reference to the graph of y = cos x.

5. Graph on the same coordinate system (label each graph):

(a) y = cos x   (b) y = cos(x - $\frac{\pi}{4}$)   (c) y = cos(x + $\frac{\pi}{4}$)

6. On the same coordinate system, graph and label:

(a) y = cos x   (b) y = 2 - cos x

7. On <u>separate</u> coordinate systems, graph and label each of the following:

(a) y = cos(2x)   (b) y = sin(2x)   (c) y = tan(2x)

Hint: Remember some identities.

8. Would you find it easy or difficult to graph each of the following?

(a) $y = \cos^2 x - \sin^2 x$

(b) $y = 2 \sin x \cos x$

(c) $y = \dfrac{2 \tan x}{1 - \tan^2 x}$

(d) $y = \cos^2 x + \sin^2 x$

9. Describe the graphs of each of the following with respect to the graph of $y = \cos x$:

(a) $y = -2 \cos x$

(b) $y = 3 - 2 \cos x$

(c) $y = 3 - 2\cos(4x)$

(d) $y = 3 - 2 \cos(4x + \pi)$

10. OPTIONAL. Sketch the graph of Exercise 9(d)

* * * ANSWERS TO GRAPHING AND RADIANS REVIEW * * *

1. (a)

degrees	90°	60°	180°	1°	27°	135°
radians	$\frac{\pi}{2}$	$\frac{\pi}{3}$	π	$\frac{\pi}{180}$	$\frac{3\pi}{20}$	$\frac{3\pi}{4}$

(b)

degrees	45°	30°	210°	$\frac{720°}{\pi}$	$\frac{180°}{\pi}$	300°
radians	$\frac{\pi}{4}$	$\frac{\pi}{6}$	$\frac{7\pi}{6}$	4	1	$\frac{5\pi}{3}$

2. (a) $r = \dfrac{21}{\pi}$

(b) $s = xr$

3. (a) inverted reflection on the x axis

(b) Amplitude = 3 times that of $y = \sin x$

(c) Amplitude = 3 times that of $y = -\sin x$, inverted and reflection with respect to the x-axis and $y = \sin x$

(d) Amplitude = $\frac{1}{2}$ that of $y = \sin x$

4. Amplitude = 5, but upside-down; Period = $\frac{2}{3}\pi$; Phase shift $\frac{\pi}{3}$ to right.

5. graphs (phase shift)   6. graphs (axis shift)

7. graphs (period change)

8. (a) $y = \cos 2x$ (b) $y = \sin 2x$ (c) $y = \tan 2x$

(d) $y = 1$

9. (a) Amplitude = 2; inverted with respect to the x-axis and $y = \cos x$

(b) Axis shift to $y = 3$; Amplitude = 2, inverted with respect to the x-axis and $y = \cos x$

(c) same as above; Period = $\frac{\pi}{2}$

(d)   same as above;   Phase shift = $-\frac{\pi}{4}$ (to the left)

10.   Optional graph

* * * GRAPHING AND RADIANS REVIEW SELF TEST * * *

1.   Find the radius of the circle if s = 15 centimeters and the measure

of θ is

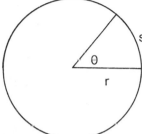

(a) $\frac{3}{4}$ radian   (b) π radians   (c) 120°   (d) 30°

2.   Sketch each of the following on the same coordinate axes and label

each graph:

(a)   y = sin x   (b) $y = \frac{1}{3} \sin 3x$ (c) $y = 3 \sin \frac{1}{3}x$

3.   Consider the graphs of each of the following:

$$y = \tan x \text{ and } y = \frac{1}{4} + \tan x$$

Do the two curves ever intersect?

4.   (a)   Sketch the curve y = cos x.

(b)   On the above sketch draw the line $y = \frac{1}{2}$. In how many points

does $y = \frac{1}{2}$ intersect y = cos x? Find the x-coordinate of six

of the points.

Hint: find x if $\cos x = \frac{1}{2}$

5.   Sketch y = -cos 2x.

6.   Graph $y = 2 \sin(2x - \frac{\pi}{3})$.

Careful! What is the phase?

7.   Describe the relation between the graph of y = cos x and the graph

of $y = 7 - 4 \cos(4x - \frac{\pi}{3})$.

ANSWERS TO SELF TEST

1.   (a)   20   (b) $\frac{15}{\pi}$ (c) $\frac{45}{2\pi}$ (d) $\frac{90}{\pi}$

2.   (b)   Amplitude = $\frac{1}{3}$; Period = $\frac{2\pi}{3}$ (c)   Amplitude = 3;   Period = 6π

3.   No; for any x, $\frac{1}{4} + \tan x > \tan x$

4. (b)   an endless number

$\frac{\pi}{3} \pm 2n\pi$ and $\frac{5\pi}{3} \pm 2n\pi$, where n is an integer

5. graph   6.   graph

7. Period = $\frac{\pi}{2}$; Phase shift = $\frac{\pi}{12}$ to the right; it is upside down with

amplitude 4, raised 7 units.

* * * GRAPHING AND RADIANS PROJECTS * * *

1. (a)   On the same coordinate system graph

y = sin x and y = 5 + sin x.

(b)   Describe how the graph of y = 5 + sin x can be obtained from
the graph of y = sin x.

2. (See Project 1)   How is the graph of

y = D + A sin B(x - C)

related to the graph of

y = A sin B(x - C)?

3. (See Projects 1 and 2)   In graphing y = D + A sin B(x - C), which
constant determines

(a)   the horizontal shift?

(b)   the vertical shift?

(c)   the horizontal squeeze or stretch?

(d)   the vertical squeeze or stretch?

*Use a trigono-
metric identity
to relate (b) to
(a)*

4. (a)   Graph y = $\frac{1}{2} - \frac{1}{2}$ cos 2x.   (b)   Graph y = $\sin^2 x$

5. (a)   Show that

sin x cos x = $\frac{1}{2}$ sin 2x.

(b)   Use (a) to graph

y = sin x cos x

6   (a)   Graph y = sin x + cos x.

(b)   Graph y = $\sqrt{2}$ sin$(x + \frac{\pi}{4})$

(c)   Use a trigonometric identity to relate (a) and (b).

7. (a)   What is the period of y = tan x?

(b)   Let B be a positive number. What is the period of y = tan Bx?

(c)   Furthermore, if A and C are fixed numbers, describe the graph of

y = A tan B(x - C).

8. Let a and b be numbers such that
$$a^2 + b^2 = 1.$$

*Projects 8-10
constitute a unit.*

Show that there is an angle θ such that
$$\cos \theta = a \text{ and } \sin \theta = b$$

9. (a) Sketch the graph of y = 3 sin x + 4 cos x.

(b) Estimate the largest value of 3 sin x + 4 cos x for all x.

10. (a) Rewrite 3 sin x + 4 cos x as
$$5(\frac{3}{5} \sin x + \frac{4}{5} \cos x)$$

Show that there is a __fixed__ angle θ such that
$$3 \sin x + 4 \cos x = 5 \sin(x + \theta)$$

for all x.

(b) From (a) deduce the maximum possible value of 3 sin x + 4 cos x.

11. In the military service a third way of measuring angles is used. The unit is a "mil," and 1600 mils describes a right angle. Find out why this method is used.

12. Radio signals are transmitted AM (amplitude modulation) and FM (frequency modulation). Frequency is "number of cycles per second," hence is inversely proportional to the period. FM tends to be relatively static-free as most electrical interference affects amplitude, not frequency, and the FM receiver rejects all signals not of a certain amplitude. For more information on FM see the McGraw Hill Encyclopedia of Science, vol. 5, pp. 511-514.

13. The period of y = sin 2πnx is $\frac{2\pi}{2\pi n} = \frac{1}{n}$. Thus, when x changes by $\frac{1}{n}$, one complete cycle is swept out. In other words, there are n cycles in "unit time." Thus,
$$y = \sin 880\pi x$$
corresponds to a pitch of 440 cycles per second. Similarly,
$$y = \sin 886\pi x$$
represents a pitch of 443 cycles per second. When two such nearby tones are sounded at one time, "beats" are heard. The explanation for this is outlined as follows:

(a) The function that records air pressure as a function of time when both notes are sounded is
$$y = \sin 880\pi x + \sin 886\pi x.$$

Using an identity from the Projects of the preceding chapter, show that the above equation is equivalent to

$$y = 2(\sin 883\pi x)(\cos 3\pi x).$$

(b)   Graph $y = 2 \cos 3\pi x$ and $y = -2 \cos 3\pi x$ on the same coordinate system.

(c)   Show that the graph of the function in (a) looks like the shaded blur in this diagram

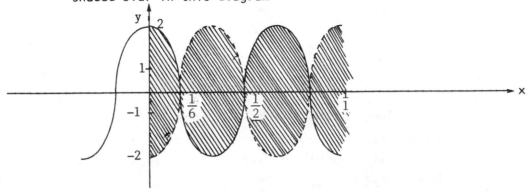

The resulting graph, which corresponds to three "pressure pulses" per second is heard as 3 "beats" per second.   (Beats are used in tuning a piano.)   See the <u>McGraw Hill</u> <u>Encyclopedia</u> <u>of</u> <u>Science</u> <u>and</u> <u>Technology</u>, vol. 2, pp. 11-15, 81, 126, 734-736, or the <u>Harvard</u> <u>Dictionary</u> <u>of</u> <u>Music</u> by Willi Apel, Harvard Univ. Press, or an elementary physics text that includes an analysis of sound.

14.   Show that the radian measure of an angle <u>equals</u> the arc length on the <u>unit</u> <u>circle</u> cut off by the angle.   (Note that r = 1.)

First check it for $\theta = 2\pi$.

15.   Show that the area of the shaded sector is

$$\frac{\theta r^2}{2}:$$

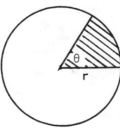

5
Trigonometry

Inverse Trigonometric Functions

* * * CENTRAL SECTION * * *

When a space capsule returns from a mission it reenters the Earth's atmosphere at some angle θ:

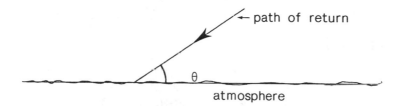

path of return

θ

atmosphere

To prevent the capsule from <u>bouncing off</u> the atmosphere it is necessary that

$$\sin \theta \geq 0.026 .$$

To prevent the capsule from entering too quickly and <u>burning up</u>, it is necessary that

$$\sin \theta \leq 0.105 .$$

To prevent the capsule from entering so quickly that the <u>deceleration</u> <u>injures</u> the astronauts it is necessary that

$$\sin \theta \leq 0.035 .$$

Fortunately these three demands can all be met; but, as Exercises 1 and 2 show, the angle θ is quite restricted.

EXERCISES

1. How small can the angle θ of reentry be? Express the answer in degrees and in radians.

2. How large can the angle θ of reentry be? Express the answer in degrees and in radians.

*Hint: If
? $\leq \sin \theta \leq$?
then
? $\leq \theta \leq$?
Use the inverse
sine key on your
calculator to find
θ.*

* * *

The opening problem requires finding

"the angle whose sine is 0.105"

and

"the angle whose sine is 0.026 ."

This short chapter is concerned with this way of looking at a trigonometric function. This "inverse" point of view was frequently required when you were solving for the angles of a triangle. Knowing cos θ or sin θ , you can estimate the angle θ by using the inverse sine key on your calculator.

THE ANGLES WHOSE SINE IS x

There are many angles whose sine is $\frac{1}{2}$. For instance,

$$\sin \frac{\pi}{6} = \frac{1}{2}$$

and $$\sin \frac{5\pi}{6} = \frac{1}{2} .$$

Since sin $(\theta + 2\pi) = \sin \theta$,

then $$\sin(\frac{\pi}{6} + 2\pi) = \frac{1}{2}$$

and $$\sin(\frac{5\pi}{6} + 2\pi) = \frac{1}{2}$$

Indeed, continued addition of 2π (or subtraction of 2π) produces an endless supply of angles whose sine is $\frac{1}{2}$:

$$\sin(\frac{\pi}{6} + 2n\pi) = \frac{1}{2}$$

and $$\sin(\frac{5\pi}{6} + 2n\pi) = \frac{1}{2}$$

for any integer n.

The most convenient angle to choose is $\frac{\pi}{6}$. More generally, the following choice is customarily made when we wish to speak of <u>the</u> angle whose sine is x, if $0 \le x \le 1$.

Thinking of "Arcsin x" as "the angle whose sine is x" is less abstract and more helpful to most students at this stage.

> Let x be a number, $0 \le x \le 1$. There is a unique number y,
>
> $$0 \le y \le \frac{\pi}{2}$$
>
> such that sin y is equal to x. This number is denoted
>
> Arcsin x.
>
> ("Arcsin x" is read "Arc sine of x" or "the angle whose sine is x.")

For instance,

$$\text{Arcsin } \frac{1}{2} = \frac{\pi}{6} .$$

(a) 0 radians

(c) $\frac{\pi}{3}$ radians

3. Evaluate the following without using a calculator:

   (a) Arcsin 0   (b) Arcsin $(\frac{\sqrt{2}}{2})$

   (c) Arcsin $(\frac{\sqrt{3}}{2})$ (d) Arcsin 1

Some scientific calculators have "arcsin" keys, others have "inv" (inverse) keys that are to be used with the sine key. Another notation is the following:

$$\sin^{-1}$$

The notation $\sin^{-1}$ does not mean $\frac{1}{\sin}$ (i.e. $\sin^{-1}x$ does not mean $\frac{1}{\sin x}$). The notations $\sin^{-1}x$, inv sin x, and Arcsin x all have the same meaning on the respective calculators.

4. Use your calculator to evaluate the following:

(a) Arcsin 0.035   (b) Arcsin 0.105

(c) Arcsin 0.423   (d) Arcsin 0.946

(a) 0.035 radians
(c) 0.436 radians

* * *

As x (the sine of the angle) goes from 0 to 1, Arcsin x (the angle whose sine is x) goes from 0 to $\frac{\pi}{2}$. The next exercise develops the graph that shows how Arcsin x varies as x goes from 0 to 1.

5. (a) Copy and fill in this table

x	0	$\frac{1}{2}$	$\frac{\sqrt{2}}{2}$	$\frac{\sqrt{3}}{2}$	1
y = Arcsin x					

(b) Copy this coordinate system accurately and plot the five points in (a). (Use $\sqrt{3} \doteq 1.7$ and $\sqrt{2} \doteq 1.4$.)

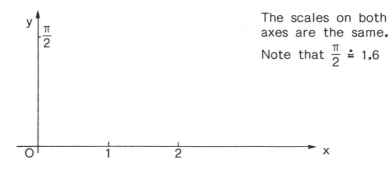

The scales on both axes are the same.
Note that $\frac{\pi}{2} \doteq 1.6$

(c) Use the five points in (b) to sketch the graph of

$$y = \text{Arcsin } x$$

for $0 \le x \le 1$.

The next exercise gives practice in working with the concept "Arcsin x."

6. Evaluate:

(a) sin (Arcsin $\frac{1}{2}$)   (b) sin (Arcsin $\frac{\sqrt{2}}{2}$)

(c) sin (Arcsin 0.574)   (d) sin (Arcsin x) for $0 \leq x \leq 1$

7. Which of these four expressions is (are) meaningful? Explain.

(a) Arcsin 1   (b) Arcsin 2   (c) Arcsin 3   (d) Arcsin 0

*   *   *

So far only angles whose sine is nonnegative have been considered. But the sine of an angle can be negative. For instance, there is an endless supply of angles whose sine is $-\frac{1}{2}$:

$$\sin (-\frac{\pi}{6}) = -\frac{1}{2}$$

and

$$\sin (\frac{7}{6}\pi) = -\frac{1}{2} ;$$

also for <u>any</u> integer n

$$\sin (-\frac{\pi}{6} + 2n\pi) = -\frac{1}{2}$$

$$\sin (\frac{7}{6} + 2n\pi) = -\frac{1}{2} .$$

It is customary to single out $-\frac{\pi}{6}$ and call it Arcsin $(-\frac{1}{2})$.   The following definition, which includes the earlier definition of Arcsin x for a positive x, defines Arcsin x for $-1 \leq x \leq 1$.

Arcsin x

Let x be a number, $-1 \leq x \leq 1$. There is a unique number y,

$$-\frac{\pi}{2} \leq y \leq \frac{\pi}{2},$$

such that sin y is equal to x. The number y is denoted

Arcsin x

8. Evaluate without using a calculator:

(a) Arcsin (-1)   (b) Arcsin $(-\frac{\sqrt{3}}{2})$.

(c) Arcsin $(-\frac{\sqrt{2}}{2})$.   (d) Arcsin $(-\frac{1}{2})$.

(a) $\frac{1}{2}$

How large can the sine of an angle be?

Again, the sine of the angle is x, and y = Arcsin x, or "y is the angle whose sine is x."

(a) $-\frac{\pi}{2}$

(c) $-\frac{\pi}{4}$

The next exercise concerns the graph of y = Arcsin x.

9. (a) Copy and fill in this table for y = Arcsin x:

x	-1	$-\frac{\sqrt{3}}{2}$	$-\frac{\sqrt{2}}{2}$	$-\frac{1}{2}$	0	$\frac{1}{2}$	$\frac{\sqrt{2}}{2}$	$\frac{\sqrt{3}}{2}$	1
y = Arcsin x									

(b) Copy this coordinate system accurately and plot the nine points in part (a):

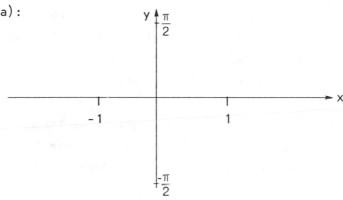

(c) Draw a smooth curve through the nine points in (b) to graph

y = Arcsin x

for -1 ≤ x ≤ 1. Label the graph.

Exercise 9 gives the graph of y = Arcsin x where -1 ≤ x ≤ 1 and

$-\frac{\pi}{2} \le y \le \frac{\pi}{2}$. This graph is the reflection across the line y = x of the

graph of y = sin x where $-\frac{\pi}{2} \le x \le \frac{\pi}{2}$:

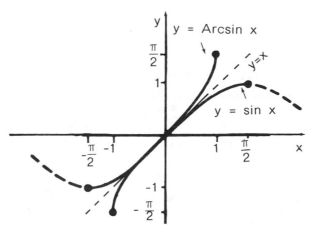

This is just another way of showing how we choose to define
y = Arcsin x (see box before Exercise 8). We want to use the inverse of
the sine function. The entire sine graph does <u>not</u> have an inverse, because when it is reflected across the line y = x that graph has more than
one output for each input. To solve the difficulty, we choose only the
portion of the graph of y = sin x where $-\frac{\pi}{2} \leq x \leq \frac{\pi}{2}$ and reflect that part
to get the graph of y = Arcsin x.

This completes the description of the Arcsin function. It is also
called the "inverse" of the sine function. In calculus it is used
frequently, almost as much as the sine function itself.

Associated with each of the six trigonometric functions is an
inverse function. In applications the two most important are Arcsin x and
Arctan x.

The third in importance is Arccos x. The three others, Arcsec x,
Arccot x, and Arccsc x are seldom used; for their definition see the
Projects. The chapter now moves to defining the function Arctan x.

THE Arctan x FUNCTION

There is an endless supply of angles whose tangent is 1. For
instance

$$\tan \frac{\pi}{4} = 1$$

and

$$\tan \frac{5}{4}\pi = 1;$$

also

$$\tan(\frac{\pi}{4} + n\pi) = 1$$

$$\tan(\frac{5}{4}\pi + n\pi) = 1$$

for any integer n, since tan $(\theta + n\pi)$ = tan θ. It is customary to single
out $\frac{\pi}{4}$ as the most convenient angle whose tangent is 1 and denote it

$$\text{Arctan 1 .}$$

This is the definition of the Arctan function, or inverse of the tangent
function.

Arctan x

Let x be any number. There is a unique number y

$$-\frac{\pi}{2} < y < \frac{\pi}{2}$$

such that tan y is equal to x. This number y is denoted

y = Arctan x.

(Arctan x is read "Arc tangent of x," hence "y is the angle whose tangent is x.")

For instance,

$$\tan \frac{\pi}{3} = \sqrt{3} \ .$$

Thus,

$$\text{Arctan } \sqrt{3} = \frac{\pi}{3} \ .$$

NOTE. While Arcsin x is defined only for $-1 \leq x \leq 1$, observe that Arctan x is defined for all numbers x.

10. Evaluate (do not use a calculator):

(a) Arctan 1   (b) Arctan 0   (a) $\frac{\pi}{4}$

(c) Arctan (-1)   (d) Arctan $(-\sqrt{3})$ (c) $-\frac{\pi}{4}$

11. Use a calculator to estimate

(a) Arctan 3   (b) Arctan (-3).

12. Use a calculator to estimate

(a) Arctan 2   (b) Arctan 4   (c) Arctan 10   (a) 1.107 radians

(d) Arctan (-2)   (e) Arctan (-4)   (f) Arctan (-10)   (c) 1.471 radians

Exercise 12 suggests the following: As x gets very large, Arctan x gets very near $\frac{\pi}{2} \doteq 1.57$ radians. A sketch of the unit circle (in terms of which the tangent function was defined) shows that this is the case. The diagram illustrates Arctan 4:

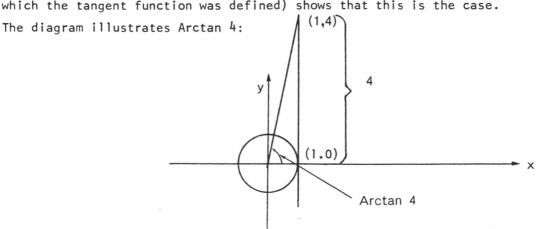

The next exercise obtains the graph of y = Arctan x. This graph does _not_ resemble the one for y = Arcsin x.

13. (a) Copy and fill in this table:

x	negative but of large absolute value	$-\sqrt{3}$	-1	$-\dfrac{1}{\sqrt{3}}$	0	$\dfrac{1}{\sqrt{3}}$	1	$\sqrt{3}$	positive and very large
y = Arctan x	near $-\dfrac{\pi}{2}$								near $\dfrac{\pi}{2}$

(b) Copy this coordinate system and guidelines accurately and plot the points in (a):

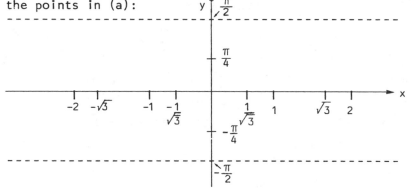

NOTE. When |x| is large (that is, far to the right or to the left) the graph of y = Arctan x gets close to the horizontal lines y = $\dfrac{\pi}{2}$ or y = $\dfrac{\pi}{2}$. These lines are the _asymptotes_ of the graph.

(c) Use the points and asymptotes in (b) to sketch the graph of y = Arctan x. Label the graph. (To check your graph, compare it with the one shown in the Summary at the end of the chapter.)

Since the graph of y = tan x has a repeating period of π, the total graph does not have an inverse. Hence we choose the part where $-\dfrac{\pi}{2} < x < \dfrac{\pi}{2}$ and reflect that across the line y = x to get the graph of the inverse of the tangent, written

$$y = \text{Arctan } x.$$

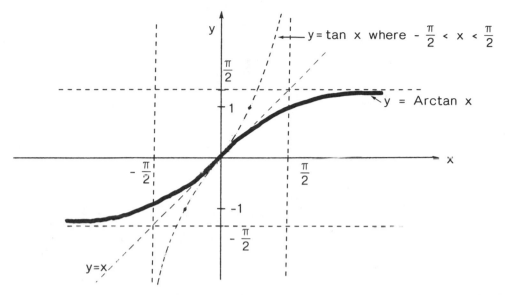

$y = \tan x$ where $-\dfrac{\pi}{2} < x < \dfrac{\pi}{2}$

$y = \text{Arctan } x$

$y = x$

PRACTICE WITH Arcsin x AND Arctan x

The two most important of the six inverse functions have now been described and their graphs sketched. Before going on to Arccos x it may be well to pause and become more familiar with Arcsin x and Arctan x. The next few exercises offer some practice in working with those two concepts.

14.   Evaluate without using a calculator:

$$\cos \left(\text{Arcsin } \dfrac{\sqrt{3}}{2}\right).$$

First find
Arcsin $\left(\dfrac{\sqrt{3}}{2}\right)$

15.   Evaluate without a calculator (show your reasoning):

   (a) $\cos (\text{Arcsin } 0)$

   (b) $\cos \left(\text{Arcsin } \dfrac{1}{2}\right)$

   (c) $\cos \left(\text{Arcsin } \dfrac{\sqrt{2}}{2}\right)$

   (d) $\cos (\text{Arcsin } 1)$

(a) 1

(c) $\dfrac{\sqrt{2}}{2}$

16.   Evaluate without a calculator (show your reasoning):

   (a) $\sin \left(\text{Arcsin } \left(-\dfrac{1}{2}\right)\right)$

   (b) $\cos \left(\text{Arcsin } \left(-\dfrac{1}{2}\right)\right)$

   (c) $\tan \left(\text{Arcsin } \dfrac{\sqrt{2}}{2}\right)$

   (d) $\tan \left(\text{Arcsin } \dfrac{\sqrt{3}}{2}\right)$

(a) $-\dfrac{1}{2}$

(c) 1

17.   Explain why each of the following steps is true.

   (a)   If $-1 \le x \le 1$, then $y = \text{Arcsin } x$ is defined and $-\dfrac{\pi}{2} \le y \le \dfrac{\pi}{2}$.

   (b)   Then $\sin y = x$, for $-\dfrac{\pi}{2} \le y \le \dfrac{\pi}{2}$

   (c) $\cos^2 y + \sin^2 y = 1$

   (d)   Use parts (a) - (c) to explain the following:

   Show that for $-1 \le x \le 1$, then $\cos (\text{Arcsin } x) = \sqrt{1 - x^2}$.

18. Evaluate without a calculator:

(a) sin (Arctan 1)   (b) sin (Arctan $\sqrt{3}$)

(c) sin (Arctan ($-\sqrt{3}$))   (d) cos (Arctan (-1))

19. Evaluate without a calculator (show your reasoning):

(a) tan (Arctan 1)   (b) tan (Arctan $\sqrt{3}$)

(c) tan (Arctan 5)   (d) tan (Arctan x)

20. Evaluate without a calculator:

(a) Arcsin (sin $\frac{\pi}{4}$)   (b) Arctan (tan 0)

(c) Arcsin (sin $\frac{\pi}{2}$)   (d) Arctan (tan ($-\frac{\pi}{3}$))

21. Evaluate without a calculator:

(a) sin (Arctan -1)   (b) sin (Arctan x)

(a) $\frac{\sqrt{2}}{2}$

(c) $-\frac{\sqrt{3}}{2}$

(a) 1

(c) 5

(a) $\frac{\pi}{4}$

(c) $\frac{\pi}{2}$

(a) $-\frac{\sqrt{2}}{2}$

THE Arccos x FUNCTION

The equation

$$\cos x = -\frac{1}{2}$$

has an endless supply of solutions. For instance,

$$\cos \frac{2\pi}{3} = -\frac{1}{2}$$

and

$$\cos \frac{4\pi}{3} = -\frac{1}{2} .$$

Since cos ($\theta + 2\pi$) = cos θ, it follows that for any integer n,

$$\cos \left(\frac{2}{3}\pi + 2n\pi\right) = -\frac{1}{2}$$

and

$$\cos \left(\frac{3}{4}\pi + 2n\pi\right) = -\frac{1}{2} .$$

Of all these solutions, only one is distinguished and labelled Arccos ($-\frac{1}{2}$).

That is the smallest positive solution, $\frac{2\pi}{3}$. Arccos x is defined as follows.

Arccos x

Let x be a number, $-1 \leq x \leq 1$. There is a unique number y,

$$0 \leq y \leq \pi$$

such that cos y is equal to x. This number y is denoted

Arccos x.

(Read "Arc cosine of x" or "the angle whose cosine is x.")

---

22. Evaluate:

(a) Arccos 0

(b) Arccos $\frac{1}{2}$

(c) Arccos $\left(-\frac{\sqrt{2}}{2}\right)$

(d) Arccos (-1)

(a) $\frac{\pi}{2}$

(c) $\frac{3\pi}{4}$

23. (a) Copy and fill in this table:

x	-1	$-\frac{\sqrt{3}}{2}$	$-\frac{\sqrt{2}}{2}$	$-\frac{1}{2}$	0	$\frac{1}{2}$	$\frac{\sqrt{2}}{2}$	$\frac{\sqrt{3}}{2}$	1
y = Arccos x									

(b) Copy the following coordinate system accurately and plot the nine points in (a).

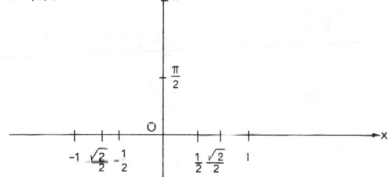

(c) Use the nine points in (b) to sketch the graph of

$$y = \text{Arccos } x.$$

for $-1 \leq x \leq 1$. Label the graph.

NOTE. Observe that

$$0 \leq \text{Arccos } x \leq \pi,$$

but

$$-\frac{\pi}{2} \leq \text{Arcsin } x \leq \frac{\pi}{2}.$$

The reason for this distinction is essentially the following: Recall from the unit circle that as θ goes from 0 to π, cos θ goes through all numbers from 1 to -1. However, as θ goes from 0 to π, sin θ does <u>not</u> go through all numbers from -1 to 1. Instead, it goes through all numbers from 0 to 1 twice. On the other hand, as θ goes from $\frac{-\pi}{2}$ to $\frac{\pi}{2}$, then sin θ <u>does</u> go through all numbers from -1 to 1.

For both y = Arccos x and y = Arcsin x the requirement is that for each x there is a unique y.

Thus, we define for $-1 \leq x \leq 1$,

$$-\frac{\pi}{2} \leq \text{Arcsin } x \leq \frac{\pi}{2} \qquad \text{and} \qquad 0 \leq \text{Arccos } x \leq \pi.$$

Because the total graph of y = cos x does not have an inverse, we choose the part where $0 \leq x \leq \pi$ and reflect that part across the line y = x to get the inverse of the cosine,

y = Arccos x.

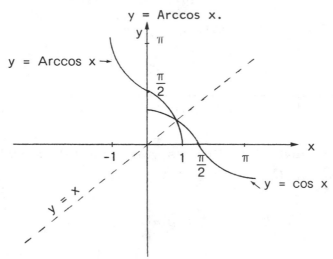

24.   Evaluate without a calculator:

(a)   sin (Arccos 1)   (b)   sin (Arccos $\frac{1}{2}$)

(c)   sin (Arccos ($-\frac{1}{2}$))   (d)   sin (Arccos x), $-1 \leq x \leq 1$.

With the aid of identities from Chapter 3, more complicated expressions can be evaluated, as Example 1 illustrates:

Example 1.   Evaluate:

$$\sin \left(\text{Arctan } \frac{1}{2} + \text{Arctan } \frac{1}{3} \right)$$

Solution.   Let $\alpha = \text{Arctan } \frac{1}{2}$ and $\beta = \text{Arctan } \frac{1}{3}$

then, $\tan \alpha = \frac{1}{2}$ and $\tan \beta = \frac{1}{3}$.

Also

$$\sin \alpha = \frac{1}{\sqrt{5}}, \quad \cos \alpha = \frac{2}{\sqrt{5}} \quad \text{and} \quad \sin \beta = \frac{1}{\sqrt{10}}, \quad \cos \beta = \frac{3}{\sqrt{10}}$$

Thus,

$$\sin \left(\text{Arctan } \frac{1}{2} + \text{Arctan } \frac{1}{3} \right) = \sin (\alpha + \beta)$$
$$= \sin \alpha \cos \beta + \cos \alpha \sin \beta$$
$$= \left(\frac{1}{\sqrt{5}} \right) \left(\frac{3}{\sqrt{10}} \right) + \left(\frac{2}{\sqrt{5}} \right) \left(\frac{1}{\sqrt{10}} \right)$$
$$= \frac{5}{\sqrt{50}} \quad \text{or} \quad \frac{1}{\sqrt{2}}$$

25. Evaluate in the manner of Example 1.

(a) $\sin\left(\text{Arcsin }\frac{3}{5} + \text{Arccos }\frac{7}{25}\right)$ (b) $\text{Sin}\left[\text{Arcsin }\frac{4}{5} + \text{Arctan }\left(-\frac{8}{15}\right)\right]$.

26. Evaluate:

(a) $\sin\left(2\text{ Arcsin }\frac{3}{5}\right)$ (b) $\tan\left(2\text{ Arctan }3\right)$.

* * *

The following example presents another type of proof.

Example 2. Prove:

$$\text{Arcsin }\frac{1}{3} - \text{Arccos }\frac{7}{11} = \text{Arcsin }\left(-\frac{17}{33}\right)$$

Solution: Take the sine of both sides:

$$\sin\left(\text{Arcsin }\frac{1}{3} - \text{Arccos }\frac{7}{11}\right) \overset{?}{=} \sin\left(\text{Arcsin }\left(-\frac{17}{33}\right)\right)$$

Let $\alpha = \text{Arcsin }\frac{1}{3}$ and $\beta = \text{Arccos }\frac{7}{11}$

then

$\sin\alpha = \frac{1}{3}$, $\cos\alpha = \frac{2\sqrt{2}}{3}$ and $\cos\beta = \frac{7}{11}$, $\sin\beta = \frac{6\sqrt{2}}{11}$ $-\frac{17}{33}$

Thus, $\sin(\alpha - \beta)$ becomes

$\sin\alpha\cos\beta - \cos\alpha\sin\beta$

$\left(\frac{1}{3}\right)\left(\frac{7}{11}\right) - \left(\frac{2\sqrt{2}}{3}\right)\left(\frac{6\sqrt{2}}{11}\right)$

or $-\frac{17}{33}$

27. Prove:

(a) $\text{Arctan }\frac{2}{3} + \text{Arctan }\frac{1}{5} = \frac{\pi}{4}$

(b) $\text{Arcsin }\frac{3}{5} + \text{Arcsin }\frac{12}{13} = \text{Arccos }\left(-\frac{16}{65}\right)$

28. (a) Find $\tan\left(\text{Arctan }\frac{1}{2}\right)$ and also $\tan\left(\text{Arctan }\frac{1}{3}\right)$.

(b) Use the identity

$$\tan(x + y) = \frac{\tan x + \tan y}{1 - \tan x \tan y}$$

to evaluate

$$\tan\left(\text{Arctan }\frac{1}{2} + \text{Arctan }\frac{1}{3}\right).$$

(c) From (b) deduce that

$$\text{Arctan }\frac{1}{2} + \text{Arctan }\frac{1}{3} = \frac{\pi}{4}.$$

29. Show that

$$2\text{ Arccos }(t^2) = \text{Arccos }(2t^4 - 1).$$

Hint: (a) Use the sin ($\alpha + \beta$) identity

(a) $\frac{117}{125}$

Hint: Recall the identities for sin 2x and tan 2x

(a) $\frac{24}{25}$

Hint: (a) Take the tangent of each side
(b) Take the cosine of each side.

Hint: (b) let Arctan $\frac{1}{2}$ = x and Arctan $\frac{1}{3}$ = y, then tan x = $\frac{1}{2}$ and tan y = $\frac{1}{3}$

Take the cosine of both sides.

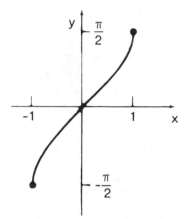

Hint:

*Let α = Arctan x.
Then consider the
other acute angle.*

30.   For any positive number x show that

$$\text{Arctan } x + \text{Arctan } \left(\frac{1}{x}\right) = \frac{\pi}{2} \, .$$

SUMMARY

This chapter consists essentially of the definitions of three functions,

   Arcsin x,   Arctan x,   and   Arccos x.

Arcsin x means "the angle between $-\frac{\pi}{2}$ and $\frac{\pi}{2}$ whose sine is x."

Arcsin x is defined for $-1 \le x \le 1$ and $-\frac{\pi}{2} \le \text{Arcsin } x \le \frac{\pi}{2}$:

y = Arcsin x

Arctan x is short for "the angle between $-\frac{\pi}{2}$ and $\frac{\pi}{2}$ whose tangent is x." Arctan x is defined for all values of x and $-\frac{\pi}{2} < \text{Arctan } x < \frac{\pi}{2}$:

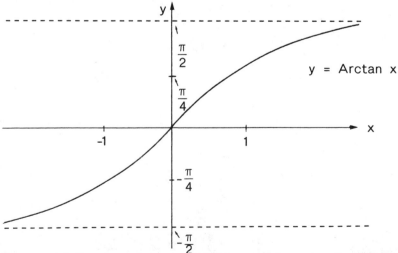

y = Arctan x

The dashed lines, which are quite useful in graphing the function, are called (horizontal) asymptotes.

The functions Arcsin x and Arctan x will be encountered by any student of calculus. They are frequently denoted by the symbols $\sin^{-1}x$ or $\text{Sin}^{-1}x$ and $\tan^{-1}x$ or $\text{Tan}^{-1}x$. They are sometimes called "the inverse sine and the "inverse tangent." For instance,

Note that the symbol '-1' in $\sin^{-1}x$ is not an exponent.

$$\sin^{-1}1 = \frac{\pi}{2} \qquad \text{and} \qquad \tan^{-1}1 = \frac{\pi}{4} \ .$$

Arccos x means "the angle between 0 and π whose cosine is x." Arccos x is defined for $-1 \leq x \leq 1$ and $0 < \text{Arccos } x \leq \pi$:

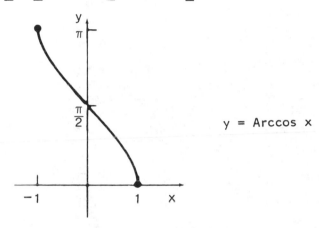

$$y = \text{Arccos } x$$

The other three trigonometric functions have inverses, but they are of less importance. See the Projects for a discussion of these functions.

WORDS AND SYMBOLS

Arcsin x	Arccos x
Arctan x	Inverse trigonometric function

CONCLUDING REMARKS

In 1674 Leibniz showed that for $-1 \leq x \leq 1$,

(1) $\text{Arctan } x = x - \dfrac{x^3}{3} + \dfrac{x^5}{5} - \dfrac{x^7}{7} + \ldots$ (angle in radians)

Replacing x in both sides by 1 yields

$$\text{Arctan } 1 = 1 - \frac{1}{3} + \frac{1}{5} - \frac{1}{7} + \ldots$$

or

$$\frac{\pi}{4} = 1 - \frac{1}{3} + \frac{1}{5} - \frac{1}{7} + \ldots$$

Thus, π is closely connected with the reciprocals of all the odd numbers.

The equation

$$\frac{\pi}{4} = 1 - \frac{1}{3} + \frac{1}{5} - \frac{1}{7} + \cdots$$

provides a way of estimating π. The more terms that are computed on the right side of the equation, the closer the estimate that is obtained for $\frac{\pi}{4}$, hence for π. The first eight terms yields this estimate:

$$\frac{\pi}{4} \doteq 1 - 0.333 + 0.200 - 0.143 + 0.111 - 0.091 + 0.077 - 0.067$$

$$\pi \doteq 0.754$$

Hence,

$$\pi \doteq 3.016 .$$

The use of more terms will give estimates that are accurate to as many decimal places as desired.

One reason that Arctan x is of importance in calculus is that the area under the graph of

$$y = \frac{1}{1 + x^2}$$

and above the x-axis between 0 and a is given by Arctan a (a > 0).

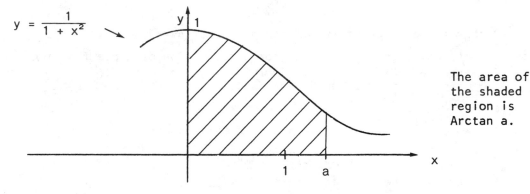

$y = \dfrac{1}{1 + x^2}$

The area of the shaded region is Arctan a.

This fact is used in physics and statistics.

INVERSE TRIG FUNCTIONS SELF QUIZ

1. For what values of x is each of the following defined?

    (a) Arcsin x   (b) Arccos x   (c) Arctan x

2. The chapter shows that $-\frac{\pi}{2} \le$ Arcsin x $\le \frac{\pi}{2}$. State similar inequalities for each of the following, if they exist:

    (a) $\underline{\ ?\ } \le$ Arccos x $\le \underline{\ ?\ }$ (b) $\underline{\ ?\ } \le$ Arctan x $\le \underline{\ ?\ }$

3. Compute:

(a) $\sin (\text{Arctan } \frac{3}{4})$ (b) $\cos (\text{Arctan } (-\frac{5}{12}))$

(c) $\tan (\text{Arccos } (-\frac{1}{3}))$ (d) $\cot (\text{Arcsin } (\frac{3}{5}))$ *Experiment!*

4. Compute:

(a) $\cos (\text{Arcsin } \frac{4}{5} - \text{Arccos } \frac{5}{13})$ (b) $\sin (\frac{1}{2} \text{Arctan } (-\frac{7}{24}))$ *Experiment!*

5. Prove:

$$\text{Arccos } (-\frac{2}{3}) - \text{Arccos } \frac{2}{3} = \text{Arccos } \frac{1}{9}$$

ANSWERS

1. (a), (b) $-1 \leq x \leq 1$ (c) all x   2. (a) $0 \leq \text{Arccos } x \leq \pi$

(b) $-\frac{\pi}{2} < \text{Arctan } x < \frac{\pi}{2}$ 3. (a) $\frac{3}{5}$ (b) $\frac{12}{13}$ (c) $-\sqrt{8}$

(d) $\frac{4}{3}$ 4. (a) $\frac{63}{65}$ (b) $-\frac{\sqrt{2}}{10}$ 5. Proof

INVERSE TRIG FUNCTIONS REVIEW

1. Define each of the following:

(a) $y = \text{Arcsin } x$ (b) $y = \text{Arccos } x$ (c) $y = \text{Arctan } x$

2. Sketch the graph of each of the following:

(a) $y = \text{Arcsin } x$ (b) $y = \text{Arccos } x$ (c) $y = \text{Arctan } x$

3. If you graph

$$y = \sin x \ (-\frac{\pi}{2} \leq x \leq \frac{\pi}{2})$$

and the ink is still wet on the paper, how could you fold the paper to make the graph of

$$y = \text{Arcsin } x?$$

4. (a) Make an accurate graph of

$$y = \sin x \ (-\frac{\pi}{2} \leq x \leq \frac{\pi}{2}).$$

(b) Bend a piece of wire in the shape of the above graph.

(c) Without bending the wire further, what could you do with the wire to get the graph of $y = \text{Arcsin } x$?

5. Compute:

(a) $\cos (\text{Arctan } (\frac{5}{12}))$ (b) $\sin (\text{Arctan } (-\frac{3}{4}))$

(c) $\sin (\text{Arctan } (-\frac{3}{2}))$ (d) $\cos (\text{Arctan } \frac{1}{2})$

6. Compute:

(a) $\sin (\text{Arcsin } \frac{3}{5})$ (b) $\tan (\text{Arcsin } \frac{3}{5})$

(c) $\cos (\text{Arcsin } \frac{8}{17})$ (d) $\sin (\text{Arctan } 1 + \text{Arctan } \frac{1}{3})$

7. Show that

(a) $\text{Arcsin } \frac{1}{2} + \text{Arccos } \frac{1}{2} = \text{Arcsin } 1$

(b) $\text{Arcsin } \frac{12}{13} + \text{Arccos } \frac{3}{5} = \text{Arccos } (-\frac{33}{65})$

(c) $\text{Arctan } \frac{2}{3} + \text{Arctan } \frac{1}{5} = \frac{\pi}{4}$ (d) $\text{Arctan } \frac{4}{3} - \text{Arctan } \frac{1}{7} = \frac{\pi}{4}$

ANSWERS:

1.-2. See Summary   3. Fold along the line y = x.   4. Rotate it

180° about the line y = x.   5.   (a) $\frac{12}{13}$ (b) $-\frac{3}{5}$ (c) $-\frac{3}{\sqrt{13}}$ (d) $\frac{2}{\sqrt{5}}$

6.   (a) $\frac{3}{5}$ (b) $\frac{3}{4}$ (c) $\frac{15}{17}$ (d) $\frac{2}{\sqrt{5}}$ 7. Proofs

INVERSE TRIG FUNCTIONS REVIEW SELF TEST

1. On the same coordinate axes, sketch the graphs of

(a) $y = \sin x \ (-\frac{\pi}{2} \le x \le \frac{\pi}{2})$ (b) $y = \text{Arcsin } x$

2. On the same axes, sketch the graph of

(a) $y = \cos x \ (0 \le x \le \pi)$ (b) $y = \text{Arccos } x$

3. On the same axes, sketch the graph of

(a) $y = \tan x \ (-\frac{\pi}{2} < x < \frac{\pi}{2})$ (b) $y = \text{Arctan } x$

4. Find:

(a) $\text{Arccos } \frac{\sqrt{3}}{2}$ (b) $\text{Arcsin } (-\frac{1}{2})$ (c) $\text{Arctan } (-0.577)$

5. Find:

(a) $\sin (\text{Arccos } \frac{1}{2} + \text{Arcsin } \frac{\sqrt{3}}{2})$ (b) $\cos (\text{Arcsin } \frac{5}{13} - \text{Arccos } \frac{3}{5})$

(c) $\sin (\text{Arctan } \frac{1}{2})$

6. Prove: $\text{Arctan } \frac{4}{3} - \text{Arcsin } (-\frac{8}{17}) = \text{Arctan } \frac{84}{13}$

ANSWERS

1.-3. graph   4. (a) $\frac{\pi}{6}$ (b) $-\frac{\pi}{6}$ (c) $-\frac{\pi}{6}$

5. (a) $\frac{\sqrt{3}}{2}$ (b) $\frac{56}{65}$ (c) $\frac{1}{\sqrt{5}}$ 6. Proof

INVERSE TRIGONOMETRIC FUNCTIONS PROJECTS

1. Write the angle A in the form Arcsin x for a suitable number x:

    (a)   (b)

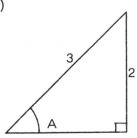 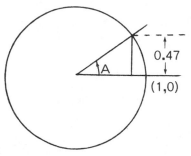

These Projects contain many important problems. Try some of them!

2. The functions Arcsin x, Arctan x, and Arccos x are special cases of inverse functions. Fill in this table:

    See Chapter 5 of Algebra II for a reference on inverse functions.

Function (and domain)	Inverse function (and domain)
x^3 (all x)	$\sqrt[3]{x}$ (all x)
10^x (all x)	$\log_{10}x$ (positive x)
$\log_{10}x$ (positive x)	10^x (all x)
x^2 $(x \geq 0)$	$\sqrt{x}$ $(x \geq 0)$
$\sin x$ $(-\frac{\pi}{2} \leq x \leq \frac{\pi}{2})$	_____
$\tan x$ $(-\frac{\pi}{2} < x < \frac{\pi}{2})$	_____
$\cos x$ $(0 \leq x \leq \pi)$	_____

3. This project describes the function Arccot x. Recall the definition of cot θ:

$$\cot \theta = \frac{1}{\tan \theta}$$

    (a) Show that if cot θ = 2, then tan θ = $\frac{1}{2}$

    (b) Show that Arctan $\frac{1}{2}$ is an angle whose cotangent is 2.

    (c) Show that for x ≠ 0, then Arctan $(\frac{1}{x})$ is an angle whose cotangent is x.

    Project 3 (c) suggests the following definition for the Arccot x function:

$$\text{Arccot } x = \text{Arctan } \left(\frac{1}{x}\right).$$

4.   This Project describes the function Arcsec x.   Recall the definition
of sec θ:
$$\sec \theta = \frac{1}{\cos \theta} \ .$$

(a)   Show that if sec θ = 2, then cos θ = $\frac{1}{2}$.

(b)   Show that Arccos $\frac{1}{2}$ is an angle whose secant is 2.

(c)   Show that for |x| ≥ 1
$$\sec \left(\text{Arccos} \left(\frac{1}{x} \right) \right) = x.$$

Project 4 (c) suggests the definition:
$$\text{Arcsec } x = \text{Arccos } \left(\frac{1}{x} \right) , \quad \text{for } |x| \geq 1 .$$

5.   (See Projects 3 and 4.)   Show that a reasonable definition of
Arccsc x when |x| ≥ 1 is
$$\text{Arccsc } x = \text{Arcsin } \left(\frac{1}{x} \right) .$$

6.   Express each of the following as simply as possible for 0 ≤ x ≤ 1 .

(a)   sin ($\frac{1}{2}$ Arccos x)   (b)   sin (Arcsin x^2 + Arccos x^2)

7.   Express in terms of x:
$$\tan (2 \text{ Arcsin } x^2)$$

Hint: Let
α = Arctan x and
β = Arctan 3x.

Take the tangent
of both sides.

8.   Solve for x:
$$\text{Arctan } x + \text{Arctan } 3x = \text{Arctan } 2$$

9.   Solve for x (x > 0):
$$\text{Arccos } x + 2 \text{ Arcsin } x = \text{Arccos } \left(-\frac{1}{2} \right) .$$

10.   Show that
$$\text{Arctan } \frac{4}{3} - \text{Arctan } \frac{1}{7} = \frac{\pi}{4}$$

11.   Read about the reentry problems of space capsules.

12.   In the Concluding Remarks it was stated that for -1 ≤ x ≤ 1,
$$\text{Arctan } x = x - \frac{x^3}{3} + \frac{x^5}{5} - \frac{x^7}{7} + \dots$$
Replace x by $\frac{1}{\sqrt{3}}$ and deduce that
$$\frac{\pi\sqrt{3}}{6} = 1 - \frac{1}{3 \cdot 3} + \frac{1}{5 \cdot 3^2} - \frac{1}{7 \cdot 3^3} + \dots$$

13. In the Concluding Remarks an estimate of π was obtained with eight summands of a certain sequence of numbers. This Project describes a much better estimate of π obtained with the same number of summands.

In Exercise 28 of the main section it was shown that

$$\text{Arctan } \frac{1}{2} + \text{Arctan } \frac{1}{3} = \frac{\pi}{4} \text{ .}$$

(a) Use the formula in Project 12 to show that

$$\text{Arctan } \frac{1}{2} = \frac{1}{2} - \frac{(\frac{1}{2})^3}{3} + \frac{(\frac{1}{2})^5}{5} - \frac{(\frac{1}{2})^7}{7} + \ldots$$

and

$$\text{Arctan } \frac{1}{3} = \frac{1}{3} - \frac{(\frac{1}{3})^3}{3} + \frac{(\frac{1}{3})^5}{5} - \frac{(\frac{1}{3})^7}{7} + \ldots$$

(b) Combining (a) with the result of Exercise 28 shows that

$$\frac{\pi}{4} = \left[\frac{1}{2} - \frac{(\frac{1}{2})^3}{3} + \frac{(\frac{1}{2})^5}{5} - \frac{(\frac{1}{2})^7}{7} + \ldots\right] + \left[\frac{1}{3} - \frac{(\frac{1}{3})^3}{3} + \frac{(\frac{1}{3})^5}{5} - \frac{(\frac{1}{3})^7}{7} + \ldots\right]$$

(c) With the aid of (b) show that

$$\pi \doteq 3.14$$

14. In calculus it is proved that

$$\text{Arcsin } x = x + \frac{1}{2} \cdot \frac{x^3}{3} + \frac{1 \cdot 3}{2 \cdot 4} \cdot \frac{x^5}{5} + \frac{1 \cdot 3 \cdot 5}{2 \cdot 4 \cdot 6} \cdot \frac{x^7}{7} + \ldots$$

Replace x in both sides by 1 and use four summands on the right hand side of the equation to obtain an estimate of π.

15. In Exercise 14 replace x by $\frac{1}{2}$ and obtain an estimate of π by using four summands on the right hand side of the equation.

16. Sketch the graph of y = Arccot x.

17. Sketch the graph of y = Arcsec x.

18. Sketch the graph of y = Arccsc x.

We are used to thinking of numbers as describing points on a straight line, the so-called x-axis. But the real number system is just a small part of a larger number system on the plane. As we shall see, the points of the (x,y)-plane can be added and multiplied in much the same way as we add and multiply the real numbers. Moreover, this addition and multiplication obeys the customary rules of algebra, such as the distributive rule, U(V + W) = UV + UW. The most remarkable property of this system, called the system of <u>complex numbers</u>, is that -1 has a square root.

THE COMPLEX NUMBER PLANE

The <u>complex number plane</u>, or simply <u>complex plane</u>, is just the ordinary (x,y)-plane equipped with a method of adding and multiplying points. In order to describe how points are added and multiplied, it will be convenient to use the following notation. Given a point P = (a,b) in the plane, we will denote it by the symbol a + bi. For example, in place of P = (3,4) we will write P = 3 + 4i.

STUDY THIS SECTION
VERY CAREFULLY!
There may be ideas
here that seem
subtle to you.
ASK QUESTIONS!

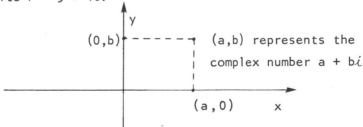

(0,b) ⌐ - - - - - ⌐ (a,b) represents the
                   complex number a + bi

(a,0)   x

If P lies on the x-axis, so P = (a,0), we shall simply write P = a rather than P = a + 0i. Thus the points on the x-axis will play their usual role as the real numbers. If P lies on the y-axis, so P = (0,b), we shall write P = bi rather than P = 0 + bi.

Given points a + bi and c + di , we <u>add</u> them as follows:

$$(a + bi) + (c + di) = (a + c) + (b + d)i$$

Example 1. Add 2 + 3i and -3 + i and sketch the result.
Solution. (2 + 3i) + (-3 + i) = -1 + 4i.

The sketch below shows the two given points and their sum:

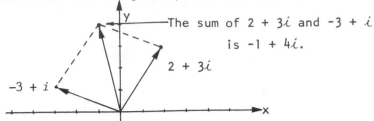

The sum of 2 + 3i and -3 + i
is -1 + 4i.

2 + 3i

-3 + i

Note that the origin, together with the two given points and their sum, form the four vertices of a parallelogram. If we represent the complex number a + bi as an "arrow" (also called a "vector") with the "tail" at the origin and the "head" at the point (a,b), then the arrows representing a + bi and c + di are adjacent sides of a parallelogram and their sum is represented by the arrow pointing along the diagonal between those two sides. This sum is sometimes called the "resultant" of the two arrows to be added.

EXERCISES

1. Compute the sums of the given complex numbers and plot the two numbers and the sum as in Example 1.

    (a) 1 + 2i and 2 + i (b) 1 and i

    (c) -3 and 3 + 2i (d) -1 - 2i and 1 + 2i

    (e) cos 45° + $i$sin 45° and cos 135° + $i$sin 135°

(a) 3 + 3i
(c) 2 i
(e) $i\sqrt{2}$

*   *   *

NOTATION. We prefer to write $i\sqrt{2}$ rather than $\sqrt{2}\,i$ in order to avoid the confusion of i being inside the radical.

2. In each case below, find real numbers x and y that make the given equation correct.

    (a) (x + yi) + (1 - 2i) = 4 + i (b) (x + yi) + (3 + 5i) = 0

(a) x = 3; y = 3

*   *   *

To <u>multiply</u> complex numbers, we proceed as in ordinary algebra, except that we apply the following rule.

DEFINITION OF i^2

Multiplication is based on this rule: whenever you see the product $i \cdot i$, or i^2, replace it by -1. That is,

$$i^2 = -1.$$

Example 2.   Compute and sketch the product $(2 + i)(1 + i)$.

Solution.   Following the usual rules of arithmetic and algebra (and remembering that $i^2 = -1$), we obtain:

$$(2 + i)(1 + i) = (2 + i)1 + (2 + i)i$$
$$= 2 + i + 2i + i^2$$
$$= 2 + i + 2i - 1$$
$$= 1 + 3i$$

When i^2 occurs, change it to -1 in the next step.

We sketch the three points:

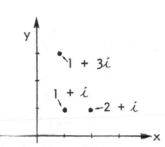

NOTE: We prefer not to draw arrows here because we are multiplying (not adding). As you will see, it is a more complicated process.

3.   Compute the product of the given numbers and sketch them and the product.

   (a) $2 + i$ and $1 + 2i$

   (b) $1 + i$ and $2 + 2i$

   (c) i and i

   (d) 3 and $2 + 3i$

   (e) $1 + i$ and $-2 - 2i$

(a) $5i$
(c) -1
(e) $-4i$

4.   If $U = 3 + 4i$ and $V = 2 - i$, compute:

   (a) UV

   (b) $(U + V)V$

   (c) $UV + V^2$

   (d) $U^2 - V^2$

   (e) $(U - V)(U + V)$

(a) $10 + 5i$
(c) $13 + i$
(e) $-10 + 28i$

5.   Let $U = \frac{1}{2} + i\frac{\sqrt{3}}{2}$.   Compute and sketch:

   (a) U^2 (b) U^3 (c) U^4 (d) U^5

Suggestion: Do successively U^2, $U^3 = U \cdot U^2$, $U^4 = U \cdot U^3$, etc.

(a) $-\frac{1}{2} + i\frac{\sqrt{3}}{2}$

(c) $-\frac{1}{2} - i\frac{\sqrt{3}}{2}$

6.   Let $U = \frac{\sqrt{2}}{2} + i\frac{\sqrt{2}}{2}$.   Compute and sketch:

   (a) U^2 (b) U^3 (c) U^4 (d) U^5

(a) i
(c) -1

7.   Let $U = 2 + 2i$.   Compute and sketch:

   (a) U^2 (b) U^3 (c) U^4 (d) U^5

   [Suggestion: You may save some labor by observing that
$$2 + 2i = 2(1 + i).]$$

(a) $8i$
(c) -64

*   *   *

As you have observed, multiplying complex numbers in a + bi form can be quite tedious. We now show another method.

MAGNITUDE (RADIUS) AND ARGUMENT (ANGLE)

STUDY THIS SECTION VERY CAREFULLY! ASK QUESTIONS!

While the addition of two complex numbers can most easily be described in terms of parallelograms, their product is easily described in terms of angles and lengths. To do this, we introduce two definitions.

The <u>magnitude</u> (or radius) of a + bi is $\sqrt{a^2 + b^2}$. The magnitude of a complex number tells the distance the point (a,b) is from the origin. It is denoted $|a + bi|$:

Do you see the Pythagorean Theorem here?

$$|a + bi| = \sqrt{a^2 + b^2}$$

NOTE: This is comparable to using the absolute value, |x|, on the number line to describe the distance the number x is from zero.

The <u>argument</u> (or angle) of a + bi is the angle from the positive x-axis to the ray from the origin through the point (a,b) that represents the number a + bi.

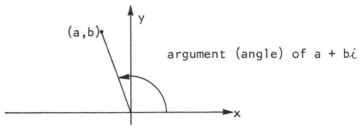

A <u>counterclockwise</u> rotation is <u>positive</u>, while a <u>clockwise</u> rotation is <u>negative</u>. It is important for later work to observe that the argument of a complex number is <u>not unique</u>. If θ is an argument, then so is θ + n · 360° for any integer n, positive or negative.

In other words, (2,60°), (2,-300°), and (2,420°) all describe the same point.

The complex number 0, which is at the origin, has magnitude (or radius) equal to 0. It is not assigned an argument (or angle).

Note that if θ is an argument (angle) of a + bi , and a ≠ 0, then tan θ = $\dfrac{b}{a}$. This permits the estimation of θ with the aid of the inverse tangent key of a calculator.

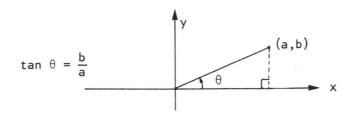

$$\tan \theta = \frac{b}{a}$$

Example 3. Compute the magnitude (or radius) of $1 - i$ and find three possible arguments (or angles).

Solution. $|1 - i| = \sqrt{(1) + (-1)^2} = \sqrt{2}$

$$\tan \theta = -\frac{1}{1}$$

$$= -1$$

$$\theta = -45°, \ 315°, \ \text{or} \ -405°$$

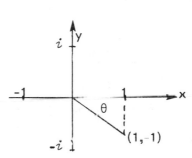

8. Find the magnitude (radius) of each of these complex numbers. Also give three possible arguments (angles) in each case. Sketch the numbers in the (x,y)-plane (complex number plane).

(a) $2 + 2i$

(b) $3 + 4i$

(c) $-2 + 2i$

(d) $-1 + i\sqrt{3}$

(e) $-1 - i\sqrt{3}$

(f) $\frac{1}{2} - i\frac{\sqrt{3}}{2}$

(g) $\cos 22° + i\sin 22°$

(h) $3\cos 22° + i(3\sin 22°)$

(a) $2\sqrt{2}$; $45°$, $405°$, $-315°$

(c) $2\sqrt{2}$; $135°$, $-225°$, $495°$

(e) 2; $240°$, $-120°$, $600°$

(g) 1; $22°$, $382°$, $-338°$

* * *

POLAR (TRIGONOMETRIC) FORM OF A COMPLEX NUMBER

The picture below shows the complex number $a + bi$ with magnitude $r = \sqrt{a^2 + b^2}$ and argument θ.

STUDY THIS SECTION CAREFULLY!
The point (a,b) and the complex number $a + bi$ have the same meaning.

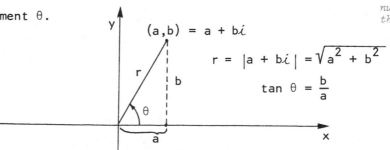

$$r = |a + bi| = \sqrt{a^2 + b^2}$$

$$\tan \theta = \frac{b}{a}$$

Since

$$\cos \theta = \frac{a}{\sqrt{a^2 + b^2}} = \frac{a}{r} \quad \text{and} \quad \sin \theta = \frac{b}{\sqrt{a^2 + b^2}} = \frac{b}{r},$$

we have

$$a = r\cos \theta \quad \text{and} \quad b = r\sin \theta.$$

Therefore,

$$a + bi = r\cos \theta + (r\sin \theta)i = r(\cos \theta + i\sin \theta).$$

This gives the following:

POLAR FORM OF A COMPLEX NUMBER

Given a complex number $a + bi$ with magnitude r and

argument θ, we have

$$a + bi = r(\cos \theta + i\sin \theta)$$

This is a very important exercise.

9. Study the above explanation carefully. Write a short explanation of your own, using a diagram and complete sentences, explaining how the polar form is obtained.

MAKE SURE YOU UNDERSTAND EXERCISE 10.

10. This exercise concerns the multiplication of complex numbers of magnitude 1. Such numbers lie on a circle with radius 1 and center at the origin.

    (a) Show that any such number is of the form $\cos \alpha + i\sin \alpha$ for some angle α.

    (b) Let $\cos \alpha + i\sin \alpha$ and $\cos \beta + i\sin \beta$ be two complex numbers on the unit circle. Compute their product,
    $$(\cos \alpha + i\sin \alpha)(\cos \beta + i\sin \beta).$$

    (c) Use trigonometric identities to simplify the result and show that:
    $$(\cos \alpha + i\sin \alpha)(\cos \beta + i\sin \beta) = \cos(\alpha + \beta) + i\sin(\alpha + \beta).$$

    (d) Explain why (c) shows that when you multiply two complex numbers, each represented by a point on the unit circle, you add the angles to get the angle of the product.

*   *   *

This means that the <u>angle</u> (<u>argument</u>) <u>of</u> <u>the</u> <u>product</u> is simply the <u>sum</u> <u>of</u> <u>the angles</u> (<u>arguments</u>) <u>of</u> <u>the</u> <u>two</u> <u>factors</u>. This holds for the product of any two complex numbers, even if they are not on the unit circle, as Exercise 11 shows.

11. (a) Using trigonometric identities, show that the product of the complex numbers

$$r_1(\cos \theta_1 + i\sin \theta_1) \text{ and } r_2(\cos \theta_2 + i\sin \theta_2)$$

is

$$r_1 r_2[\cos(\theta_1 + \theta_2) + i\sin(\theta_1 + \theta_2)].$$

(b) Use (a) to justify the following rule for multiplying complex numbers in polar form:

> To multiply two complex numbers, just add the angles (arguments) and multiply the radii (magnitudes).

12. Sketch U, V and UV if

(a) $U = 2(\cos 100° + i\sin 100°)$, $V = 3(\cos 80° + i\sin 80°)$

(b) $U = 5(\cos 180° + i\sin 180°)$, $V = 2(\cos 180° + i\sin 180°)$

13. Compute the product $(\cos 22° + i\sin 22°)(\cos 68° + i\sin 68°)$ as simply as possible.

14. State the magnitude and argument of UV if

(a) U has argument 30°, magnitude 2 and V has argument 45°, magnitude 3.

(b) U has argument 135°, magnitude $\frac{1}{2}$ and V has argument 450°, magnitude 2.

(c) Draw U, V, and UV for (a) and (b).

15. Let U, V, and W be nonzero complex numbers. If you know the magnitudes and arguments of U, V, and W, how would you find them for the product UVW?

16. State the magnitude and argument for each of the following complex numbers:

(a) 1   (b) i (c) $i^2 = -1$ (d) $i^3 = -i$ (e) $i^4 = 1$

17. Observe what happens when you keep multiplying by i for each case in Exercise 16.

(continued)

(a)   Does the radius change? Explain.

(b)   Does the angle change? How?

Hint: $i = i$
$i^2 = -1$
$i^3 = -i$
$i^4 = 1$
etc.
(a) $-i$
(c) i

18.   Simplify each of the following to the form of 1, -1, i, or $-i$.

(a) i^7 (b) i^{49} (c) i^{101} (d) i^{2000}

(e)   Explain to another person (or write a short paragraph) what you learned in Exercises 16-18. (Hint: what is a shortcut?)

19.   (a)   Let U be the complex number of magnitude 2 and argument $\frac{\pi}{3}$.

Show that there is exactly one complex number V such that UV = 1. (Hint: What must the magnitude be? The argument?)

(b)   Draw U and V.

20.   Let U = cos 72° + $i$sin 72°.

(a)   Draw U, U^2, U^3, U^4, U^5, and U^6.

(b)   What do you observe about U and U^6?

21.   (a)   Give the magnitude and all possible arguments for the complex number i.

(b)   If $U^2 = i$, what is the magnitude of U?

(c) $45° + n(360°)$
or $225° + n(360°)$
where n is an
integer .

(c)   If $U^2 = i$, give all possible arguments of U.

(d)   Sketch all complex numbers U such that $U^2 = i$.

22.   (a)   If U = r(cos θ + $i$sin θ), find the magnitude and argument of U^2.

(b)   Write U^2 in polar form.

23.   If U = r(cos θ + $i$sin θ), explain why $U^3 = r^3$(cos 3θ + $i$sin 3θ).

* * *

DE MOIVRE'S FORMULA

MEMORIZE
DE MOIVRE'S
FORMULA!

As Exercises 22 and 23 suggest, it is easy to find U^n if we are lucky enough to know the magnitude r and argument θ of U. If

U = r(cos θ + $i$sin θ), then $U^n = r^n$(cos nθ + $i$sin nθ).

(This is known as De Moivre's formula.)

Example 4. Find $(\frac{1}{2} + \frac{\sqrt{3}}{2} i)^3$.

Solution. We could grind out the answer by brute multiplication. But instead, let us sketch $\frac{1}{2} + \frac{\sqrt{3}}{2}i$, as in the following figure.

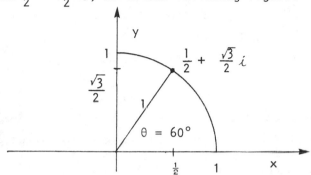

We see that $\frac{1}{2} + \frac{\sqrt{3}}{2}i$ has argument 60° and magnitude 1, since $\frac{1}{2} = \cos 60°$ and $\frac{\sqrt{3}}{2} = \sin 60°$. Thus,

$$\frac{1}{2} + \frac{\sqrt{3}}{2}i = 1(\cos 60° + i \sin 60°).$$

Cubing gives

$$1^3(\cos 180° + i \sin 180°)$$
$$= 1 (-1 + i0)$$
$$= -1$$

24. Find (a) $\left(\frac{1}{\sqrt{2}} + i\frac{1}{2}\right)^4$ (b) $\left(\frac{1}{\sqrt{2}} + \frac{i}{\sqrt{2}}\right)^8$ (c) $\left(-\frac{1}{2} + \frac{i\sqrt{3}}{2}\right)^6$ *Hint: Change to polar (Trig) form.*

* * *

Of course, if we are not dealing with a well known angle, such as 30°, 45°, or 60°, we will have to make some trigonometric estimations using the calculator.

Example 5. Find $(3 + 4i)^{10}$.

Solution. Since $\sqrt{3^2 + 4^2} = 5$, r = 5 for $3 + 4i$. How do we find an argument of $3 + 4i$? A glance at the figure below shows that $\theta = \tan^{-1}\left(\frac{4}{3}\right) \doteq 53.13°$.

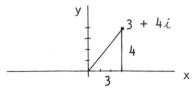

Compute $\tan^{-1}\left(\frac{4}{3}\right)$ using the inverse tangent key on your calculator.

Hence, the angle (argument) of our answer is approximately
$(10)(53.13°) = 531.3°$. The radius (magnitude) of $3 + 4i$ is
$|3 + 4i| = \sqrt{3^2 + 4^2} = 5$. Thus, the magnitude of our answer is 5^{10}.
The final answer may be written as

$$(3 + 4i)^{10} = 5^{10}(\cos 531.3° + i\sin 531.3°)$$
$$= 5^{10}(\cos 171.3° + i\sin 171.3°).$$

Hint:
531.3° – 360° =
171.3°

NOTE: We could convert our answer above to a + bi (rectangular) form if
we had reason to do so. The above form is acceptable.

25.   Compute in polar form:

(a) 32(cos 150°+
sin 150°)
(c) 125(cos 200.4°
+ i sin 200.4°)
approximately.

(a) $(\sqrt{3} + i)^5$ (b) $(1 + i)^8$ (c) $(-2 + i)^6$ (d) $(\frac{1}{2} + \frac{\sqrt{3}}{2}i)^{10}$

26.   Write the answers for Exercises 25 (a) and (d) in a + bi form.

(a) −16√3 + 16i
(c) −117.2 − 43.6i

FINDING THE nth ROOT OF A COMPLEX NUMBER

Just as raising to a power is easy in polar form, so is taking the
n^{th} roots of a complex number (it will turn out that a nonzero complex
number has n different n^{th} roots). The next few exercises pursue this
idea.

27.   The complex number i has magnitude 1 and argument 90°. Thus we may
write

$$i = \cos 90° + i\sin 90°.$$

(a)   Show that $\cos 45° + i\sin 45°$ is a square root of i.

(b)   Show that $\cos 225° + i\sin 225°$ is another square root of i.

(c) Hint: How are
the roots (points)
spaced about the
unit circle.

(c)   Sketch the roots in (a) and (b). Observe their special
positions on the unit circle.

28.   Let $U = 16(\cos 240° + i\sin 240°)$.

(a)   Show that $V = 2(\cos 60° + i\sin 60°)$ is a fourth root of U.

(b)   Consider the following facts about angles:
$\cos 240° = \cos(240° + 360°) = \cos(240° + 2\cdot360°)$
$= \cos(240° + 3\cdot360°)$ and similarly
for sin 240°. Now find the other fourth roots of U.

(c)   Sketch the roots. (Observe that they all lie on a circle of
radius 2 and are equally spaced.)

29. Let U = r(cos θ + $i$sin θ) and let n be a positive integer. Let

V = $r^{\frac{1}{n}}$[cos $(\frac{\theta}{n})$ + $i$sin $(\frac{\theta}{n})$]. Show why V is an nth root of U.

* * *

In general, if n is a positive integer, then r(cos θ + $i$sin θ) has n^{th} roots of the form:

$$r^{\frac{1}{n}}[\cos (\frac{\theta}{n}) + i\sin (\frac{\theta}{n})]$$

$$r^{\frac{1}{n}}[\cos (\frac{\theta + 360°}{n}) + i\sin (\frac{\theta + 360°}{n})]$$

$$r^{\frac{1}{n}}[\cos (\frac{\theta + 2(360°)}{n}) + i\sin (\frac{\theta + 2(360°)}{n})]$$

. . . and so on to get n roots.

Observe the roots have the same radius (magnitude) and are equally spaced on a circle.

STUDY CAREFULLY THIS GENERAL STATEMENT.

30. Find as many fifth roots of 32 as you can. Hint: What is the magnitude of 32? What are possible arguments of 32? (Remember that an angle of 0° is only one possibility.)

One root is $2(\cos 72° + i\sin 72°)$

31. Compute and sketch all solutions of the equation

(a) $U^4 = 1$ (b) $U^5 = 1$ (c) $U^5 = \frac{1}{32}$ (d) $U^4 = -1$

Explain your answers to another student.

32. Compute and sketch all roots of the equation $U^7 = 1$. (There are seven.)

33. Find and sketch all complex roots of the equations:

(a) $x^2 = 4$ (b) $x^2 = -4$ (c) $x^2 + 2x + 5 = 0$

Hint: (c) Recall the quadratic formula.

34. Explain why every nonzero complex number has two complex square roots.

35. (a) Let a, b, and c be complex numbers, with a ≠ 0. Explain why the equation $ax^2 + bx + c = 0$ has at least one solution among the complex numbers.

Hint: Consider the quadratic formula.

(b) When will it have two different solutions?

* * *

CONJUGATE OF A COMPLEX NUMBER

Before describing how we divide complex numbers, it will be useful to define the <u>conjugate</u> of a complex number.

The <u>conjugate</u> of the complex number $a + bi$ is the complex number $a - bi$. For example,

$$3 - 4i \text{ is the conjugate of } 3 + 4i$$
$$2 + i \text{ is the conjugate of } 2 - i$$
$$-i \text{ is the conjugate of } i$$
$$5 \text{ is the conjugate of } 5.$$

The picture below shows the geometrical relationship between $a + bi$ and its conjugate $a - bi$:

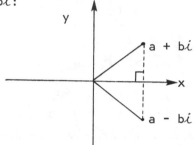

Note that $a - bi$ is the "reflection" of $a + bi$ across the x-axis.

The notation for the conjugate is

$$\overline{a + bi} = a - bi.$$

Similarly,

$$\overline{a - bi} = a + bi.$$

You should check that

$$(a + bi)\overline{(a + bi)} = a^2 + b^2 = |a + bi|^2.$$

Observe that the sum of two squares is factorable using complex numbers!

DIVISION OF COMPLEX NUMBERS

Let U, V, and W be complex numbers, with $U \neq 0$. Assume that $UV = W$. Then V is called the quotient of W divided by U, and is written $V = \dfrac{W}{U}$. It is easy to find the quotient in trigonometric (or polar) form, as the next exercise shows.

36. Show that

$$\frac{r_1(\cos\theta_1 + i\sin\theta_1)}{r_2(\cos\theta_2 + i\sin\theta_2)} = \frac{r_1}{r_2}\left[\cos(\theta_1 - \theta_2) + i\sin(\theta_1 - \theta_2)\right]$$

That is,

> "To divide complex numbers, divide the radii (magnitudes) and subtract the angles (arguments)."

Hint: Consider conjugates, then multiply by 1 in the form:

$$\frac{\cos\theta_2 - i\sin\theta_2}{\cos\theta_2 - i\sin\theta_2}$$

37. Compute and sketch $\frac{W}{U}$ if

(a) $W = 6(\cos 135° + i\sin 135°)$ and $U = 2(\cos 45° + i\sin 45°)$

(b) $W = 6(\cos 240° + i\sin 240°)$ and $U = 3(\cos 120° + i\sin 120°)$

(c) $W = 8(\cos 300° + i\sin 300°)$ and $U = 4(\cos 120° + i\sin 120°)$

(a) $3i$

DIVISION IN a + bi FORM

Division is also easy to carry out when the numbers are given in rectangular form. One simply "conjugates" the denominator. The method depends on the fact that $(a + bi)(a - bi) = a^2 + b^2$, a fact that can be checked by brute force.

Example 6. Compute $\frac{2 + 3i}{1 + 2i}$.

Solution.

$$\frac{2 + 3i}{1 + 2i} = \frac{2 + 3i}{1 + 2i} \cdot \frac{1 - 2i}{1 - 2i}$$

$$= \frac{(2 + 3i)(1 - 2i)}{1^2 + 2^2}$$

$$= \frac{8 - i}{5}$$

$$= \frac{8}{5} - \frac{i}{5}$$

Note the use of the conjugate in the form:

$$1 = \frac{1 - 2i}{1 - 2i}$$

The result is expressed in rectangular form.

38. Compute the quotients:

(a) $\frac{2 - i}{1 + i}$ (b) $\frac{-1 + 2i}{3 - 4i}$ (c) $\frac{2 + 3i}{3 - 2i}$ (d) $\frac{-2 - i}{2 + i}$

(a) $\frac{1}{2} - \frac{3}{2}i$

(c) i

39. Compute $\dfrac{2 + 2i}{1 - i}$ two ways:

(a) by conjugating the denominator in rectangular form,

(b) by putting numerator and denominator in polar or trigonometric form.

* * * SUMMARY * * *

In this chapter we considered the properties of complex numbers and their connection with trigonometry in the polar form representation.

We can think of the complex number $a + bi$ geometrically as a point (a,b) in the (x,y)-plane. We add and multiply complex numbers using the ordinary rules of algebra, except that we substitute -1 wherever we see i^2.

The magnitude (or radius) of $a + bi$ is $|a + bi| = \sqrt{a^2 + b^2}$, representing the distance of $a + bi$ from the origin. The argument (or angle) of $a + bi$ is the angle θ from the x-axis to the ray from the origin to $a + bi$. If $a \neq 0$, then $\tan \theta = \dfrac{b}{a}$.

If r is the magnitude and θ the argument of $a + bi$, then we have the polar form,

$$a + bi = r(\cos \theta + i\sin \theta).$$

Using the addition formulas for $\cos \theta$ and $\sin \theta$, we found that the product of the complex numbers

$$r_1(\cos \theta_1 + i\sin \theta_1) \text{ and } r_2(\cos \theta_2 + i\sin \theta_2)$$

is

$$r_1 r_2(\cos(\theta_1 + \theta_2) + i\sin(\theta_1 + \theta_2)).$$

Thus, the magnitude of the product is the product of the magnitudes and the argument of the product is the sum of the arguments. This leads to DeMoivre's formula:

$$[r(\cos \theta + i\sin \theta)]^n = r^n[\cos(n\theta) + i\sin(n\theta)].$$

A complex number $r(\cos \theta + i\sin \theta) \neq 0$ has two square roots:

$$r^{1/2}(\cos \frac{\theta}{2} + i\sin \frac{\theta}{2}) \text{ and}$$

$$r^{1/2}[\cos(\frac{\theta + 360°}{2}) + i\sin(\frac{\theta + 360°}{2})] = -r^{1/2}[\cos \frac{\theta}{2} + i\sin \frac{\theta}{2}].$$

More generally, the complex number $r(\cos \theta + i\sin \theta) \neq 0$ has n different
n^{th} roots, given by

$$r^{\frac{1}{n}}(\cos \frac{\theta}{n} + i\sin \frac{\theta}{n})$$

$$r^{\frac{1}{n}}(\cos (\frac{\theta + 360°}{n}) + i\sin(\frac{\theta + 360°}{n}))$$

$$r^{\frac{1}{n}}(\cos (\frac{\theta + 2\cdot360°}{n}) + i\sin(\frac{\theta + 2\cdot360°}{n}))$$

$$\cdot$$
$$\cdot$$
$$\cdot$$

$$r^{\frac{1}{n}}[\cos (\frac{\theta + (n-1)360°}{n}) + i\sin(\frac{\theta + (n-1)360°}{n})].$$

To divide complex numbers, we divide their magnitudes and subtract
their arguments:

$$\frac{r_1 (\cos \theta_1 + i\sin \theta_1)}{r_2 (\cos \theta_2 + i\sin \theta_2)} = \frac{r_1}{r_2} [\cos (\theta_1 - \theta_2) + i\sin(\theta_1 - \theta_2)].$$

* * * WORDS AND SYMBOLS * * *

Complex number

Magnitude(radius)

Conjugate

DeMoivre's formula

Complex plane

Argument(angle)

Polar(trigonometric)
   form

* * * CONCLUDING REMARKS * * *

The great eighteenth century mathematician Leonard Euler noticed a
remarkable relationship connecting the exponential, sine and cosine
functions. We begin with the series representation of $\sin \theta$, where θ is
in radian measure. (See the CONCLUDING REMARKS of Chapter 4):

$$\sin \theta = \theta - \frac{\theta^3}{3!} + \frac{\theta^5}{5!} - \frac{\theta^7}{7!} + \ldots ,$$

and the series that represents $\cos\theta$:

$$\cos\theta = 1 - \frac{\theta^2}{2!} + \frac{\theta^4}{4!} - \frac{\theta^6}{6!} + \ldots .$$

Now we consider the series representation for e^x (See the CONCLUDING REMARKS of Chapter 4):

$$e^x = 1 + x + \frac{x^2}{2!} + \frac{x^3}{3!} + \frac{x^4}{4!} + \ldots$$

In this last equation we let $x = i\theta$ to obtain

$$e^{i\theta} = 1 + i\theta + \frac{(i\theta)^2}{2!} + \frac{(i\theta)^3}{3!} + \frac{(i\theta)^4}{4!} + \ldots$$

Using $i^2 = -1$, $i^3 = -i$, $i^4 = 1$, etc. we change this last equation to

$$e^{i\theta} = 1 + i\theta - \frac{\theta^2}{2!} - \frac{i\theta^3}{3!} + \frac{\theta^4}{4!} + \frac{i\theta^5}{5!} - \frac{\theta^6}{6!} - \frac{i\theta^7}{7!} + \ldots$$

$$= (1 - \frac{\theta^2}{2!} + \frac{\theta^4}{4!} - \frac{\theta^6}{6!} + \ldots) + i(\theta - \frac{\theta^3}{3!} + \frac{\theta^5}{5!} - \frac{\theta^7}{7!} + \ldots)$$

This equation contains the series representing $\sin \theta$ and $\cos \theta$, so we have Euler's wonderful formula:

$$\boxed{e^{i\theta} = \cos \theta + i\sin \theta.}$$

If in Euler's formula we let $\theta = \pi$, we obtain (since $\cos \pi = -1$, $\sin \pi = 0$) the almost mystical relationship

$$e^{i\pi} = -1,$$

connecting the four important numbers e, π, i, and -1.

* * * TRIGONOMETRY AND COMPLEX NUMBERS SELF QUIZ * * *

1. Compute the sum of the complex numbers $-2 + 3i$ and $5 + 4i$, and show in a sketch how the sum is a vertex of a parallelogram in the (x,y)-plane having the given numbers as two of its vertices.

2. Write each of the following in the form $a + bi$:

    (a) $(2 + i) + (7 - 6i)$ (b) $(-2 + 5i)(3 - i)$

    (c) $(3 + 4i)^2$ (d) $\frac{1}{3 + 4i}$

    (e) $(\frac{1}{2} + \frac{\sqrt{3}}{2} i)^3$ (f) $(\frac{1 + i}{\sqrt{2}})^2$

    (g) $(\frac{1 + i}{\sqrt{2}})^4$ (h) $(2 - i)(\overline{2 - i})$

3. Find the magnitude (radius) and argument (angle) of each of the following:

(a) $1 + i$ (b) $\frac{1}{2} + \frac{\sqrt{3}}{2} i$ (c) $\frac{1 - i}{\sqrt{2}}$ (d) $-5i$

4. Describe all the possible arguments that can be assigned to the complex number $-1 + i$.

5. Find and sketch all complex numbers U such that

$$U^4 = 1.$$

6. Find and sketch all the cube roots of 8.

7. If C is a real number, show that both $1 + Ci$ and $1 - Ci$ are solutions of the equation

$$x^2 - 2x + 1 + C^2 = 0$$

ANSWERS TO SELF QUIZ

1. The sum is $3 + 7i$

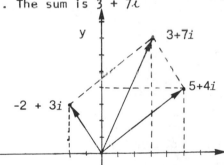

2. (a) $9 - 5i$ (b) $-1 + 17i$

(c) $-7 + 24i$ (d) $\frac{3}{25} - \frac{4}{25} i$

(e) -1 (f) i

(g) $i^2 = -1$ (h) 5

3. (a) magnitude: $\sqrt{2}$, argument: $45°$

(b) magnitude: 1, argument: $60°$

(c) magnitude: 1, argument: $315°$

(d) magnitude: 5, argument: $270°$

4. $135° + n \cdot 360°$, n = 0, ±1, ±2, ±3, ...

5. $1, -1, i, -i,$

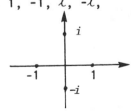

6. $2, -1 + i\sqrt{3}, -1 - i\sqrt{3}$

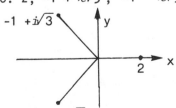

7. If $x = 1 + Ci$, then $x^2 - 2x + 1 + C^2 = (1 + Ci)^2 - 2(1 + Ci) + 1 + C^2$

$$= 1 + 2Ci - C^2 - 2 - 2Ci + 1 + C^2$$

$$= 0.$$

Similar calculation for $x = 1 - Ci$.

* * *   TRIGONOMETRY AND COMPLEX NUMBERS REVIEW   * * *

1. If $a + bi$ is a complex number, define

    (a) the magnitude of $a + bi$ (b) the argument of $a + bi$

    (c) the conjugate of $a + bi$

2. Write each of the following in the form $a + bi$:

    (a) $(6 + i)+(4 + 3i)$ (b) $(-5 + 2i) - (3 - i)$

    (c) $(1 + i)(1 - i)$ (d) $(\frac{1}{1 + i})^2$

    (e) $[\cos(30°) + i\sin(30°)]^5$ (f) $\frac{3 + 2i}{5 - 4i}$

3. Show that if U is a complex number such that $U^2 = 16$, then the magnitude of U is 4. [Suggestion: If in polar form $U = r(\cos\theta + i\sin\theta)$, then what is the polar form of U^2? From the polar form of a complex number how do we find the magnitude?]

4. Compute and sketch $\frac{W}{U}$ if

    (a) $W = 2 - i$ and $U = 1 + i$.

    (b) $W = 4(\cos 80° + i\sin 80°)$ and $U = 2(\cos 50° + i\sin 50°)$.

5. Find the magnitude and argument of $\frac{i}{1 + i}$.

6. Derive the trigonometric identities

    $$\cos 2\theta = \cos^2\theta - \sin^2\theta$$

    $$\sin 2\theta = 2\sin\theta \cos\theta$$

    from DeMoivre's formula by comparing the two sides of:

    $$(\cos\theta + i\sin\theta)^2 = \cos 2\theta + i\sin 2\theta.$$

7. Show that

    $$\frac{1}{\cos\theta + i\sin\theta} = \cos\theta - i\sin\theta$$

8. Find all the

    (a) square roots of $2 - 2i\sqrt{3}$ (b) cube roots of 1

    (c) fourth roots of 1

9. Use the polar form and DeMoivre's formula to find

    (a) $(1 + i)^6$ (b) $(\sqrt{3} - i)^4$ (c) $(i - 1)^{10}$

ANSWERS TO TRIGONOMETRY AND COMPLEX NUMBERS REVIEW

1. (a) $\sqrt{a^2 + b^2}$ (b) The angle from the positive x-axis   (c) $a - bi$
                               to the ray from the origin through
                               the point (a,b).

2.   (a) $10 + 4i$ (b) $-8 + 3i$ (c) 2 (d) $-\frac{1}{2}i$ (e) $-\frac{\sqrt{3}}{2} + \frac{1}{2}i$ (f) $\frac{7}{41} + \frac{22}{41}i$

3. If $U = r(\cos\theta + i\sin\theta)$, then $U^2 = r^2(\cos 2\theta + i\sin 2\theta)$.
   Thus if $U^2 = 16$, then $r^2 = 16$, and $r = 4$.

4. (a) $\dfrac{2 - i}{1 + i} = \dfrac{(2 - i)(1 - i)}{(1 + i)(1 - i)} = \dfrac{1 - 3i}{2} = \dfrac{1}{2} - \dfrac{3}{2}i$

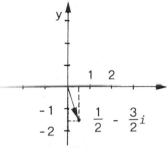

   (b) $\dfrac{W}{U} = \dfrac{4(\cos 80° + i\sin 80°)}{2(\cos 50° + i\sin 50°)} = 2(\cos 30° + i\sin 30°) = \sqrt{3} + i$

5. $\dfrac{i}{1 + i} = \dfrac{i(1 - i)}{(1 + i)(1 - i)} = \dfrac{i + 1}{2}$ Magnitude: $\dfrac{\sqrt{2}}{2}$; argument: $45°$

6. $\cos 2\theta + i\sin 2\theta = (\cos\theta + i\sin\theta)^2$

$$= \cos^2\theta - \sin^2\theta + i(2\sin\theta\cos\theta)$$

$\therefore \cos 2\theta = \cos^2\theta - \sin^2\theta$ and $\sin 2\theta = 2\sin\theta\cos\theta$

7. $\dfrac{1}{\cos\theta + i\sin\theta} = \dfrac{\cos\theta - i\sin\theta}{(\cos\theta + i\sin\theta)(\cos\theta - i\sin\theta)} = \dfrac{\cos\theta - i\sin\theta}{\cos^2\theta + \sin^2\theta} = \cos\theta - i\sin\theta$

8. (a) $\sqrt{3} - i$ and $-\sqrt{3} + i$ (b) $1, -\frac{1}{2} + \frac{\sqrt{3}}{2}i, -\frac{1}{2} - \frac{\sqrt{3}}{2}i$.   (c) $1, i, -1, -i$.

9. (a) $1 + i = \sqrt{2}(\cos 45° + i\sin 45°)$

   $\therefore (1 + i)^6 = (\sqrt{2})^6(\cos 270° + i\sin 270°) = -8i$

   (b) $\sqrt{3} - i = 2(\cos 330° + i\sin 330°)$

   $\therefore (\sqrt{3} - i)^4 = 16[\cos(4\cdot330°) + i\sin 4\cdot330°] = 16[-\frac{1}{2} - \frac{\sqrt{3}}{2}i]$

   $$= -8[1 + i\sqrt{3}]$$

   (c) $i - 1 = \sqrt{2}(\cos 135° + i\sin 135°)$

   $\therefore (i - 1)^{10} = 32(\cos 1350° + i\sin 1350°) = -32i$

* * *   TRIGONOMETRY AND COMPLEX NUMBERS REVIEW SELF TEST   * * *

1. If $U = \frac{\sqrt{2}}{2} + \frac{\sqrt{2}}{2} i$, in $a + bi$ form

    (a) find V such that $U + V = 0$

    (b) find W such that $UW = 2$

2. Find all roots of the equation:

    (a) $x^2 = 16$ (b) $x^3 = 8$ (c) $x^2 - 4x + 5 = 0$ (d) $x^2 + 4 = 0$

3. Use two different methods to compute

$$\left(-\frac{1}{2} + \frac{\sqrt{3}}{2} i\right)^3 .$$

4. Show that if $a + bi \neq 0$ then

$$\frac{1}{a + bi} = \frac{a}{a^2 + b^2} - \frac{b}{a^2 + b^2} i$$

5. Show that the product of a complex number and its conjugate is the magnitude squared.

6. State DeMoivre's formula.

7. Explain why every nonzero complex number has three different cube roots.

8. If U has magnitude 6 and argument 120° and V has magnitude 2 and argument 15°, find the magnitude and argument of

    (a) UV (b) $\frac{U}{V}$

ANSWERS TO REVIEW SELF TEST

1. (a) $V = \frac{-\sqrt{2}}{2} - \frac{\sqrt{2}}{2} i$ (b) $W = \frac{2}{U} = \sqrt{2}(1 - i)$

2. (a) 4, -4   (b) 2, $-1 + i\sqrt{3}$, $-1 - i\sqrt{3}$ (c) $2 + i$, $2 - i$ (d) $2i$, $-2i$

3. Direct multiplication, or else use DeMoivre's formula together with the fact that $-\frac{1}{2} + \frac{\sqrt{3}}{2} i = \cos 120° + \sin 120°$, to obtain

    $\left(-\frac{1}{2} + \frac{\sqrt{3}}{2} i\right)^3 = 1$.

4. $\dfrac{1}{a + bi} = \dfrac{a - bi}{(a + bi)(a - bi)} = \dfrac{a - bi}{a^2 + b^2} = \dfrac{a}{a^2 + b^2} - \dfrac{b}{a^2 + b^2} i$

5. $(a + bi)\overline{(a + bi)} = (a + bi)(a - bi) = a^2 + b^2 = |a + bi|^2$

6. For any positive **integer** n,

    $[r(\cos\theta + i\sin\theta)]^n = r^n(\cos n\theta + i\sin n\theta)$

7. If the complex number U has polar form
$$U = r(\cos\theta + i\sin\theta),$$
then each of the complex numbers

$$r^{\frac{1}{3}}(\cos\frac{\theta}{3} + i\sin\frac{\theta}{3}), \quad r^{\frac{1}{3}}[\cos(\frac{\theta + 360°}{3}) + i\sin(\frac{\theta + 360°}{3})],$$

$$r^{\frac{1}{3}}[\cos(\frac{\theta + 2\cdot360°}{3}) + i\sin(\frac{\theta + 2\cdot360°}{3})] \text{ is a cube root of U.}$$

8. (a) magnitude 12; argument 135°   (b) magnitude 3; argument 105°

* * * TRIGONOMETRY AND COMPLEX NUMBERS PROJECTS * * *

1. Prove that for any complex numbers, the magnitude of their product
    is the product of their magnitudes. In other words, check that
$$|(a + bi)(c + di)| = |a + bi||c + di|$$

Hint: $|a + bi| = \sqrt{a^2 + b^2}$

2. (a) Show that the result in the preceding exercise is equivalent to
    the identity
$$(a^2 + b^2)(c^2 + d^2) = (ac - bd)^2 + (ad + bc)^2$$
    (b) Some positive integers can be represented as the sum of two
    squares of positive integers, and some cannot. For example,
    5 is the sum of two squares:
$$5 = 1^2 + 2^2,$$
    but 6 cannot be so represented. Use part (a) to show that if the
    positive integers M and N are each the sum of two squares of
    positive integers, then so is MN.

3. Show that if a and b are positive integers, then there exist
    positive integers A and B such that
$$(a^2 + b^2)^3 = A^2 + B^2$$

4. Use DeMoivre's formula to prove the following trigonometric
    identities:
    (a) $\cos3\theta = \cos^3\theta - 3\cos\theta\sin^2\theta$ (b) $\sin3\theta = 3\cos^2\theta\sin\theta - \sin^3\theta$

5. Deduce from the preceding that
    (a) $\cos3\theta = 4\cos^3\theta - 3\cos\theta$ (b) $\sin3\theta = 3\sin\theta - 4\sin^3\theta$

6. Show that if $x = \cos20°$, then x is a root of the cubic equation
$$8x^3 - 6x - 1 = 0$$

(Use Projects exercise 5 (a))

7. An "n^{th} root of unity" is any complex number U such that $U^n = 1$.

   (a) If $U = \cos\left(\frac{2\pi}{n}\right) + i\sin\left(\frac{2\pi}{n}\right)$, show that U is an nth root of unity.

   (b) If U is the complex number in part (a), show that
   $U^2, U^3, U^4, \ldots, U^{n-1}$ are also n^{th} roots of unity.

   (c) There are n different n^{th} roots of unity, namely
   $$1, U, U^2, U^3, \ldots, U^{n-1}.$$
   Show that their sum is 0, that is,
   $$1 + U + U^2 + U^3 + \ldots + U^{n-1} = 0.$$
   Hint: $[x^n - 1 = (x - 1)(x^{n-1} + x^{n-2} + \ldots + x + 1)]$

   (d) Draw a sketch to illustrate the result of part (c).

8. Let U be an nth root of unity (see Projects exercise 7). If $U \neq 1$, show that
   $$1 + 2U + 3U^2 + 4U^3 + \ldots + nU^{n-1} = \frac{n}{U-1}$$

9. What is wrong with the following?
   $$1 = \sqrt{1} = \sqrt{(-1)(-1)} = \sqrt{(-1)}\ \sqrt{(-1)} = i \cdot i = -1$$

10. (a) Show that if a complex number is multiplied by i, the effect is a 90° rotation about the origin in the positive direction.

    (b) What if a complex number is multiplied by $\cos\theta + i\sin\theta$, where θ is a given angle? Describe what happens geometrically, in a fashion similar to part (a).

* * * TRIGONOMETRY CUMULATIVE REVIEW * * *

1. In terms of the unit circle, how do we define the following?
   (a) $\cos\theta$ (b) $\sin\theta$ (c) $\tan\theta$

2. Write in terms of $\sin\theta$ or $\cos\theta$ (use a unit circle for a reference):
   (a) $\sin(90° - \theta)$ (b) $\cos(90° + \theta)$ (c) $\cos(90° - \theta)$
   (d) $\sin(180° - \theta)$ (e) $\cos(180° + \theta)$ (f) $\sin(90° + \theta)$

3. If $\sin\theta = -\frac{5}{13}$ and $270° < \theta < 360°$, find the following.
   (a) $\tan\theta$ (b) $\sec\theta$ (c) $\csc\theta$ (d) $\cot\theta$ (e) $\cos\theta$

4. How do we define the trigonometric functions in terms of an acute angle of a right triangle?

5. Find x:

(a)

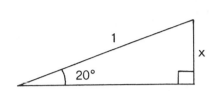

(b)

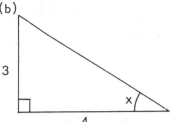

(c)

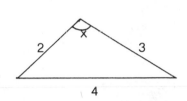

(d)

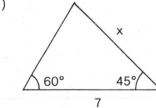

6. If $(7,-5)$ is a point on the second arm of θ, when θ is an angle in the coordinate plane with the vertex at the origin, then find

(a) $\sec \theta$ (b) $\csc \theta$ (c) $\tan \theta$

(d) $\cot \theta$ (e) $\cos \theta$ (f) $\sin \theta$

7. Prove the identity:

(a) $\cos^2 x - \sin^2 x = 2 \cos^2 x - 1$ (b) $\csc^2 x \tan^2 x = \tan^2 x + 1$

(c) $\dfrac{\tan x - 1}{\tan x + 1} = \dfrac{1 - \cot x}{1 + \cot x}$ (d) $\dfrac{\sin x - \cos x}{\sin x + \cos x} = \dfrac{\tan x - 1}{\tan x + 1}$

8. Prove the identity:

(a) $\tan x = \dfrac{\sin 2x}{1 + \cos 2x}$ (b) $(\sin x + \cos x)^2 = 1 + \sin 2x$

9. Solve for x such that $0° \leq x \leq 360°$

(a) $2 \cos^2 x + \sin x = 1$ (b) $\tan x = 3 \cot x$

10. Make a rough sketch of the graph of

(a) $y = \sin x$ (b) $y = \cos x$ (c) $y = \tan x$

(d) $y = \csc x$ (e) $y = \sec x$ (f) $y = \cot x$

11. How do we define radian measure?

12. Fill in the table:

θ in degrees	70°		640°			1°
θ in radians		1		$\dfrac{7\pi}{6}$	π	

13. Describe in words how $y = \cos x$ is modified to get the graph
    $y = 3 + 2 \cos (3x - \pi)$

14. Define:

    (a) Arcsin x   (b) Arccos x   (c) Arctan x

15. Sketch the graphs:

    (a) $y =$ Arcsin x   (b) $y =$ Arccos x   (c) $y =$ Arctan x

16. Discuss the relation between

    (a) $y = \sin x$ and $y =$ Arcsin x   (b) $y = \cos x$ and $y =$ Arccos x

    (c) $y = \tan x$ and $y =$ Arctan x

17. Evaluate:

    (a) $\sin \left(\text{Arccos } \frac{3}{5}\right)$ (b) $\tan \left(\text{Arcsin } \left(-\frac{24}{25}\right)\right)$ (c) $\cos \left(\text{Arcsin } \left(-\frac{15}{17}\right)\right)$

18. Prove:

    (a) $\text{Arcsin } \frac{1}{4} + \text{Arccos } \frac{7}{8} = \text{Arcsin } \frac{11}{16}$ (b) $\text{Arctan } \frac{4}{3} - \text{Arctan } \frac{1}{7} = \frac{\pi}{4}$

19. Prove:

    (a) $(1 + \cos \theta)(\csc \theta - \cot \theta) = \sin \theta$

    (b) $(\csc \theta - \sin \theta) \tan \theta \sec \theta = 1$

    (c) $(\cot \theta + 1)^2 + (\cot \theta - 1)^2 = 2 \csc^2 \theta$

    (d) $(1 + \sin x - \cos x)(1 - \sin x + \cos x) = \sin (2x)$

20. Prove:

    (a) $\sqrt{\dfrac{1 - \cos\theta}{1 + \cos\theta}} = \dfrac{\sin \theta}{1 + \cos\theta}$ if $0° \leq \theta < 180°$

    (b) $\sqrt{\dfrac{1 + \sin\theta}{1 - \sin\theta}} = \dfrac{\cos \theta}{1-\sin \theta}$ if $-90° < \theta < 90°$

21. Let $P = \cos \theta + i\sin \theta$. Express each of the following in
    $r(\cos \theta + i\sin \theta)$ form.

    (a) 3P   (b) P^2 (c) $\overline{P}$ (d) P^{45}

    (e) $\frac{1}{P}$ (f) 2P   (g) P^3 (h) $-P$

22. Explain using a diagram that if $P = a + bi$ and has radius r and
    angle θ, then $a = r \cos \theta$ and $b = r \sin \theta$. Hence, the complex
    number $P = a + bi$ may be written as $P = r(\cos\theta + i\sin \theta)$.

23. Change to $r(\cos \theta + \sin \theta)$, or polar form. [Hint: Make a sketch;
    Show the point (a,b) that represents $a + bi$, then find (r,θ).]

    (a) $\sqrt{2} + i\sqrt{2}$ (b) $1 - i\sqrt{3}$ (c) $-\sqrt{3} + i$ (d) $-1 - i$

24. Write in $a + bi$ form: $29(\cos 50° - i\sin 50°)$.

25. If $Z = 12(\cos 165° + i\sin 165°)$ and $W = 3(\cos 135° + i\sin 135°)$

    compute in polar form:

    (a) WZ   (b) $\dfrac{Z}{W}$ (c) W^3

26. Do Exercise 25 for $Z = 20(\cos 100° + i\sin 100°)$ and

    $W = 4(\cos 40° + i\sin 40°)$.

27. Find the two square roots of $9(\cos 100° + i\sin 100°)$.

28. Find the three cube roots of $8(\cos 240° + i\sin 240°)$.

29. Find the four fourth roots of $5(\cos 200° + i\sin 100°)$.

ANSWERS TO TRIGONOMETRY CUMULATIVE REVIEW

1. See Chapter 1.   2. (a) $\cos θ$ (b) $-\sin θ$ (c) $\sin θ$ (d) $\sin θ$

    (e) $-\cos θ$ (f) $\cos θ$ 3.(a) $-\dfrac{5}{12}$ (b) $\dfrac{13}{12}$ (c) $-\dfrac{13}{5}$ (d) $-\dfrac{12}{5}$ (e) $\dfrac{12}{13}$

4. (a) $\sin θ = \dfrac{\text{opp}}{\text{hyp}}$ (b) $\cos θ = \dfrac{\text{adj}}{\text{hyp}}$ (c) $\tan θ = \dfrac{\text{opp}}{\text{adj}}$ (d) $\cot θ = \dfrac{\text{adj}}{\text{opp}}$

    (e) $\sec θ = \dfrac{\text{hyp}}{\text{adj}}$ (f) $\csc θ = \dfrac{\text{hyp}}{\text{opp}}$ 5.(a) 0.342   (b) 36.9°   (c) 104.5°

    (d) 6.28   6.(a) $\dfrac{\sqrt{74}}{7}$ (b) $-\dfrac{\sqrt{74}}{5}$ (c) $-\dfrac{5}{7}$ (d) $-\dfrac{7}{5}$ (e) $\dfrac{7}{\sqrt{74}}$

    (f) $-\dfrac{5}{\sqrt{74}}$ 7. and 8. Proofs   9. (a) 210°, 330°, 90°

    (b) 60°, 120°, 240°, 300°   10. Sketch.   11. See Chapter 4.

12.

70°	$\left(\dfrac{180°}{π}\right)$	640°	210°	180°	1°
$\dfrac{7π}{18}$	1	$\dfrac{32π}{9}$	$\dfrac{7π}{6}$	$π$	$\dfrac{π}{180}$

13. (i) raised 3 units   (ii) amplitude is 2   (iii) period is $\dfrac{2}{3}π$

    (iv) phase shift to the right $\dfrac{π}{3}$ 14. and 15. See Chapter 5

16. (a) inverses   (b) inverses   (c) inverses   17. (a) $\dfrac{4}{5}$ (b) $-\dfrac{24}{7}$

    (c) $\dfrac{8}{17}$ 18. - 20. Proofs   21. (a) $3(\cos θ + i\sin θ)$ (b) $\cos 2θ + i\sin 2θ$

    (c) $\cos θ - i\sin θ$ (d) $\cos 45θ + i\sin 45θ$ (e) $\cos θ - i\sin θ$

    (f) $2(\cos + i\sin θ)$ (g) $\cos (3θ) + i\sin (3θ)$ (h) $-\cos θ - i\sin θ$.

22. See text between Exercises 8 and 9 in the Central Section.

23.　(a) $2(\cos 45° + i\sin 45°)$　　　(b) $2(\cos 300° + i\sin 300°)$

　　　(c) $2(\cos 150° + i\sin 150°)$　　(d) $\sqrt{2}(\cos 225° + i\sin 225°)$

24.　Approximately $18.6 - 22.2i$

25.　(a) $36(\cos 300° + i\sin 300°)$　　(b) $4(\cos 30° + i\sin 30°)$

　　　(c) $27(\cos 45° + i\sin 45°)$

26.　(a) $80(\cos 140° + i\sin 140°)$　　(b) $5(\cos 60° + i\sin 60°)$

　　　(c) $64(\cos 120° + i\sin 120°)$

27.　$3(\cos 50° + i\sin 50°)$ and $3(\cos 230° + i\sin 230°)$

　　　The roots are 180° apart.

28.　$2(\cos 80° + i\sin 80°)$; $2(\cos 200° + i\sin 200°)$;

　　　$2(\cos 320° + i\sin 320°)$. The roots are 120° apart.

29.　$\sqrt[4]{5}(\cos 50° + i\sin 50°)$; $\sqrt[4]{5}(\cos 140° + i\sin 140°)$;

　　　$\sqrt[4]{5}(\cos 230° + i\sin 230°)$; $\sqrt[4]{5}(\cos 320° + i\sin 320°)$. The roots

　　　are 90° apart.

THESE TABLES ARE INCLUDED FOR STUDENTS WHO DO NOT HAVE ACCESS TO A CALCULATOR.

N	N^2	$\sqrt{N}$	N	N^2	$\sqrt{N}$
1	1	1.000	51	2601	7.141
2	4	1.414	52	2704	7.211
3	9	1.732	53	2809	7.280
4	16	2.000	54	2916	7.348
5	25	2.236	55	3025	7.416
6	36	2.449	56	3136	7.483
7	49	2.646	57	3249	7.550
8	64	2.828	58	3364	7.616
9	81	3.000	59	3481	7.681
10	100	3.162	60	3600	7.746
11	121	3.317	61	3721	7.810
12	144	3.464	62	3844	7.874
13	169	3.606	63	3969	7.937
14	196	3.742	64	4096	8.000
15	225	3.873	65	4225	8.062
16	256	4.000	66	4356	8.124
17	289	4.123	67	4489	8.185
18	324	4.243	68	4624	8.246
19	361	4.359	69	4761	8.307
20	400	4.472	70	4900	8.367
21	441	4.583	71	5041	8.426
22	484	4.690	72	5184	8.485
23	529	4.796	73	5329	8.544
24	576	4.899	74	5476	8.602
25	625	5.000	75	5625	8.660
26	676	5.099	76	5776	8.718
27	729	5.196	77	5929	8.775
28	784	5.292	78	6084	8.832
29	841	5.385	79	6241	8.888
30	900	5.477	80	6400	8.944
31	961	5.568	81	6561	9.000
32	1024	5.657	82	6724	9.055
33	1089	5.745	83	6889	9.110
34	1156	5.831	84	7056	9.165
35	1225	5.916	85	7225	9.220
36	1296	6.000	86	7396	9.274
37	1369	6.083	87	7569	9.327
38	1444	6.164	88	7744	9.381
39	1521	6.245	89	7921	9.434
40	1600	6.325	90	8100	9.487
41	1681	6.403	91	8281	9.539
42	1764	6.481	92	8464	9.592
43	1849	6.557	93	8649	9.644
44	1936	6.633	94	8836	9.695
45	2025	6.708	95	9025	9.747
46	2116	6.782	96	9216	9.798
47	2209	6.856	97	9409	9.849
48	2304	6.928	98	9604	9.899
49	2401	7.000	99	9801	9.950
50	2500	7.071	100	10000	10.000

De-grees	Radians	Sine	Cosine	Tan-gent	De-grees	Radians	Sine	Cosine	Tan-gent
0	.000	0.000	1.000	0.000					
1	.017	.018	1.000	.018	46	0.803	0.719	0.695	1.036
2	.035	.035	0.999	.035	47	.820	.731	.682	1.072
3	.052	.052	.999	.052	48	.838	.743	.669	1.111
4	.070	.070	.998	.070	49	.855	.755	.656	1.150
5	.087	.087	.996	.088	50	.873	.766	.643	1.192
6	.105	.105	.995	.105	51	.890	.777	.629	1.235
7	.122	.122	.993	.123	52	.908	.788	.616	1.280
8	.140	.139	.990	.141	53	.925	.799	.602	1.327
9	.157	.156	.988	.158	54	.942	.809	.588	1.376
10	.175	.174	.985	.176	55	.960	.819	.574	1.428
11	.192	.191	.982	.194	56	.977	.829	.559	1.483
12	.209	.208	.978	.213	57	.995	.839	.545	1.540
13	.227	.225	.974	.231	58	1.012	.848	.530	1.600
14	.244	.242	.970	.249	59	1.030	.857	.515	1.664
15	.262	.259	.966	.268	60	1.047	.866	.500	1.732
16	.279	.276	.961	.287	61	1.065	.875	.485	1.804
17	.297	.292	.956	.306	62	1.082	.883	.470	1.881
18	.314	.309	.951	.325	63	1.100	.891	.454	1.963
19	.332	.326	.946	.344	64	1.117	.899	.438	2.050
20	.349	.342	.940	.364	65	1.134	.906	.423	2.145
21	.367	.358	.934	.384	66	1.152	.914	.407	2.246
22	.384	.375	.927	.404	67	1.169	.921	.391	2.356
23	.401	.391	.921	.425	68	1.187	.927	.375	2.475
24	.419	.407	.914	.445	69	1.204	.934	.358	2.605
25	.436	.423	.906	.466	70	1.222	.940	.342	2.747
26	.454	.438	.899	.488	71	1.239	.946	.326	2.904
27	.471	.454	.891	.510	72	1.257	.951	.309	3.078
28	.489	.470	.883	.532	73	1.274	.956	.292	3.271
29	.506	.485	.875	.554	74	1.292	.961	.276	3.487
30	.524	.500	.866	.577	75	1.309	.966	.259	3.732
31	.541	.515	.857	.601	76	1.326	.970	.242	4.011
32	.559	.530	.848	.625	77	1.344	.974	.225	4.331
33	.576	.545	.839	.649	78	1.361	.978	.208	4.705
34	.593	.559	.829	.675	79	1.379	.982	.191	5.145
35	.611	.574	.819	.700	80	1.396	.985	.174	5.671
36	.628	.588	.809	.727	81	1.414	.988	.156	6.314
37	.646	.602	.799	.754	82	1.431	.990	.139	7.115
38	.663	.616	.788	.781	83	1.449	.993	.122	8.144
39	.681	.629	.777	.810	84	1.466	.995	.105	9.514
40	.698	.643	.766	.839	85	1.484	.996	.087	11.43
41	.716	.656	.755	.869	86	1.501	.998	.070	14.30
42	.733	.669	.743	.900	87	1.518	.999	.052	19.08
43	.751	.682	.731	.933	88	1.536	.999	.035	28.64
44	.768	.695	.719	.966	89	1.553	1.000	.018	57.29
45	.785	.707	.707	1.000	90	1.571	1.000	.000	∞

THESE TABLES ARE INCLUDED FOR STUDENTS
WHO DO NOT HAVE ACCESS TO A CALCULATOR.

Index for Trigonometry